Control of Electric Machine Drive Systems

IEEE Press
445 Hoes Lane
Piscataway, NJ 08854

Control of Electric Machine Drive Systems

Seung-Ki Sul

IEEE PRESS SERIES ON POWER ENGINEERING

Mohamed E. El-Hawary, *Series Editor*

IEEE PRESS

WILEY

A JOHN WILEY & SONS, INC., PUBLICATION

For general information on our other products and services or for technical support, please contact our Customer Care Department within the United States at (800) 762-2974, outside the United States at (317) 572-3993 or fax (317) 572-4002.

Wiley also publishes its books in a variety of electronic formats. Some content that appears in print may not be available in electronic formats. For more information about Wiley products, visit our web site at www.wiley.com.

Library of Congress Cataloging-in-Publication Data:

Sul, Seung-Ki.
Control of electric machine drive system / Seung-Ki Sul.
p. cm. – (IEEE Press series on power engineering ; 55)
Includes bibliographical references.
Summary: "This book is based on the author's industry experience. It contains many exercise problems that engineers would experience in their day-to-day work. The book was published in Korean at 500 pages as a textbook. The book will contain over 300 figures". – Provided by publisher.
Summary: "This book is based on the author's industry experience. It contains many exercise problems that engineers would experience in their day-to-day work"– Provided by publisher.
ISBN 978-0-470-59079-9 (hardback)
1. Electric driving–Automatic control. I. Title.
TK4058.S8513 2011
621.46–dc22

2010039507

Printed in the United States of America

eBook: 978-0-470-87655-8
oBook: 978-0-470-87654-1

To my father, who lived his whole life as an unknown engineer.

Contents

Preface

It has been eight years since my book, *Control Theory of Electric Machinery*, was published in Korean. In the past six years, more than 2500 copies of the book have been sold in Korea. Some of them are used as a textbook for a graduate course at several universities in Korea. But most of them are used as a reference book in the industry. After publishing the book in Korean, I received a lot of encouragement and inquiry to translate the book into English. But my tight schedule has delayed the translation. However, four years ago, several foreign students and visitors attended my graduate course class and they need some study materials so I was forced to translate the book into English. After two years of hard work, the English-version manuscript is now ready for publication. During the translation, the contents of this book was revised and upgraded. I hope that this book will be a good reference for the students and engineers in the field.

Modern technology, which today is called *information technology,* is based on the stable supply of energy, especially electric energy which is the most widely used. Many people in modern society think that electric energy can be produced for as long as we want. However, because clean water and air are growing scarce, electric energy comes to us as a very limited resource. As modern society develops, more and more electric energy is needed. But mass production, transportation, and use of environmentally friendly electric energy have become a very difficult problem. Electric energy goes through various steps of transformation from production to final use. Mechanical energy acquired from a primary energy source such as oil, gas, nuclear, and hydraulic power can be converted to electric energy through electromechanical energy conversion. More precisely, after mechanical energy from various sources is transformed to electric energy through a generator, voltage and frequency are controlled for proper purpose, and in developed countries, more than 60% of energy is transformed into mechanical energy again for later use. Hence, in the whole process of production and consumption of electric energy, the most critical fields of engineering are efficient control of voltage/current and frequency and appropriate control of electric machines. For 30 years my academic interest has been the control of electric machinery and I have dedicated myself to research and development of this field. This book has been written to share these experiences with my colleagues.

Even small research results cannot be achieved alone and I owe this book to the efforts of many others. First of all, I mention my academic advisor for Master's and Ph.D. courses taken at Seoul National University, Professor Minho Park, who opened my path in the field of power electronics and control of electric machines. Second, I recognize my honored professor and at the same time my father-in-law, Jongsoo

Won. He taught electric machine subjects and I inherited many good, everyday reference books related to that area from him. Third, I thank Professor Thomas Lipo, who accepted me as a visiting researcher at the University of Wisconsin in Madison. During my two years at the University of Wisconsin, my understanding of electric machine control jumped at quantum speed. I am also very grateful to the faculty and colleagues at the University of Wisconsin and the many other professors and students of Seoul National University and other universities I have visited for the many discussions that lead to this book. Through conversations and arguments, I can now more solidly understand the problems and solutions of electric machine control. Also, this book would not have been possible without the extremely hard work of my former Master's and Ph.D. students. Their invaluable effort and time are merged together in this book, which summarizes and reorganizes our academic research results for the past 20 years. In particular, many Ph.D. and Master's dissertations are foundations of this book. Problems in Chapters 1, 4, 5, 6, and 7, and Appendix A come from the research results of my collaboration and work with many in industry and I thank the many companies and people there.

The book is constructed in following order. Chapter 1 explains the features of the electric drive system and trends of development in related technologies. Moreover, Chapter 1 also explains basic knowledge of mechanics, which is used throughout this book, and also a description for typical characteristics of the load driven by electric machines is provided. The many end-of-chapter problems provide many examples of the drive systems I designed and tested. By solving the problems, some knowledge of the actual industry can be shared.

Chapter 2 discusses the basic structure and operation principle of the electric machine, which converts mechanical energy to electrical energy like a generator or converts electrical energy to mechanical energy, like a motor. Steady-state equivalent circuits of several machines are introduced to understand the steady-state characteristics and control of machines. Also, several examples of machine control from the motor viewpoint with control features are discussed. In addition, power converters, which convert electrical energy to another form of electrical energy based on power semiconductors, are modeled as equivalent circuits. In Chapter 3, the transformation of physical variables of AC machine using reference frame theory is introduced. The transformation makes the understanding and analysis of AC machine easy by transforming time-variant differential equations to time-invariant differential equations. Electrical variables such as voltage, current, and flux in a, b, and c phases of a three-phase system can be transformed to the variables in *d–q–n* (direct, quadrature, neutral) orthogonal axis, where the magnetic couplings between axes are zero. And, the three-phase system can be easily represented only by *d–q* components assuming a balanced three-phase operation. The *d–q* component, which is orthogonal to each other, can be expressed simply by a single complex number, where real part stands for *d*-axis components and imaginary for *q*-axis components. Transformation from the a, b, c phase to the orthogonal axis can be done easily by complex vector algebra. This complex vector concept is used in this book. Chapter 4 is the essence of this book because here several control algorithms of electric machines and power converters are discussed. At the first, the concept of active damping, which is a kind of state feedback control, is introduced. Next, how to regulate the current, speed, position in feedback manner are described. In order to regulate physical variables, sensors for the measurement of variables such as current sensors and position/speed sensors are introduced. If variables are not measurable, the principles of the state observer are introduced and used for the regulation electric machine speed. Furthermore, some tricks to enhance control performance of electric machines are introduced. Finally the algorithm to detect the phase angle of an AC source and to control the DC link voltages of power converters is discussed. Most of the end-of-chapter Problems in this chapter come from industry collaboration and solving the problems enhance the understanding of this chapter conspicuously.

Chapter 5 discusses the concept of vector control. Electric machines basically convert current (or torque) to torque (or current) under the excitation flux. In many high-precision motion control systems, where acceleration, speed, position are all regulated instantaneously according to their references, instantaneous torque control is a prerequisite. Instantaneous torque of an electric machine comes from the cross product of the flux linkage vector and the line vector, where current flows. Therefore, to control torque instantaneously, the flux linkage and the line vector should be controlled instantaneously. Hence, not only the magnitude of the current and flux linkage, but also the relative angle between two vectors should be controlled instantaneously. From this fact, the name "vector control" originated. In this chapter, the principles of instantaneous torque control are described in the case of several electric machines. In Chapter 6, control algorithms for position/speed senorless drive of AC machines applied to real industry are introduced. The back EMF-based sensorless control algorithms are widely investigated and some of them commercialized. This chapter discusses their merits and demerits. Because the magnitude of back EMF decreases as the speed decreases, the performance of sensorless control based on back EMF degrades rapidly at low speed. At zero speed and/or at zero frequency operation, AC machines cannot run in sensorless operation maintaining torque controllability. To escape from this problem inherently, the sensorless control algorithms exploiting saliency of AC machine are introduced in this chapter, too. By injecting some signals to the AC machine, variation of inductance according to the rotor flux position can be measured. From the variation, the position of rotor flux linkage or rotor position itself can be estimated. A general purpose inverter equipped the algorithm injecting fluctuating high frequency voltage signal to the permanent magnet AC machine has been marketed. Several practical problems presented in Chapter 7 which appear to implement the control algorithms described in the previous chapters are addressed and the possible solutions to the problems are suggested. First, the problems associated with a dead time or blanking time are discussed. To lessen the problems, some countermeasures are introduced. For the accurate measurement of current, the offset and scale errors of the current sensors and delays in the measurement system are investigated. Methods to reduce their negative effects to the control performance of the drive system are also described. Finally because of the digital implementation of the control algorithm, there can be delays from the sample and holder, the execution time of the algorithm, and PWM of the power converter. These delays may severely limit the control performance of the drive system. Some remedies to cope with delays from digital signal processing are discussed in this chapter. In Appendix A, several methods to identify the parameters of electric machines are introduced. To apply the control algorithms in this text to the control of the electric machines, the parameters of electric machines should be identified for setting of gains of the regulators, limiting values of limiters of the controller, reference and feed-forwarding values to the regulator, etc. The parameters of electric machines may be calculated or estimated from design data or from performance test data of the manufacturer, but these data are not easily available in the application field. Most of the methods introduced in this appendix dose not require any special measurement tool, but relies on the controller of the drive system itself. In Appendix B, the matrix algebra to model three-phase AC machine in the *d–q–n* axis is briefly described. Matrix algebra may be easier for computer simulation and real-time control programming of the electric machine than the complex vector notation mentioned in Chapter 2.

For comprehensive understanding of this book, basic knowledge of undergraduate physics and modern linear control theory are needed. Electric circuit theory and basic control theory are essential, and in addition, power electronics and electric machine theory are included to help the reader. This book is suitable as a graduate course text or as a reference for engineers who majored in related fields. As a graduate textbook, Chapters 1 to 5 are appropriate for a one-semester lecture scope. More precisely, Chapters 1, 3, and 4 should be explained in detail. After understanding Chapters 3 and 4, it is easier to handle topics in Chapter 5. Chapter 6 is a good summary and a starting point to understand sensorless control of AC machines. Chapter 7 and Appendix A would be very helpful to students and engineers who implement control algorithm. As a reference book on industrial site, not only the basic theories but also many problems in Chapters 1, 4, 5, 6, and 7, and the algorithms in Appendix A will be helpful. Most of the problems in this book have been used in midterm and final exams or as homework for the past 20 years of my teaching at Seoul National University. I give my thanks to students who may have suffered solving these problems while taking my lecture. Also, some contents and situations of problems are acquired from various research done at many corporations. These problems are very practical. To understand the contents of this book, I strongly recommend some experimental study with real electric machines and power converters. If that is not possible, al least the computer simulations considering real field situation are essential.

To my two daughters–Yoojin and Hyojin–please know that this book owes you because you understand your busy dad. In addition, my mother's care and concerns made me all I am today. Lastly, I give my special thanks and love to my wife, Miyun, who will always be at my side, and at least half of what I have achieved is hers.

SEUNG-KI SUL

Seoul, South Korea
August 2010

Chapter 1

Introduction

1.1 INTRODUCTION

As human beings have evolved from anthropoid species that lived in Africa several millions years ago, solving basic needs for food, clothing, and shelter have been the great concern for all humankind. As the living standard of humankind is being upgraded, the quantity of commodities is increased. All of these commodities are obtained by manufacturing natural resources or by recycling existing resources. In those processes, energy is indispensable. The civilization of human culture has evolved according to the source and form of energy. The farming culture was possible because of the energy and power of animals. The Industrial Revolution started with the invention of the steam engine in the 18th century. Since the invention of the internal combustion engine (ICE) in the 19th century, the productivity of the manufacturing has been greatly increased. In the 19th century, when electric machinery was invented, the mechanical power of the electric machine was the best source of the mechanical power that humankind ever had. It is the most widely used source of mechanical power excluding transportation area. Though the total efficiency of electricity from primary energy source to final stage is, at best, 40%, the electricity is the most convenient energy source to control and to convert to other form. Consequently, electromechanical power based on an electric machine is the basic source of mechanical power to support today's industrialized society. Recently, even in the transportation area, where the internal combustion engine has dominated for the past 100 years as a source of mechanical power, electric machines are applied as a main source of traction force in the electric vehicle, the hybrid vehicle, and the electrically propelled vessels. Through continuation of this trend, before the end of the first half of the 21st century, most of the mechanical power could be obtained from electromechanical power conversion.

Electric machinery has the following advantages compared to ICE and the gas turbine [1].

1. From an electric machine to run an electric watch to the electric machine to drive the pump of hydro pump storage, the power range can be extended from milliwatts to hundreds of Megawatts.

Control of Electric Machine Drive Systems, by Seung-Ki Sul

2. From a high-speed centrifugal separator machine running at over several hundred thousand revolutions per minute to a main mill machine in a steel process line generating over several tens of Mega Newton-meters, the operating range of speed and torque is very wide.
3. An electric machine can be easily adapted to any external environment such as vacuum, water, and extreme weather condition. Compared to an internal combustion engine, it is emissionfree in itself, has less vibration and audible noise, and is environmental friendly.
4. The response of an electric machine is faster than that of an internal combustion engine and a gas turbine by at least 10 times.
5. The running efficiency is higher, and no load or standby loss is smaller.
6. The direction of force (torque) and movement (rotation) can be easily changed.
7. The force (torque) can be easily controlled regardless of the direction of movement (rotation).
8. An electric machine can be designed in various shapes such as thin disk type, long cylinder type, rotating type, and linear motion type. And it can be easily attached to the right place where mechanical power should be applied.
9. Its input is electricity and the control system of an electric machine is easily compatible with modern information processing devices.

The abovementioned advantages of the electric machine over the ICE and the gas turbine have been intensified with the development of power electronics, magnetic and insulation materials, and information technology. Especially with the recent progress of the rare earth magnet such as the niodium–iron–boron magnet, the force (torque) density of the electric machine is comparable to a hydraulic system. And, the many motion control systems based on hydraulic pressure are replaced with an electric machine. Moreover, with the development of power electronics, the electric machine drive system can be easily controlled directly from the information processing system, and the drive system would be automated without additional hardware. However, regardless of these merits of the electric machine, it has been applied to a very limited extent to the traction force of the transportation system because of its continuous connection to the utility line. Recently, to lessen the pollution problem of the urban area, the electric vehicle is getting attention; but because of the limited performance of the battery as an energy storage, it would take considerable time to use the pure electric vehicle widely. In these circumstances a hybrid electric vehicle, where after getting mechanical power from an ICE a part or all of the power from ICE is converted to electric power to run the electric machine, has been developed and has had practical use in the street.

After Jacobi invented a DC machine in 1830 and Ferraris and Tesla invented an induction machine, the electric machine has been a prime source of mechanical power for the past 150 years. In modern industrialized society, more than 60% of electricity is used to run electric machines. Among them, more than 80% are used for induction machines [2]. The induction machine based on the rotating magnetic motive force

Figure 1.1 Induction machine at early stage of development.

Figure 1.2 Today's Induction machine.

during the early days of development is shown in Fig. 1.1, while in Fig. 1.2 a modern induction machine is shown. Due to the developments of the insulation materials and the magnetic materials, the power density, which is defined as the ratio between output power and weight, and the price has been remarkably improved. In 1890, the weight and price of a 5-horsepower (Hp) induction machine were 454 kg and $900; in 1957, 60 kg and $110; and in 1996, 22 kg and $50 [3].

In Fig. 1.3, there are several outer sizes of a 10-hp, 4-pole, totally enclosed induction machine according to years [4]. These trends of smaller size and less weight would be continued, and the efficiency of the machine would be improved continuously to save electricity for a better environment. Ever since the induction machine

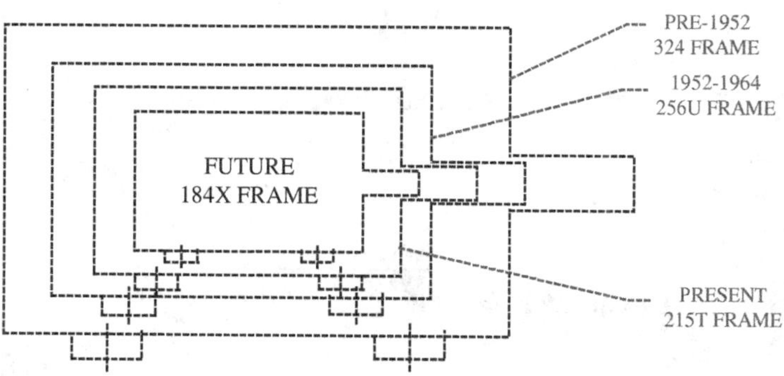

FRAME	TYPICAL STATOR DIAMETER (mm)
324	343
256	267
215	222
184	191

Figure 1.3 Outer sizes of general-purpose, 10-hp, 4-pole, totally enclosed induction machines [4].

was invented 80 years ago, it had been run by a 50-Hz or 60-Hz utility line, and its rotating speed is almost constant. However, after the invention of the thyristor in 1960, the input voltage and frequency to the machine could be changed widely, and the machine itself has been designed to adapt to these variable voltage variable frequency (VVVF) sources [5].

1.1.1 Electric Machine Drive System [6]

An electric machine drive system usually consists of several parts such as driven mechanical system, electric machine, electric power converter, control system, and so on. For the design of the drive system, several other things including the electric machine itself should be considered as shown in Fig. 1.4. As usual engineering design, the drive system showing the same performance could be implemented in various ways. The final criterion for the best design would be not only economic reasons such as initial investment, running cost, and so on, but also noneconomic reasons such as environmental friendliness, ethics, and regulations. Recently, because of the concern of engineering to the social responsibility, the noneconomic reasons are becoming important.

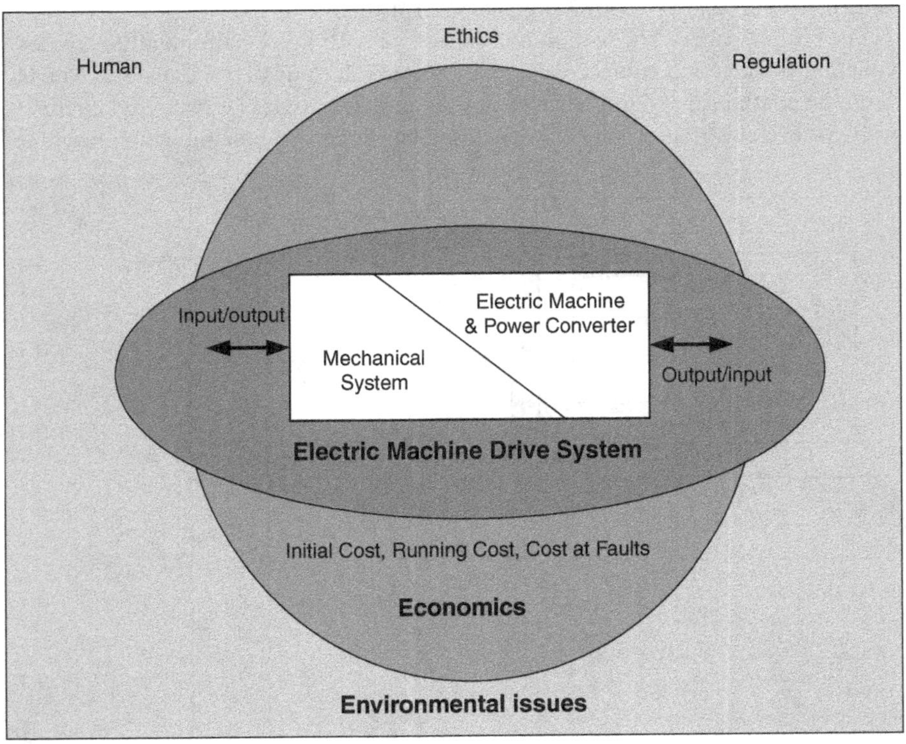

Figure 1.4 Consideration points in designing the electric machine drive system [6].

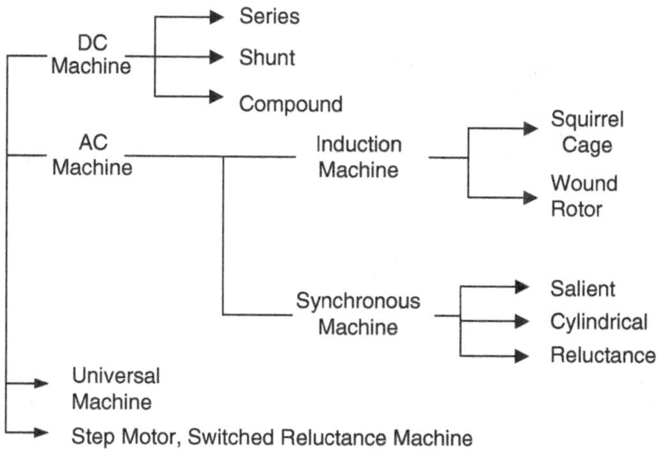

Figure 1.5 Classification of the electric machine according to power source and operating principles.

Through 100 years of development, the electric machines have diverse shapes, and a suitable shape is applied to the specific area according to the purpose of the machines. The machines can be classified as a rotary motion machine and a linear motion machine according to the motion of the rotor (mover). Also, if the machine is classified according to the electric source and operating principles of the machine, it can be classified as shown in Fig. 1.5. For typical rotational motion machine, the range of the output power and rotating speed has the relationship as shown in Fig. 1.6 [6]. If the output power of the machine is getting larger, the size, especially radius of the rotor, of the machine is larger and the centrifugal force is getting larger. Hence, a high-power and simultaneously high-speed machine is extremely difficult to make because of limited yield strength of rotor materials. Recently, with the development of computer-aided design techniques and the developments of materials, especially permanent magnet materials, a high-speed and high-power machine is appearing in some special applications such as turbo compressor [7], flywheel energy storage, and so on. And the range of the output power and speed of the permanent magnet synchrounous machine will be extended further.

1.1.2 Trend of Development of Electric Machine Drive System

In the past, due to convenience of torque and speed control, the DC machine had been used widely for adjustable speed drive (ASD). However, recently, with the development of power electronics technology, the AC machine drive system such as the induction machine and the synchronous machine driven by a variable voltage variable frequency (VVVF) inverter have been used widely. The inverter can replace the commutuator and brush of DC machine, which need regular maintenance and

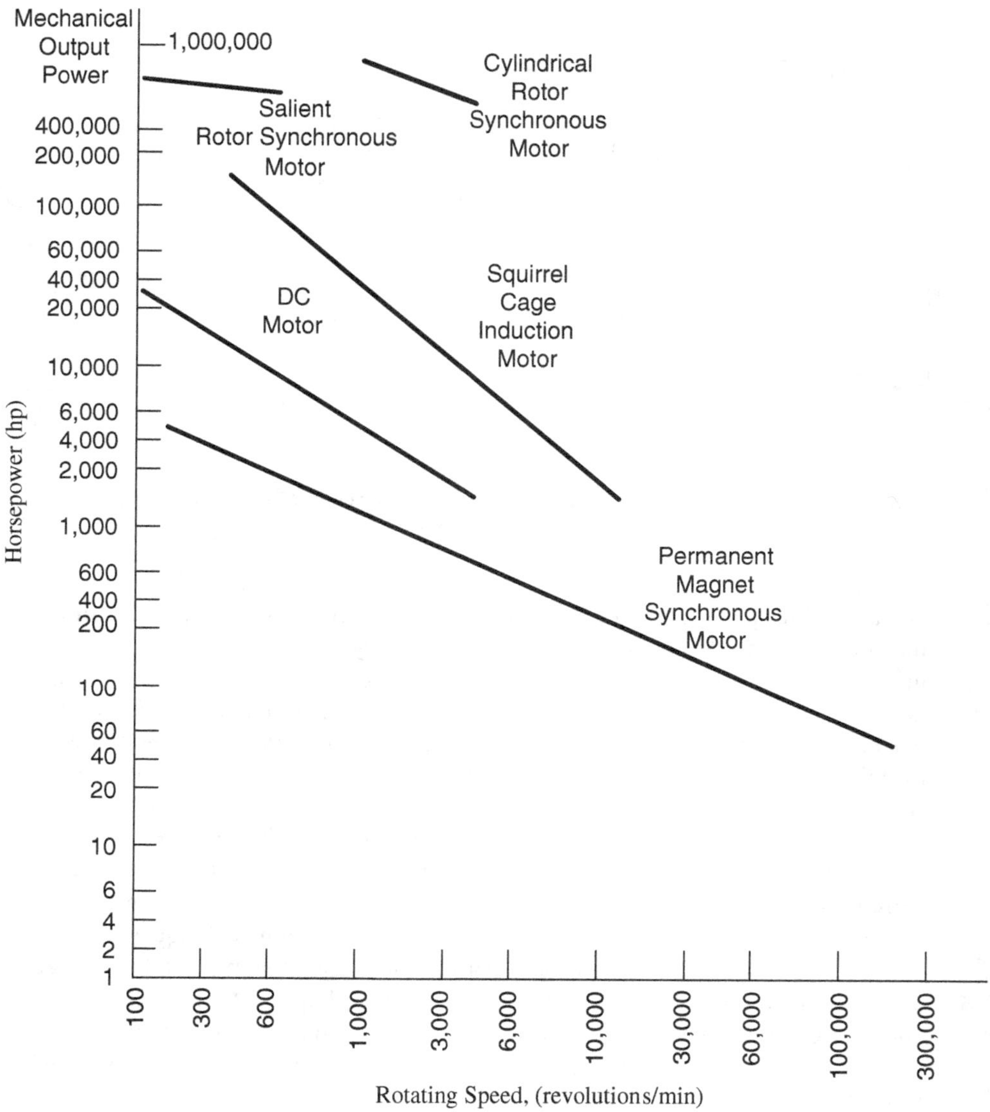

Figure 1.6 Boundary of speed and output power of rotating machines [6].

are the weak points of a DC machine. And this trend—the shift from DC machine to AC machine—would be continued because of the developmenent of not only the previously mentioned power electronics but also the control theory of AC machine such as field orientation control. In early times, DC and AC machines, both received the field flux from separated field windings. The field fluxes of both the DC machine and the synchronous machine come from the current flowing field windings, while the field flux of the induction machine comes from the part of the stator current. But with the use of a high-performance reliable permanent magnet, even in a megawatt-range machine, the field flux of the machine comes from the remanence flux of the permanent magnet. By replacing separate field winding with the permanent magnet, the torque and power density of the machine can be increased; and, simultaneously, the efficiency of the machine can be improved by eliminating the copper loss of the field winding. This trend—change from the field flux from external winding to the flux from the permanent magnet—would be continued. So, in the future, an AC machine with a permanent magnet will be used more widely.

In the beginning of the 20th century, because the price of the electric motor and its associated control system was very expensive, a large electric motor was used in the whole factory, and the mechanical power from the motor was distributed to every mechanical machine where the mechanical power is needed through gears and belts. According to the reduction of the price of the electric motor and the control system, an electric motor was used in each mechanical machine, which has several motions, and still the mechanical power from the motor was transmitted and converted to an appropriate form at each point of the motion in the machine. Recently, even in a single mechanical machine, multiple electric motors are used at each motion point. The motion required at that point could be obtained by the motor directly without speed or torque conversion from the motor. In this way, the efficiency of the system can be enhanced; furthermore, the motion control performance can be improved by eliminating all nonlinear effects and losses such as backlash, torsional oscillation, and friction. In the future, this tendency could be continued and the custom designed motor could be used widely at each moving part. For example, for high-speed operation, the high-speed motor could be used without amplification of the speed through gears. For linear motion, a linear motor can be used without a ball screw mechanism. For high-torque low-speed traction drive, the direct drive motor can be applied to reduce the size and loss of the system.

The control method of the machine drive system has been developed from manual operation to automatic control system. Recently, intelligent control techniques have been used and the control system itself can operate the system at optimal operating conditions without human intervention. Also, in the early stages of automatic control of the machine drive system, the simple supervisory control was implemented, and the control unit transferred the operating command set by the user to the machine drive system. Through the direct digital control, right now, distributed intelligent control techniques are used widely in the up-to-date motion control system.

1.1.3 Trend of Development of Power Semiconductor

In the late 1950s, with the invention of the thyristor, power electronics was born. The power semiconductor was the key of the power electronics. With the rapid improvement of performance against cost of the power semiconductors, the power electronics technology improved in a revolutionary way. The original thyristors of the 1950s and 1960s could only be turned on by an external signal to the gate but should be turned off by the external circuits. And it needs a complicated forced commutating circuit. In the

1970s, the gate turn-off (GTO) thyristor had been commercialized. And the GTO thyristor could be not only turned on but also turned off by external signal to the gate of the semiconductor. In the late 1970s, the bipolar power transistor opened a new horizon of the control of power because of its relatively simple on and off capabilities. With the transistor, general-purpose VVVF inverters had been commercialized and used in many ASD applications. Recently, with the introduction of the integrated gate controlled thyristor (IGCT) and the fifth-generation insulated gate bipolar transistor (IGBT) to the market, the performance of the electric machine drive system has been dramatically improved in the sense of output power of the system and the control bandwidth of the motion of the drive system. However, still, all the power semiconductors have been fabricated based on silicon, and its junction temperature has been limited up to 150°C in the most cases. Recently, the power semiconductor based on silicon carbide (SiC) has been introduced, and the operating temperature and operating voltage of the power semiconductor can be increased severalfold [8]. With this material, the semiconductor operating at above 300°C and at several thousand voltage can conduct several hundred amperes within one-tenth of the wafer size of the device made by silicon. In particular, the Schotky diode and field effect transistor (FET) based on SiC were the first devices in the field, and extraordinary performances of the devices have been reported.

1.1.4 Trend of Development of Control Electronics

In the early days of research and development, the control signal for the power semiconductors came from analog electronics circuits consisting of transistors, diodes, and R, L, C passive components. And, with the development of electronics technology, especially integrated circuit technology, the mixed digital and analog circuit consisting of operational amplifiers and TTL logic circuit was used. Recently, except for high-frequency switching power supplies, the major part of the power electronics system, especially the electric machine drive system, is controlled digitally by one or a few digital signal processors (DSP). Right now, a DSP chip can do over 1 gigaflop/s (one $\times$ 10^9 floating point operation per second) [9], and versatile input and output (I/O) function can be achieved by the chip without any extra hardware. In the future, this tendency of full digital control with a single chip would be widespread because of the developments of microelectronics technology. The future control electronics for the power electronics system would be on a single chip, which can execute the complex algorithm based on the modern control theory in real time with a minimized extra measurement system. And it can accomplish the user's desire with minimum energy, and simultaneously it can adapt intelligently to the change of operating conditions and parameters of the plant under control.

1.2 BASICS OF MECHANICS

Electric machines are usually connected to mechanical system, and it converts the electrical energy to mechanical energy as a motor and converts mechanical energy to electrical energy as a generator. Hence, in these energy conversion processes, understanding of mechanics is essential.

1.2.1 Basic Laws [10]

1. A physical body will remain at rest, or continue to move at a constant velocity, if net force to the body is zero.
2. The net force on a body is proportional to the time rate of change of its linear momentum:

$$f = \frac{d(Mv)}{dt} \tag{1.1}$$

   where M is the mass and v is the velocity of the body.
3. Whenever a particle A exerts a force on another particle B, B simultaneously exerts a force on A with the same magnitude in the opposite direction.
4. Between two particles, there is attractive force directly along the line of centers of the particles, and the force is proportional to the product of masses of the particles and inversely proportional to the square of distance of two particles:

$$f = G\frac{M_1 M_2}{R^2} \tag{1.2}$$

   where M_1 and M_2 are the masses of the particles, R is the distance between two particles measured from the center to center of particle, and G is a proportional constant. When a particle is on the surface of earth, the force can be represented as $f = Mg$, where M is the mass of the particle and g is a gravitational constant.

1.2.2 Force and Torque [1]

In the linear motion system as shown in Fig. 1.7, the equation of the motion with external forces can be derived as (1.3) from (1.1):

$$f_d - f_L = \frac{d}{dt}(Mv) = M\frac{dv}{dt} + v\frac{dM}{dt} \tag{1.3}$$

If there is no change of the mass for the motion, which is true in the most of cases, (1.3) can be simplified as follows:

$$f_d - f_L = M\frac{dv}{dt} = M\frac{d^2 l}{dt^2} \tag{1.4}$$

where v is the velocity of the mass, and l is the moving distance.

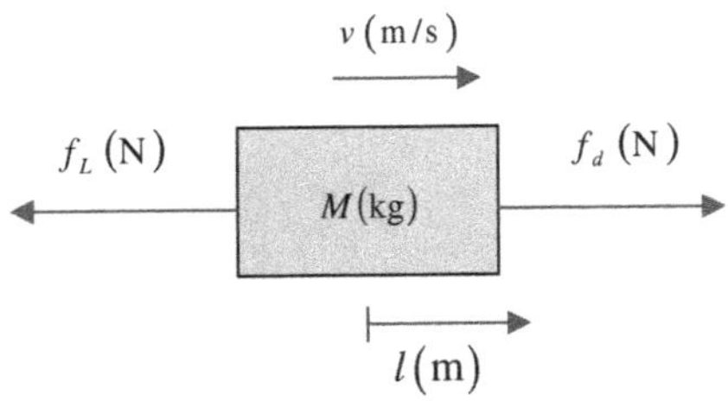

Figure 1.7 External forces in a linear motion system.

In rotating motion system as shown in Fig. 1.8, similar equation can be derived. In this equation, the rotational inertia, J, may vary according to the motion in some cases. Generally, to consider the variation of the inertia, (1.5) can be applied to the rotational motion.

$$\begin{aligned} T_d - T_L &= \frac{d}{dt}(J\omega) = J\frac{d\omega}{dt} + \omega\frac{dJ}{dt} \\ &= J\frac{d^2\theta}{dt^2} + \frac{d\theta}{dt}\frac{dJ}{dt} \end{aligned} \tag{1.5}$$

In many application cases of motion drives, as shown in Fig. 1.9 (which is a hoist drive), rotational motion and linear motion are coupled through some mechanical connections. In this system, the torque and the force have a relationship as shown in (1.6), considering gravitational force.

If there is no elongation of rope between mass, M, and sheave whose radius is r, and if the mass of rope is neglected, then (1.6) can be deduced:

$$T_d = J_{\text{sheave}}\frac{d\omega_m}{dt} + r\frac{d}{dt}(Mv) + Mgr \tag{1.6}$$

where J_{sheave} is the inertia of the sheave. The linear speed of the mass can be represented as $v = r\omega_m$. And if the radius of the sheave is constant, then from (1.6), we can derive (1.7):

$$\begin{aligned} T_d &= J_{\text{sheave}}\frac{d\omega_m}{dt} + r\frac{d(Mr\omega_m)}{dt} + Mgr \\ &= J_{\text{sheave}}\frac{d\omega_m}{dt} + Mr^2\frac{d\omega_m}{dt} + Mgr = J_{\text{sheave}}\frac{d\omega_m}{dt} + J_{\text{eq}}\frac{d\omega_m}{dt} + Mgr \end{aligned} \tag{1.7}$$

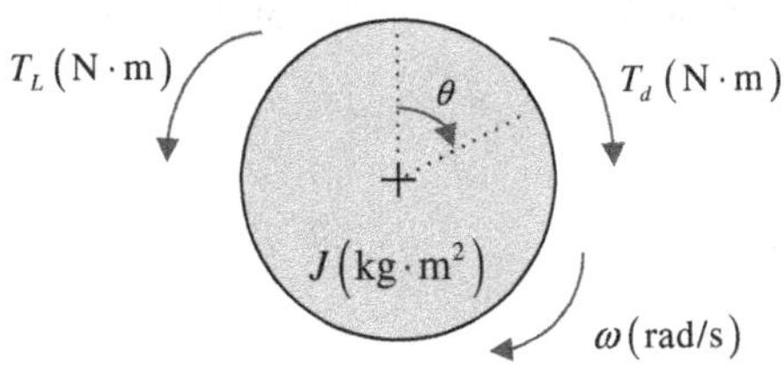

Figure 1.8 External torques in a rotating motion system.

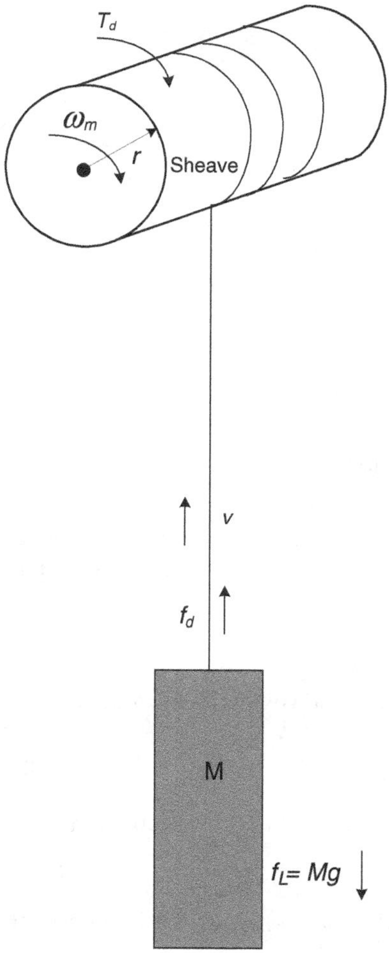

Figure 1.9 Coupling of linear and rotating motion.

where $J_{eq} = Mr^2$. From (1.7), it can be seen that the mass, M, is converted to equivalent inertia, J_{eq}, at the rotational motion of sheave. And, similarly, the inertia in the rotational motion can be converted to equivalent mass in the linear motion, and it is called *equivalent inertia mass.*

1.2.3 Moment of Inertia of a Rotating Body [11]

The moment of inertia of the rotating body asymmetry to the rotating axis as shown in Fig. 1.10 can be deduced as follows. In general, every rotating body has some asymmetry to rotating axis. Hence, to find the force to the part supporting rotating motion such as bearings, the rotating inertia of arbitrary shape should be investigated.

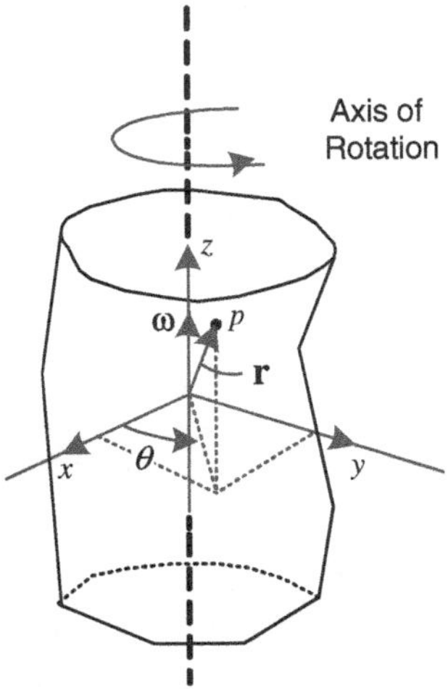

Figure 1.10 Asymmetric rigid rotating body.

$p(x, y, z)$ is the position of an infinitesimal mass, whose mass, δM, is expressed as (1.8). The position vector from the origin can be represented as (1.9). Also, the velocity vector, $\boldsymbol{\nu}$, of the mass is expressed as (1.10).

$$\delta M = \rho \delta V \tag{1.8}$$

where ρ is the density of body at $p(x, y, z)$ and δV is the volume of the infinitesimal mass at $p(x, y, z)$.

$$\mathbf{r} = \mathbf{i_x} x + \mathbf{i_y} y + \mathbf{i}_z z \tag{1.9}$$

where, $\mathbf{i_x}, \mathbf{i_y}$, and $\mathbf{i_z}$ are the unit vectors at each x, y, and z axis, respectively.

$$\boldsymbol{\nu} = \boldsymbol{\omega} \times \mathbf{r} \tag{1.10}$$

where $\boldsymbol{\omega}$ is the angular velocity vector, defined as $\boldsymbol{\omega} \equiv \frac{d\theta}{dt}\mathbf{i_z}$, of the infinitesimal mass, δM.

The accelerating force applied to the infinitesimal mass can be expressed as (1.11) from (1.1):

$$\delta \mathbf{f}_a = \frac{d}{dt}(\delta M \boldsymbol{\nu}) = \rho \delta V \frac{d\boldsymbol{\nu}}{dt} \tag{1.11}$$

By differentiating the velocity vector, $\boldsymbol{\nu}$, regarding time, (1.11) can be derived as (1.12) by using (1.10):

$$\delta \mathbf{f}_a = \rho \delta V \left[\frac{d\boldsymbol{\omega}}{dt} \times \mathbf{r} + \boldsymbol{\omega} \times (\boldsymbol{\omega} \times r) \right] \tag{1.12}$$

By using vector identity in (1.13), the force in (1.12) can be rewritten as (1.14):

$$\boldsymbol{\alpha} \times (\boldsymbol{\beta} \times \boldsymbol{\gamma}) = \boldsymbol{\beta}(\boldsymbol{\alpha} \times \boldsymbol{\gamma}) - \boldsymbol{\gamma}(\boldsymbol{\alpha} \times \boldsymbol{\beta}) \tag{1.13}$$

$$\delta \mathbf{f}_a = \rho \delta V \left[\frac{d\boldsymbol{\omega}}{dt} \times \mathbf{r} + \boldsymbol{\omega}(\boldsymbol{\omega} \cdot \mathbf{r}) - \mathbf{r}(\boldsymbol{\omega} \cdot \boldsymbol{\omega}) \right] \tag{1.14}$$

Because the torque vector is defined as the cross product of a force vector and a position vector as (1.15), the torque applied to the infinitesimal mass of the asymmetry body can be deduced as follows:

$$\delta \mathbf{T_e} = \mathbf{r} \times d\mathbf{f}_a \tag{1.15}$$

$$\delta \mathbf{T_e} = \rho \delta V \left[\mathbf{i}_z (x^2 + y^2) \frac{d^2\theta}{dt^2} - \mathbf{i}_x \left[xz \frac{d^2\theta}{dt^2} - yz \left(\frac{d\theta}{dt} \right)^2 \right] - \mathbf{i}_y \left[yz \frac{d^2\theta}{dt} + xz \left(\frac{d\theta}{dt} \right)^2 \right] \right] \tag{1.16}$$

where the inertia at each axis, with the assumption of rigid body, can be defined as

$$\begin{aligned} J_z &\equiv \int (x^2 + y^2) \rho \, dV \\ J_{xz} &\equiv \int xz\rho \, dV \\ J_{yz} &\equiv \int yz\rho \, dV \end{aligned} \tag{1.17}$$

where $\int_v \bullet \, dV$ means the integral of "•" over the entire volume, v.

Finally, the total torque vector applied to the whole body can be expressed as

$$\mathbf{T_e} = \mathbf{i}_z J_z \frac{d^2\theta}{dt^2} - \mathbf{i}_x \left[J_{xz} \frac{d^2\theta}{dt^2} - J_{yz} \left(\frac{d\theta}{dt} \right)^2 \right] - \mathbf{i}_y \left[J_{yz} \frac{d^2\theta}{dt^2} + J_{xz} \left(\frac{d\theta}{dt} \right)^2 \right] \tag{1.18}$$

If the rotating body is symmetry to the rotating axis, then $J_{xz} = J_{yz} = 0$. So, (1.18) can be simplified as $\mathbf{T_e} = \mathbf{i}_z J_z \frac{d^2\theta}{dt^2}$. And only torque in **z** axis exists. In the case of asymmetry, there is always torque at **x** and **y** axes, and such torque would be applied to the parts supporting the rotating motion. It should be noted that as seen in (1.18) the torque due to asymmetry is proportional to the square of the rotating speed. Hence keeping symmetry to the rotating axis is getting important as the rotating speed is getting higher.

1.2.4 Equations of Motion for a Rigid Body

If a rigid body is acted upon by external forces and does not have any constraints, it shows a combinational motion of translation and rotation. This combinational motion of a rigid body can be represented by equations of motion that have six degrees of freedom (DOF): three independent axes for the translational motion and three independent axes for the rotational motion in a three-dimensional space. The acceleration of a body in each axis is expressed by a nonlinear combination of the external forces. This kinematic analysis of a rigid body is commonly used in the manufacturing equipment which requires highly precise motion control.

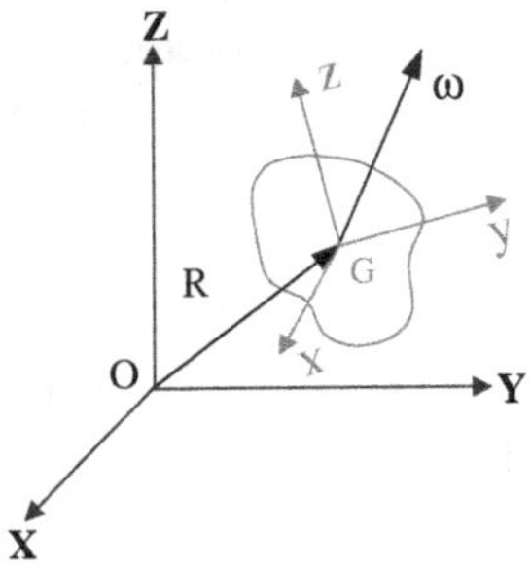

Figure 1.11 Inertial reference frame and body fixed reference frame.

In Fig. 1.11, a coordinate system for the kinematic analysis of a rigid body is shown. In the figure, *OXYZ* is an inertial reference frame that is attached to an absolute point and does not change its orientation to any external conditions. $oxyz$ is a body fixed frame that is attached to the center of the mass of a rigid body, and it changes its orientation according to the rigid body's translational or rotational motion. *G* is the center of mass and also the center of rotation of the rigid body.

Euler angles that describe the rotational motion of a rigid body with three different angles are defined in Fig. 1.12. A rotation about the *Z* axis in the *XYZ* frame is defined as angle ψ, a rotation about the y_1 axis in the x_1y_1Z frame is defined as angle θ, and a rotation about the x_2 axis in the $x_2y_1z_2$ frame is defined as angle ϕ. The reference frame x_2yz is same as the body fixed frame xyz. Hence, any arbitrary rotation of a rigid body can be represented by (ϕ, θ, ψ).

The transformation matrix representing the rotation of a rigid body with Euler angle is shown in (1.19):

$\mathbf{u_X}, \mathbf{u_Y}, \mathbf{u_Z}$: Unit vector of **OXYZ** frame

$\mathbf{u_x}, \mathbf{u_y}, \mathbf{u_z}$: Unit vector of **oxyz** frame

$$\begin{bmatrix} \mathbf{u_X} \\ \mathbf{u_Y} \\ \mathbf{u_Z} \end{bmatrix} = \begin{bmatrix} \cos\psi & -\sin\psi & 0 \\ \sin\psi & \cos\psi & 0 \\ 0 & 0 & 1 \end{bmatrix} \begin{bmatrix} \cos\theta & 0 & \sin\theta \\ 0 & 1 & 0 \\ -\sin\theta & 0 & \cos\theta \end{bmatrix} \begin{bmatrix} 1 & 0 & 0 \\ 0 & \cos\phi & -\sin\phi \\ 0 & \sin\phi & \cos\phi \end{bmatrix} \begin{bmatrix} \mathbf{u_x} \\ \mathbf{u_y} \\ \mathbf{u_z} \end{bmatrix} \tag{1.19}$$

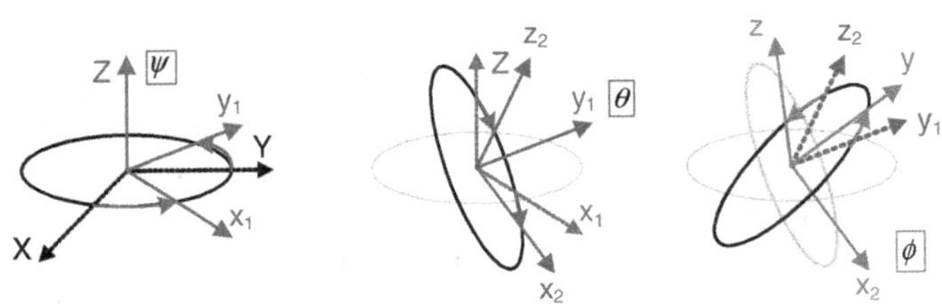

Figure 1.12 Euler angle(ϕ, θ, ψ).

From (1.19), we can derive (1.20):

$$\begin{bmatrix} \mathbf{u_X} \\ \mathbf{u_Y} \\ \mathbf{u_Z} \end{bmatrix} = \begin{bmatrix} \cos\theta\cos\psi & \cos\psi\sin\theta\sin\phi - \cos\phi\sin\psi & \cos\phi\cos\psi\sin\theta + \sin\phi\sin\psi \\ \cos\theta\sin\psi & \cos\phi\cos\psi + \sin\theta\sin\phi\sin\psi & -\cos\psi\sin\phi + \cos\phi\sin\theta\sin\psi \\ -\sin\theta & \cos\theta\sin\phi & \cos\theta\cos\phi \end{bmatrix} \begin{bmatrix} \mathbf{u_x} \\ \mathbf{u_y} \\ \mathbf{u_z} \end{bmatrix} \tag{1.20}$$

In (1.20), if all of the angles of rotation are small enough to approximate the value of sine function as the angle itself and the value of cosine function as unity, (1.20) can be approximated as a linearized matrix in (1.21):

$$\begin{bmatrix} \cos\theta\cos\psi & \cos\psi\sin\theta\sin\phi - \cos\phi\sin\psi & \cos\phi\cos\psi\sin\theta + \sin\phi\sin\psi \\ \cos\theta\sin\psi & \cos\phi\cos\psi + \sin\theta\sin\phi\sin\psi & -\cos\psi\sin\phi + \cos\phi\sin\theta\sin\psi \\ -\sin\theta & \cos\theta\sin\phi & \cos\theta\cos\phi \end{bmatrix}$$

$$\approx \begin{bmatrix} 1 & -\psi & \theta \\ \psi & 1 & -\phi \\ -\theta & \phi & 1 \end{bmatrix} \tag{1.21}$$

Using this simplified matrix, the relation between the unit vector of inertial reference frame and the unit vector of body fixed reference frame can be represented as (1.22):

$$\begin{bmatrix} \mathbf{u_X} \\ \mathbf{u_Y} \\ \mathbf{u_Z} \end{bmatrix} \approx \begin{bmatrix} 1 & -\psi & \theta \\ \psi & 1 & -\phi \\ -\theta & \phi & 1 \end{bmatrix} \begin{bmatrix} \mathbf{u_x} \\ \mathbf{u_y} \\ \mathbf{u_z} \end{bmatrix} \tag{1.22}$$

As mentioned earlier, if the center of mass is chosen as the center of rotation in a 6-DOF system, the equations of translational motion and rotational motion separately can be derived as (1.23) and (1.24):

$$\mathbf{F} = m\mathbf{a} + 2m\mathbf{v} \times \boldsymbol{\omega} \tag{1.23}$$

$$\mathbf{T_e} = \mathbf{J}\boldsymbol{\alpha} + \boldsymbol{\omega} \times \mathbf{J}\boldsymbol{\omega} \tag{1.24}$$

where $\mathbf{J} = \begin{bmatrix} J_{XX} & -J_{XY} & -J_{XZ} \\ -J_{XY} & J_{XX} & -J_{YZ} \\ -J_{XZ} & -J_{YZ} & J_{XX} \end{bmatrix}$

In (1.23), **F** stands for external force acting on a rigid body, M stands for mass of a rigid body, **a** stands for acceleration of the center of mass of a rigid body, **v** stands for velocity of the center of mass of a rigid body, $\mathbf{T_e}$ stands for external torque acting on a rigid body, **J** stands for a tensor of the moment of inertia of a rigid body against to the center of mass, $\boldsymbol{\omega}$ stands for angular velocity of a rigid body against to the center of mass, and $\boldsymbol{\alpha}$ stands for angular acceleration of a rigid body against to the center of mass, G.

In (1.23) and (1.24), there are two nonlinear terms caused by the Coriolis effect and the gyroscopic effect. If the magnitude of these terms is very small compared to the magnitude of linear terms, the equations above can be linearized by ignoring the nonlinear terms. Then the equations of motion can be expressed as linear equations:

$$\begin{bmatrix} F_x \\ F_y \\ F_z \\ \hdashline T_{ex} \\ T_{ey} \\ T_{ez} \end{bmatrix} = \left[\begin{array}{ccc:ccc} m & 0 & 0 & 0 & 0 & 0 \\ 0 & m & 0 & 0 & 0 & 0 \\ 0 & 0 & m & 0 & 0 & 0 \\ \hdashline 0 & 0 & 0 & J_{xx} & -J_{xy} & -J_{xz} \\ 0 & 0 & 0 & -J_{xy} & J_{yy} & -J_{yz} \\ 0 & 0 & 0 & -J_{xz} & -J_{yz} & J_{zz} \end{array}\right] \begin{bmatrix} a_x \\ a_y \\ a_z \\ \hdashline \alpha_x \\ \alpha_y \\ \alpha_z \end{bmatrix} \tag{1.25}$$

If several external forces are acting upon a rigid body, each of them can be decomposed into three independent components against to the axes of the body fixed frame. The resultant force of a specific axis can be represented by the summation of all the components in that axis. Also, the resultant torque can be obtained by the multiplication of the magnitude of each force and the distance from the center of the body fixed frame to the point of application of a force. For example, it is shown in Fig. 1.13 that seven different forces are acting on a rigid body parallel to the each axis of a body fixed frame.

The summation of f_1 and f_2 is the **x**-axis force, $\mathbf{F_x}$, the summation of f_3 and f_4 is the **y**-axis force, $\mathbf{F_y}$, and the summation of f_5, f_6, and f_7 is the **z**-axis force, $\mathbf{F_z}$. Because the point of application of f_1 and f_2 is not on the **x** axis, these two forces induce **y**- and **z**-axis torque. For same reason, **y**-axis forces induce **x**- and **z**-axis torque and **z**-axis forces induce **x**- and **y**-axis torque. The torque acting on each axis of the body fixed frame can be expressed as $\mathbf{r}_i = (x_i, y_i, z_i)$, $1 \leq i \leq 7$, which is the distance from the center of rotation of the rigid body to the point of application of $f_1 \sim f_7$.

$$\begin{aligned} &\mathbf{f}_i = (f_i, 0, 0),\ i = 1, 2;\ \mathbf{f}_i = (f_i, 0, 0), i = 3, 4;\ \mathbf{f}_i = (f_i, 0, 0), i = 5, 6, 7; \\ &\mathbf{T}_e = (T_{ex}, T_{ey}, T_{ez}) = \sum_{i=1}^{7} \mathbf{r}_i \times \mathbf{f}_i \end{aligned} \tag{1.26}$$

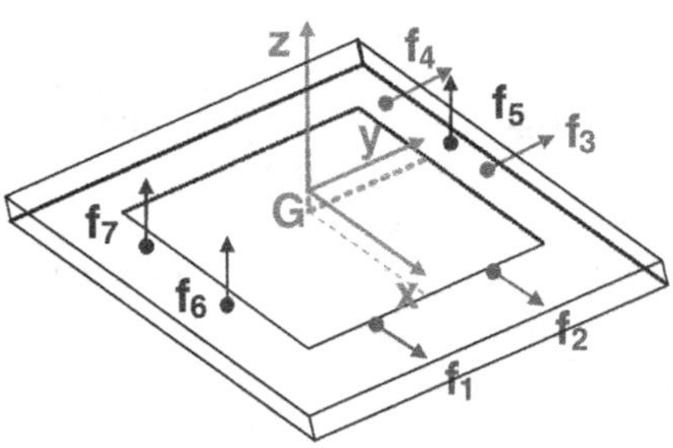

Figure 1.13 Seven external forces are acting on a rigid body.

The matrix form of the above equation is shown in (1.27)

$$\mathbf{u} = \mathbf{A}\mathbf{f} \tag{1.27}$$

$$\mathbf{A} = \left[\begin{array}{cc:cc:ccc} 1 & 1 & 0 & 0 & 0 & 0 & 0 \\ 0 & 0 & 1 & 1 & 0 & 0 & 0 \\ 0 & 0 & 0 & 0 & 1 & 1 & 1 \\ \hdashline 0 & 0 & -z_3 & -z_4 & y_5 & y_6 & y_7 \\ z_1 & z_2 & 0 & 0 & -x_5 & -x_6 & -x_7 \\ -y_1 & -y_2 & x_3 & x_4 & 0 & 0 & 0 \end{array}\right] \tag{1.28}$$

$$\mathbf{u} = [F_x, \ F_y, \ F_z, \ T_{ex}, \ T_{ey}, \ T_{ez}]^T \tag{1.29}$$

$$\mathbf{f} = [f_1 \ \ f_2 \ \ f_3 \ \ f_4 \ \ f_5 \ \ f_6 \ \ f_7]^T \tag{1.30}$$

In the above equations, $[\ldots]^T$ means transpose of vector or matrix $[\ldots]$.

In (1.27)–(1.30), the resultant forces and torques, which have 6-DOF in the body fixed frame, are represented by seven independent forces. In this case, to control the motion of the body only for a specific direction from the six different axes of motion, three or four forces among the seven independent forces should be controlled simultaneously. Then, there may be unintended forces or torques caused by the coupling effect of external forces. To implement precise control of 6-DOF rigid body motion, it should be done to decompose the motion of rigid body to the intended one and to the unintended one appropriately.

1.2.5 Power and Energy [1]

In linear motion system, the power, P, can be described as

$$P = f \cdot v \tag{1.31}$$

If the force, f, is constant during the motion, then the energy, E, is the product of the force, f, and the moving distance, l:

$$E = f \cdot l \tag{1.32}$$

Also, the energy is the integration of power regarding to the time. If the mass, M, does not vary during the motion, then the energy can be expressed as

$$E = \int P\,dt = \int f \cdot v\,dt = \int M\frac{dv}{dt} \cdot v\,dt = \frac{1}{2}Mv^2 \tag{1.33}$$

In a rotating motion, the power can expressed as

$$P = T_e \cdot \omega \tag{1.34}$$

As the linear motion, if the torque is constant during the motion, then the energy can be expressed as

$$E = T_e \cdot \theta \tag{1.35}$$

If the inertia, J, does not varies during the motion, then the energy can be expressed as

$$E = \int T_e \cdot \omega \, dt = \int J \frac{d\omega}{dt} \cdot \omega \, dt$$
$$= \int J\omega \, d\omega = \frac{1}{2} J\omega^2 \tag{1.36}$$

If the inertia, J, varies during the motion, then the variation of the inertia should be considered as (1.5).

As shown in (1.36), the energy can be stored in a rotating body. Recently, there have been many applications of the energy storage system using a high-speed rotating body under the name of flywheel energy stoage [13].

1.2.6 Continuity of Physical Variables

All physical vriables in the nature are finite, and a physical variable expressed as the time integral of another physical variable is always continuous. Because the force in the nature is always finite, the velocity and moving distance are always continuous in the linear motion and the angular velocity and moving angle are continuous in the rotating motion. In an electric machine, the thrust force and the torque are generated by the cross product of the current and its associated flux linkage. In the electromagnetic circuit, there is always inductance, and the flux and current are always continuous. Hence, the thrust force and the torque are also continuous. And, the linear acceleration and angular acceleration are also continuous. Therefore, the discontinuous function that can be implemented in reality is the jerk, which is time derivative of the acceleration. Furthermore, in the trajectory control, the planned trajectory (position or angle) can be obtained through the succesive time integration of the jerk. In this sucessive integration, the acceleration reference and velocity reference are easily obtained, and the references can be used to enhance the control performance of a postion regulation loop (see Section 4.4.2).

1.3 TORQUE SPEED CURVE OF TYPICAL MECHANICAL LOADS

The electric machines provide torque or force to operate the mechanical load or sometimes absorb torque or force from the mechanical load. The mechanical load employing the electrical machine as an actuator has its own torque–speed characteristics.

1.3.1 Fan, Pump, and Blower

Fan, pump, and blower are the loads that consume the most electricity in the developed countries. And those are used to move the fluids, and the torque of the loads in steady

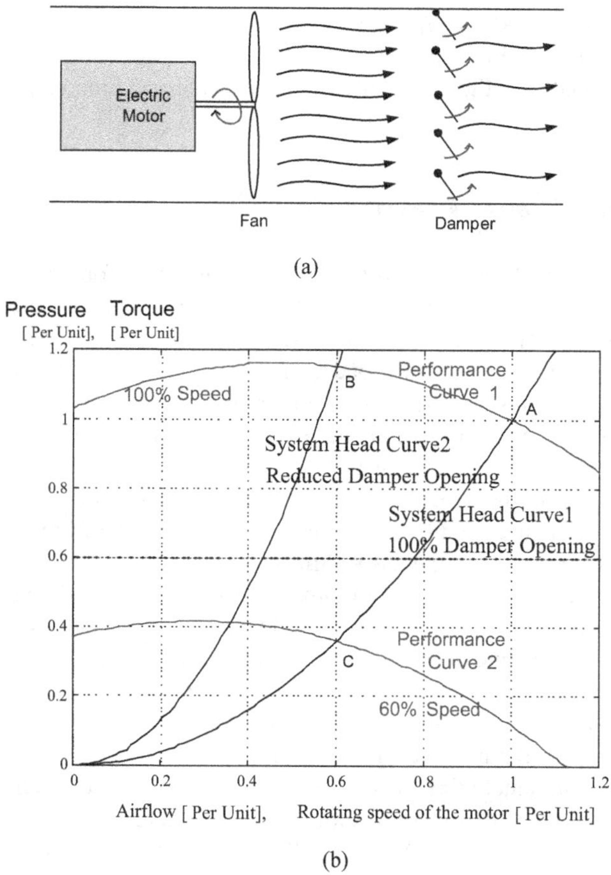

Figure 1.14 Control of airflow of a fan. (a) Control of airflow by a damper. (b) Performance curves and system head curves.

state is proportional to the square of the speed of flow of the fluid. Also, the power of the electric machine to drive the load is proportional to the cubic of the speed of flow. In Fig. 1.14a, the conceptual diagram of airflow control system by damper and a fan driven by an electric motor is shown. And in Fig. 1.14b, the typical performance curve of the fan and its torque–speed curve (system head curve) are shown. The operating point lies at the crossing point of two curves. As shown in Fig. 1.14a, if the airflow is controlled by the damper of the fan, then the flow, Q, can be reduced, but the pressure applied to the blades of the fan would be increased. In this case, as shown in Fig. 1.14b, the operating point moves from A to B, and the mechanical power by the machine changes from $P_A = H_A Q_A$ to $P_B = H_B Q_B$.

If the speed of the machine is adjusted to control the airflow, as shown in Fig. 1.14b, the operating point moves from A to C, and the mechanical power by the machine changes from $P_A = H_A Q_A$ to $P_C = H_C Q_C$. In this case, the torque to drive the fan decreases as the speed of the machine decreases, and the power by the machine would be decreased proportionally to the cubic of the speed of the machine or to the cubic of the airflow.

1.3.2 Hoisting Load; Crane, Elevator

In the steady state, the hoisting load requires torque due to gravitational force and friction force of the load. The torque against the gravitational force is independent with the moving speed of the load. However, the friction force increases as the speed increases, and the torque to drive the hoisting load could increase as the speed increases. In high-speed gearless elevator drive system or high-power crane drive system, where the friction force is negligible compared to the gravitational force or to the acceleration force, the torque is almost constant regardless of the speed. In Fig. 1.15, the torque–speed curves of the typical hoisting load are shown as a solid line and as a dashed line. The curve by the solid line curve is the case where the friction torque can be neglected, while the curve by the dashed line represents the case where the friction torque is proportional to the speed. If Coulomb friction is also considered in this case, the curve may have discontinuity at null speed. In the case of the elevator system, at the steady state the torque due the difference of the weight of the cage and counter weight is covered by the electric machine. In the high-speed elevator drive system, at acceleration and deceleration, 50% to 200% of the torque of the steady-state torque is needed to get the required acceleration and deceleration force to accelerate/decelerate the total mass including the masses of the cage and the counter weight. Hence, the electric machine to drive the elevator should have at least 150% to 300% overload capability for a short time, which is usually less than 10 s, to handle this torque. The peak motoring power of the machine occurs at just before the finishing point of the acceleration. In these hoisting loads, the electric machine should generate not only positive torque but also negative torque at either direction of rotation. Hence,

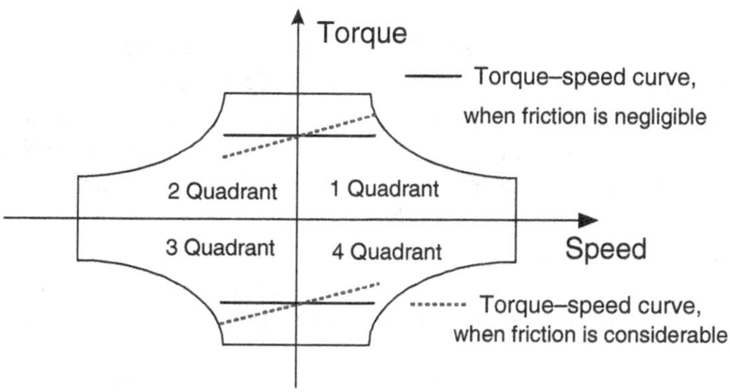

Figure 1.15 Torque–speed curve of a typical hoisting load.

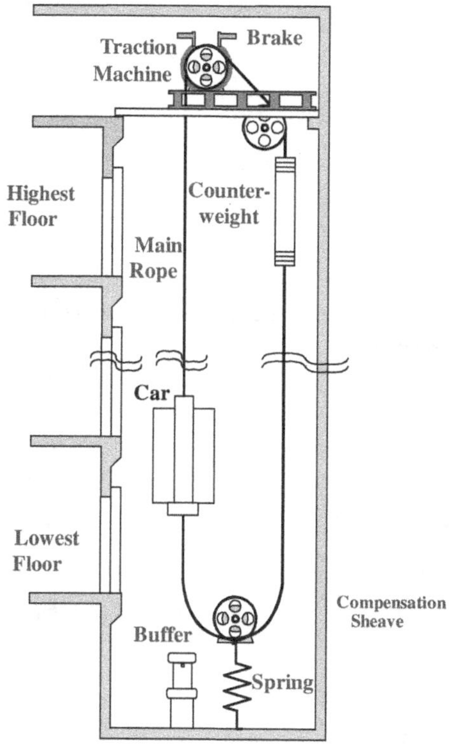

Figure 1.16 Gearless direct drive elevator system.

as shown in Fig. 1.15, the four-quadrant operation in a torque–speed plane is necessary in these hoisting loads.

In Fig. 1.16, a conceptual diagram of an elevator drive system is shown. As shown in the figure, the cage or car, where the passengers are, and a counterweight, whose weight is usually a half of the full weight of cage and passengers, are connected by a rope through the sheave of the traction machine driven by the electric machine. And by the rotation of the electric machine, the cage moves up or down.

1.3.3 Traction Load (Electric Vehicle, Electric Train)

The machine, used as the traction machine of the electric vehicle or the electric train, requires high torque at starting and low speed and requires low torque at high speed, as shown in Fig. 1.17. In the conventional internal combustion engine (ICE), the torque–speed range with reasonable efficiency is quite narrow, and the multi-ratio gear system—so-called transmission—is used to match the torque and speed of ICE to the operating condition of the vehicle. However, the electric machine can provide the required torque–speed characteristics without complex gear system. The required characteristics can be easily obtained by field (flux) weakening control of the machine. Also, the electric machine for the traction application can operate at four quadrants in a torque-speed plane contrast to ICE.

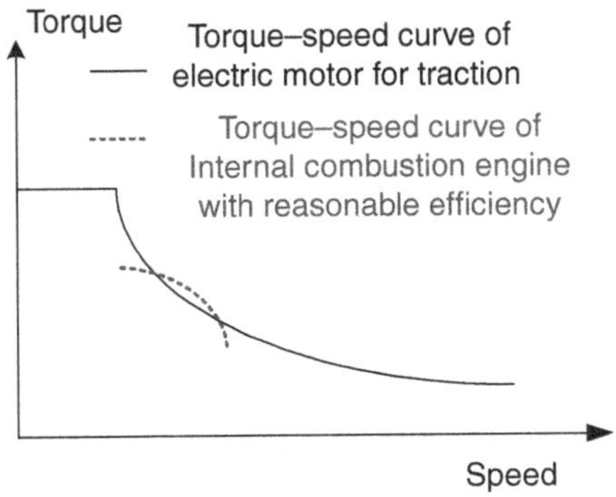

Figure 1.17 Torque–speed curves of an electric machine and internal combustion engine for traction application.

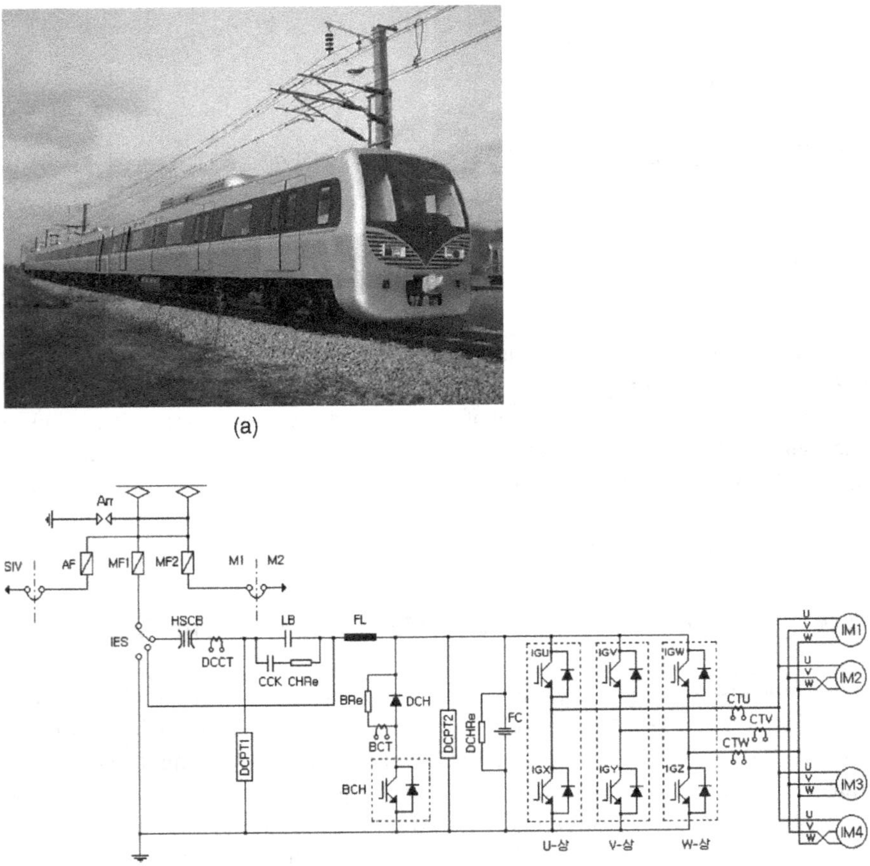

Figure 1.18 (a) Outer view of a subway train and (b) Main power circuit of a motor car of the subway train.

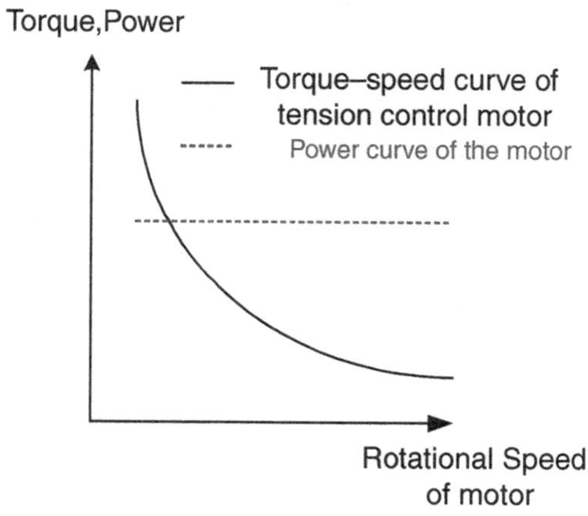

Figure 1.19 Torque–speed–power curve of tension control machine.

A circuit diagram of a motor car of the subway train is shown in Fig. 1.18, where an inverter driving four traction machines is powered by the catenary. With the development of the power electronics, to enhance performance and efficiency of the drive system, a new drive system, where each electric machine is driven by an inverter separately, is already applied in the field.

1.3.4 Tension Control Load

Usually, in the driving of the paper mill, steel mill, pay-off roll, and tension roll, the tension should be controlled as constant in the steady state. In this case, if the transportation speed of a paper sheet or a metal sheet is constant, the rotational speed of the machine decreases as the radius of the roll increases. Also, the output power of the machine is constant. However, in the acceleration or deceleration time, due to the torque for the acceleration and deceleration the constant power operation cannot be kept. In Fig. 1.19, the curves of torque and power of the electric machine driving a typical tension control system, where the metal sheet is moving at the constant speed, are shown. As an example of a tension control system, a continuous annealing line is shown in Fig. 1.20. In this line, the accuracy and bandwidth of the torque and speed control of the electric machine is crucial in the productivity of the process and the quality of the product.

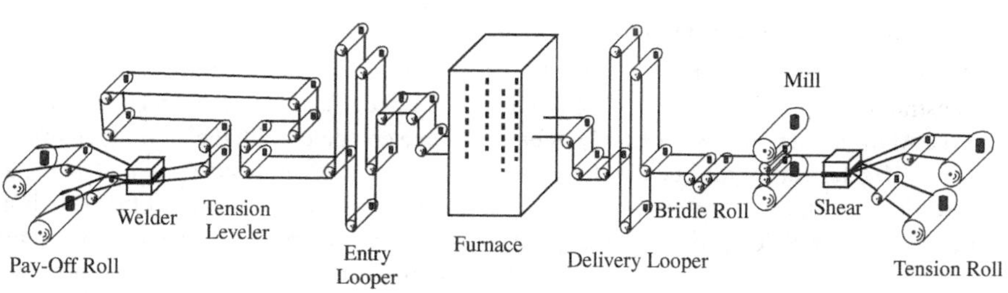

Figure 1.20 Continuous annealing processing line.

PROBLEMS

1. Calculate the moment of inertia of rotating cylinder as shown in Fig. P1.1. The density of the cylinder is ρ. Describe how to maximize the ratio, J/M under the condition of $r_1 + r_2 = \text{constant}$. Here, J stands for the inertia of the cylinder and M for the mass of the cylinder.

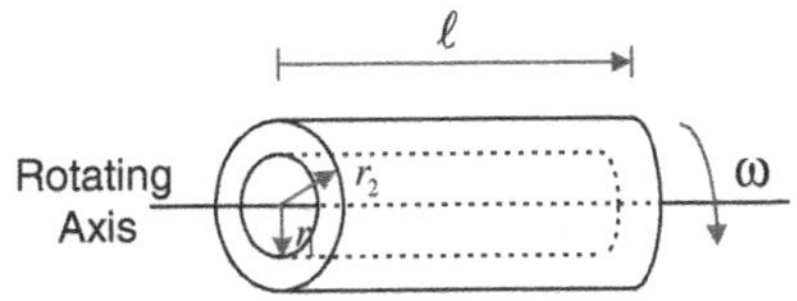

Figure P1.1 Moment of inertia of rotating cylinder.

2. As shown in Fig. P1.2, a disk is rotating regarding the z axis. The origin of the Cartesian coordinate is apart from the center of the mass, G, by 2(mm), 2(mm), 1(mm) as shown in the figure. The G can be represented by (–2 mm,–2 mm,–1 mm) in the coordinate. The radius of the disk, r, is 100(mm) and thickness of the disk is 10(mm).

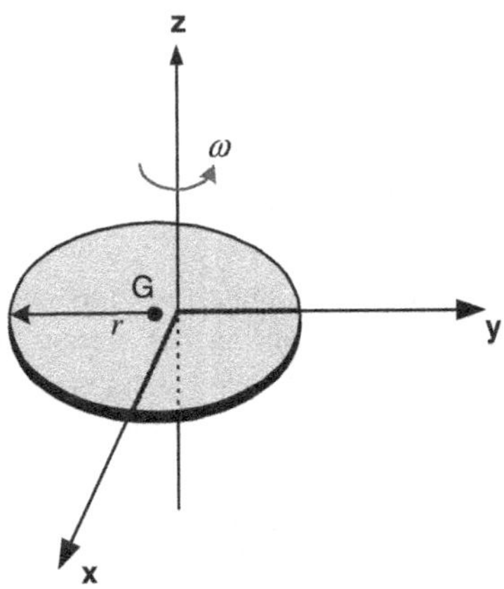

Figure P1.2 Rotating disk, whose rotating axis is slightly offset from the center of mass.

(1) In this case, calculate rotating inertia, $\mathbf{J_{xz}}$, $\mathbf{J_{yz}}$, $\mathbf{J_z}$, defined as (1.17).

(2) If the rotating speed of the disk, ω, is 1000 r/min, calculate torque at each **x, y, z** axis. Also calculate energy stored in this disk due to the rotation.

(3) Repeat problem 2 when the rotating speed is 100,000 r/min.

3. As shown in Fig. 1.13, there is a rigid body moving by seven forces. The body is not constraint in any axis of motion. The gravitational force is acting in the direction of the **z** axis. At the starting instant, the body reference frame coincides with the inertial reference frame. The lengths of the body in the **x** and **y** direction are 100 mm and the length in the **z** direction is 5 mm. The material of the body is stainless steel and the shape of the body is a rectangular parallelepiped. The initial position of the center of mass, simultaneously center of the rotation, G, is expressed in Cartesian coordinate in two reference frame as $(\mathbf{x}, \mathbf{y}, \mathbf{z}) = (\mathbf{X}, \mathbf{Y}, \mathbf{Z}) = (0, 0, 0)$, respectively. And the operating points, $\mathbf{r}_1 - \mathbf{r}_7$, of forces, $\mathbf{f}_1 - \mathbf{f}_7$, in the body reference frame are followings. All of the angles of rotation are small enough to

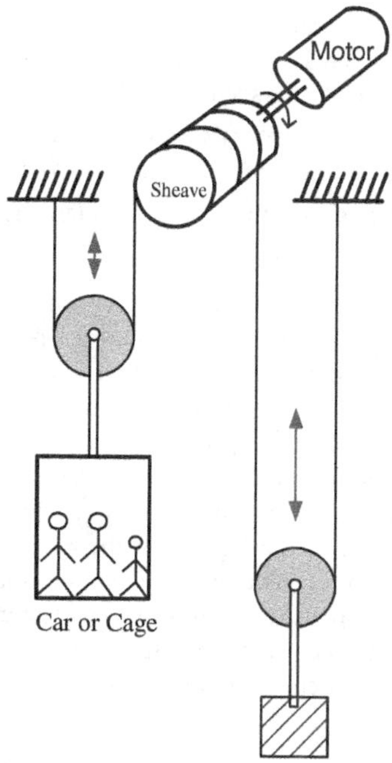

Figure P1.3 Conceptual diagram of a high-speed gearless elevator.

approximate the value of sine function as the angle itself and the value of cosine function as unity, and the nonlinear matrix in (1.20) can be linearized as (1.21).

$\mathbf{r}_1 = (40, -20, -2.5)\text{mm}, \quad \mathbf{r}_2 = (40, 20, -2.5)\text{mm}, \quad \mathbf{r}_3 = (20, 40, 2.5)\text{mm},$

$\mathbf{r}_4 = (-20, 40, 2.5)\text{mm}$

$\mathbf{r}_5 = (0, 40, 2.5)\text{mm}, \quad \mathbf{r}_6 = (20, -40, 2.5)\text{mm}, \quad \mathbf{r}_7 = (-20, -40, 2.5)\text{mm}$

(1) Find the matrix **A** in (1.27), which transforms the force, $\mathbf{f}_1$–$\mathbf{f}_7$, to the force and torque acting in each axis of the motion independently.

(2) Find the inertia matrix regarding three rotating axes.

$$J = \left[\begin{array}{ccc|ccc} m & 0 & 0 & 0 & 0 & 0 \\ 0 & m & 0 & 0 & 0 & 0 \\ 0 & 0 & m & 0 & 0 & 0 \\ \hline 0 & 0 & 0 & J_{xx} & -J_{xy} & -J_{xz} \\ 0 & 0 & 0 & -J_{xy} & J_{yy} & -J_{yz} \\ 0 & 0 & 0 & -J_{xz} & -J_{yz} & J_{zz} \end{array}\right]$$

(3) If the acceleration regarding only the **X** axis is 1 m/s^2 in the inertial reference frame and regarding all other axes there is no acceleration and movement except **X** axis translational motion, then find force $\mathbf{f}_1$–$\mathbf{f}_7$ for such a motion. *Hint*: $\mathbf{f} = \mathbf{A}^{\mathrm{T}}(\mathbf{A}\mathbf{A}^{\mathrm{T}})^{-1}\mathbf{u}$.

(4) If angular acceleration regarding only the **x** axis is 1 rad/s^2 in body fixed frame and regarding all other axes there is no acceleration and movement except **x**-axis rotational motion, then find force $\mathbf{f}_1$–$\mathbf{f}_7$ for such a motion.

(5) Describe the method regarding how to control the linear and angular acceleration independently by manipulating only the forces $\mathbf{f}_1$–$\mathbf{f}_7$.

4. In the cooling fan drive system for a thermal power plant, the airflow and air pressure has the following relationship (performance curve).

$H = 1.03N^2 + 0.56NQ - 0.59Q^2$, where N is the rotational speed of the fan, and Q stands for flow rate, H stands for air pressure, and all units are per unit (P.U.). The 1 P.U. of the speed of the electric machine corresponds to 1800 r/min, 1 P.U. flow rate corresponds to 1000 m^3/min, and 1 P.U. air pressure corresponds to 4243 N/m^2. The efficiency of the fan is given by $\eta = 0.5 + 0.3Q$, where η is per unit. The system head curve of the fan can be expressed as $H = Q^2$ when the damper is fully opened. And according to the damper opening the curve can be represented by $H = KQ^2$, where K depends on the damper opening. The operating point of the fan lies at the crossing point of the performance curve and the system head curve. The flow rate can be controlled by adjusting damper opening or by adjusting the rotating speed of the electric machine driving the fan. The required flow rate is proportional to the load factor of the generator of the power plant. If the required airflow for a year is assumed as follows, then answer the following questions:

$$50\%, 4000\ \text{hours};\quad 30\%, 2000\ \text{hours};\quad 20\%, 2000\ \text{hours}$$

(1) Select an electric machine to drive the fan from following choices. The fan should provide 100% flow rate. The rated speed of the all machines at following choices is 1800 r/min.
(a) 100 Hp (b) 125 hp (c) 150 Hp (d) 200 Hp (e) 250 Hp (f) 300 Hp

(2) Calculate total electricity (kWh) to drive the fan during a year for the following cases to control flow rate. It is assumed that the efficiency of the electric machine is 90% constant regardless of the load factor.

- (a) Calculate the electricity consumed for a year in the case of control of the damper opening.
- (b) Calculate the electricity consumed for a year in the case of adjusting speed of the electric machine. In this case, it is assumed that the efficiency of the VVVF system for the adjustable speed drive of the electric machine is 95% constant regardless of load factor, and damper is fully opened.
- (c) If the rate for the electricity is 0.1$/kWh, how much is the cost of electricity saved for a year by the adjustable speed control compared to damper opening control?

(3) Describe the advantages and disadvantages of adjustable speed control compared to damper opening control.

(4) Describe the reason why the flow rate of the fan is equal to or less than 50% in the case of the normal operation.

5. The high-speed elevator system, shown in Fig. P1.3, has following specifications. Answer the following questions.

- Rated speed: 240 m/min, 24 passengers (weight of payload: 1600 kg), maximum number of floor for movement: 30 floors

- Distance between floors: 3 m
- Mass of counterweight: 3134 kg
- Mass of cage: 2345 kg
- Inertia of sheave: 95 kg-m^2
- Inertia of machine: 25 kg-m^2
- Diameter of sheave: 710 mm

The mass of rope and inertia of the pulley on the cage can be neglected. The jerk profile at ascending operation is shown in Fig. P1.4.

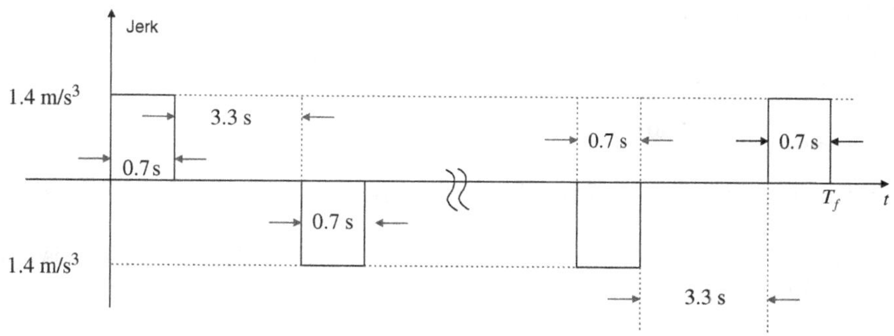

Figure P1.4 Jerk profile of the elevator at ascending operation.

(1) When the elevator moves from the first floor to the thirty-first floor (30 floors × 3 *m*/floor) with the jerk profile as shown in Fig. P1.4, calculate the total travel time, T_f.

(2) In the case of part 1, plot the acceleration, velocity, and position of the cage according to the time.

(3) If the traction machine of the elevator can withstand 200% overload for 10 s, select the minimum capacity of an electric machine to drive the elevator from the following choices:
(a) 40 Hp (b) 50 Hp (c) 60 Hp (d) 70 Hp (e) 80 Hp

(4) From the following choices, what are the number of poles and rated frequency of the machine selected from part 3? Here, there is no slip between machine, sheave, and rope. The maximum speed of the machine is decided by "rated frequency × 60/(number of poles/2)" (revolutions/min).
(a) 15 Hz, 4 poles (b) 15 Hz, 8 poles (c) 30 Hz, 4 poles (d) 30 Hz, 8 poles (e) 60 Hz, 8 poles (f) 60 Hz, 16 poles

(5) Plot the output power of the motor according to the time in case of part 1. In this problem, the slip between rope, sheave, and pulley can be neglected and the friction of all moving parts also can be ignored. The weight of the payload is 1600 kg.

6. Answer the following questions for a crane, shown in Fig. P1.5 under the following assumptions and specification.

- The torque transfer efficiency of each pulley is 98%.
- The gear ratio is 400:1 and efficiency is 90%. And the inertia of gear itself can be neglected. There is no slip between sheaves, rope, and pulleys.

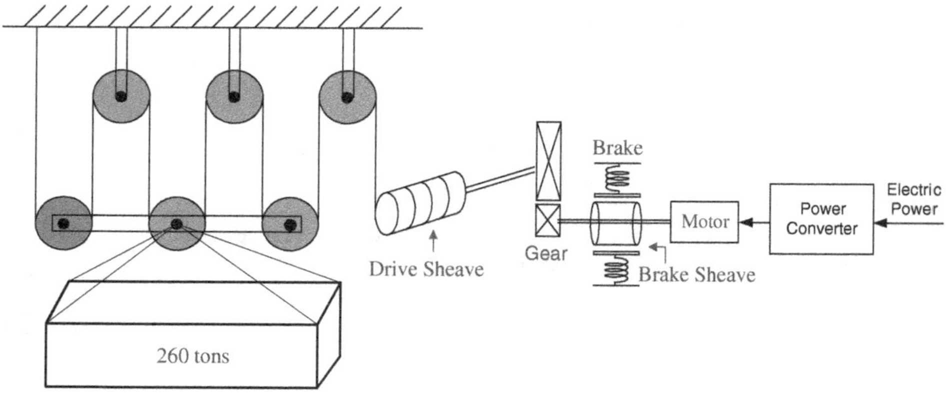

Figure P1.5 Conceptual block diagram of a crane.

- The drive sheave is a solid cylinder whose length is 1.2 m; its diameter is 1 m and its density is 7800 kg/m^3.
- The density of the steel part of the brake sheave, which is a hollow cylinder as shown in Fig. P1.6, is 7800 kg/m^3.

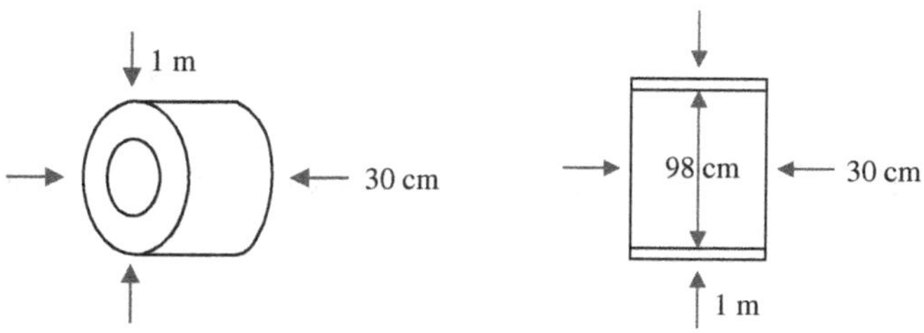

Figure P1.6 Structure of the brake sheave.

- The mass of the rope and the elongation of the rope can be neglected.
- The maximum speed of the machine is decided by "rated frequency × 60/(number of poles/2)" (revolutions/min).

The speed pattern to move a 260-metric-ton payload is shown in Fig. P1.7.

(1) Plot the rotational speed of the machine according to the time for the given speed pattern.

(2) Plot the torque of the machine according to the time for the given speed pattern.

(3) It is assumed that the total efficiency of the electric power conversion system (VVVF drive ASD sytem) including electric motor is 90% regardless of the load factor of the system. Plot the power to the electric power conversion system according to the time.

(4) Select a suitable electric motor from following choices. The motor can withstand 150% overload, which means 2.5 times of rated power of the machine, for 30 seconds. The pole numbers of the machine is four and the rated frequency of the machine is 60 Hz. There is no slip between pulleys, machine, and sheave. The maximum speed of the

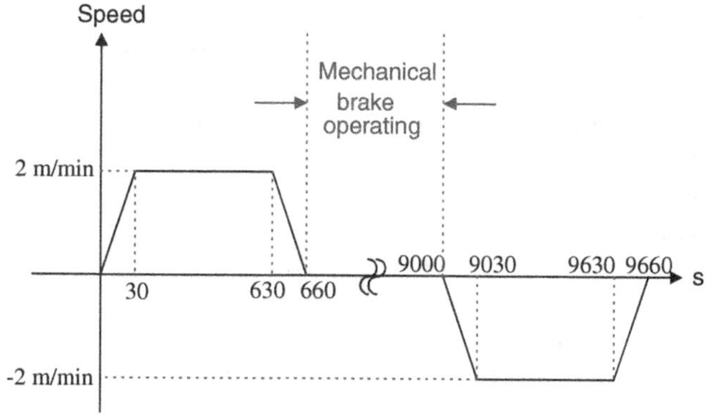

Figure P1.7 Speed pattern of a 260 ton payload

machine is decided by "rated frequency × 60/(number of poles/2)" (revolutions/min).
(a) 75 kW (b) 90 kW (c) 110 kW (d) 132 kW

7. As shown in Figure 1.P8, there is a commuter train consisted with four motor cars (M car) and four trailer cars (T car).

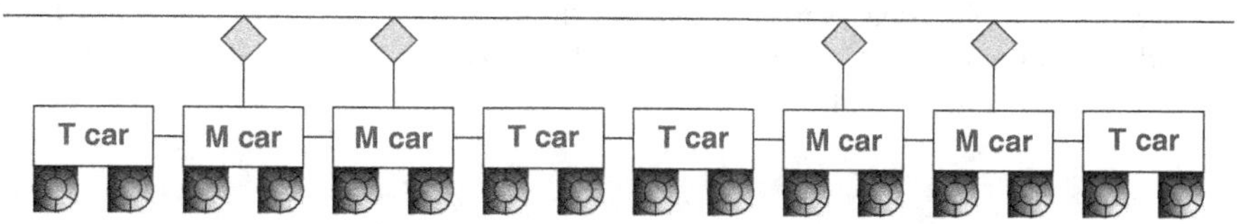

Figure P1.8 Typical configuration of the commutating train

The mass of T car is 28 metric tons, and that of M car is 30 metric tons. The maximum payload of each car is 20 metric tons. The rolling resistance of the train is given by the following equation:

$$R = 1.867 + 0.0359\,v_a + 0.000745\,v_a^2$$

where the unit of R is kg-G/ton = 9.8 N/ton, v_a is the speed of the train, and its unit is km/h.

The train accelerates up to 35 km/h with the maximum acceleration, 3.0 km/h/s. The maximum speed of the train is 100 km/h. During acceleration, the inertia of the rotating parts of the train such as wheels, gears, and electric machines can be converted to an equivalent mass in calculating of the traction force. The converted inertia to the mass is called "inertia mass." The compensation factor of the inertia mass to consider the traction force due to the rotating parts of each car is 6% in the case of a T car and 14% in the case of an M car. Hence, the equivalent total mass (including the inertia mass) to consider the torque due to the rotating parts of the car can be given by following equation.

Equivalent mass at acceleration = Total mass of car including payload + (Compensation factor for the inertia mass * Mass of each car itself)

Only 97% of the mechanical torque from the machine is transfered to the rail and that is used as the tractive force to move the train. Each M car has four traction motors.

(1) Calculate the power rating of the motor in kilowatts. The power rating of the motor is decided at the operating speed where the tractive force is the maximum (in this problem it is 35 km/h).

(2) The gear ratio between motor and wheel is 7.07:1 and the gear is reducing gear. When the average speed of the train is 48 km/h, the rotating speed of the motor is the rated value. The diameter of the wheel is 0.82 m constant.

(a) What is the rated rotating speed (revolutions/min) of the traction motor?

(b) At maximum train speed, 100 km/h, what is the rotating speed (revolutions/min) of the motor?

(c) At the operating condition given in part 1, if the efficiency of the AC traction motor is 92% and the power factor is 85%, then calculate the apparent input power (kVA) to the machine.

8. In Fig. P1.29, a conceptual diagram of the recoil line of steel mill processing system.

The total inertia of the rotor of the motor including gearbox, referred to the axis of the rotor of the motor, is $J = 200$ kg-m^2. The density of steel web is 7870 kg/m^3, and the web is tightly wound on the drums and there is no empty space in the steel roll. The density of the drum, whose minimum diameter is 0.61 m, is the same with that of steel web. And the drum is a solid cylinder. Also, the minimum diameter of the roll is 0.61 m when no web on the drum, and the maximum diameter of the roll is 2.6 m when the web is fully wound on the drum. The thickness of the steel web is 2 mm, the width, 900 mm, and the profile of the speed of the web is shown in Fig. P1.10. The constant 64,000 N tension is always applied to the steel web by pay-off and tension rolls.

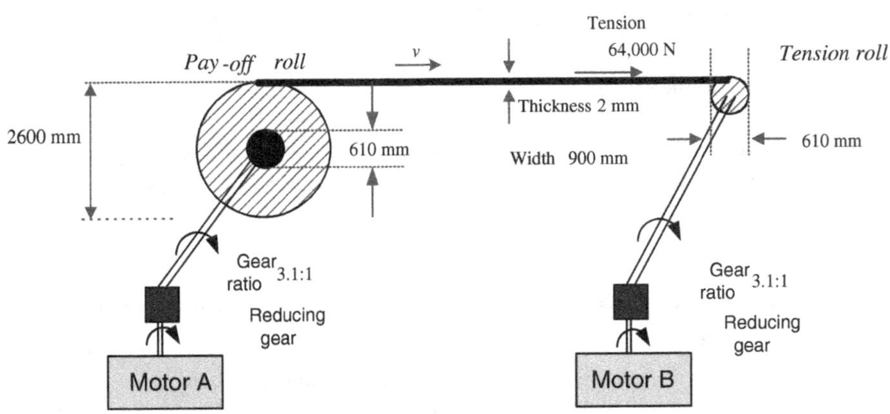

Figure P1.9 Conceptual diagram of a recoil line of a steel mill processing system.

(1) Plot the radius of the tension roll according to the time from minimum value to the maximum value.

(2) With consideration of gear ratio, plot the rotational speed (r/min) of the motor B according to the time.

(3) Plot the torque (N-m) of the motor B according to the time.

(4) Plot output power (kW) of the motor B according to the time.

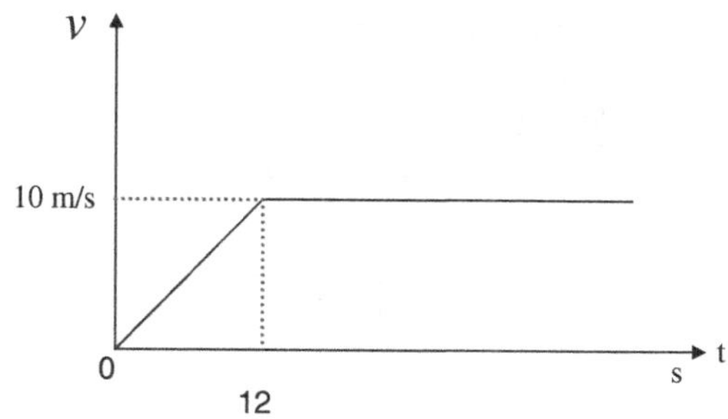

Figure P1.10 Speed profile of the movement of steel web.

9. In Fig. P1.11 a photo of a rubber-tired gantry crane (RTGC) used for handling containers in a port is shown. The mass of a spreader, which is used to catch the container, is 11 metric tons. The mass of trolley is 23 metric tons [14].

Figure P1.11 Photo of a rubber-tired gantry crane.

The mechanical system of RTGC including an electric machine is similar to that shown in Fig. P1.25. At hoist motion the total inertia, including gears, brake drum, rope, sheave, and machine itself reflected to the rotating axis of electric machine is 11 kg-m^2. Regardless of torque and direction of motion, the total efficiency of the mechanical system for hoist motion is 86% constant and that of the electric machine is 95% constant. And there is no slip between all moving components. The weight of the rope can be neglected. The speed, 800 r/min, of the hoist motor, used for the vertical motion of the container, means 60 m/min of hoisting speed of the container.In trolley motion, the total inertia including gears, brake drum, rope, sheave, and electric machine itself reflected to the rotor of the machine is 0.4 kg-m^2. Regardless of torque and direction, the total efficiency of the mechanical system for trolley motion is 90% constant and that of the electric machine is 92% constant. The speed, 1750 r/min, of the trolley motor, used for the horizontal motion of the container, means 70 m/min of hoisting speed of the container. In trolley motion, the hoist motor is locked by the mechanical brake, and the power to the hoist motor is null.

(1) The spreader of the RTGC is moving from Layer 4, Column 6 to Layer 4, Column 1, shown in Fig. 1.12, without a container. First (1) the spreader moves vertically up from

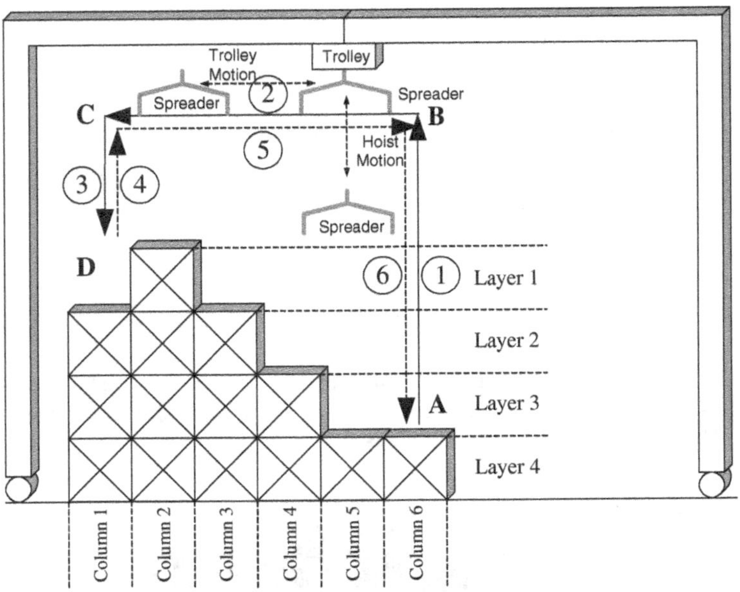

Figure P1.12 Operation of RTGC (rubber-tired gantry crane).

Layer 4 to the top (line between point A and point B), for 7.95 m, and then (2) the spreader moves horizontally to Column 6 (line between point B and point C), for 17.24 m; finally (3) the spreader moves vertically down to Layer 4 (line between point C and point D). In the hoist motion, the acceleration is (60 m/min)/2.5 s constant and deceleration is (60 m/min)/3.5 s constant. The maximum speed of the hoist motion is 60 m/min. In the trolley motion, the acceleration and deceleration are both (70 m/min)/5.5 s constant. The maximum speed of the trolley motion is 70 m/min. During trolley motion, the spreader and trolley can be considered as a single mass. Plot the torque and speed of the hoist machine and trolley machine according to the time, respectively. Also plot the sum of input power to the machines according to the time.

(2) The spreader of the crane is now holding a container whose mass is 40 metric tons, and moves in reverse (from point D, C, B, and A as the motion described in part 1). Plot the torque and speed of the hoist machine and trolley machine according to the time, respectively. Also, plot the sum of input power to the machines according to the time. During trolley motion, the spreader, trolley, and container can be considered as a single mass. The acceleration and deceleration of the hoist motion is the same with the case of part 1, but the maximum speed of the hoist motion is 24 m/min.

10. For RTGC in problem 9, the electric power to the machine is provided by a 400-kW engine generator set as shown in Fig. P1.13.

(1) When the hoist machine is a six-pole, permanent-magnet synchronous machine, plot the input frequency to the hoist machine according to the time at the motion described in problem 9.
In trolley motion, the machine for hoist motion is stopped. (The speed of the machine is decided by "input frequency × 60/(number of poles/2)" (r/min).)

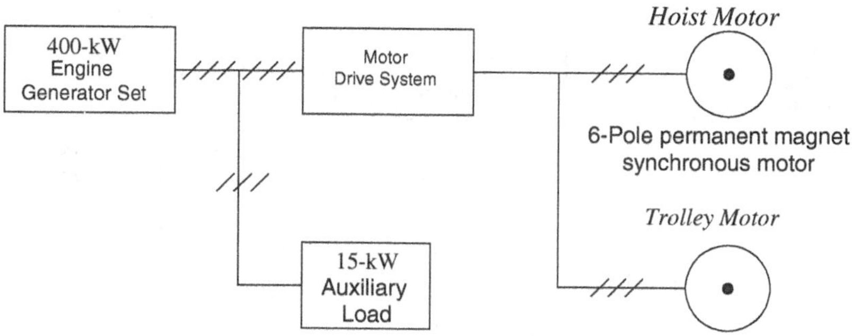

Figure P1.13 Electric power supply system of RTGC based on an engine generator set.

(2) The fuel consumption of the engine generator is given by $y = \frac{14.16}{3600} + \frac{0.163}{3600}x$, where "$x$" means output power of the generator in kW and "y" means consumption of diesel fuel in liter/s. The auxiliary load is a 15-kW constant, and the regenerated power at the vertical down motion of the load (the spreader and the container) cannot be transferred to the auxiliary load or not to the generator. Hence, the regenerated power is dissipated at the resistor box, which is separately installed in the drive system. The efficiency of the motor drive system is 90% constant regardless of the load factor.

- **(a)** For the motion described in problem 9, part 1, calculate total quantity of the consumed diesel fuel.
- **(b)** For the motion described in problem 9, part 2, calculate total quantity of the consumed diesel fuel.
- **(c)** For an hour, 8 times of motion described in problem 9, part 1 and 8 times of the motion described in problem 9, part 2 have been done, and for other times of the hour the generator only supplies the electricity to the auxiliary load. For such operation, calculate total quantity of the consumed diesel fuel for an hour.

(3) The electric power system for RTGC has been changed to the system shown in Fig. P1.14.

The engine generator can supply up to 150 kW, and its response is fast enough. The response of the power converter is also fast enough. And the response time of the engine and power converter can be neglected. The efficiency of the power converter is 90% constant regardless of the load factor and direction of power flow. Before the hoist motion the super-capacitor, C_s, is charged to 600 V, always. And the capacitance of the capacitor is 25 F. The power needed for the electric machine drive system is, at the first, supplied from the engine generator set as much as possible, and then the additional power to the drive system is supplied from super-capacitor through the power converter. The efficiency of the electric machine drive system is 90% constant regardless of the load factor and direction of power flow. The regenerated power at the vertical down motion of the load (the spreader and the container) can be used to charge the super-capacitor. And if the voltage of the super-capacitor is less than 600 V after the hoist motion, the capacitor should be charged to 600 V using the power from the generator through the power converter. The super-capacitor, C_s, can be assumed as

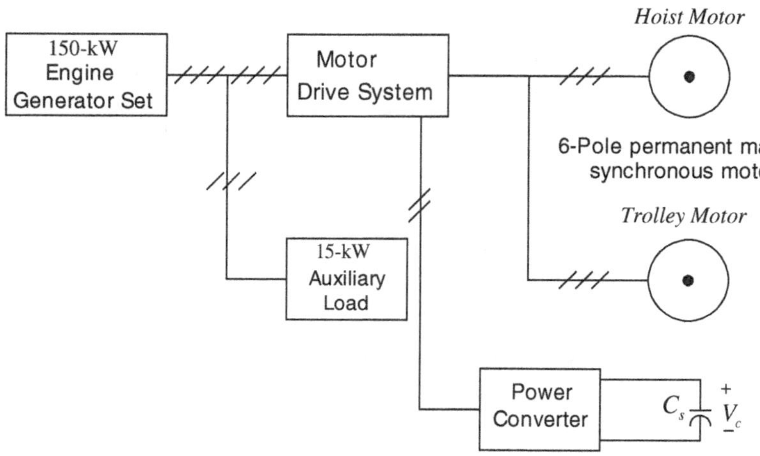

Figure P1.14 Power supply system of RTGC based on engine generator and an energy storage system based on the super-capacitor.

an ideal capacitor. The stored energy in the capacitor and the power to the capacitor are given by following equations.

$$E = \frac{1}{2} C_s V_c^2 [\text{J}]$$

$$P = C_s V_c \cdot \frac{dV_c}{dt} [\text{W}]$$

Also, the power converter for charging and discharging of the super-capacitor operates only for hoisting motion. At the other operation, the power converter turns off and there is no loss at the power converter and at super-capacitor. The fuel consumption of the 150-kW engine–generator set is given by $y = \frac{7.76}{3600} + \frac{0.156}{3600}x$, where "$x$" means output power of the generator in kilowatts and "y" means consumption of diesel fuel in liters per second.

(a) For the motion described in problem 9, part 1, calculate total quantity of the consumed diesel fuel.

(b) For the motion described in problem 9, part 2, calculate total quantity of the consumed diesel fuel

(c) For the motion described in problem 9, part 2, plot the voltage of the super-capacitor according to the time. In here, the initial and final value of the voltage of the capacitor should be 600 V.

(d) For an hour, 8 times of motion described in problem 9, part 1 and 8 times of motion described in problem 9, part 2 have been done, and for other times of the hour the generator only supplies the electricity to the auxiliary load. For such operation, calculate total quantity of the consumed diesel fuel for an hour.

(4) For a year, calculate the saving of diesel fuel cost of the system shown in Fig. P1.14 compared to the system shown in Fig. P1.13 under the following assumptions.

The price of diesel fuel is \$1/liter, and the operation time of RTGC is 5000 h/year, the operating patterns are given by parts 2(c) and 3(d), respectively.

(5) Compare the advantages and disadvantages of the system shown in Fig. P1.14 against the system shown in Fig. P1.13 in the following viewpoints:

(a) Total life cycle cost (Initial cost and running cost)

(b) Effects to environment

(c) Control performance

REFERENCES

1. W. Leonhard, *Control of Electrical Drives, Introduction,* 2nd edition, Springer, Berlin, 1996, Chapter 1.
2. A. Bonnett, Understanding efficiency in squirrel-cage induction machines, *IEEE Trans. Ind. Appl.*, Vol. IA-16, No. 4, July/August 1980, pp. 476–483.
3. R. Browning, Evolution of induction machines the ever-shrinking machine, *IEEE Ind. Appl. Mag.*, January/February 1997, pp. 16–18.
4. A. Bonnett, An update on AC induction machine efficiency, *IEEE Trans. Ind Appl.*, Vol. 30, No. 5, September/October 1994, pp. 1362–1372.
5. H. C. J. de Jong, *AC Machine Design,* Hemisphere Publishing Corporation, New York, 1989.
6. K. K. Schwarz, *Design of Industrial Electric Machine Drives,* Butterworth-Heinemann Ltd., Oxford, U.K., 1991, Chapters 1 and 2.
7. B. Bae et al., Implementation of sensorless vector control for super-high-speed PMSM turbo-compressor, *IEEE Trans. Ind. Appl.*, Vol. 39, No. 5, September/October 2003, pp. 811–818.
8. A. Elasser et al., A comparative evaluation of new silicon carbide diodes and state-of the art silicon diodes for power electronic applications, in *IEEE Industrial Applications Society Annual Meeting Conference Record,* 1999, pp. 341–345.
9. Texas Instruments *TMS6701 Reference Manual,* 1999.
10. J. R. Calvert, *Mechanics for Electrical and Electronics Engineers,* Ellis Horwood Ltd., 1992.
11. H. H. Woodson and J. R. Melcher, *Electromechanical Dynamics,* John Wiley & Sons, New York, 1968, Section 2.2.1.
12. J. Park, Development of 6 DOFs ultra precision positioning system using the PZT actuators and elastic hinges, Ph.D thesis, School of Mechanical Engineering, Seoul National University, 2003.
13. www.upei.ca/~physics/p261/projects/flywheel1/flywheel1.htm.
14. S. Kim et al., control of rubber tyred gantry crane with energy storage based on supercapacitor bank, *IEEE Trans. Power Electron.*, Vol. 21, No. 5, September/October 2006, pp. 1420–1427.

Chapter 2

Basic Structure and Modeling of Electric Machines and Power Converters

In this chapter, the basic structure and the operation principle of the electric machine, which converts mechanical energy to electrical energy as a generator or converts electrical energy to mechanical energy as a motor, is described. And the steady-state equivalent circuits of several machines are introduced to understand the steady-state characteristics and control of the machines. Also, several examples to control the machine with regard to a motor are discussed with its control features. In addition, the power converters, which convert the electrical energy to another form of the electrical energy based on power semiconductors, are modeled as the equivalent circuits. The brief description of the operation principle of each power converter is given. The models of the machine and the power converter will be used as a plant model in Chapter 4, where the design of the controller is described in detail.

2.1 STRUCTURE AND MODELING OF DC MACHINE

The DC machine had been used for 100 years for adjustable speed drive (ASD), and the variable torque control had been implemented based on the DC machine. However, since the late 1980s with the developments of power electronics and control technologies of the AC machine, the DC machine is getting out of date. But, still in many traditional industries, the DC machine is operating.

As shown in Fig. 2.1, the DC machine has a field winding for the excitation flux and armature winding to generate torque through the interaction with the flux. As shown in Fig. 2.2, the machine can be classified as shunt, series, and compound type according to the connection of a field winding and an armature winding. Each type of the machine has its own torque–speed characteristics. In early times, a specific type of the machine had been used for the specific application, but with the development of

Control of Electric Machine Drive Systems, by Seung-Ki Sul

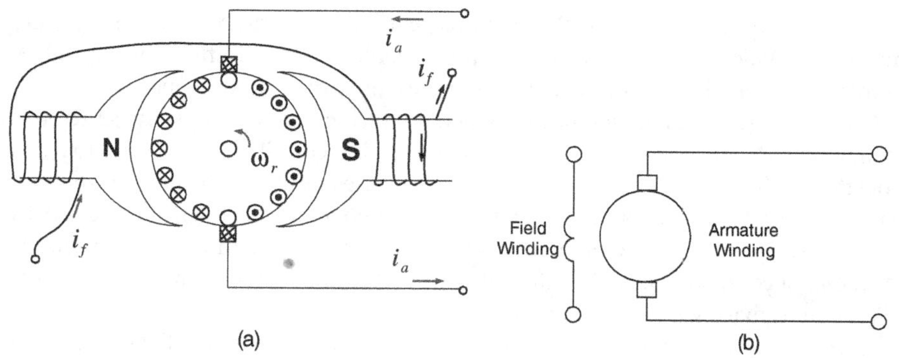

Figure 2.1 Basic structure of DC machine and its circuit symbol. (a) Field winding (current, i_f) and armature winding (current, i_a). (b) Symbol of DC machine.

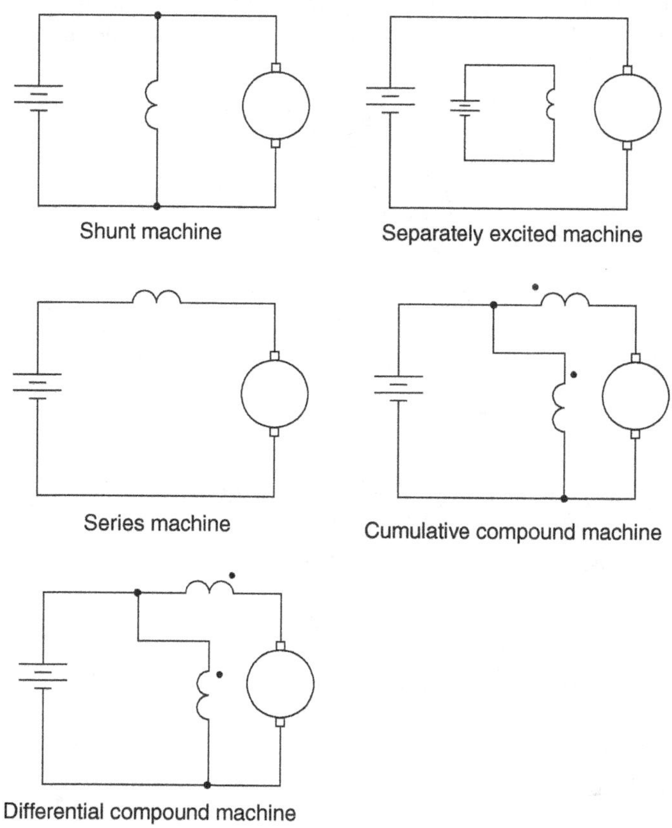

Figure 2.2 Classification of DC machine according to the connection of a field winding and an armature winding.

power electronics technology, the shunt (especially a separately excited shunt) DC machine is widely used in conjunction with the control of both the field current and the armature current. By the control of both currents, the various torque–speed curves can be implemented, and it has been applied to various applications. In series and compound DC machine, the armature current also contributes to excitation flux together with its own field winding. And in the series machine, the field current and armature current are the same. In the cumulative compound machine, the field flux by the armature current is the same direction of the flux by the field winding current. In differential compound machine, the field flux by the armature current is the opposite direction to that by the field winding current.

To get the continuous torque, the DC motor needs electric rectification devices consisting of a commutator and a brush, because the induced voltage at each turn of the armature winding by the excitation flux is AC voltage whose frequency is proportional to the rotating speed of the armature. To connect AC voltage of the armature winding to a DC source, the AC voltage should be rectified. In a DC machine the commutator and the brush rectify AC voltage to DC voltage.

In Fig. 2.3, a simple circuit diagram of a two-pole DC machine is shown with armature and field winding. The armature winding is connected to an external DC

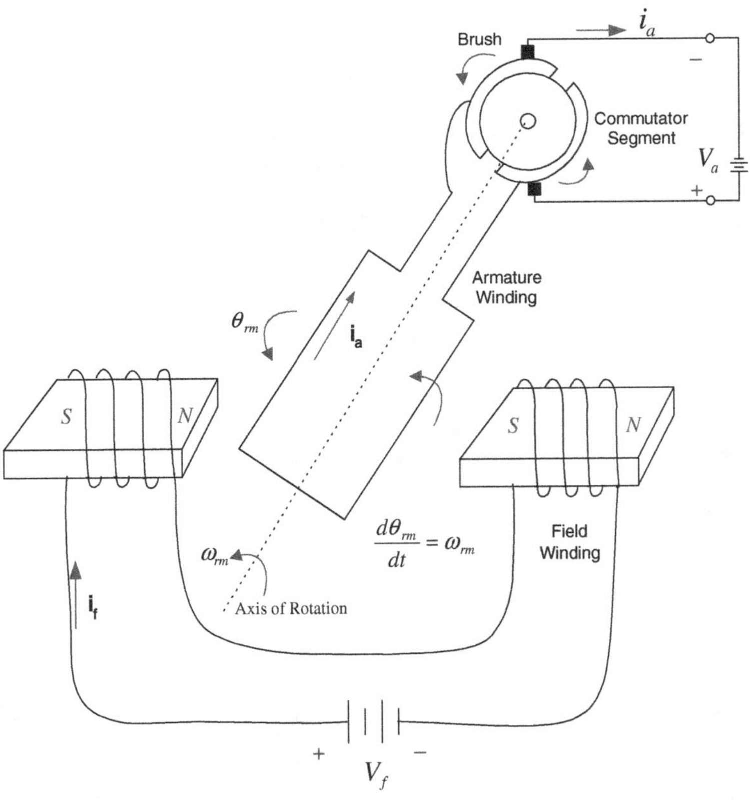

Figure 2.3 Commutator and brush of DC machine.

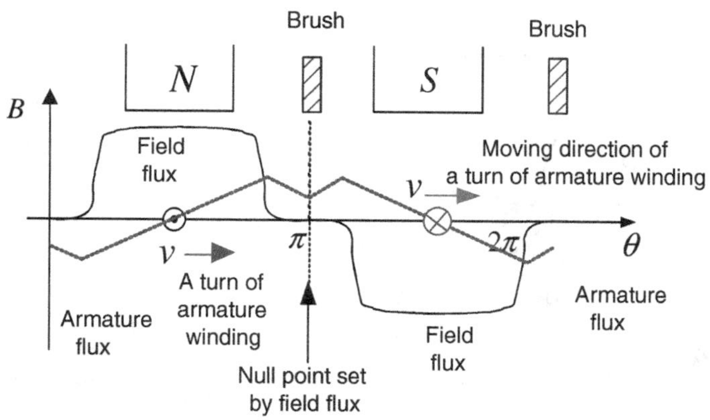

Figure 2.4 Field flux and armature flux.

source, V_a, through a commutator and a brush. As shown in Fig. 2.4, the brush locates at the position, where the flux density is null. And through the commutator and the brush, the turn of the armature winding, whose polarity of the induced voltage is positive, is always connected to the positive terminal of the DC source feeding DC machine. The current flows in the armature winding, as shown in Fig. 2.4, induces a flux by the current itself. Because of the flux by armature current the spatial flux distribution, originally set by the excitation flux from the field winding current, is distorted. The spatial distribution of the air gap flux is the vector sum of the flux by armature winding current and the field winding current. Also, the distortion of the spatial distribution of the flux is getting worse as the armature current is getting larger in magnitude. The distortion of the flux by the armature current, referred to as an armature reaction, results in the shift of the null point of the flux density distribution, and the commutating of the armature current from one turn to another turn becomes difficult. In the worst case, the commutator segments experience flashover and over-current due to the short circuit of the induced voltage at the commutating instants. To reduce these harmful effects on the commutation from the armature reaction, commutating poles and a compensation winding can be installed as shown in Fig. 2.5. In the case of a larger DC machine or a highly dynamic DC machine, the commutating poles and the compensation winding are indispensable.

Under the assumption that the armature reaction is perfectly canceled out by the commutating poles and the compensation winding, a DC machine can be modeled using the following equations:

$$V_f = R_f i_f + \frac{d\lambda_f}{dt} \tag{2.1}$$

$$e = K_e \lambda_f \omega_{rm} \tag{2.2}$$

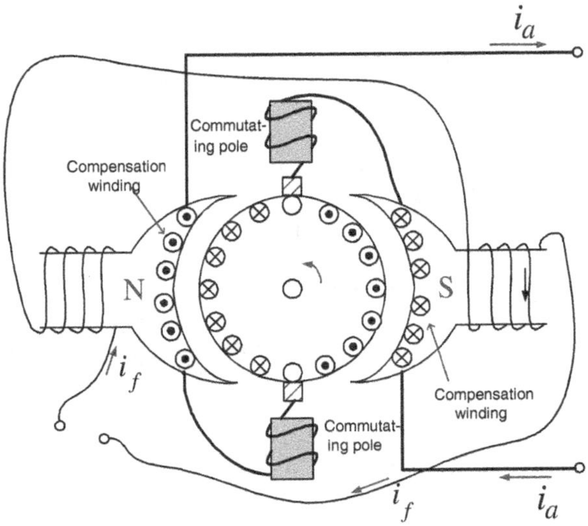

Figure 2.5 Commutating poles and a compensation winding.

$$V_a = R_a i_a + L_a \frac{di_a}{dt} + e + V_b \tag{2.3}$$

$$T_e = K_T \lambda_f i_a \tag{2.4}$$

$$\lambda_f = L_f i_f \tag{2.5}$$

$$T_e = J\frac{d\omega_{rm}}{dt} + \omega_{rm}\frac{dJ}{dt} + B\omega_{rm} + T_L + K_{sh}\theta_{rm} \tag{2.6}$$

where e stands for back electromotive force (EMF) in volts(V), V_a stands for terminal voltage in V, T_e stands for torque in N·m, λ_f stands for the linkage flux to the armature current by the field winding current in Wb·turn, V_f stands for applied voltage to the field winding in V, K_e stands for the back EMF constant in V/(Wb·turn·rad/s), ω_{rm} stands for the rotating angular speed of the rotor in (rad/s), R_a stands for an equivalent armature winding resistance in Ω, and L_a stands for an equivalent armature winding inductance in H. V_b stands for the voltage drop by the brush in V, which depends on the rotating speed of the rotor and the wear-out of the brush. The voltage drop shows nonlinear characteristics and is usually less than 1% of the rated armature voltage of DC machine, and it can be easily neglected. However, in the low-speed operation or in stalled operation the voltage drop by the brush should be considered. K_T stands for the torque constant in N·m/(Wb·turn·A), R_f stands for an equivalent field winding resistance in Ω, and L_f stands for an equivalent field winding inductance in H.

If the magnetic saturation of the field flux by the field current on the magnetic circuit of DC machine occurs, the field flux should be expressed as a nonlinear function of the field current as $\lambda_f = f(i_f)$. In the case of series DC machine where the armature current flows in the field winding, the saturation should be considered.

In (2.1)–(2.5), the armature reaction is neglected. However, if the flux by the field winding is much smaller (less than one-third of the rated value) than its rated value or the armature current is much larger (larger than several times of the rated value) than its rated value, then the armature reaction should be considered. In this case, torque constant and back EMF constant, K_T and K_e, may be reduced considerably due to the armature reaction.

Equation (2.6) is an equation of the motion of the rotor, where T_L means load torque. The torque usually reveals nonlinear characteristics according to the rotating speed, and it can be expressed as a nonlinear function of the speed, $T_L(\omega_{rm})$. J stands for the total inertia of the rotating part of the drive system, referred to as the rotor of the motor in $kg \cdot m^2$ driven by DC machine including the inertia of machine itself. B stands for friction coefficient in N·m/(rad/s), and K_{sh} stands for a stiffness coefficient in N·m/rad. When a DC machine rotates continuously and the inertia of the rotating part of the drive system is constant during the motion, then $K_{sh} = 0$ and $dJ/dt = 0$. And, (2.6) can be simplified as (2.7). In addition, the friction coefficient, B, also varies nonlinearly according to the speed and direction of the rotation, and if the torque due to the friction is included in the term of the load torque, then (2.7) can be further simplified as (2.8).

$$T_e = J\frac{d\omega_{rm}}{dt} + B\omega_{rm} + T_L(\omega_{rm}) \tag{2.7}$$

$$T_e = J\frac{d\omega_{rm}}{dt} + T_L(\omega_{rm}) \tag{2.8}$$

2.2 ANALYSIS OF STEADY-STATE OPERATION

In the steady state, the rotating speed of a DC machine is constant as $d\omega_{rm}/dt = 0$, and the armature and field current are also constant as $di_a/dt = di_f/dt = 0$. And if the voltage drop by the brush is neglected, (2.1)–(2.5), (2.8) can be rewritten as (2.9)–(2.14) in the steady state.

$$e = K_e\lambda_f\omega_{rm} \tag{2.9}$$

$$V_a = R_a i_a + e \tag{2.10}$$

$$T_e = K_T\lambda_f i_a \tag{2.11}$$

$$\lambda_f = L_f i_f \tag{2.12}$$

$$V_f = R_f i_f \tag{2.13}$$

$$T_e = T_L \tag{2.14}$$

From (2.9)–(2.14), the steady-state characteristics of several different types of DC machine can be derived.

2.2.1 Separately Excited Shunt Machine

In a separately excited shunt machine, the voltage applied to the field winding and the current flowing armature winding can be controlled separately, and hence versatile torque–speed control characteristics can be obtained. From (2.9), (2.10), and (2.11), after eliminating e and i_a, the relationship between V_a, λ_f, and ω_{rm} can be deduced as (2.15):

$$T_e = \frac{K_T \lambda_f V_a}{R_a} - \frac{K_T K_e \lambda_f^2 \omega_{rm}}{R_a} \tag{2.15}$$

At (2.15), if λ_f is constant, then the torque–speed curves varies according to V_a as shown in Fig. 2.6.

The operating point of the machine in torque–speed plane locates at the crossing point of the torque–speed curve of the machine and the torque–speed curve of the load. Whether the continuous operation at the crossing point is possible or not is decided by the stability of the system at the point. For a stable operating, point the following condition should be met: $\partial T_e/\partial\omega_{rm} < \partial T_L/\partial\omega_{rm}$. Otherwise, even with a very small disturbance the operating point moves to a different stable point. In the case shown in Fig. 2.7, as the armature voltage, V_a, increased, the operating point would move from A to B to C in the steady state, because all points are stable. However, while it is certain that when the voltage, V_a, varied from V_{a1} to V_{a2}, the operating point would move finally from A to B, it is uncertain with regard to what trajectory the operating point would move from A to B and also with regard to when the operating point would arrive at B from A. As shown in Fig. 2.7, there are infinite cases of trajectory from A to B, and the trajectory can be decided by the transient analysis described in Section 2.3.

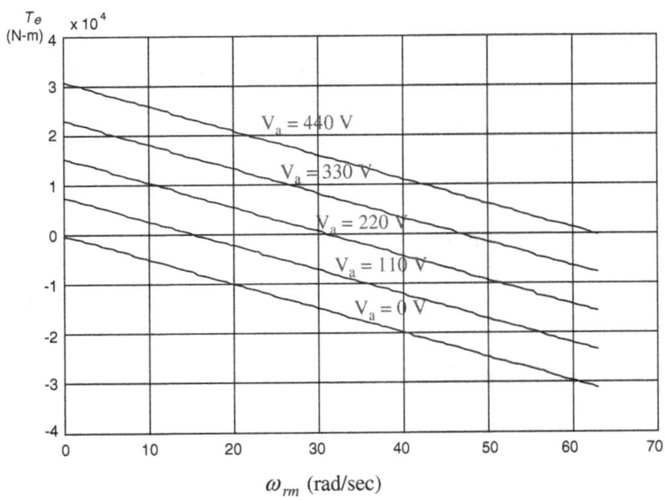

Figure 2.6 Torque–speed curves of a separately excited shunt DC machine according to V_a with constant λ_f (rated armature voltage 440 V, rated power 110 kW, rated speed 560 r/min).

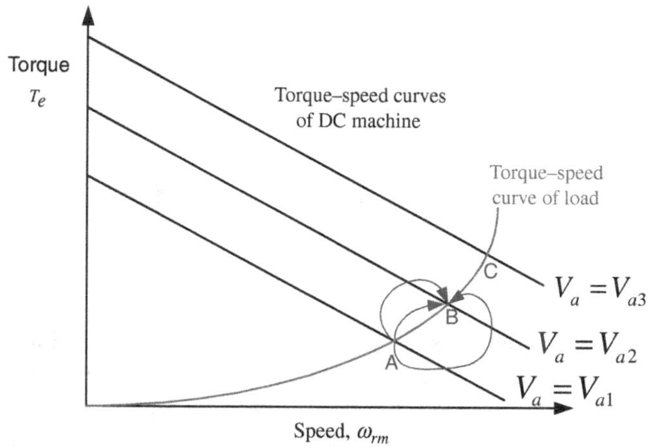

Figure 2.7 Migration of the operating point according to variation of the armature voltage.

In (2.15), if V_a is constant, then by varying field current, i_f, and hence by varying λ_f, the torque–speed curves as shown in Fig. 2.8 can be obtained.

With these torque–speed curves, if the load torque–speed curve looks like the curve shown in Fig. 2.8, then the operating points would move from A to B to C as the field flux decreases. Also, the operating speed increases. Usually, the rated field flux is set as the maximum value to get the maximum torque constant, where the magnetic saturation does occur slightly. Because the rated field flux is the available maximum value to the machine, the variation of the field flux means the weakening of the flux. So, if λ_{f1} is the rated value, then λ_{f2} and λ_{f3} are the values in which the flux are reduced (field weakening). Through the field weakening control, the operating speed can be increased, but as given in (2.11), to get the same torque as the torque without the field weakening, the armature current should be increased by the amount by which the flux reduced. Also, the increased armature current results in increased copper losses. The maximum armature current is usually limited due to the thermal limit of the machine itself or due to the other parts of the drive system. Hence, if the speed is increased by the field weakening, then the available maximum torque at the increased speed is

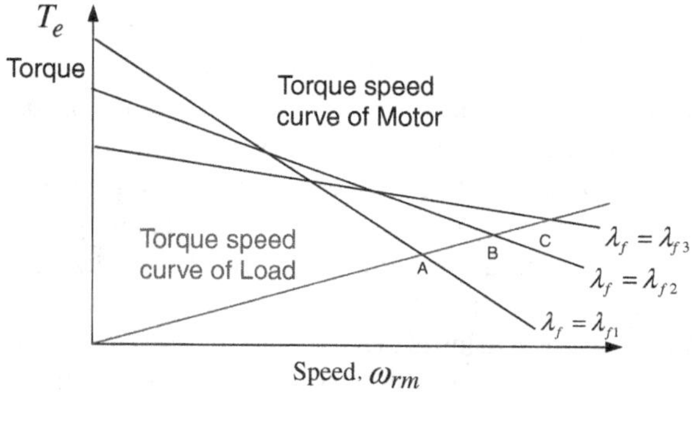

Figure 2.8 Speed–torque curves with the variation of the field flux.

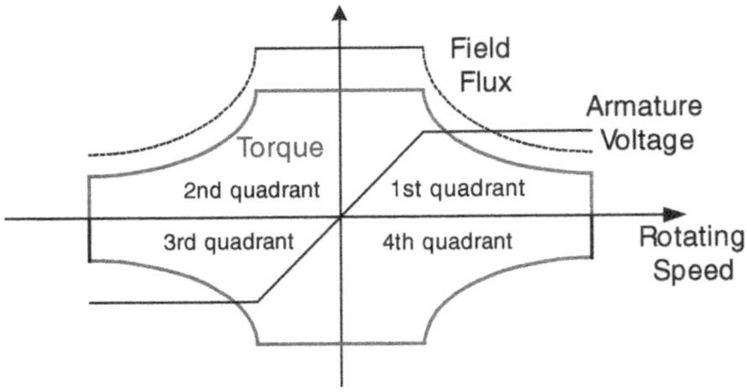

Figure 2.9 Operating area (capability curve) with the control of armature voltage and field weakening control.

reduced proportional to the reduced field flux. Moreover, in the field flux weakening operation, the armature reaction is relatively larger and commutation problems may occur. Because of the higher operating speed by the field weakening control, the mechanical problems in the commutator and the brush of the machine may also occur. Hence, the minimum flux level for the field flux weakening control is usually limited down to a certain value—for example, one-third of the rated value.

Through the flux weakening control along with armature voltage control, a DC machine can be operated in a wide region of torque–speed plane as shown in Fig. 2.9. In Fig. 2.9, the boundary of the region where a DC machine can operate under limited voltage and limited current condition is defined as the capability curve of a DC machine. The capability curve can be specified in the first quadrant of the torque–speed plane or in all four quadrants of the plane. The capability curve shown in Fig. 2.9 is obtained after neglecting the armature reaction, voltage drop due to the armature winding and the brush, and friction and windage losses of the machine itself. With consideration of the losses and the voltage drops, the capability curve would be asymmetric with regard to the vertical axis of the torque–speed plane. In this case, the area in the second and the fourth quadrant of the torque–speed plane, where the machine works as a generator, is larger than the area in the first and the third quadrant, where the machine operates as a motor.

The control method to increase the operating speed of the electric machine through the field weakening can be applicable to not only a DC machine but also to an AC machine. The detailed description about the flux weakening control of AC machine will be in Sections 2.8, 2.15, and 5.4. The field weakening control to get a higher operating speed can be compared with the gearing of the machine to get a higher speed. In the case of the gearing, the speed is decided by gear ratio, which is mostly discontinuous. In addition, the gear itself adds rotating inertia to the drive system, and the acceleration performance would be degraded. However, by the field weakening, the speed can be increased continuously up to the maximum speed. The maximum output power of the machine by the field weakening control is the same as the output power of the machine through gearing at the same speed if the loss of gear itself is neglected. With the gearing and with the field weakening, the speed can be increased but the available torque at the increased speed is reduced, and the power, which is the product of the speed and the torque, is always constant regardless of the speed. One advantage of the gearing is that the speed can be increased to more than several times of rated speed, while the speed control by the field weakening is limited up to a certain point—for example, three times the rated speed—due to above-mentioned commutating and mechanical problems.

If the machine operates in a light load, then the load torque can be matched by the machine with the reduced flux in the cost of increased the armature current. While the increased current results in more copper loss, the iron loss would be decreased due to the reduced flux level. The iron loss is roughly proportional to the square of the flux level, and there should be an optimum flux level where the total loss, which is the sum of copper loss and iron loss, is minimized at the given load torque [1]. By this kind of field weakening control, the efficiency of the machine can be improved in the light load condition. However, if the flux level is too low and an impact load is suddenly applied, then the machine would stall because the flux cannot increase rapidly due to the field circuit time constant $\left(\tau_f = \frac{L_f}{R_f}\right)$. So, the machine might be permanently damaged with the excessive armature current at the stalled condition. Thus, to apply this field weakening control to improve operating efficiency of the machine, the dynamics of the drive system should be considered carefully.

To control a DC machine in four quadrants of the torque–speed plane, both the torque and the speed of the machine should be controlled in positive and negative direction independently. As seen in (2.11), because the product of the flux and the armature current represents the torque, and to change the direction of the torque, for four-quadrant operation the direction of the flux or the direction of the armature current should be changed. The direction of the flux can be changed by changing the direction of the field current, but the time constant of the field circuit is quite large, and it would take more than several hundred milliseconds in the case of above several-kilowatt-range machines. Moreover, in the course of changing the direction of the flux, the flux level would be zero at a certain instant. Near at that instant, the armature reaction, if the armature current flows, could be prohibitive, and armature current should be controlled as zero during field current reversal. Therefore, during the field reversal, DC machine loses the torque controllability over the load. This is a severe demerit of the field reversal control for the reversal of the torque. However, the power rating of the field circuit is only a small percentage of the armature circuit, and the field reversal can be easily implemented without much cost. That is a merit of the field reversal control. Another method to change direction of the torque is to change the direction of the armature current. For four-quadrants operation by armature current control, the voltage and current of the armature should be controlled in the positive and the negative direction. The detailed implementation of the control of armature voltage and current will be described in Section 2.4.

2.2.2 Series Excited DC Machine

Nowadays, the series excited DC machine is not used very much in newly designed system, but it is still used in the limited application area such as a traction motor or as a universal motor for home appliance. From (2.9)–(2.13), the voltage equation of a

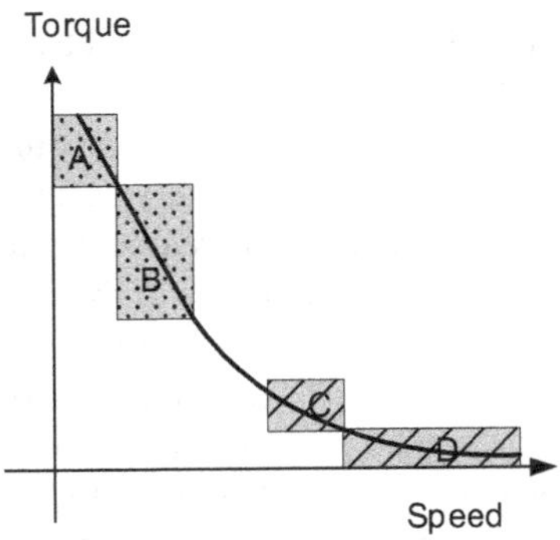

Figure 2.10 Torque–speed curve of a series excited DC machine under constant terminal voltage.

series machine driven by DC source can be expressed as (2.16) with the constraint of $i_a = i_f$ after eliminating e and i_a.

$$V_a = \sqrt{\frac{T_e}{K_T L_f}} R_a + \sqrt{\frac{T_e}{K_T L_f}} R_f + K_e L_f \sqrt{\frac{T_e}{K_T L_f}} \omega_{rm} \tag{2.16}$$

In Fig. 2.10, there is a torque–speed curve of a series motor in the first quadrant of the torque–speed plane under the constant terminal voltage.

In the figure, regions C and D are the field weakening area, and especially in the case of the operation in region D, the flux is too small to commutate current smoothly. Also, the operation at D region is normally not allowed because of the armature reaction. In regions A and B, the magnetic circuit of the machine is saturated due to the excessive field current, where $i_f = i_a$, and the flux is almost constant as λ_{sat} regardless of the increase of the current. Hence, the torque would vary linearly with the current at those regions. Especially at region A where back EMF of the machine is too small, the continuous operation may result in permanent damage to the machine because of the excessive current. The torque–speed curve of the series machine, as shown in Fig. 2.10, reveals larger torque at lower speed and smaller torque at higher speed without any special control means. So, the machine had been used widely in the traction application, where such torque–speed characteristics are requisite. But, nowadays, due to the developments of power electronics, the application of a series machine to that application area is becoming rare.

2.3 ANALYSIS OF TRANSIENT STATE OF DC MACHINE [2]

Without considering armature reaction, magnetic saturation, and voltage drop due to the brush and the commutator, the equivalent circuit of a separately excited DC machine can be represented as shown in Fig. 2.11. Also, (2.1)–(2.7) can be simplified as (2.17)–(2.22). Here, the inertia and the friction coefficient are assumed to be constant regardless of the rotational speed. In the transient state, the characteristics of a DC machine can be expressed by solving the equations. The three differential equations, (2.17), (2.18), and (2.22), in general, are nonlinear equations, and it is very difficult to find general analytical solutions. Furthermore, the analysis of the stability

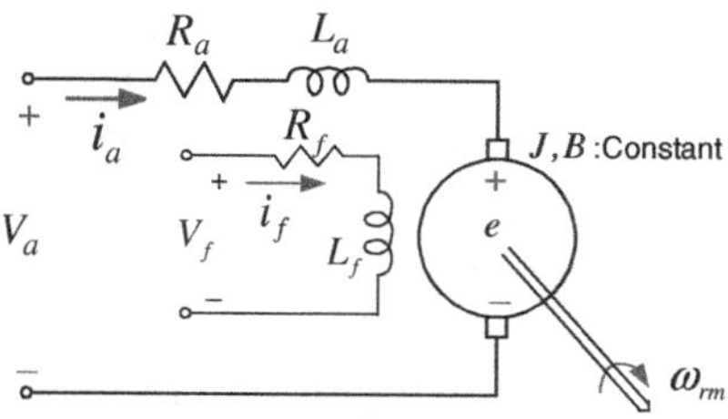

Figure 2.11 Equivalent circuit of a separately excited DC machine.

of the nonlinear system in overall operating region is quite difficult. At a given operating point, by using small signal analysis the system can be linearized, and the stability at that operating point can be evaluated under the assumption of the small disturbance. Also, by using numerical analysis based on the computer simulation, the operating characteristics of the DC machine can be evaluated in the overall operating region.

$$V_a = R_a i_a + L_a \frac{di_a}{dt} + e \tag{2.17}$$

$$V_f = R_f i_f + L_f \frac{di_f}{dt} \tag{2.18}$$

$$e = K_e \lambda_f \omega_{rm} \tag{2.19}$$

$$T_e = K_T \lambda_f i_a \tag{2.20}$$

$$\lambda_f = L_f i_f \tag{2.21}$$

$$T_e = J \frac{d\omega_{rm}}{dt} + B\omega_{rm} + T_L \tag{2.22}$$

2.3.1 Separately Excited Shunt Machine

If the field winding of machine is excited by a constant current source or if the field flux is provided by a permanent magnet, then the excitation flux can be considered as constant. Under this assumption, (2.17) and (2.22) can be rewritten as the linear differential equations. If MKS units are used in (2.17)–(2.22), then back EMF constant, K_e, and torque constant, K_T, are identical. If the products of the constant and field flux, λ_f, is expressed as (2.23), then (2.17), (2.20), and (2.22) can be transformed to (2.24)–(2.26) through Laplace transformation with zero initial conditions. In these equations, the unit of K is $\mathrm{N \cdot m/A}$ or V/(rad/s), which are identical to each other.

$$K_e \lambda_f = K_T \lambda_f = K \tag{2.23}$$

$$V_a = (R_a + sL_a) I_a + K\omega_{rm} \tag{2.24}$$

$$T_e = K I_a \tag{2.25}$$

$$T_e = (Js + B)\omega_{rm} + T_L \tag{2.26}$$

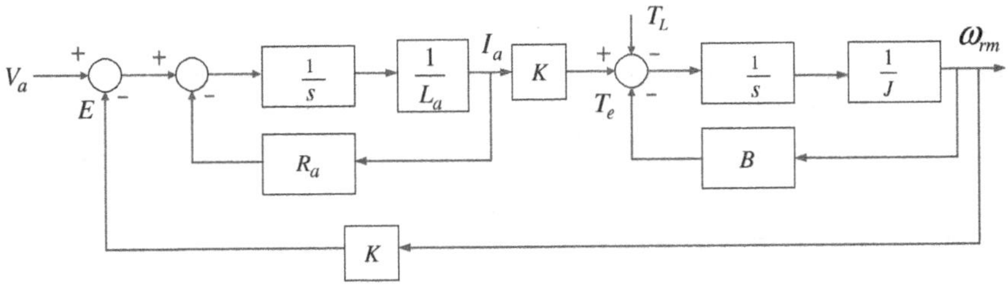

Figure 2.12 Control block diagram of a separately excited DC machine.

Equations (2.24)–(2.26) can be formulated as a control block diagram as shown in Fig. 2.12. In the figure, the overall system consists of the cascade connection of the electrical part and the mechanical part. And there is a feedback loop given by back EMF.

In Fig. 2.12, R_a is the armature winding resistance and it is a damping component of the electric part. Similarly, B is the friction coefficient, and it is a damping component of the mechanical part. Both damping components degrade the efficiency of the system, but they may contribute to enhance the stability of the system. If there are no damping components, that means $R_a = 0$ and $B = 0$, as shown in Fig. 2.13, the system would oscillate without decaying and its oscillation angular frequency is $K/\sqrt{L_a/J}$. By introduction of the damping components into the mechanical part and/or electrical part, though the efficiency of the system is getting worse, the oscillation decays out.

The transfer function regarding speed to the terminal voltage of the system in Fig. 2.12 is given in (2.27).

$$\frac{\omega_{rm}}{V_a} = \frac{K}{(sL_a + R_a)(Js + B) + K^2} \tag{2.27}$$

As K is getting larger, the eigenvalues, which are roots of denominator of (2.27), have imaginary parts, and their magnitudes are getting larger. And, it can be said that as K is getting larger, the damping is getting smaller. Hence, the current, torque, and speed responses are all getting more oscillatory according to the variation of the armature voltage.

In (2.27), if the torque due to the friction is included in load torque, T_L, or the friction torque is neglected and if the transient phenomena by the armature

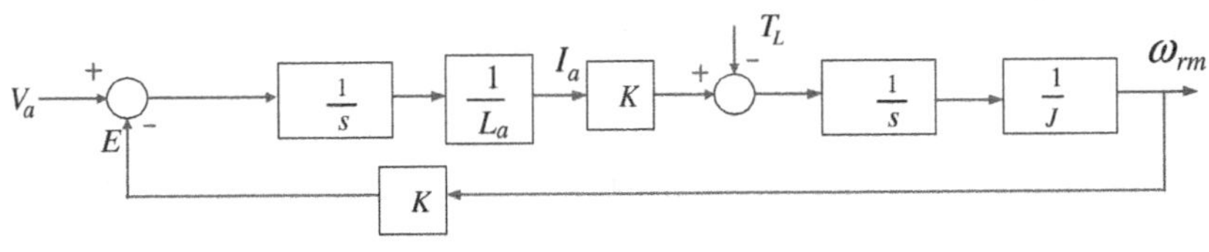

Figure 2.13 Control block diagram of a separately excited DC machine with no damping component.

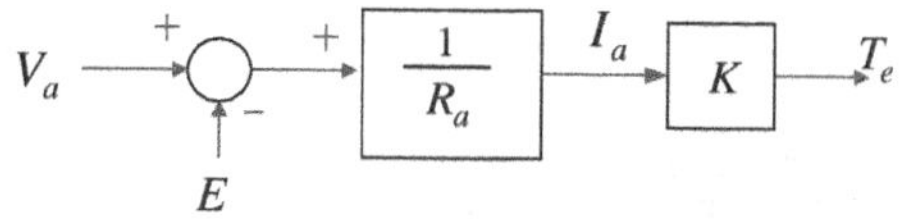

Figure 2.14 Block diagram of a DC machine with neglecting of the armature inductance.

inductance, L_a, is neglected, then (2.27) can be simplified as follows:

$$\frac{\omega_{rm}}{V_a} \approx \frac{\frac{K}{J}}{R_a s + \frac{K^2}{J}} = \frac{\frac{K}{JR_a}}{s + \frac{K^2}{JR_a}} = \frac{1}{K}\frac{\omega_c}{s+\omega_c} \tag{2.28}$$

As seen from (2.28), the speed response due to the terminal voltage variation can be modeled as the output of the first-order low-pass filter of the terminal voltage. In here the angular cutoff frequency of the filter is $\omega_c = K^2/JR_a$. The time constant of the filter, which is the inverse of the cutoff frequency, JR_a/K^2, is called as electromechanical time constant, T_m. In this case, the machine can be modeled as the first-order system, whose pole lies at $-\omega_c$. The torque from the machine after neglecting L_a can be expressed as shown in Fig. 2.14, where the torque can be obtained instantaneously according to the variation of the terminal voltage. As seen from this approximation, if the rotating inertia is large enough and the armature circuit response is fast enough ($K^2/JR_a \ll R_a/L_a$), then the torque is directly proportional to the voltage applied to the armature winding without any time delay.

The eigenvalues of (2.27) varies according to the electrical and mechanical parameters of the drive system such as $L_a, R_a, J,$ and B. It can be noted that the speed responses against the variation of the terminal voltage are decided as underdamping, critical-damping, or overdamping according to the eigenvalues. After finding eigenvalues, it can be said that in the case of (2.29) the responses are nonoscillatory, which means that the system is overdamping. Otherwise, the system is critical-damping or underdamping.

$$\frac{\left(R_a + \frac{BL_a}{J}\right)^2}{\frac{L_a}{J}(R_a B + K^2)} > \frac{1}{4} \tag{2.29}$$

It can be seen from (2.29) that as K is larger, the response is getting oscillatory (under-damping). By reducing the field current, i_f, K can be reduced, and then the system can be changed to an overdamped case from an underdamped case. In (2.29), if the friction torque is neglected or if it is included in the load torque, then (2.29) can be simplified as follows:

$$\frac{JR_a^2}{L_a K^2} > \frac{1}{4} \tag{2.30}$$

From (2.30) it can be said that to prevent the oscillatory responses the armature resistance or the inertia should be increased. However, the increased inertia degrades the acceleration characteristics of the machine, and the increased resistance degrades system efficiency because of the increased copper loss. Hence, by increasing these

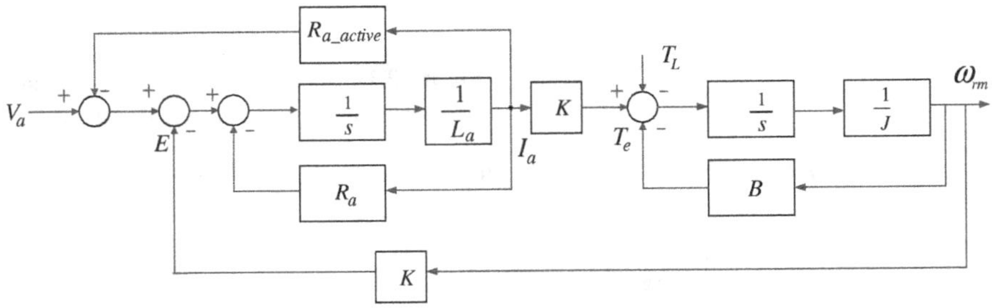

Figure 2.15 Control block diagram of a separately excited DC machine with active damping term.

passive elements such as the inertia and the resistance, the improvement of the control performance is very limited, and the implementation is also costly. However, by a feedback control after measuring or estimating the control variables such as acceleration and current, an artificial damping can be added to the control loop as if the added damping is the physical one. This artificial damping can be said to be an active damping, in contrast to the passive damping such as resistance and friction. In this way, without any modification in the physical system such as the increase of inertia and/or the increase of the resistance, the control performance can be easily improved. Moreover, there are no additional losses for the improvement of the performance by these acting damping components. A simple example of an active damping is shown in Fig. 2.15. In here, after measuring the current, by a negative feedback of the current with gain, which is an active damping resistance, to the armature voltage as shown in Fig. 2.15, it may generate the same control performance as if the armature resistance increases. Like this example, by feedback control of the measurable states or observable states, the eigenvalue of the physical system can be adjusted to get better control performance. This is a kind of state feedback control. The detailed explanation of the active damping will be given in Section 4.1.

2.4 POWER ELECTRONIC CIRCUIT TO DRIVE DC MACHINE

Even before the power semiconductors were available, a DC machine had been widely used for the application area where adjustable torque or speed is essential. At that time, the adjustable speed drive of the DC machine was implemented in electromechanical means based on multiple electric machines as shown in Fig. 2.16. The system shown in the figure, the so-called Ward–Leonard system, consists of an induction machine and two DC machines, where one is used as a motor and the other as a generator.

In this system, the speed of the DC motor is adjusted by controlling the terminal voltage of the motor, V_t, through the control of the field current of the DC generator, and simultaneously by controlling the field current of DC motor. With control of V_t and V_{fm} in Fig. 2.16, the operating region of the DC motor can be expanded to all quadrants of the torque–speed plane as shown in Fig. 2.9. Especially in the application

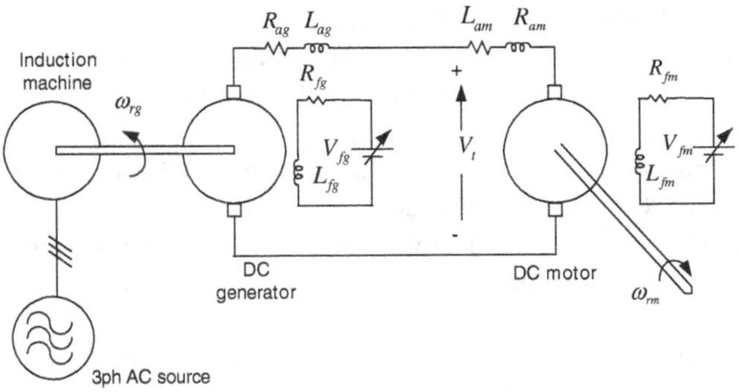

Figure 2.16 Ward–Leonard system.

of a steel mill process line, where normally impact loads are frequently applied to the motor, a large inertia can be attached to the axis that connects the induction machine and the DC generator. With the help of the energy $\left(\frac{1}{2}J\omega_{rm}^2\right)$ stored in the inertia, it should be noted that the speed drop, the variation of the terminal voltage, and the torque of the motor can be minimized against the impact loads. This kind of Ward–Leonard system that has a large inertia is called an Ilgner system. Though the Ward–Leonard system has some advantages such as no harmonic currents to an AC source and the higher over-load rating, the efficiency of the system is poor and the weight and volume of the system, because of multiple machines to drive a motor, is prohibitive in the modern manufacturing facilities. Moreover, due to the regular maintenance of the machines and expensive running cost, a traditional Ward–Leonard system is no longer installed as the adjustable speed drive system.

2.4.1 Static Ward–Leonard System

As shown in Fig. 2.17 [3], a variable voltage DC source can be easily obtained by thyristors, which were invented in the late 1950s. The average DC voltage of the motor terminal can be controlled from –135% to 135% of line-to-line rms voltage of a three-phase AC source by manipulating the gating signals to the thyristors. The field current

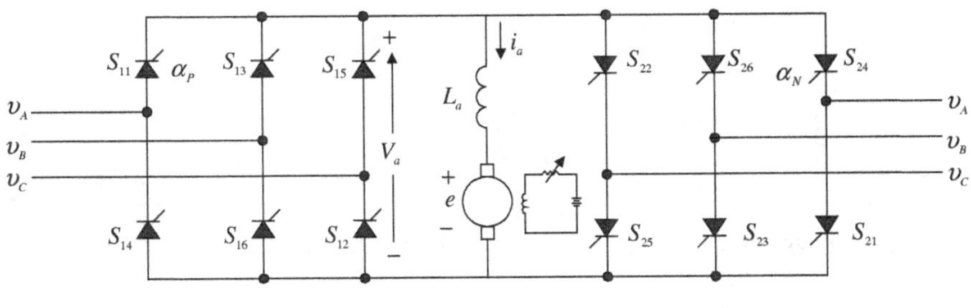

Figure 2.17 Static Ward–Leonard system based on the power semiconductors.

of the motor can also be controlled by a semiconverter, where only the upper or lower three thyristors are controlled and the others are replaced by diodes. In the case of a small motor, the field current can be controlled simply by adjusting the value of the resistance, which is inserted in series to the field winding. By this static variable DC voltage source and field current control, the same operating region with the region by Ward–Leonard can be obtained. Compared to the traditional Ward–Leonard system, this static Ward–Leonard system has several advantages such as faster current regulation (in the case of a 60-Hz, three-phase AC source, several tens of rad/s of current regulation bandwidth is possible), easy maintenance, and smaller size and weight. However, there are still some disadvantages such as smaller overload rating, harmonics to AC source, sensitivity to the power quality of the AC source, and poor power factor at light load condition. Regardless of these shortcomings, since the late 1960s, in above several 100-kW variable-speed DC drive systems the static Ward–Leonard had been a mainstream. Recently, with developments of power electronics technology, the DC machine drive system is becoming obsolete and the AC machine drive system is replacing the DC machine drive system. Hence, the static Ward–Leonard system is not installed in the newly designed adjustable speed and torque control system.

2.4.2 Four-Quadrant Chopper System

A four-quadrant chopper circuit shown in Fig. 2.18 can easily control the DC machine whose rated power is below several tens of kilowatts. In particular, a DC servomotor whose rating is less than a few kilowatts is usually controlled by this circuit.

In the circuit shown in Fig. 2.18, by turning on and off the four power semiconductor switches, T_1, T_2, T_1', and T_2', the terminal voltage to the motor, V_a, can be varied from $-V_d$ to $+V_d$, which is normally almost the peak of the line-to-line voltage of AC source. When a DC machine operates as a generator, it means that when a DC motor is operating in regenerative braking mode, the link voltage, V_d, will increase due to the energy accumulated in the DC link capacitor. To prevent the overvoltage of a DC link, the discharging switch implemented by a power

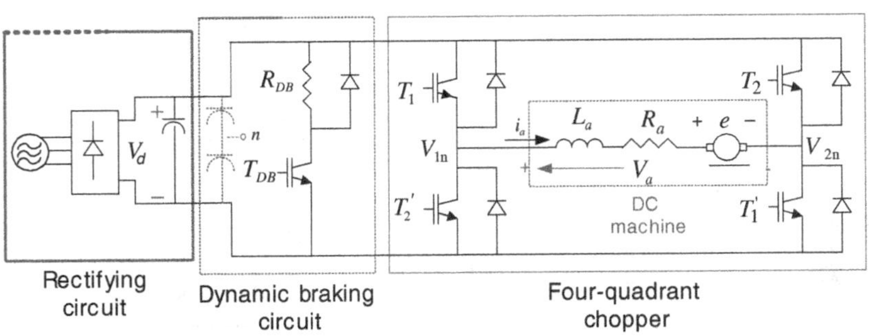

Figure 2.18 Four-quadrant chopper.

semiconductor turns on and the regenerated energy is dissipated at a resistor, R_{DB}, as a form of the heat. This kind of braking energy absorbing method is called *dynamic braking*. Through the four-quadrant chopper circuit, the current regulation bandwidth can be easily extended up to several thousand radians per second.

2.5 ROTATING MAGNETIC MOTIVE FORCE [4]

AC machines powered by an AC voltage source have been widely used to get the mechanical power for the past hundred years after they were invented, because it can be directly connected to an AC utility grid. Among AC machines, the squirrel cage induction machine has consumed more than 90% of the total electricity to drive an AC machine due to its simplicity, ruggedness, and virtually no maintenance as described in Section 1.1. The operation principle of all AC machines is based on a rotating magnetic motive force (MMF). In this section the principle of rotating MMF is described. An AC machine rotates with the interaction of the magnetic materials of the rotor and the rotating MMF generated by a stator winding current. As shown Fig. 2.19, by applying balanced three-phase AC voltage to the three sets of windings spatially apart by 120°, each winding generates MMF. The vector sum of an MMF by three sets of winding is equivalent to an MMF by a rotating magnet.

If the numbers of turns of three windings are the same and the magnitude of the current in each winding is identical but its phase is different by 120°, then the vector sum of an MMF by three windings can be described as (2.31) at the orientation with an MMF by a.

$$\mathbf{F}_{abc} = N i_a + N i_b\left(-\frac{1}{2} + j\frac{\sqrt{3}}{2}\right) + N i_c\left(-\frac{1}{2} - j\frac{\sqrt{3}}{2}\right) \tag{2.31}$$

where N is the effective number of turns of each winding. i_a, i_b, and i_c current flows in a, b, and c, phase winding, respectively. If the current is sinusoidal and its peak value is

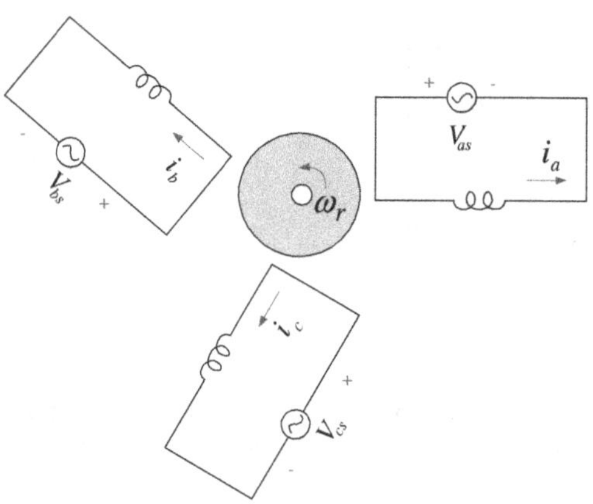

Figure 2.19 Rotating magnetic motive force (MMF) by three-phase windings.

I and if the angular frequency of the current is ω_e (rad/s) and all currents are balanced, then (2.31) can be rewritten as

$$\mathbf{F}_{abc} = NI\cos\omega_e t + NI\left(-\frac{1}{2}+j\frac{\sqrt{3}}{2}\right)\cos\left(\omega_e t - \frac{2\pi}{3}\right) + NI\left(-\frac{1}{2}-j\frac{\sqrt{3}}{2}\right)\cos\left(\omega_e t + \frac{2\pi}{3}\right) = \frac{3}{2}NI[\cos\omega_e t + j\sin\omega_e t] \tag{2.32}$$

The MMF in (2.32) is equivalent to the MMF by two windings spatially apart by 90° whose number of turn is $\frac{3}{2}N$ as shown in Fig. 2.20a. The current in each winding is $i_\alpha = I\cos\omega_e t$ and $i_\beta = I\sin\omega_e t$, respectively. The MMF by these two sets of winding is equivalent to the MMF by a winding, which is rotating with angular speed, ω_e, and the current in the winding is constant DC and its magnitude is I as shown in Fig. 2.20b.

Also, the MMF by a rotating winding can be equalized to the MMF by a rotating permanent magnet as shown in Fig. 2.21. Such a rotating MMF can be understood

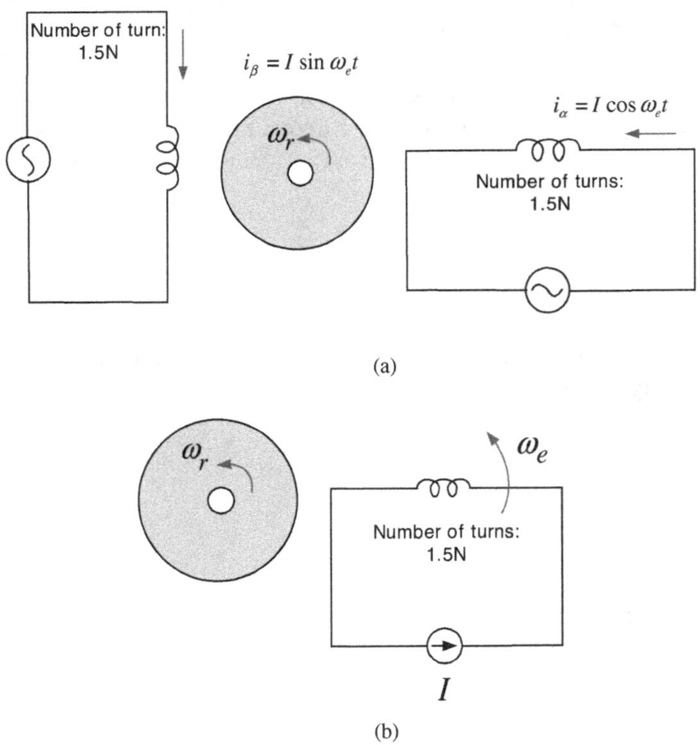

Figure 2.20 Equivalence of MMF by three-phase winding. (a) Equivalence of MMF by three-phase winding to two-phase windings. (b) Equivalence of MMF of three-phase winding by a rotating winding powered by a DC current source.

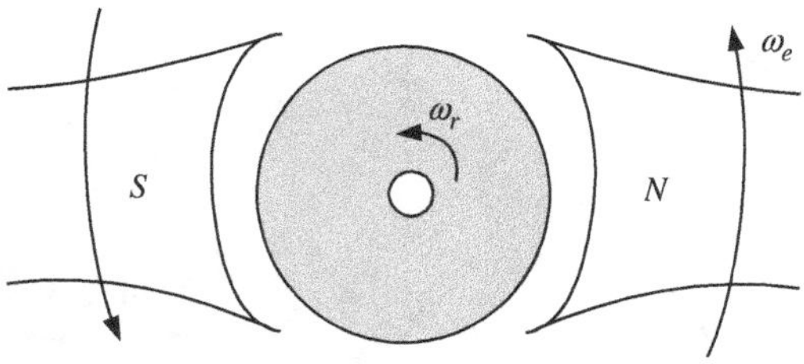

Figure 2.21 Equivalence of MMF by a rotating permanent magnet.

easily with a mechanical analogy as shown in Fig. 2.22. In the figure, three persons are holding one end of a line spatially 120° apart. All the other ends of the line are tied together to a small disc-shaped body. If each person pulls the line sequentially, then the disc would rotate. If the force applied to the body by each person is ideal sinusoidal and exactly spaced in time by 120° apart, then the motion of the disc would be a perfect circle. But if each force is not sinusoidal and not 120° apart in time, the motion would be a distorted circular shape. Like this motion of the disc, the MMF of an AC machine would rotate, which is the reason for the rotation of the rotor in all AC machines.

In Fig. 2.21, if a rotor has an asymmetric shape with respect to the rotating MMF and is made by ferromagnetic material as shown in Fig. 2.23, then the magnetic reluctance is different along the angle, θ_e. If there is no external force to the rotor, then the rotor rotates to minimize the reluctance. In this case, the instantaneous angle of the rotating MMF, θ_e and the angle of the rotor, θ_r, would be the same $(\theta_e - \theta_r = \theta_\delta = 0)$.

As the dragging force to the rotor, which is a load torque, increases, the angle difference, θ_δ, increases. If the magnetic saturation is ignored, then torque generated by the rotor and the rotating MMF for the minimization of the reluctance is maximum when θ_δ is equal to 45°. Above 45°, the torque would decrease. Hence, if the load torque including friction torque continues to increase above $\theta_\delta = 45°$, the rotor

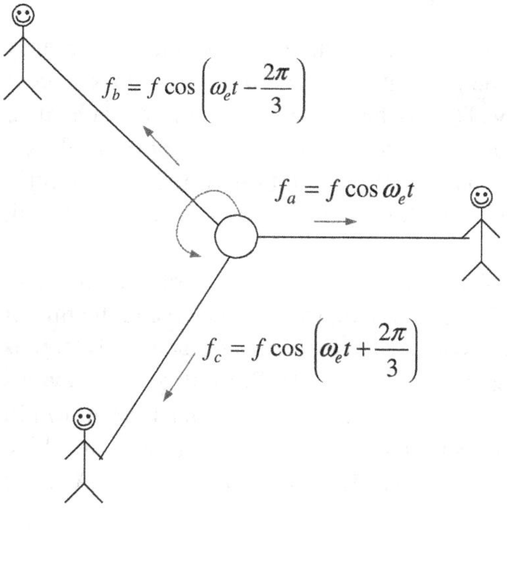

Figure 2.22 Analogy of rotating MMF to rotating movement by three forces.

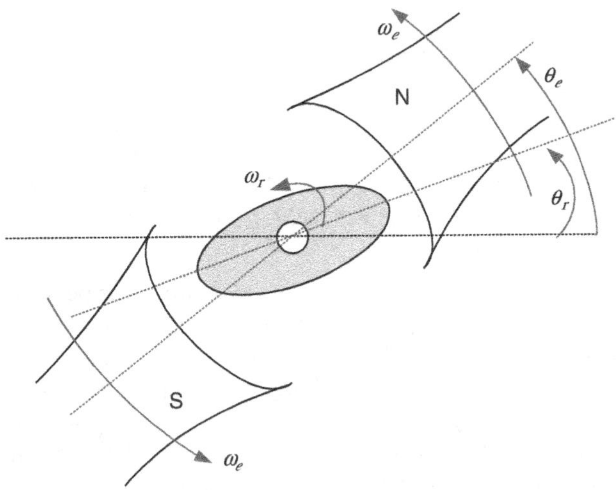

Figure 2.23 Operating principle of a synchronous reluctance machine (SynRM).

eventually stops. In this case the rotor could not generate DC torque in average, but only generates ripple torque whose average is zero. The machine exploiting the reluctance minimization principle under the rotating MMF is called a synchronous reluctance machine (SynRM). If the rotor of the machine is driven by an external prime mover such as an engine and a turbine, then three-phase AC voltage would be induced by the residual flux of the rotor. If the stator winding is connected to the outer circuit and the current flows, then rotating MMF would formulate. With the interaction with this rotating MMF and the rotor driven by the prime mover, the electric power is generated in the stator winding continuously. In this generating operation, the angle difference, θ_δ, is negative as in (2.33):

$$\theta_e - \theta_r = \theta_\delta < 0 \tag{2.33}$$

Like this example, in all electric machines based on the electromagnetic energy conversion principles, the motor and the generator are inherently the same except for the direction of energy (power) flow. Due to the virtue of the simple mechanical structure of the rotor, the synchronous reluctance generator is widely used as a generator in aero and military applications, where the reliability is the first concern. The analysis of transient characteristics of this type of machine is described in Section 3.3.4.

If a set of permanent magnets is mounted on the surface of the rotor as shown in Fig. 2.24, the rotor rotates through the interaction with rotating MMF and the flux of the permanent magnet. In this case, like a synchronous reluctance machine, if there is no external force, then the angle difference is $\theta_e - \theta_r = 0$. But unlike the previous machine, the maximum torque occurs not at 45° but at 90°. The machine shown in Fig. 2.24 is called a surface-mounted permanent magnet synchronous machine (SMPMSM) because the magnet is mounted on the surface of the rotor. Also, in

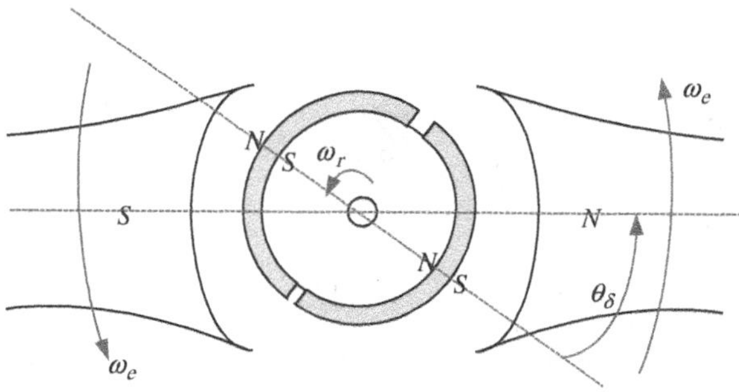

Figure 2.24 Operating principle of a surface-mounted permanent magnet synchronous machine (SMPMSM).

this machine, if the rotor is driven by an external prime mover, then the machine acts as a generator. The analysis of transient characteristics of this type of machine is described in Section 3.3.3.1.

As shown in Fig. 2.25, if the permanent magnet is inserted inside of the rotor, then the reluctance of the rotor would be asymmetry because the relative permeability of the permanent magnet is almost unity like air. This type of permanent magnet machine is called an interior permanent magnet synchronous machine (IPMSM).

In this type of machine, as a motor, not only torque due to the permanent magnet but also torque due to the difference of the reluctance can be obtained. Hence the versatile operating characteristics can be obtained. In this case the angle difference, θ_{δ_max}, where maximum torque occurs, is between 90° and 135°. If the torque due to the reluctance torque is getting larger relative to the torque due to the magnet, then θ_{δ_max} is getting nearer to 135°. The detailed description of this type of the machine is given in Section 3.3.3.2.

In the case of the above-several-megawatt synchronous machine, where the permanent magnet cannot provide enough excitation flux, or even in the case of smaller power machines, the excitation flux can be obtained from a separately wound winding, where excitation current flows. The winding is called a field winding, and as shown in Fig. 2.26 the machine can be classified as a round rotor machine or a salient rotor machine according to the shape of the rotor. In the case of above-multi-megawatt and the above-several-thousand revolution/minute rotor speed machines, the round rotor machine is usually used due to even distribution of a centrifugal force. The synchronous machine, which has a field winding, is called a wound rotor synchronous machine, and the torque and power factor of the machine as a motor, as well as the generated voltage and power factor as a generator, can be manipulated by adjusting

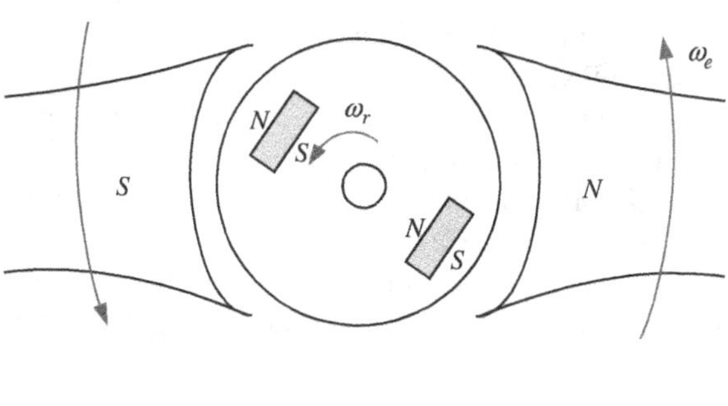

Figure 2.25 Operating principle of an interior permanent magnet synchronous machine (IPMSM).

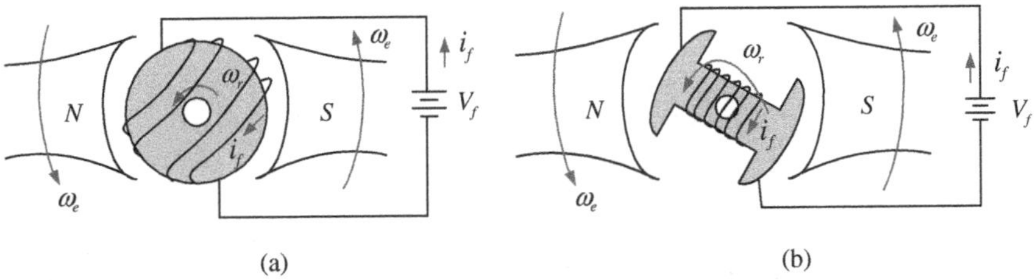

Figure 2.26 Rotor structures of a wound synchronous machine. (a) Round rotor synchronous machine. (b) Salient rotor synchronous machine.

the field current, i_f. In the case of the salient rotor machine, there is the torque due to the difference of the reluctance like the case of the interior permanent magnet (IPM) machine. But unlike the IPM machine, as shown in Fig. 2.26b, the reluctance at the direction of the flux by field winding, which is the direct axis direction, is smaller than the quadrature direction of the flux. In the wound rotor synchronous machine, the instantaneous speed difference between the rotating MMF and the rotor induces alternating voltage to the field winding, which results in a large AC current to the field winding. To prevent such a current and to improve the transient characteristics of the machine, damper windings are usually installed in the rotor of the large wound rotor synchronous machine. The analysis of the machine with the damper windings is described in detail in Section 3.3.

2.6 STEADY-STATE ANALYSIS OF A SYNCHRONOUS MACHINE [5]

In the analysis of a three-phase synchronous machine drive system, the balanced steady state means that all current and voltages of each phase is sinusoidal and apart by 120° in phase, the phase difference between voltage and current at each phase is constant, and the speed of the rotor and rotating MMF is constant: $d\theta_e/dt = d\theta_r/dt = \text{const} = \omega_e$, $\theta_e - \theta_r = \theta_\delta = \text{const}$. In this case, even if there are damper windings, no current flows through the damper windings in the steady states. Also, the damper windings do not affect the analysis of the steady-state characteristics of the machine. In the balanced steady-state analysis, like analysis of the steady-state analysis of a three-phase AC circuit, the system is analyzed as if there is a single-phase system such as only the "a" phase. And other phase ("b" and "c" phase) responses can be obtained by considering simply a $\pm 120°$ phase difference. In the case of the salient rotor machine including SynRM and IPMSM, it is difficult to describe the steady-state characteristics by only a simple equivalent circuit. The analysis of this type of machine is in Section 3.3, including the transient state.

In the cases of the round rotor and SMPM synchronous machine, the reluctance is symmetry to the rotating axis, and the induced EMF at the stator winding is synchronous to the rotor position. If the air gap of the machine is smooth, and there is no eccentricity and no magnetic saturation, then the induced voltage would be

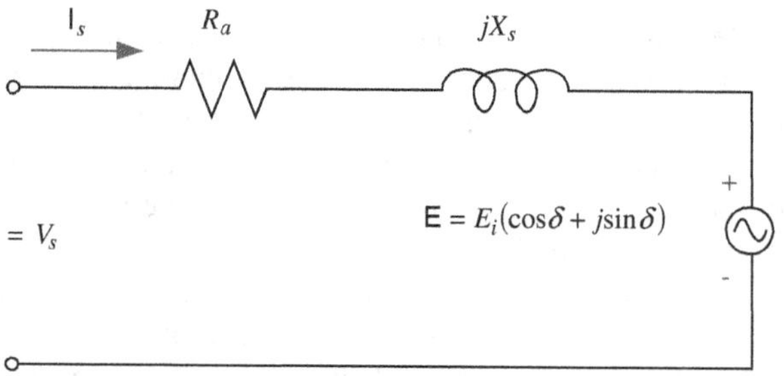

Figure 2.27 Equivalent circuit of round rotor synchronous machine at balanced steady state.

sinusoidal. In Fig. 2.27, an equivalent circuit is shown under the above assumptions using a so-called "phasor." Usually, the armature (stator) winding resistance, R_a, is much less than the synchronous reactance, X_s, at the rated speed as $R_a \ll X_s$. X_s of the wound rotor synchronous machine is around 1 per unit (see Section 2.19), while that of the SMPMSM is typically several tens of a percentage per unit because of a large equivalent air gap of the machine, where the thickness of the permanent magnet acts as an air gap.

In Fig. 2.27, the angular frequency of the voltage and the current is the same with the rotating speed of MMF, ω_e, which is also the same with the rotating speed of the rotor, ω_r. In the circuit, if the magnitude of R_a is much smaller than that of X_s and R_a can be neglected, then the load angle, δ, has the following relationship with the angle difference previously mentioned:

$$\delta = -\theta_\delta \tag{2.34}$$

Under the above assumptions, the current in the stator winding can be deduced as follows:

$$\mathbf{I}_s = \frac{\mathbf{V}_s - E_i\cos\delta - jE_i\sin\delta}{jX_s} \tag{2.35}$$

Under the assumption of a balanced three-phase system, the input power to the machine can be described by

$$P_s = 3\mathrm{Re}\left[\mathbf{V}_s \cdot \mathbf{I}_s^*\right] = \frac{-3V_sE_i\sin\delta}{X_s} \tag{2.36}$$

where * stands for complex conjugate and Re stands for real part of the complex number. As seen from (2.36), if $\delta < 0$, then the machine acts as a motor, and if $\delta > 0$, then it acts as a generator. If it works as a motor, the torque including friction torque of motor itself in the balanced steady state can be given as (2.37) after neglecting the internal losses of the machine such as the copper loss, the iron loss, and the stray loss.

$$T_e = T_L \approx -\frac{P}{2}\frac{3V_sE_i\sin\delta}{\omega_eX_s} \tag{2.37}$$

where P is the number of pole of the machine. It can be seen that the magnitude of the load angle, $|\delta|$, increases as the load torque increases. The maximum value occurs when $\delta = -90°$. If δ is less than $-90°$ (i.e., the magnitude of δ is larger than 90°), then the machine would be out of synchronism, and it cannot generate average DC torque. If the rotating speed of the machine is low, where the voltage drop due the stator winding resistance is almost same magnitude compared to that due to X_s, then the effect due to R_a cannot be ignored. In this case, the maximum torque occurs when $\delta > -90°$. As mentioned before, the magnitude of X_s is almost unity per unit, and the magnitude of the short circuit ($\mathbf{V}_s = 0$) current of the machine is almost unity per unit, which means that the short current level in the steady state is almost the order of the rated current of the machine. However, if short circuit occurs, due to the damper winding, and if there is damper winding, then the transient value of X_s, namely as X'_sor X''_s, is much smaller than the steady-state value of X_s. And the short current after the brief moment of the short circuit would be much larger than the steady-state short-circuit current. But in the case of an SMPMSM, even the magnitude of the short-circuit current in the steady state is several times the rated current due to its smaller value of X_s. So, this excessive short-circuit current may result in demagnetization of the permanent magnet at the short-circuit faults of an SMPMSM, and the magnet of the machine can be damaged irreversibly. Normally, at the designing stage of an SMPMSM, the short-circuit current level is investigated, and the machine is designed not to be demagnetized even at the worst-case short-circuit fault. In the equivalent circuit at Fig. 2.27, E_i is proportional to the product of the rotating speed and the flux linkage to the stator winding by the field current on the rotor as follows:

$$E_i = \lambda_f \cdot \omega_r = K_f \cdot I_f \cdot \omega_r \tag{2.38}$$

where I_f is the current flowing in the field winding of the wound rotor synchronous machine, and K_f is a proportional constant decided by the number of turns of the field winding and the structure of the magnetic circuit. If the saturation of the magnetic circuit is neglected, K_f is a constant regardless of the operating condition. As seen from (2.35) and (2.38), by changing the magnitude and direction of I_f, E_i can be adjusted and the phase and magnitude of the stator current can be controlled even under the constant stator voltage. Hence the power factor of the machine can be adjusted by controlling the field current. In Fig. 2.28, the phasor diagrams of

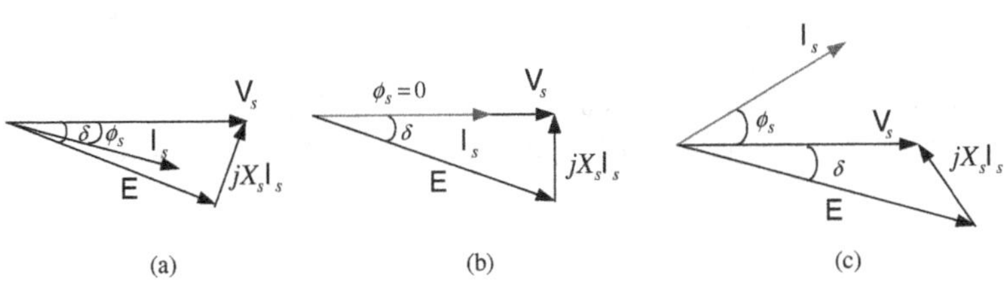

Figure 2.28 Phasor diagram of the terminal voltage and current of a round rotor synchronous machine with different field current at motoring operation. (a) Lagging power factor operation. (b) Unit power factor operation. (c) Leading power factor operation.

the motoring operation of the synchronous machine after neglecting R_a are shown in the case of leading, unity, and lagging power factor, respectively. In the figure, ϕ_s is a power factor angle.

Like the above phasor diagrams, by controlling I_f, the power factor of the machine can be adjusted under the same terminal voltage and the same output power. As I_f increases, the power factor varies from lagging to leading in the case of motoring operation. Using this feature of the machine, the large power round rotor synchronous machine had been used as a device to improve the power factor of the distribution network of the power system. The machine for this application is called as a synchronous phase shifter. But with the development of the power electronics, the synchronous phase shifter was replaced by a static var compensator consisted with power semiconductors such as thyristors or IGBTs.

By exploiting the characteristics that the terminal power factor can be a leading power factor with overexcitation of the round rotor synchronous machine whose field flux is supplied by external field winding, the speed of the machine can be controlled by a simple thyristor bridge circuit as shown in Fig. 2.29. The thyristors are commutated sequentially by turning on the next thyristors due to the back EMF voltage, **E**, of the machine (load commutation) in the case of leading power factor operation. Such a variable-speed drive system using a large wound rotor synchronous machine fed by the thyristor bridge is called a load commutated inverter (LCI) system or a thyristor motor, and it had been used widely from the 1970s until the 1980s. But, with the development of self-commutated power semiconductors such as IGCT, IEGT, and IGBT, the system had been faded out faded out except extremely large motor drive.

In the case of an SMPMSM, because the excitation flux is provided by the magnet and it is constant, the magnet can be modeled by an ideal current source. Usually to get the maximum torque per ampere (MTPA), the phase of the stator current phasor is kept to be the same with the phase of EMF phasor as shown in Fig. 2.30. In this kind of MTPA operation, the power factor of an SMPMSM is always lagging.

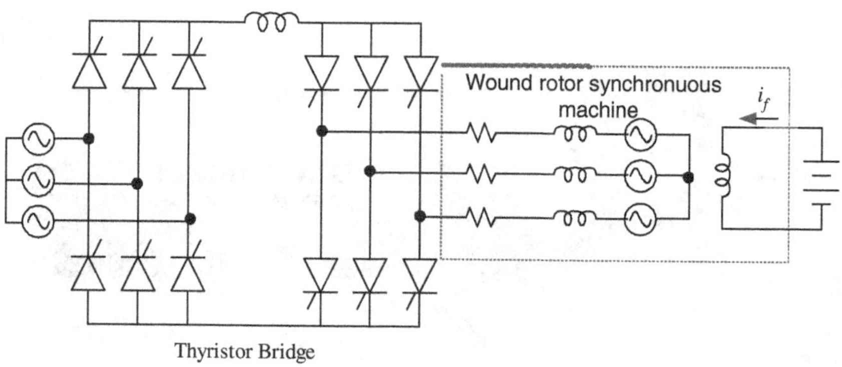

Figure 2.29 Circuit diagram of a thyristor motor (or LCI system).

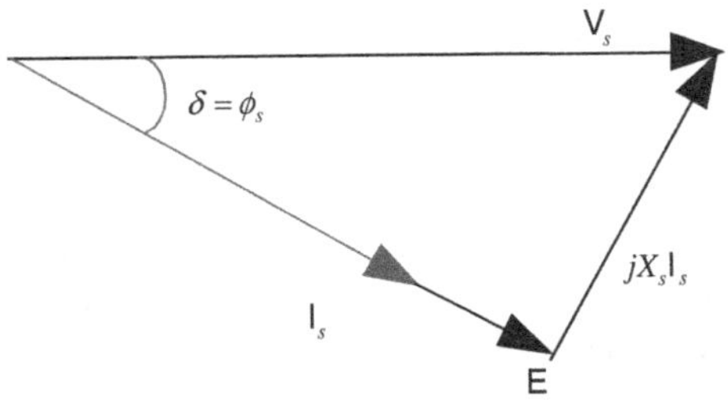

Figure 2.30 Phasor diagram in the case of MTPA operation of SMPMSM.

2.7 LINEAR ELECTRIC MACHINE

For the application where the linear motion is required, the rotary motion of an electric machine is converted to the linear motion with a proper mechanism such as ball and screw, wheel and rail. However, the conversion of the motion through the mechanical devices has several demerits such as increased weight and volume, poor efficiency, less stiffness, and less accuracy in positioning due to the backlash and the Coulomb friction. Because of these demerits, the linear machine is getting widely applied to the application where highly accurate and fast positioning of the linear motion is the first concern. All the electric machines mentioned previously—DC machine, induction machine, permanent magnet AC machine, and wound rotor AC machine—can be implemented as a linear machine, but an SMPMSM is widely implemented as a linear electric machine. In this case the control characteristics are almost the same as the rotary motion SMPMSM except for a slight difference due to asymmetry of the magnetic circuit. In Fig. 2.31, the photo of a linear SMPMSM is shown. The linear machine can be classified by a moving magnet type and a moving coil type. The machine in the figure is a moving coil type, where the magnet is fixed and it is installed at the stator, and the coil on the mover is moving according to the motion. Hence, the electric power line should be connected to the mover via a flexible wire harness. In the

Figure 2.31 Moving-coil-type linear permanent magnet synchronous machine.

case of a less-than-several-meter stroke, a moving-coil-type linear machine is widely used because of its simplicity of the control.

2.8 CAPABILITY CURVE OF SYNCHRONOUS MACHINE

2.8.1 Round Rotor Synchronous Machine with Field Winding

From the equivalent circuit at Fig. 2.27, if the magnitudes of the terminal voltage, $\mathbf{V}_s$, and line current, $\mathbf{I}_s$, are limited, then the capability curve of the synchronous machine is similar to that of the separately excited DC machine. In the region where the speed is less than the base speed, ω_b, referred to as the constant torque operation region, if $R_a \ll \omega_e L_s$, then the same phasor diagram shown in Fig. 2.30 can be drawn for the round rotor machine with the constant field winding current for MTPA operation. L_s stands for synchronous inductance, which is the sum of leakage inductance, L_{ls}, and magnetizing inductance, L_m. The rotational angular speed in electrical angle, ω_r, is the same with the angular speed of rotating MMF, ω_e, in the steady state.

In Fig. 2.30 we have $X_s = j\omega_e L_s$, and EMF can be represented as $\mathbf{E} = K_f I_f \omega_e \angle\delta$ in phasor. (For motoring operation, δ is negative.). If $\mathbf{V}_s$ is controlled proportionally to ω_e—that is, $\mathbf{V}_s = K_a \cdot \omega_e$—then the shape of the phasor diagram in Fig. 2.30 does not vary regardless of the variation of ω_e. Here, K_a is a proportional constant. The shape of the phasor diagram can be kept until $\mathbf{V}_s$ reaches to its limiting value, $\mathbf{V}_s = V_{\text{lim}}$. The torque in this operation mode can be represented as (2.39) from the phasor diagram and (2.35), (2.37), and (2.38):

$$T_e = \frac{3X_s|\mathbf{I}_s| \cdot K_f I_f \cdot \omega_e}{\frac{2}{P} X_s \omega_e} = \frac{3}{2} P \cdot |\mathbf{I}_s| \cdot K_f I_f \tag{2.39}$$

Hence, the field current, I_f, is kept as constant, which is the rated value, then torque, T_e, can be maximized under the limiting condition of the armature (stator) current, $|\mathbf{I}_s| = I_{\text{lim}}$. In this operating region, the maximum torque with the current limitation, $|\mathbf{I}_s| = I_{\text{lim}}$, is always the same regardless of the operating speed, ω_e, as seen from (2.39). Also, this operation region is called a constant torque operation region. If ω_e is quite small and the magnitude of R_a is comparable to $\omega_e L_s$, then the voltage drop due the resistance of the stator winding, R_a, should be considered. So, in this case the armature voltage should be controlled as $\mathbf{V}_s \approx R_s \mathbf{I}_{\text{lim}} + K_a \omega_e$.

Above the base speed, ω_b, if I_f is controlled as the rated value, then the terminal voltage would be above the rated value, which is the limiting value of the machine. So, to keep the terminal voltage equal to the limited value as $|\mathbf{V}_s| = V_{\text{lim}}$, the field current should be decreased according to the speed above the base speed. In this operating region where the field current should be reduced, the R_a is sufficiently small compared to the magnitude of $\omega_e L_s$, and the voltage drop due to R_a can be neglected. Also, the copper loss in Fig. 2.27 can be neglected. Furthermore, if the iron loss and the stray loss of the machine are neglected, the output power of the machine is the same as the

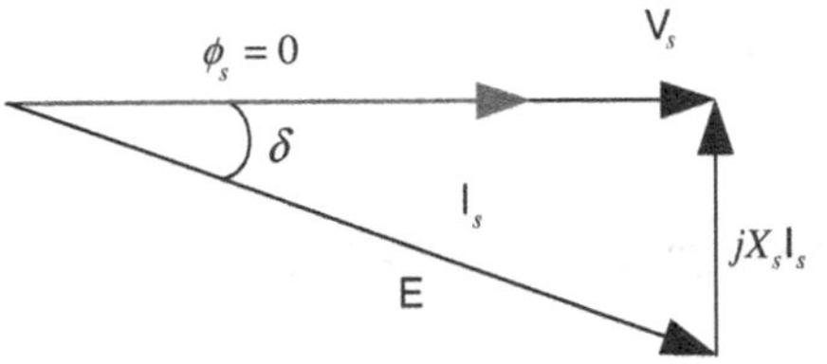

Figure 2.32 Maximum torque per ampere (MTPA) operation of a round rotor synchronous machine with field winding under limited voltage and current condition after neglecting the voltage drop of stator winding resistance.

input power of the machine. Under these assumptions, to get maximum torque per ampere (MTPA) with both the limitation of the voltage and current, the power factor of the machine should be kept as unity. The phasor diagram in this operation is shown in Fig. 2.32. With the increase of ω_e, I_f should be reduced to keep the shape of the phasor diagram in Fig. 2.32. In this operation, the air gap flux of the machine decreases as the speed increases, and the available maximum torque also decreases exactly proportional to the inverse of the speed as shown in (2.40). This operation region is called the field weakening region.

$$T_e = \frac{3}{2}P\frac{V_{\lim}I_{\lim}}{\omega_e} \tag{2.40}$$

Between this field-weakening region and the previously described constant-torque operating region, there is a brief operating region where the power factor angle, ϕ_s, decreases from $|\delta|$ to null. In this operating region, the voltage and current are both at their limiting values, $\mathbf{V}_s = V_{\lim}$ and $|\mathbf{I}_s| = I_{\lim}$ while the power factor angle varies according to the operating speed, ω_e. The torque in this operating mode decreases according to the speed, ω_e, but is not exactly inversely proportional to the speed.

Unlike a separately excited DC machine, because there are no commutation problems in the synchronous machine, the torque inversely proportional to the speed can be obtained up to the speed where the mechanical problems occur due to centrifugal forces. After considering the above characteristics of the constant torque control and field weakening control, the capability curve of a round rotor synchronous machine with separate field winding can be drawn as shown in Fig. 2.33.

Because the mechanical output power is the product of torque and speed, the output power is constant in the overall field-weakening region. In the case of a salient rotor synchronous machine, the similar capability curve can be drawn, but in this case it should be noted that there is another torque that comes from the difference of reluctance.

2.8.2 Permanent Magnet Synchronous Machine

In the case of the permanent magnet synchronous machine, where the reluctance difference is almost none or small (less than 20%), most of the torque occurs through the interaction with rotating MMF and the flux of the permanent magnet. In this case the air gap flux is almost constant, and the flux-weakening region is very narrow. So, the capability curve of such a machine can be drawn as shown in Fig. 2.34, where

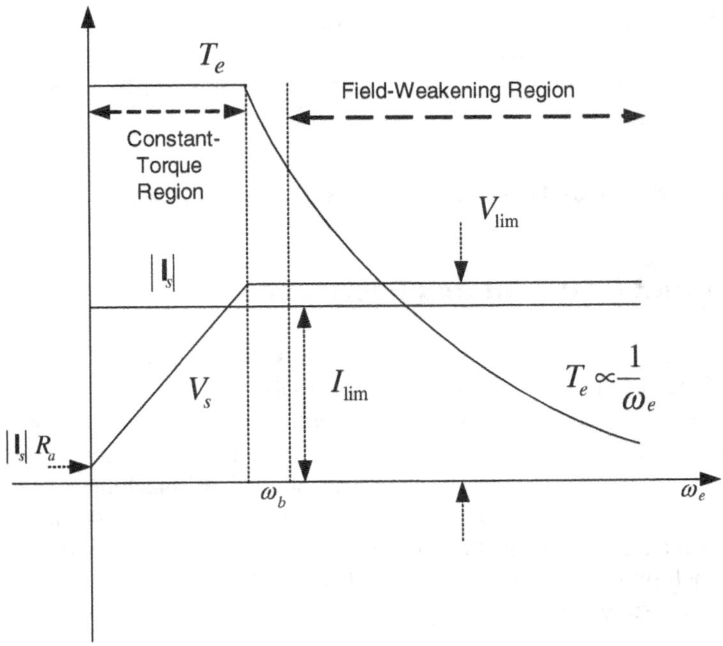

Figure 2.33 Capability curve of a of round rotor synchronous machine with field winding.

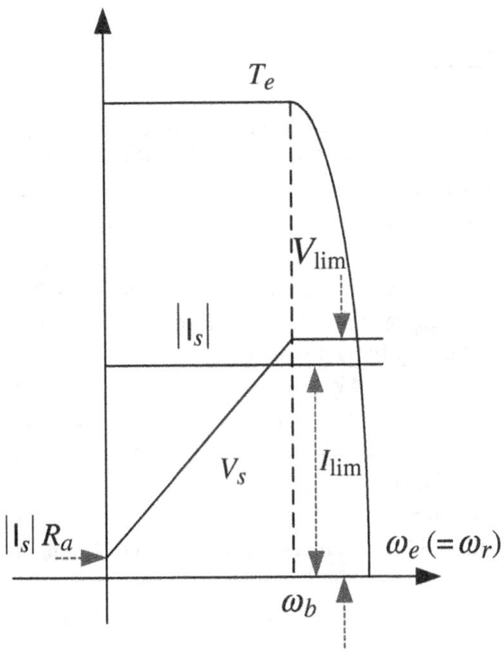

Figure 2.34 Capability curve of a synchronous machine where the air gap flux is dominated by the permanent magnet.

the torque decreases rapidly above the base speed. Even in this type of the machine, the flux-weakening region can be extended by improving the control methods. The detailed discussion of the improved flux-weakening control method for this type of machine is given in Section 5.4.2.

2.9 PARAMETER VARIATION OF SYNCHRONOUS MACHINE

2.9.1 Stator and Field Winding Resistance

The stator and field winding of the synchronous machine is usually made by copper wire, and the resistance of the copper varies according to the temperature as shown in (2.41). Also, the resistance of the copper varies according to the frequency of the current flowing in the winding because of the skin effect. The normal frequency range of the current flowing in the stator winding of the conventional AC machine is less than a couple hundred hertz, and the stator winding is usually made by a stranded copper wire for easy manufacturing. So, the skin effect can be neglected in most of the general-purpose synchronous machines. Even in the field winding, the field current is normally DC, and the skin effect can be neglected.

$$R_{r\text{ at }T} = R_{r75^\circ\text{C}}\left(\frac{235+T}{310}\right) \qquad (2.41)$$

where T stands for temperature in °C, and $R_{r75^\circ\text{C}}$ for the resistance value of the winding at 75°C.

2.9.2 Synchronous Inductance

The synchronous inductance, which is the sum of a mutual inductance and a leakage inductance, is dominated by the mutual inductance, which varies according to the air gap flux, which is a vector sum of the flux by field winding and the flux by the stator winding. Because of the relatively large air gap in the q axis, in the case of the salient rotor synchronous machine with separate field winding, the variation of the synchronous inductance at the d axis, L_d, where the flux by field winding lies, is larger than that of the synchronous inductance at the q axis, L_q, which is the quadrature to the d axis. That is because the saturation occurs more easily in the magnetic circuit where the air gap is smaller. However, as shown in Fig. 2.35, in the case of an interior permanent magnet synchronous machine (IPMSM), where the magnet lies at the d axis and the permeability of the magnet is almost same as that of the air, the equivalent air gap at the d axis is much larger than that of the q axis. Hence, the variation of the inductance at the d axis is smaller than that of the inductance at the q axis. The equivalent circuit of this type of machine is considered in Section 3.3.2.2. The q-axis inductance, L_q, of an IPMSM for a traction application can vary up to several times according to the operating condition. The inductance decreases as the magnitude of the current at the q axis increases.

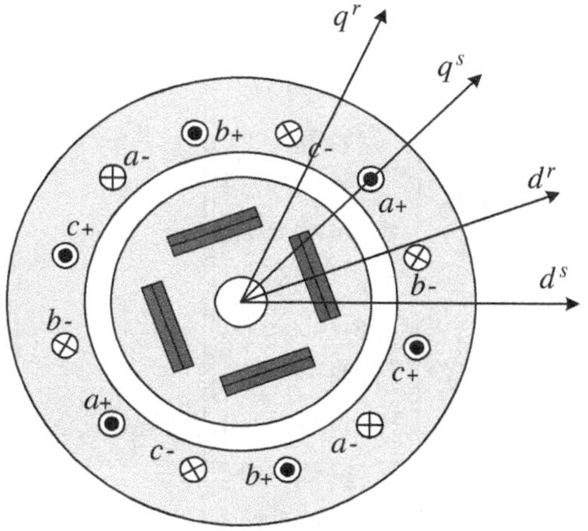

Figure 2.35 Stator winding and magnet structure of a four-pole interior permanent magnet synchronous machine (IPMSM).

In Fig. 2.36, the variation of L_d and L_q of a typical IPMSM for the traction application according to the magnitude of the current and to the angle of the current, β, at the d–q current plane, where angle 0° stands for the d axis and angle 180° stands for $-d$ axis. Hence, in the figure, β equals 90° means that all current is in the q axis, which is the quadrature to the d axis where flux by the magnet lies. Also, β equals 180° means that all current is in the d axis, meaning that the flux by the stator current is against the flux by the magnet, which means flux weakening by the stator current. In the measurement of L_d, when β is near 90°, due to the difficulties in the control of the current, the accuracy of the measurement degrades rapidly. If the values of L_d and L_q in the range of $\beta = 90°–100°$ are neglected, then L_q varies about 2.5 times and L_d varies about 1.5 times in this specific IPMSM. These kinds of the variations of the synchronous inductances should be considered in the design of the high-performance IPMSM drive system. In Fig. 2.37, the capability curve of the above-mentioned IPMSM is shown. From the figure, it can be seen that from 2000 r/min to 9000 r/min the constant power operation by flux weakening control is possible under the voltage and current constraints. The detailed description of the flux weakening control of IPMSM is seen in Sections 5.4.2 and 5.4.3.

2.9.3 Back EMF Constant

In the case of a wound rotor synchronous machine, where the excitation flux is provided by field winding current, the back EMF constant is mainly decided by the field current. However, as mentioned in Section 2.5, in the case of the permanent magnet synchronous machine where the excitation flux is provided by the magnet, the field current, I_f, in (2.38) can be modeled as a constant value set by a current source. Usually, the remanence of the magnet varies according to the temperature. Hence, the back EMF constant, K_f, in (2.38), varies according to the temperature of the magnet itself. The temperature coefficient of a ferrite magnet is typically −0.2%/°C, and that of a NdFeBr (neodymium–iron–boron) magnet is typically −0.1%/°C. So, in

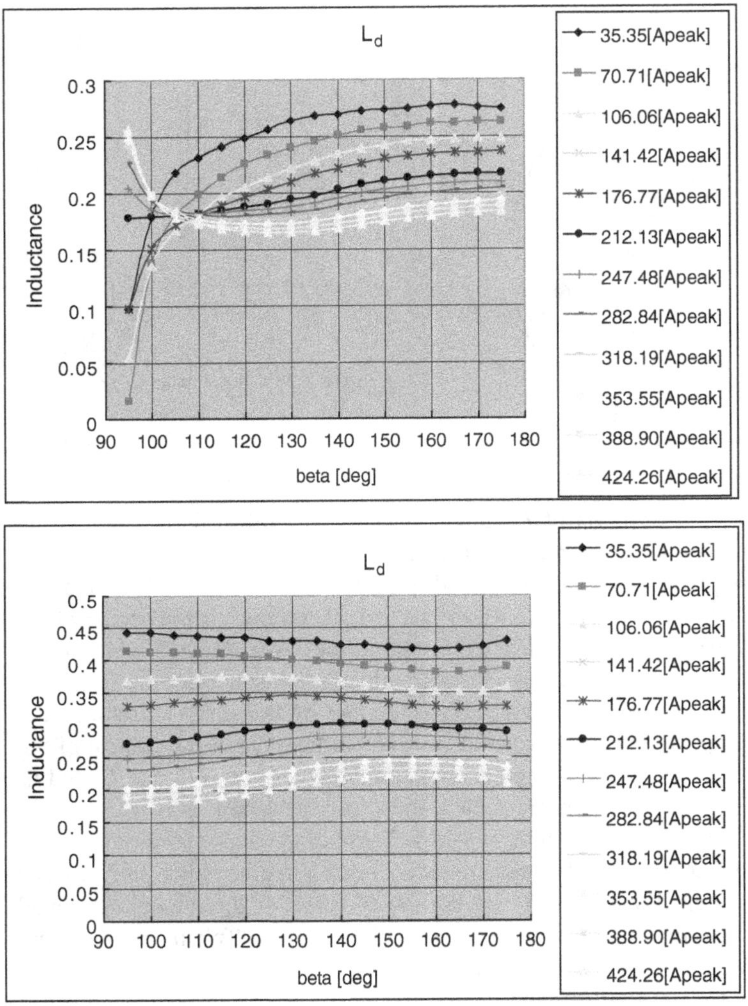

Figure 2.36 Variation of synchronous inductances, L_d and L_q of 50-KW, 220-V, 20-pole IPMSM for traction application.

the case of a NdFeBr permanent magnet synchronous machine, the back EMF constant would decrease by 10% according to a 100°C increase of the temperature of the magnet. In Fig. 2.38, the variation of the flux density according to the temperature of NEOMAX-32H, one of the NdFeBr magnets, is shown. From the figure, the temperature coefficient, α, is about –0.11%/°C, and the flux can be represented as the function of temperature in (2.42)–(2.44).

$$B_r(T) = B_{r0}\left(1 + \alpha\frac{\Delta T}{100}\right) \tag{2.42}$$

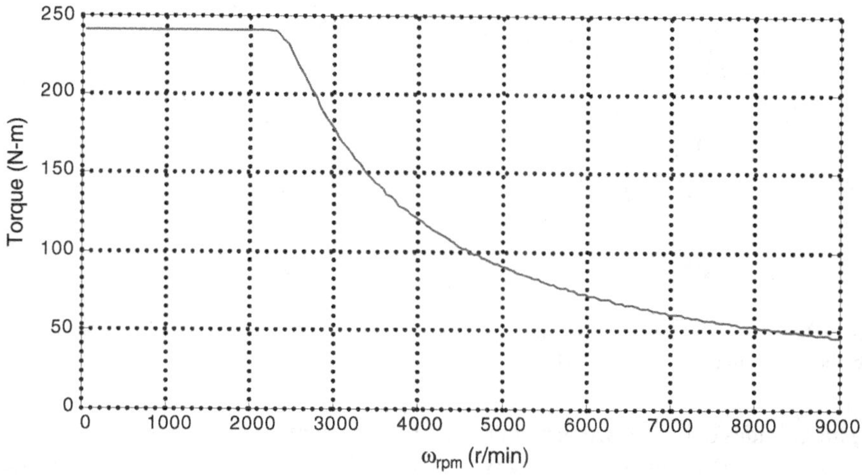

Figure 2.37 Capability curve of 50-KW, 220-V, 20-pole IPMSM for traction application.

$$\lambda_f = \int B_r \, dA \tag{2.43}$$

$$\lambda_f(T) = \lambda_{f0}\left(1 + \alpha\frac{\Delta T}{100}\right) \tag{2.44}$$

where B_{r0} stands for remanence of the magnet at 30°C, λ_{f0} for flux of the magnet at 30°C, ΔT for difference between 30°C and operating temperature T (°C), and dA for infinitesimal area where the flux passes. In Fig. 2.39, the back EMF waveforms of IPMSM using NEOMAX-32H magnet at two different operating temperatures are shown. In the figure, the operating speed is 1605 (r/min), constant at both waveforms.

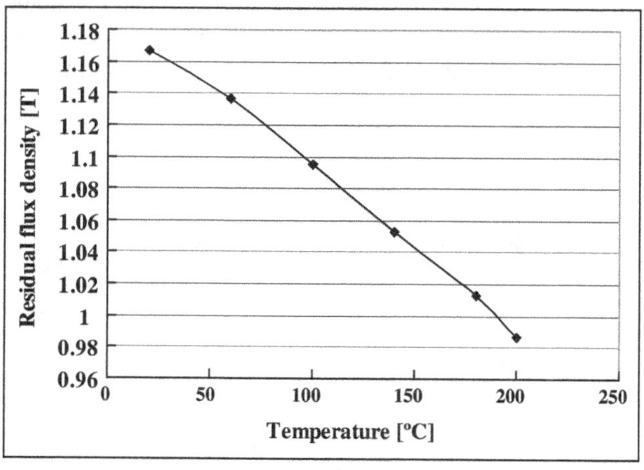

Figure 2.38 Variation of remanence of NEOMAX-32H magnet according to the temperature.

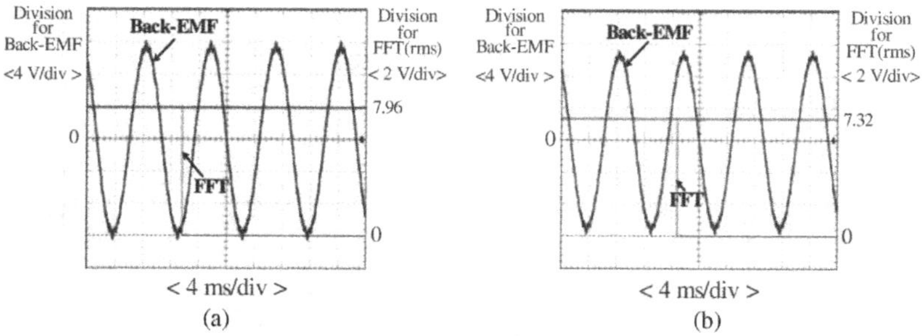

Figure 2.39 Back EMF waveforms of 4-kW, 6-pole IPMSM made by a NEOMAX-32H magnet at different temperatures. (a) 30°C. (b) 110°C.

The magnitudes of back EMF in rms through fast Fourier transformation (FFT) at two temperatures, 30°C and 110°C are 7.62 (V) and 7.32 (V), respectively. Because, the magnitude is set by the product of the speed and the flux linkage, λ_f, it is concluded that λ_f varies according to (2.42)–(2.44).

Such a variation of the flux according to the temperature may also affect the variation of the synchronous inductances through the variation of the level of the magnetic saturation. But the variation of the inductance according to the temperature variation would be minor compared to that according to the magnitude of the stator current.

2.10 STEADY-STATE ANALYSIS OF INDUCTION MACHINE

The induction machine is rotating based on the rotating MMF by the stator winding current like the synchronous machine; but unlike the synchronous machine, the induction machine has no reluctance difference and no excitation flux on the rotor by the permanent magnet or by a separate field winding. The induction machine can be classified according to the structure of the rotor as a wound rotor induction machine as shown in Fig. 2.40, where three-phase windings are in the rotor and the terminals of

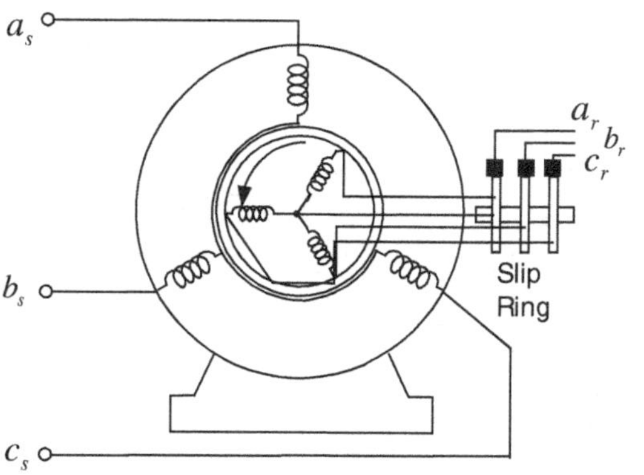

Figure 2.40 Conceptual diagram of a wound rotor induction machine.

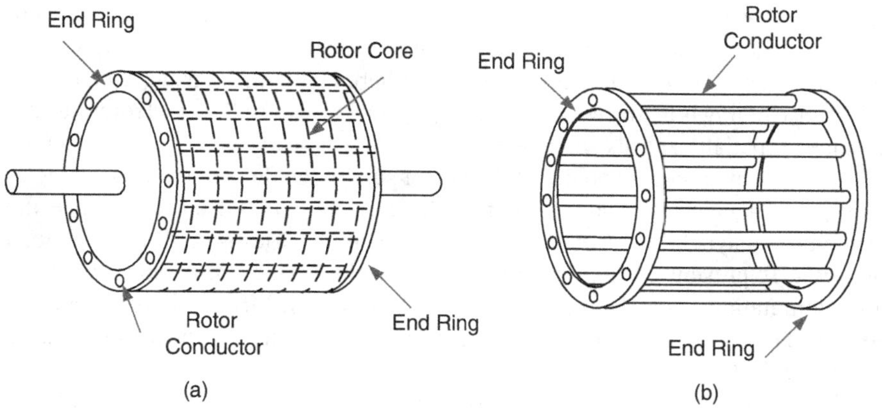

Figure 2.41 (a) Rotor structure of a squirrel cage rotor induction machine with rotor conductors and iron core. (b) Rotor structure of a squirrel cage rotor induction machine with rotor conductors only.

the windings are brought out, and a squirrel cage rotor induction machine as shown in Fig. 2.41. In the 20th century, the wound rotor induction machine had been used for an adjustable speed drive system because of its simplicity of the speed control through rotor winding terminals. But recently, due the development of power electronics, the application of the wound rotor induction machine is limited except for a high-power wind generation system. In this section, mainly the operation principle and the steady-state characteristics of the squirrel cage machine will be discussed. But the same principle works for a wound rotor induction machine. When the rotating MMF is applied to the squirrel cage rotor, EMF is induced at the conductors of the rotor, and EMF let the current flow in the conductor because the rotor circuit is short circuited by the end rings as shown in Fig. 2.42.

The current in the rotor conductors, by the induced EMF, again generates a rotating MMF. With the interaction between the rotor current and the rotating MMF by stator current, the torque of the induction machine is generated. When the load torque applies, the flux by the load current of the rotor is canceled out by the additional stator current except for the leakage flux of rotor flux. Hence, the air gap flux of the induction machine is constant regardless of the load condition if the excitation current of the stator winding is constant. This is the same with the case of a

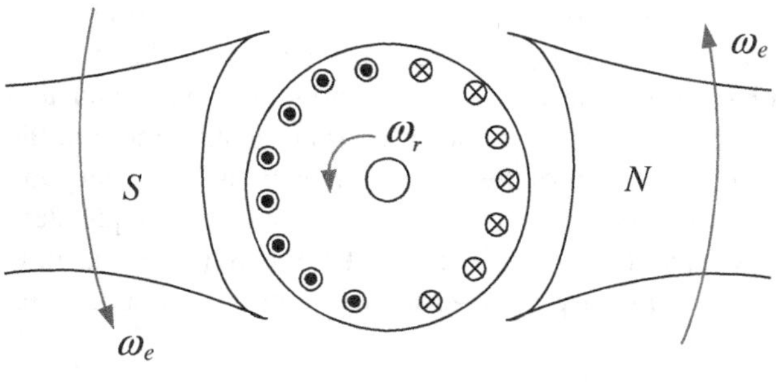

Figure 2.42 Current flowing in the conductors of the cage rotor with a rotating MMF.

transformer, where the flux by the load current of secondary winding is canceled out by the flux generated by the additional primary winding current. So, there is no armature reaction in the induction machine, inherently. If the leakage flux at rotor and stator winding is neglected, then the air gap flux, the stator flux, and rotor flux are the same in spatially and also in time. However, in the case of a synchronous machine and a DC machine the air gap flux is the vector sum of the flux by a rotor winding or by the permanent magnet and the stator flux by the stator winding current. Hence, the air gap flux of synchronous machine and DC machine varies according to the load condition. To prevent this flux variation, a DC machine has compensating windings and commutating poles. However, in the case of an SMPMSM, where the air gap flux is usually dominated by the permanent magnet, the flux by the stator winding may be neglected.

The frequency of the current and voltage of the rotor conductor of the induction machine is the difference between the angular frequency of the rotating MMF and that of the rotating speed of the rotor. And, the slip angular frequency, which is the frequency of the electrical variables of the rotor circuit, is given by

$$\omega_{sl} = \frac{d}{dt}(\theta_e - \theta_r) \tag{2.45}$$

where θ_r is the instantaneous position of the rotor in electric angle, and it can be expressed as

$$\theta_r = \frac{P}{2}\int_0^t \omega_{rm}\, d\tau + \theta_{ro} \tag{2.46}$$

where θ_{ro} stands for the rotor position at $t = 0$ s, ω_{rm} stands for the mechanical angular frequency of the rotor in rad/s, and P stands for the number of pole of the induction machine. The angular frequency of the rotor can be expressed as $\omega_r = \frac{P}{2}\omega_{rm}$ in electric angle.

2.10.1 Steady-State Equivalent Circuit of an Induction Machine [6]

The operation principle of the induction machine is the same as that of the transformer, which is that the rotor current flows through the induced voltage from the variation of the stator flux with regard to the rotor conductors. Similarly, in the case of the transformer the secondary voltage is induced through the flux variation of the primary flux with regard to the secondary conductors, whereas in the case of transformer there is no relative motion between primary and secondary conductors compared to the case of the induction machine. Because of no relative motion, the angular frequency of the electrical variables of the transformer in both the primary and the secondary windings are always the same. By using the steady-state equivalent circuit of the transformer, the steady-state equivalent circuit of the induction machine can be easily understood as follows by only considering the relative motion between the stator and rotor conductors.

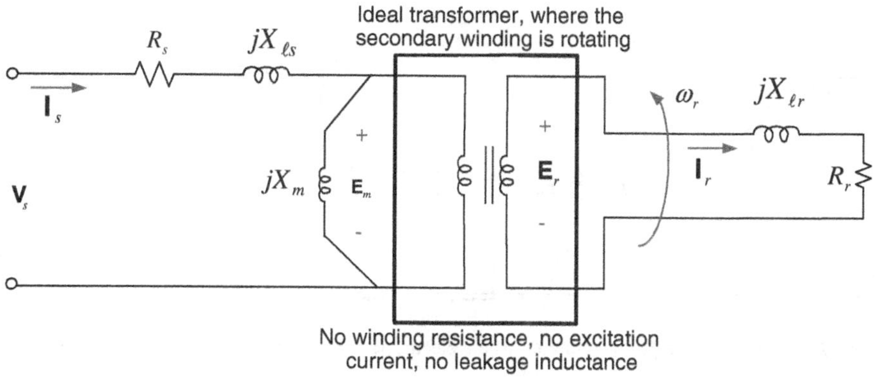

Figure 2.43 Steady state per phase equivalent circuit of a squirrel cage rotor induction machine.

The slip of the induction machine can be defined as

$$S \equiv \frac{\omega_e - \omega_r}{\omega_e} \tag{2.47}$$

where $\omega_e = \frac{d\theta_e}{dt}$ and $\omega_r = \frac{d\theta_r}{dt} = \frac{P}{2} \cdot \omega_{rm}$.

The equivalent circuit of the induction machine can be drawn as Fig. 2.43 using the equivalent circuit of an ideal transformer.

In the figure, R_s stands for the stator winding resistance, X_{ls} stands for the stator winding leakage reactance, X_m stands for the excitation reactance, X_{lr} stands for the rotor winding leakage reactance, and R_r stands for the rotor winding resistance.

From the circuit, the excitation voltage, $\mathbf{E}_m$, can be represented by (2.48) by using the phasor method, which is usually used to analyze linear AC circuit in the steady state.

$$\mathbf{E}_m = \mathbf{V}_s - (R_s + jX_{ls})\,\mathbf{I}_s \tag{2.48}$$

where $\mathbf{E}_m$ stands for the voltage in the phasor applied to mutual inductance, $\mathbf{V}_s$ stands for the phase voltage in the phasor between the stator terminal and the neutral point of Y connection of three phase stator winding, and $\mathbf{I}_s$ stands for the current in the phasor flowing in a phase of the stator winding.

If the induced voltage to the rotor circuit is $\mathbf{E}_r$ and the turn ratio between the stator circuit and rotor circuit is 1:1, then (2.49) can be deduced.

$$|\mathbf{E}_r| = S|\mathbf{E}_m| \tag{2.49}$$

If there is no slip, $S = 0$, which means that the speed of rotating MMF and the rotor speed is the same, then the induced voltage to the rotor circuit is $|\mathbf{E}_r| = 0$. In this case, no current flows in the rotor conductors and no torque can be generated. That can be easily understood from the principle of the induction machine mentioned previously. And, when $S = 1$, which means that the rotor is stationary, then the induction machine

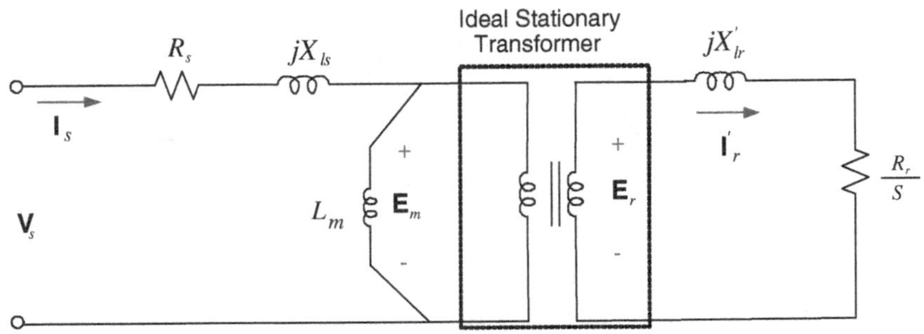

Figure 2.44 Per phase equivalent circuit, where electric variables in the rotor circuit is referred to stator side.

is the exactly same to the transformer. Because of the motion of the rotor, there is a difference between the transformer and the induction machine in the analysis using the phasor. While the frequency of the variables in the stator circuit is ω_e, that of the variables in the rotor circuit is $\omega_{sl}(\equiv S\omega_e)$, and the reactance at each circuit is $jX_m = j\omega_e L_m$, $jX_{ls} = j\omega_e L_{ls}$, and $jX_{lr} = j\omega_{sl} L_{lr}$ respectively. Here, L_m means excitation inductance of the machine, and L_{ls} and L_{lr} are leakage inductances of stator and rotor winding, respectively. Such an inconsistency of the frequency makes it difficult to simultaneously analyze both stator and rotor circuit. In particular, because the current and voltage of the squirrel cage rotor cannot be measured directly, the variables in the rotor circuit can be referred to the stator side, like in the analysis of the transformer where the secondary variables are usually referred to the primary side. If the variables at the rotor circuit are referred to the stator circuit after considering slip, then an equivalent circuit as shown in Fig. 2.44 can be drawn, where the ideal transformer is stationary. Instead, the rotor resistance varies according to the slip.

In Fig. 2.44, the angular frequency of both stator and rotor circuits is the same as ω_e. And if the magnitude of the referred rotor current, $\mathbf{I}'_r$ is the same as that of the actual rotor current, $\mathbf{I}_r$, then the impedance of the rotor circuit referred to the stator side should be increased to $1/S$ of its actual value because the induced voltage referred to the stator side is increased to $1/S$ of its actual value. So, the rotor resistance should be $1/S$ of its actual value as shown in the figure. But the leakage reactance referred to stator side, X'_{lr}, is already increased because the frequency of the rotor circuit has been increased from ω_{sl} to $\omega_e[= (1/S)\omega_{sl}]$. The actual parameters of the rotor circuit such as L_{lr}, R_r cannot be easily measured. Instead, it is deduced from the tests such as the no-load test and the locked rotor test done in the stator side, and the deduced rotor parameters and the calculated rotor current from the tests are inherently the values referring to the stator side such as X'_{lr} and I'_r. Moreover, the control of the squirrel cage induction machine is always done in the stator side, and the equivalent circuit can be simplified by omitting the ideal transformer as shown in Fig. 2.45, where all rotor variables are expressed at the stator side without the prime symbol ($'$), which means the referred variable.

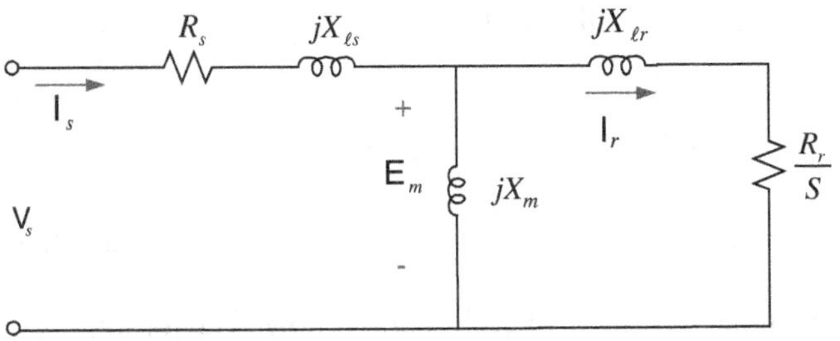

Figure 2.45 Simplified per phase equivalent circuit of an induction machine.

Using the circuit in Fig. 2.45, the characteristics of the induction machine can be investigated by using the phasor method like the case of a simple AC circuit.

From the circuit and phasor of the stator and currents, the losses due to the winding resistances—that is, the so-called copper losses—can be represented as (2.50) in the case of a three-phase induction machine.

$$P_{\text{copper}} = 3\left(R_s|\mathbf{I}_s|^2 + \frac{R_r}{S}|\mathbf{I}_r|^2\right) \tag{2.50}$$

The air gap power, P_{gap}, defined as the power transmitted from stator to rotor through the air gap can be represented as $\text{Re}\left[3\mathbf{E}_m \cdot \mathbf{I}_s^*\right]$ using the complex power theory of the phasor method in the case of three-phase balanced circuit. To consider the copper loss by the rotor resistance, R_r, in the rotor circuit, the resistance can be segregated to R_r and $\frac{1-S}{S}R_r$. Then the circuit in Fig. 2.45 can be redrawn as the circuit in Fig. 2.46, where the power consumed at the resistance, $\frac{1-S}{S}R_r$, can be considered as the mechanical output power, P_m, and the power consumed at the resistance R_r can be considered as the rotor copper loss.

Then, P_m can be expressed as

$$P_m = 3 \cdot |\mathbf{I}_r|^2 \cdot \frac{1-S}{S}R_r \tag{2.51}$$

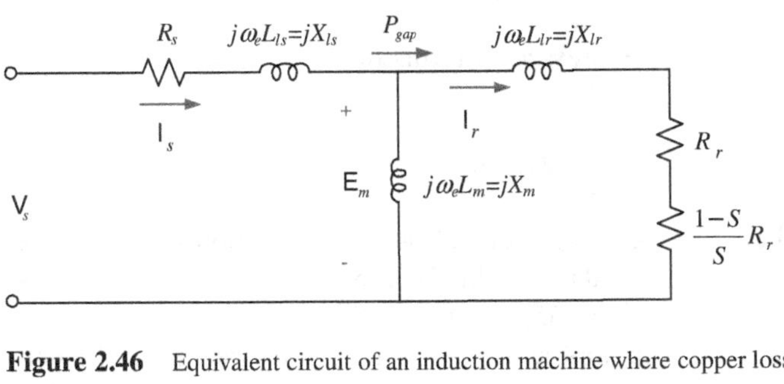

Figure 2.46 Equivalent circuit of an induction machine where copper loss due to the rotor resistance is separately considered.

From (2.51), the torque of the machine at the operating speed, ω_{rm}, can be deduced as

$$T_e = \frac{P_m}{\omega_{rm}} = \frac{P_m}{(1-S)\omega_e \frac{2}{P}} = \frac{P}{2}\frac{P_{\text{gap}}}{\omega_e} = 3\frac{P}{2}\frac{R_r}{S\omega_e}|\mathbf{I}_r|^2 = 3\frac{P}{2}\cdot\frac{R_r}{\omega_{sl}}|\mathbf{I}_r|^2 \tag{2.52}$$

As seen in (2.52), the torque can be expressed using the air gap power, P_{gap}, and the synchronous speed, $\omega_{rm} = \omega_e/(P/2)$.

The input power to the machine is $\text{Re}[3\mathbf{V}_s\mathbf{I}_s^*]$, and the power factor angle is defined as the difference of the phase of $\mathbf{V}_s$ and $\mathbf{I}_s$. The mechanical power, P_m, and the torque, T_e, at synchronous speed, $\omega_e = \omega_r(S=0)$, is null, because there is no current at the rotor circuit, $|\mathbf{I}_r| = 0$. However, at zero speed, $\omega_r = 0(S=1)$, while mechanical power is zero, the torque can be represented as (2.53) by inserting $S=1$ in (2.52).

$$T_{e\_st} = \frac{P}{2}3|\mathbf{I}_r|^2 \cdot R_r \cdot \frac{1}{\omega_e} \tag{2.53}$$

From the circuit in Fig. 2.45, the air gap voltage, $\mathbf{E}_m$, can be expressed as

$$\mathbf{E}_m = \left(jX_{lr} + \frac{R_r}{S}\right)\mathbf{I}_r \tag{2.54}$$

The power factor at the air gap can be represented by

$$\cos\phi_r = \frac{\frac{R_r}{S}}{\sqrt{X_{lr}^2 + \left(\frac{R_r}{S}\right)^2}} \tag{2.55}$$

The torque can be expressed in terms of $\mathbf{I}_r$, ϕ_r, and $\mathbf{E}_m$ as

$$T_e = 3\cdot\frac{P}{2}\cdot|\mathbf{E}_m|\cdot\cos\phi_r|\mathbf{I}_r|\frac{1}{\omega_e} \tag{2.56}$$

The air gap flux, $\boldsymbol{\lambda}_m$, can be defined as

$$\boldsymbol{\lambda}_m \equiv \frac{\mathbf{E}_m}{j\omega_e} \tag{2.57}$$

And, the torque in (2.56) can be expressed in terms of the air gap flux and rotor current as

$$T_e = 3\cdot\frac{P}{2}\cdot|\boldsymbol{\lambda}_m|\cdot|\mathbf{I}_r|\cos\phi_r \tag{2.58}$$

The slip of a standard NEMA B-type general-purpose induction machine (see Section 2.13), if the output power of the machine is above several kilowatts, is less than 5%, and $X_{lr} \ll R_r/S$. Hence, in (2.55), the power factor at the air gap can be approximated as $\cos\phi_r \approx 1$. With this approximation, the torque in (2.58) can be expressed as simply the product of the magnitude of the air gap flux and that

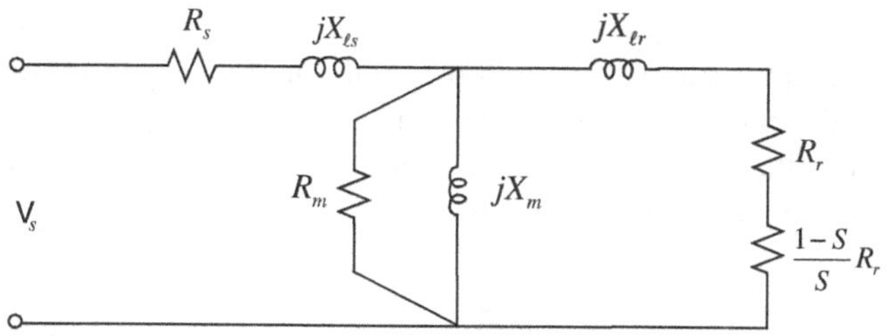

Figure 2.47 Equivalent circuit considering iron loss.

of the rotor current. This expression for the torque of the induction machine is very similar to that of DC machine in (2.4).

At the circuit in Fig. 2.46, a resistance paralleled to excitation (magnetizing) inductance, L_m, can be added to express iron loss of the induction machine as shown in Fig. 2.47. The effects of the resistance, R_m, for the iron loss, to the torque is less than a few percentages, and it can be easily ignored. However, to evaluate the efficiency and the no-load power factor, it should be considered. Usually, in the case of the standard general-purpose induction machine the iron loss is almost same as the copper loss at around 80% load condition at the rated speed, where the efficiency is maximum. So, the efficiency evaluated from the circuit in Fig. 2.46 is unrealistically high compared to that from the circuit in Fig. 2.47, where the iron loss is considered together with copper loss. In addition, the power factor at the stator terminal would be increased remarkably due to the iron loss by R_m in the case of no-load operation. However, the power factor at the rated load condition by the circuit in Fig. 2.46 would be virtually the same as the power factor in Fig. 2.47.

2.10.2 Constant Air Gap Flux Operation

To control the air gap flux, $\boldsymbol{\lambda}_m$, as constant regardless of the variation of the stator frequency, ω_e, the air gap voltage, $\mathbf{E}_m$, in Fig. 2.45 should be adjusted as follows:

$$\mathbf{E}_m = j\omega_e\boldsymbol{\lambda}_m = j\omega_e L_m \mathbf{I}_m \tag{2.59}$$

The rotor current can be expressed by the voltage, $\mathbf{E}_m$, and the slip angular frequency, ω_{sl}, as

$$\mathbf{I}_r = \left(\frac{|\mathbf{E}_m|}{\omega_e}\right)\frac{\omega_{sl}}{\sqrt{R_r^2 + (\omega_{sl}L_{lr})^2}} \tag{2.60}$$

where the slip angular frequency is given by $\omega_{sl} = S \cdot \omega_e$.

Equation (2.60) can be rewritten as (2.61) in terms of the air gap flux and slip angular frequency.

$$|\mathbf{I}_r| = |\boldsymbol{\lambda}_m|\frac{\omega_{sl}}{\sqrt{R_r^2 + (\omega_{sl} \cdot L_{lr})^2}} \tag{2.61}$$

Usually in the case of the standard general-purpose induction machine, when the load torque is less than rated value and the air gap flux is constant, the magnitude of rotor current, $|\mathbf{I}_r|$, is proportional to ω_{sl} as seen in (2.61), because $R_r \gg \omega_{sl}L_{lr}$ at this operating condition. By substituting (2.61) into (2.52), the torque of the induction machine can be represented as (2.62) in terms of the air gap flux and slip angular frequency.

$$T_e = 3\frac{P}{2}|\boldsymbol{\lambda}_m|^2 \cdot \frac{\omega_{sl}R_r}{R_r^2 + (\omega_{sl}L_{lr})^2} \tag{2.62}$$

Equation (2.62) can be approximated as (2.63) in the normal operating region where $R_r \gg \omega_{sl}L_{lr}$.

$$T_e \approx 3\frac{P}{2} \cdot |\boldsymbol{\lambda}_m|^2 \cdot \frac{\omega_{sl}}{R_r} \tag{2.63}$$

Equation (2.63) means that the torque of the induction machine is proportional to ω_{sl}, and proportional to the square of the air gap flux.

By differentiating (2.62) regarding ω_{sl}, the slip frequency, where maximum or minimum torque occurs, can be found as

$$\omega_{sl\_pk} = \pm\frac{R_r}{L_{lr}} \tag{2.64}$$

The torque at ω_{sl_pk} is called as the pull-out torque of the induction machine with the given constant air gap flux. The pull-out torque can be calculated by substituting (2.64) into (2.63) as (2.65) in both motoring and generating cases.

$$T_{e\_pk} = \pm 3\frac{P}{2} \cdot |\boldsymbol{\lambda}_m|^2 \cdot \frac{1}{L_{lr}} \tag{2.65}$$

It can be noted that the pull-out torque is independent on the rotor resistance, and it is decided only by the rotor leakage inductance under the constant air gap flux operation. If the air gap flux is kept as constant, then the stator voltage per phase, $\mathbf{V}_s$, can be expressed as

$$\mathbf{V}_s = j\omega_e\boldsymbol{\lambda}_m + (R_s + jX_{ls})\mathbf{I}_s \tag{2.66}$$

If ω_e is large enough, then the following inequality holds.

$$|\boldsymbol{\lambda}_m| \cdot \omega_e = |\mathbf{E}_m| \gg |(R_s + jX_{ls})\mathbf{I}_s| \tag{2.67}$$

Then the phase voltage can be approximated by the air gap voltage as

$$|\mathbf{V}_s| \approx |\boldsymbol{\lambda}_m| \cdot \omega_e \tag{2.68}$$

Hence, in the operation region where (2.67) holds, if the phase voltage is adjusted as (2.69), then the air gap flux can be maintained as constant. In the case of the standard induction machine, if rated output power is above several kilowatts

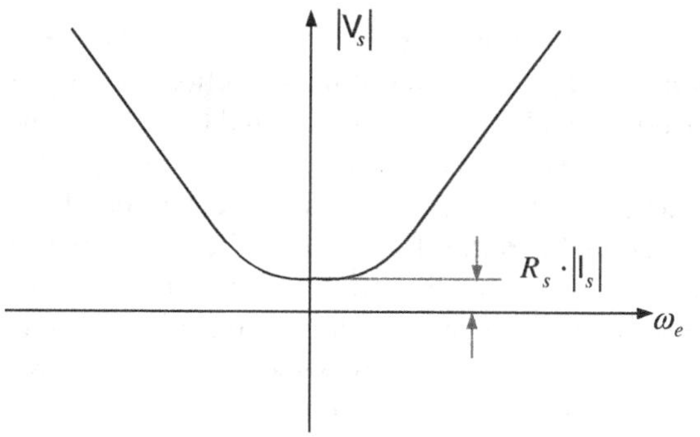

Figure 2.48 Relationship between the magnitude of the stator voltage and the frequency under constant air gap flux operation.

and the stator frequency is above 20% of the rated value, then (2.67) generally holds.

$$\frac{\mathbf{V}_s}{\omega_e} = |\boldsymbol{\lambda}_m| = \text{const} \tag{2.69}$$

Equation (2.69) means that only by adjusting the phase voltage proportional to the stator frequency the air gap flux can be controlled as approximately constant. This kind of approximated constant air gap flux operation is called as V/F control of the induction machine. However, the frequency, ω_e, is getting smaller and the magnitude of the air gap voltage, $|\boldsymbol{\lambda}_m| \cdot \omega_e = |\mathbf{E}_m|$, is comparable to or less than the magnitude of the voltage drop of the stator impedance, $|(R_s + jX_{ls})\mathbf{I}_s|$; thus the voltage drop should be considered in the control of the stator voltage. Also, if the frequency is small enough, then $R_s \gg X_{ls}$. In this operating condition, the voltage drop of the stator impedance is almost the same as the drop by stator resistance as $|(R_s + jX_{ls})\mathbf{I}_s| \approx R_s|\mathbf{I}_s|$. Considering the overall frequency range, to keep air gap flux as constant, the magnitude of the stator voltage should have the relationship with the frequency as shown in Fig. 2.48. The voltage at zero frequency, $R_s|\mathbf{I}_s|$, is referred to as the voltage for torque boost or simply as the voltage for IR compensation.

2.11 GENERATOR OPERATION OF AN INDUCTION MACHINE

From the circuit in Fig. 2.46, If slip, S, is negative, which means that the rotating speed of the rotor is larger than the speed of the rotating MMF, then the mechanical output, P_m, is negative. And, in this case the induction machine operates as a generator. If the induction machine is driven by a prime mover to set the slip as negative, then the resistance $\frac{1-S}{S}R_r$ in the equivalent circuit, which represents mechanical output, is negative. In this case if the current flows through the negative resistance, the power is not consumed but generated. Unlike the synchronous machine where always excitation flux is provided by the field current or the permanent magnet, the induction machine cannot be operated as the generator if there is no excitation current at the stator winding provided from the external circuit. As understood from the equivalent circuit, if the stator terminal of the machine is not connected to the external circuit, then rotating MMF cannot exist. Hence, there is no induced voltage to the stator winding by simply rotating the rotor by the prime mover. If the induction machine is connected to the external source, whose angular frequency is ω_e and the rotor of the machine is driven as $\omega_r > \omega_e$, then the source would provide the reactive power (VAR), $3X_m \cdot |\mathbf{I}_m|^2$, to the machine, and the prime mover, which rotates the rotor of the machine, provides active power (Watts) to the source. Here, a part of the active power provided by the prime mover through the rotor axis is used to cover the losses of the induction machine itself, and the remaining active power is transmitted to the source. The active power to the source can be expressed as (2.70), neglecting the iron loss, windage loss, friction loss, and stray loss of the induction machine.

$$P_{\text{out}} = T_e \cdot \omega_{rm} - 3|\mathbf{I}_s|^2 \cdot R_s - 3|\mathbf{I}_r|^2 R_r = 3\text{Re}\left[\mathbf{V}_s \cdot \mathbf{I}_s^*\right] \tag{2.70}$$

where T_e is the torque of the prime mover. In addition, for the real machine there is the iron loss, windage loss, friction loss, and stray loss, which are not considered in (2.70). So, the actual power transmitted to the external circuit would be less than the power calculated by (2.70).

In Fig. 2.49, the variations of the magnitude of the stator current and mechanical power of a 22-kW, 60-Hz, four-pole induction machine are drawn with the variation of the slip from −1 to 1. In the figure, iron loss, windage loss, friction loss and stray loss is not considered. When the slip is negative, the machine operates as a generator, where the mechanical power means input power to the machine. In the figure, the asymmetry regarding the vertical axis comes from the internal copper loss of the machine. When it operates as a motor, the loss is covered by the electric input power and the mechanical power is less than the electrical input power. However, when it operates as a generator,

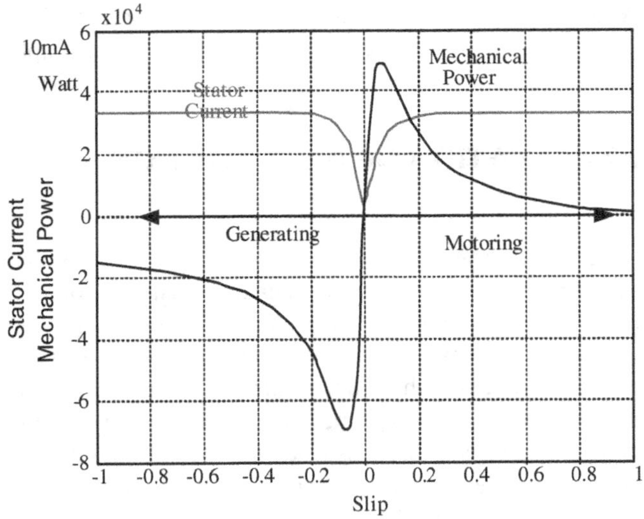

Figure 2.49 Steady-state characteristics of a four-pole, 220-V, 75-A, 60-Hz, 22-kW induction machine. $R_s = 0.044\,\Omega$, $R_r = 0.0252\,\Omega$, $L_{ls} = 0.55$ mH, $L_{lr} = 0.47$ mH, $L_m = 12.9$ mH.

the mechanical input power covers the loss and the mechanical power is always larger than the electrical output power.

2.12 VARIATION OF PARAMETERS OF AN INDUCTION MACHINE [7]

Like the synchronous machine, the parameters of the equivalent circuit of an induction machine in Fig. 2.47, R_s, R_r, L_m, L_{ls}, L_{lr}, R_m, vary according to the operating conditions such as temperature, operating frequency, air gap flux, and the magnitude of the current. Also, the parameters of all electric machines including DC machine, AC machine, and transformer vary very widely according to the rated voltage, speed, and frequency. Hence, it is very difficult to compare the characteristics of the machines when their size, voltage, and frequency are different. For the comparison, the per unit method could be a helpful tool. By using the method, the parameters of the electric machine can be easily compared even if their operating voltage, frequency, and power are different. Also, the trends of the parameters according to the power or the frequency can be easily understood. The detailed description about per unit method is in Section 2.19. The power, voltage, current, frequency, speed, and impedance in per unit have no physical unit and they are relative values to the base values. In addition, the measurement and/or the estimation of the parameters of the electric machine drive system including the inertia and the friction coefficient is described in Appendix A.

2.12.1 Variation of Rotor Resistance, R_r

The rotor resistance varies according to the frequency, ω_{sl}, of the current flowing in the rotor conductors. Usually, the rotor conductor is made by aluminum- or copper-based alloy, and the resistance varies by the skin effect, which means that the current tends to flow on the top surface of the conductor as the frequency of the current increases. So, the resistance increases as the frequency,ω_{sl}, increases. By exploiting the variation of the rotor resistance due to the skin effect, the starting torque of the induction machine can be enhanced without any penalty in the running efficiency. In the starting, the slip is unity and $\omega_{sl} = \omega_e$. At normal running condition, the slip is quite small, and $\omega_{sl} \ll \omega_e$. So, the rotor resistance may be set to be large at the starting and small in the running. From (2.53), the starting torque is decided by the rotor resistance, and the larger resistance means higher starting torque. But the running efficiency can be improved with the reduced rotor resistance due to the reduced frequency of the current flowing in the rotor conductors. To maximize the skin effect, the rotor bar of the squirrel cage rotor induction machine can be designed as shown in Fig. 2.50, where the resistance varies remarkably according to ω_{sl}.

In addition to the skin effect, the resistance varies according to the temperature of the conductor as mentioned in Section 2.9.1. If the rotor conductor is made by a copper, then the rotor resistance varies according to the temperature coefficient of copper as (2.41).

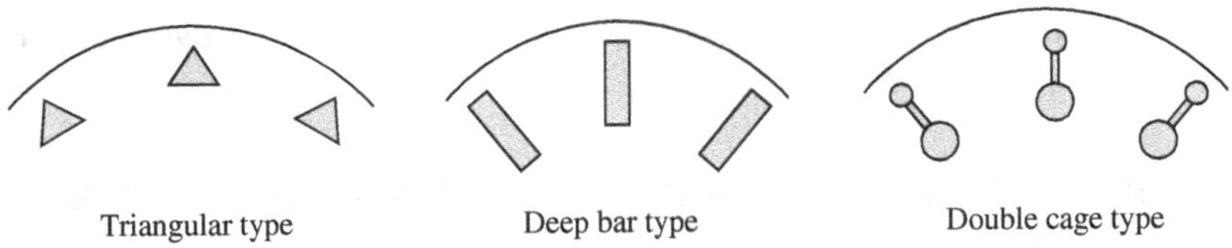

Figure 2.50 Shapes of rotor conductors of the cage rotor of the induction machine to maximize skin effect for improvement of the starting characteristics.

2.12.2 Variation of Rotor Leakage Inductance, L_{lr}

The rotor leakage inductance, L_{lr}, is affected by the skin effect, and the inductance decreases as the frequency, ω_{sl}, increases. In addition to the skin effect, the leakage inductance is affected by the magnetic saturation of the leakage flux in the rotor core. This variation due the magnetic saturation is severe in the case of the closed slot cage rotor, where the narrow area (the so-called bridge) in the front of the conductor can be easily saturated even with a moderate rotor current as shown in Fig. 2.51.

The typical variation of L_{lr} according to the magnitude of the rotor current is shown in Fig. 2.52, where the value varies by severalfold. In the case of open slot or semi-closed slot rotor shown in Fig. 2.53, the variation of the inductance is not severe as the case of the closed slot. But there are still some variations even in the case of the open slot or semi-closed slot.

2.12.3 Variation of Stator Resistance, R_s

Because the stator conductors are normally made by a stranded wire, the skin effect can be neglected. However, the variation of the resistance due to the temperature should be considered as (2.41).

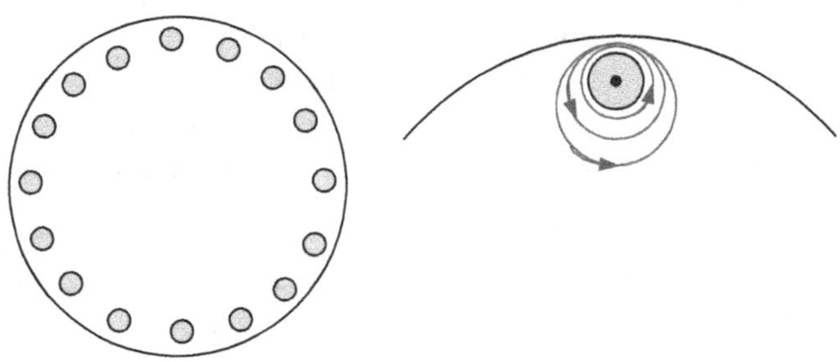

Figure 2.51 Distribution of the leakage flux of a closed slot cage rotor.

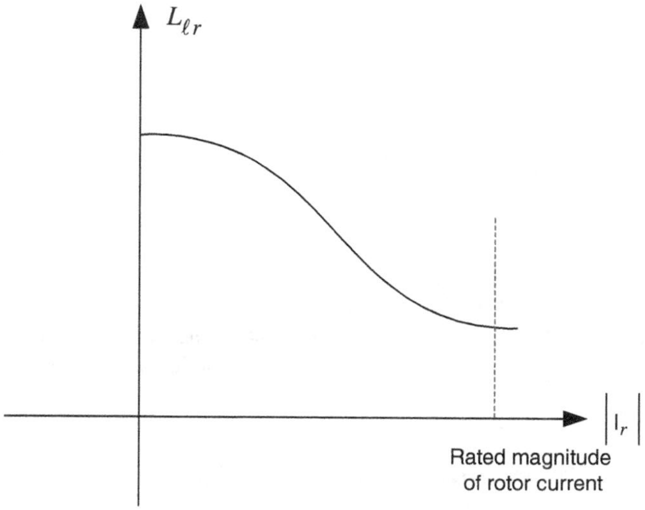

Figure 2.52 Variation of rotor leakage inductance according to the magnitude of the rotor current.

2.12.4 Variation of Stator Leakage Inductance, L_{ls}

The slot of the stator is normally open magnetically, and the variation of the leakage inductance due to the magnetic saturation of the leakage flux is not severe compared to that of the rotor leakage inductance of the closed slot rotor. But, still due to the magnetic saturation, according to the magnitude of the stator current, several tens percentages of the variation can be expected.

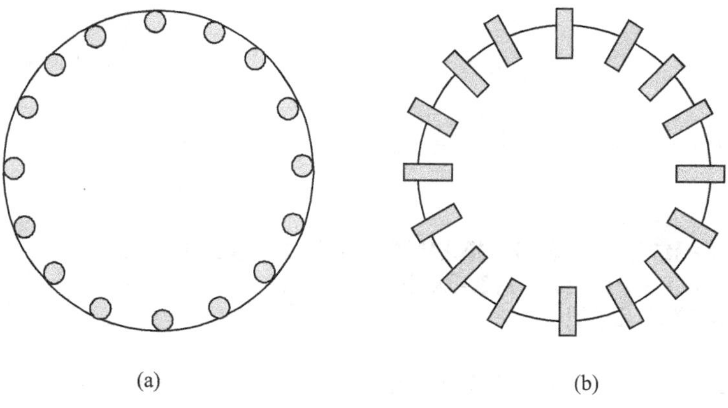

Figure 2.53 Rotor slot structure that reveals less variation of the leakage inductance. (a) Semi-closed slot. (b) Open slot.

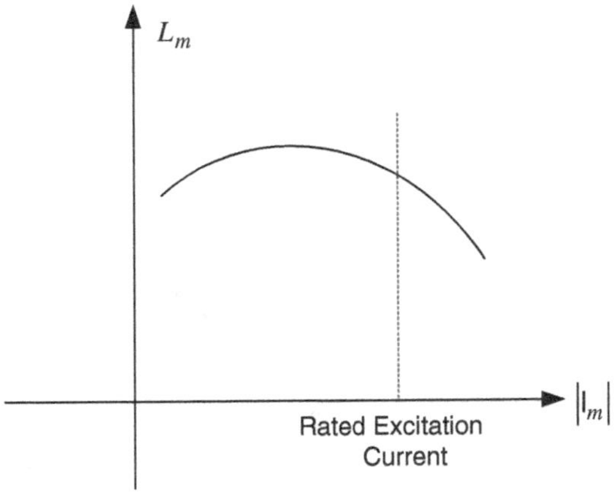

Figure 2.54 Variation of L_m according to the magnitude of the excitation current, $|I_m|$.

2.12.5 Variation of Excitation Inductance, L_m

The excitation (magnetizing) inductance varies according to the excitation current–air-gap flux level, by several tens of percentages as shown in Fig. 2.54. If the excitation current reduced from its rated value, the inductance increases slightly and decreases with further decrease of the current.

2.12.6 Variation of Resistance Representing Iron Loss, R_m

In Fig. 2.47, the resistance, R_m, represents iron loss varies according to the air-gap flux, $\boldsymbol{\lambda}_m$, and the excitation frequency, ω_e. The iron loss expressed by R_m comes from eddy current loss and hysteresis loss. The former is approximately proportional to $\omega_e^2|\boldsymbol{\lambda}_m|^2$ and the latter approximately proportional to $\omega_e^2|\boldsymbol{\lambda}_m|^{1.6}$. Hence, R_m can be modeled as (2.71) as a function of the air gap flux and the frequency.

$$R_m = \left(K_e + \frac{K_h}{\omega_e|\boldsymbol{\lambda}_m|^{0.4}}\right) \tag{2.71}$$

where K_e and K_h are coefficients for eddy current loss and hysteresis loss, respectively.

2.13 CLASSIFICATION OF INDUCTION MACHINES ACCORDING TO SPEED–TORQUE CHARACTERISTICS [8]

The induction machine is classified as A, B, C, D, or F type according to the torque at the operating speed. This classification is prepared by NEMA (National Electric Manufacturer's Association) of the United States. The NEMA F type is a special one that has small pull-out torque and small starting torque, and its application is very

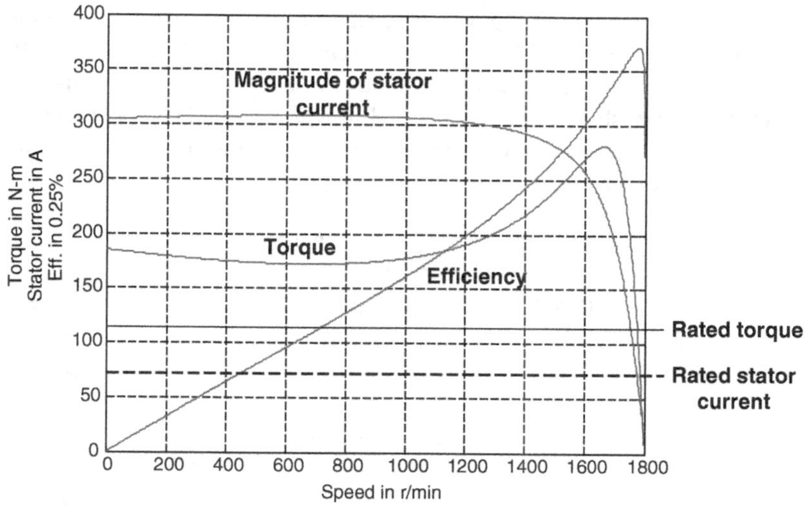

Figure 2.55 Torque–speed curve of a typical 22-kW NEMA B-type induction machine.

limited. The most widely used machine is NEMA B type, which reveals starting torque more than 150% of rated torque of the machine and pull-out torque more than 200%. The typical torque–speed curve of NEMA B-type induction machine is shown in Fig. 2.55 along with the efficiency and the magnitude of the stator current, where the parameter variations, except for the rotor resistance, according to the operating conditions are ignored. The rotor resistance varies due to skin effect as mentioned in Section 2.12.1. And that variation is important with regard to increasing the starting torque while decreasing starting current.

In the case of a NEMA D-type machine, the slip where pull-out torque occurs is quite large and sometimes it would be unity, which means that the peak torque occurs at the start. And the starting current of NEMA D-type machine is limited, and starting torque is quite high. The operating speed of this type of machine can be easily controlled by simply adjusting terminal voltage of the machine. However, the efficiency of the induction motor is always less than 1-slip, and the running efficiency of NEMA D type of machine is poor, especially when the slip is large. NEMA A-type machines reveal higher starting current and less starting torque, but the running efficiency is the best among all types because of the smaller rotor resistance compared to other types. For this reason, an A-type induction machine would be the best choice if the induction machine is driven by a variable-voltage variable-frequency (VVVF) inverter, where the starting current and starting torque can be regulated by a VVVF inverter itself.

If the parameters of the standard NEMA B-type machine are represented in per unit method (see Section 2.19), then they can be formulated empirically as (2.72)–(2.76). These equations come from the physical limitations of the materials of the machine such as current density of the conductors, the flux density of the core materials, cooling conditions, and so on. If the pole pitch of the machine is set by τ_p in meters, then the mechanical output power of a standard NEMA B-type induction machine in horsepower (1 Hp = 746 W) has the relationship with pole pair (PP) as (2.72) [9]. From the pole pitch, other parameters of a NEMA B-type induction machine have the relationship as (2.73)–(2.76). However, it should be noted that the parameters in (2.72)–(2.76) are typical ones and because of the special design and special application, the range of the parameters are quite wide.

$$\tau_p = 0.084\left(\mathrm{Hp}/\mathrm{PP}^2\right)^{6/23} \qquad \text{(meters)} \tag{2.72}$$

$$R_s = 0.0033\tau_p^{-1} \qquad \text{(per unit)} \tag{2.73}$$

$$R_r = 0.004\tau_p^{-1} \qquad \text{(per unit)} \tag{2.74}$$

$$X_\sigma = X_{ls} + X_{lr} = 0.2 \qquad \text{(per unit)} \tag{2.75}$$

$$X_m = 10\left(\tau_p/\mathrm{PP}\right)^{1/2} \qquad \text{(per unit)} \tag{2.76}$$

The parameters given by the above equations in the case of a four-pole machine, PP = 2, can be expressed as shown in Fig. 2.56.

From the figure, it can be seen that X_m increases and R_s and R_r decrease as output power of the induction machine increases. Hence, as the power of the machine is getting larger, the power factor and efficiency are getting better. And, at the same output power, as the number of the poles of the machine increases, X_m decreases as (2.76) and the power factor decreases.

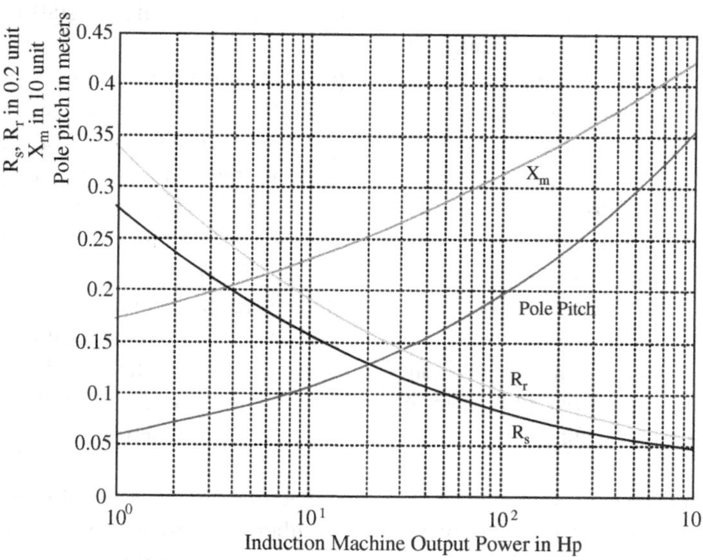

Figure 2.56 Parameter variations of a four-pole NEMA B-type induction machine according to output power.

2.14 QUASI-TRANSIENT STATE ANALYSIS [6]

The transient response of the induction machine due to the input voltage transient for short period, where short period means that for that period the variation of the rotating speed and that of rotor flux can be negligible, can be easily found without the full analysis of the transient state described in Section 3.2. The variation of the rotating speed is resulted from the variation of torque at the transient, but the variation of the speed is quite slow compared to the variation of the torque due to the low-pass filtering action of the inertia of the system. Also, the variation of the rotor flux depends on the rotor time constant $\left(\tau_r = \frac{(L_m + L_{lr})}{R_r}\right)$. Hence, in the case of above several-hundred-kilowatt induction machine, several milliseconds can be considered as the short period.

If the induction machine was operated in the steady state just before the transient, the machine can be represented as a circuit based on the phasor method as shown in Fig. 2.57 during the quasi-transient.
where R_s stands for the stator winding resistance as shown in Fig. 2.45, and X_K can be represented as $\omega_e L_K$, where $L_K = L_{ls} + \frac{L_m \cdot L_{lr}}{L_{lr} + L_m}$, approximately $L_K \approx L_{ls} + L_{lr}$. And $\mathbf{E}_s$ in Fig. 2.57 is called the phasor voltage behind transient reactance which is decided by the rotor flux and rotor inductance just before the transient as expressed by

$$\mathbf{E}_s = j\omega_e \frac{L_m}{L_m + L_{lr}} \cdot \boldsymbol{\lambda}_{ro} = \frac{L_m}{L_m + L_{lr}} \cdot \mathbf{I}_{ro} \cdot \frac{R_r}{S} \tag{2.77}$$

where $\mathbf{I}_{ro}$ is the phasor rotor current just before the transient. If the angular frequency of the phasor of the stator voltage, $\mathbf{V}_s$, is ω_e, then the operating frequency of the circuit in Fig. 2.57 is ω_e. And if the stator circuit is opened suddenly and no current flows in the circuit, $\mathbf{I}_s = 0$ in the quasi-transient, then the operating frequency of the circuit in Fig. 2.57 is the same as the rotating speed of the rotor in electrical angle, $\omega_r = \frac{P}{2}\omega_{rm}$. In this situation, it is assumed that the current flows through X_K disappeared instantaneously through the sudden discharge of the current in X_K during the opening of the stator circuit. This quasi-transient analysis is helpful to deduce the transient responses of the machine at the open circuit of the stator or at several induction machines tied to the same input line against the input line voltage transient.

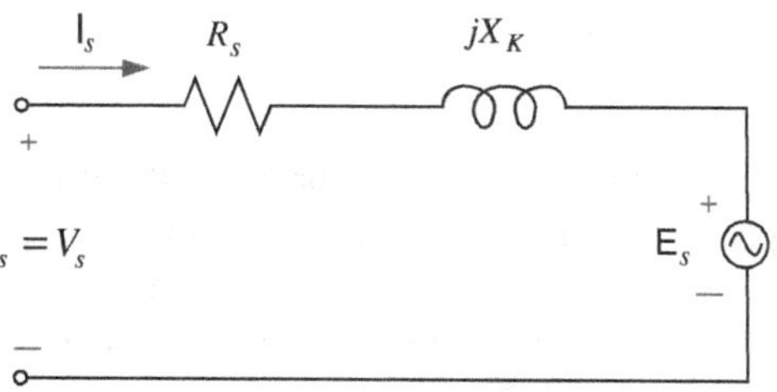

Figure 2.57 Quasi-transient equivalent circuit of the induction machine.

2.15 CAPABILITY CURVE OF AN INDUCTION MACHINE

The available maximum torque and speed range of the induction machine in the torque–speed plane under the limited stator voltage and current magnitude ($|\mathbf{V}_s| \le V_{\text{lim}}, |\mathbf{I}_s| \le I_{\text{lim}}$), which is defined as the capability curve, can be obtained by controlling the air gap flux in constant up to the base speed and by reducing the air gap flux proportional to the speed above the base speed. The reason to control the air gap flux as the constant value, which is normally the rated value, is that the rated value maximizes the torque per ampere without the saturation of the magnetic circuit of the machine. If the flux is set to above the rated value, then due to the magnetic saturation the iron loss increases rapidly and efficiency drops with a little increase of the torque. So, the benefits of the slightly increased torque at the same current are further offset by the increased iron loss. As a reverse case, if the flux is set to below the rated value, then to get the rated torque more current should flow into the machine and the copper loss increases rapidly. So, again efficiency drops.

Above the base speed, if the flux is kept as the rated value, then the magnitude of the stator voltage, $\mathbf{V}_s$, increases above the rated value, normally V_{lim}, which is the maximum value accommodated by the machine itself or by the electric power supply of the machine. This situation is exactly the same to the case of DC machine or a wound rotor synchronous machine. While in a DC machine and a wound rotor synchronous machine the field winding current can be reduced for the field weakening control, in the case of the induction machine and a permanent magnet synchronous machine the rotating MMF should be adjusted with regard to the flux of the rotor to reduce the air-gap flux. Because of no field winding in the case of the induction machine and permanent magnet synchronous machine, the air gap flux should be controlled by the stator current. Hence, in this case it is reasonable to refer to such a weakening of the excitation flux as not the field weakening control but as the flux weakening control. The flux weakening control of an AC machine is described in detail in Section 5.4. Above the base speed, if the $|\boldsymbol{\lambda}_m|$ is controlled as $1/\omega_e$, because $|\mathbf{E}_m| = \omega_e|\boldsymbol{\lambda}_m|$ in the circuit at Fig. 2.46, $|\mathbf{E}_m|$ would be constant in the entire flux weakening region. However, the impedance due to the leakage inductance of the rotor, $X_{lr} = \omega_e L_{lr}$, would increase as the frequency increases. Hence, if the slip is kept as constant, then the magnitude of the rotor current, $|\mathbf{I}_r|$, decreases and that of the stator current, $|\mathbf{I}_s|$, also decreases. To keep the magnitude of the stator current as I_{lim}, the slip should be increased according to the increase of ω_e in order to keep $|\mathbf{I}_r|$ as a constant decided by the magnitude of the rotor impedance $|Z_r| = |\frac{R_r}{S} + j\omega_e L_{lr}|$:

$$|\mathbf{I}_r| = \frac{|E_m|}{\sqrt{\left(\frac{R_r}{S}\right)^2 + (\omega_e L_{lr})^2}} \tag{2.78}$$

In the operating condition, where the magnitude of stator current is the limited value, $|\mathbf{I}_s| = I_{\text{lim}}$, and under the assumption that the limited value is equal or larger than the rated value of the stator current, then $\mathbf{I}_s = \mathbf{I}_m + \mathbf{I}_r \approx \mathbf{I}_r$. Also, in the case of the standard general-purpose induction machine above several-kilowatt output power range, at the normal running condition we have $R_r/S \gg \omega_e L_{lr}$. To keep $|\mathbf{I}_s|$ as I_{lim}, the slip should be constant and the slip angular frequency increases proportional to ω_e. If the slip angular frequency, ω_{sl}, increases up to ω_{sl_pk}, which is the value where the pull-out torque occurs, then the torque would not further increase with the increase of ω_{sl}. Hence, above that speed, the slip angular frequency, ω_{sl}, should be kept as constant as ω_{sl_pk}, and the magnitude of the rotor and the stator current decreases with the further increase of the speed above that speed. It is referred to as a flux weakening region 1, where $|\mathbf{I}_s|$ and approximately $|\mathbf{I}_r|$ can be kept as a constant like $|\mathbf{I}_s| \approx I_{\text{lim}}$ by controlling the slip angular frequency as the speed increases. And it is referred to as flux weakening region 2 or as a characteristic region, where $|\mathbf{I}_s|$ decreases as speed increases. The available maximum torque in the region 1 is almost inversely proportional to ω_e as seen from (2.57) and (2.58), where the variation of the power factor of the rotor circuit is negligible. Hence, the slip angular frequency is small enough compared to the stator frequency as $\omega_e \gg \omega_{sl}$ and $\omega_e \approx \frac{P}{2}\omega_{rm}$, then the output power of the machine, given by $P_m = T_e \cdot \omega_{rm}$, can be maintained as a constant. Hence, the flux weakening region 1 is called the constant power region. However, in the flux weakening region 2, T_e decreases inversely proportional to the square of ω_e because $|\mathbf{I}_r|$ decreases inversely proportional to ω_e in addition to the decrease of the air-gap flux, $\boldsymbol{\lambda}_m$. Hence the output power of the induction machine at the region 2 decreases inversely proportional to ω_e. The existence of this characteristic region in a flux weakening control is an inherent difference between the induction machine and other machines such as a DC machine or a synchronous machine.

In Fig. 2.58, the magnitude of the stator current, the torque, the slip angular frequency, and the output power is shown under the limitation of the air-gap flux, the stator current, and the stator voltage neglecting the voltage drop according to the leakage inductances of the machine. From the figure, the variation of the torque and output power in the constant torque region, as well as the flux weakening region including a characteristic region, can be seen. If the voltage drop due to the stator leakage inductance is considered, the stator voltage would increase under the

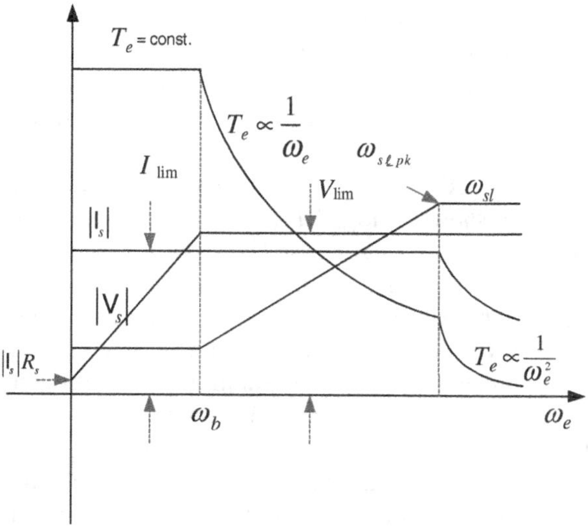

Figure 2.58 Capability curve of an induction machine in the first quadrant of torque–speed plane.

constant air-gap voltage. Otherwise, the air gap voltage decreases under the constant stator voltage, and the torque would be smaller than the value given by $1/\omega_e$ in region 1 and by $1/\omega_e^2$ in region 2. The detailed analysis including the effects of the leakage inductance can be understood by solving problem 15 in this chapter. The flux weakening control of the induction machine with the vector control is described in Section 5.4.4.

2.16 COMPARISON OF AC MACHINE AND DC MACHINE

2.16.1 Comparison of a Squirrel Cage Induction Machine and a Separately Excited DC Machine [11]

The maximum stator current slew rate, di_s/dt, of the induction machine is much larger than that of even a specially designed DC machine for the variable torque operation. While the slew rate of a DC machine is limited with the commutator and brush and it would be less than 30 per unit/s, that of the usual induction machine for the variable speed drives is over 200 per unit/s. Because the torque of the electric machine under the constant flux is decided instantaneously by the stator current, the higher slew rate of the stator current means higher slew rate of the torque, and finally higher torque regulation bandwidth, which is a key factor for the high-performance servo drive system. Furthermore, because of the mechanical reason from the commutator and brush, the line speed of the commutator given by $V_{com} = \omega_{rm} \cdot r_{com}$, where r_{com} is the radius of the commutator, is limited. The product of the output power and the rotation speed of a DC machine is usually limited to $<2.6 \times 10^6$ kW-r/min, which can be easily understood as following physical facts [10]. To increase the output power of the machine, the physical size of the machine should be increased by extending axial length and/or radius of the machine. However, the axial length is limited by the bending of the axis, and the increased radius results in commutation problems due to the increased line speed of the commutator. With these reasons, as shown in Fig. 1.6, high-power and high-speed DC machine is very difficult to manufacture. But in the case of the squirrel cage rotor induction machine, the power and rotating speed can be increased as much as the stiffness of the rotor allows against the centrifugal force.

In the viewpoint of the power factor, $\cos\phi_s$, if a DC machine is controlled by a three-phase thyristor full bridge circuit shown in Fig. 2.17, the power factor varies widely according to the delay angle, α, because $\cos\phi_s < \cos\alpha$ [3]. However, the power factor of a standard general-purpose induction machine varies from 0.2 to 0.9 according to the load factor. If the machine is controlled by a VVVF inverter, then the power factor at the input utility line is decided independently with the power factor of the machine itself. If the rectifying circuit of the inverter is a diode full bridge circuit with some reactors, then the power factor at the utility line would be in the range of 0.9 regardless of the load factor of the induction machine.

Also, in the viewpoint of the inertia of the machine itself, the inertia of the induction machine is generally smaller that that of a DC machine at the same rated power and at the same rated speed. With the smaller inertia, the slew rate of the speed of the induction machine is larger than that of a DC machine and results in higher acceleration capability. And the speed regulation bandwidth of the induction machine can be extended. Again this is another key factor for the high-performance variable-speed drive system. In the viewpoint of the protection gears such as circuit breakers and overload relays, the breakers for a DC machine to cut off DC current of the armature circuit at overload and overcurrent faults are expensive and bulky compared to the breakers to cut off AC current of the induction machine at the same current level. In the maintenance point, while a DC machine needs the regular maintenance due to the commutator and the brush, the induction machine operating in the usual application environment does not need any maintenance for several years or even more. At stalled operation or at the extreme low-speed rotation, while the current of the DC machine is concentrated to a few segments of the commutator and results in local heating problem, the current of the induction machine is distributed to all the stator and rotor conductors due to slip frequency even at stalled condition. Also in the sense of the weight and volume, the induction machine is advantageous over the DC machine at the same rated power and at the same rated speed.

In addition, the induction machine is advantageous over the DC machine in the cabling. The input power to the DC machine and the induction machine can be represented as (2.79) and (2.80), respectively.

$$P_{in\_DC} = I_a \cdot V_a \cdot \eta_{DC} \tag{2.79}$$

$$P_{in\_AC} = \sqrt{3} V_{e-l} |\mathbf{I}_s| \cos\phi \eta_{AC} \tag{2.80}$$

The efficiency of the induction machine, η_{AC}, is generally equal to or higher than that of the DC machine, η_{DC}. The power factor of a four-pole standard NEMA B-type general-purpose induction machine above several-kilowatt power range is equal to or larger than $\frac{3}{2}\frac{1}{\sqrt{3}}$ ($\approx$0.866). Hence, the area of the copper of the cable, which is proportional to the current, and the insulation level of the cable, proportional to the voltage, for the armature winding terminal is almost the same as that of the cable for the induction machine if the number of cables—three cables for AC machine and two cables for DC armature circuit)—is considered. However, in the case of a separately excited DC machine, it needs two more lines for the field winding, though its current rating is quite small compared to cable to the armature winding. One disadvantage of the induction machine is the cost of the power converter to control the machine. In the case of the variable-speed drive of an induction machine, a VVVF inverter is generally used as the power converter. But in the case of a DC machine, a simple thyristor bridge can be used, though its performance is poor. Still, a VVVF inverter is quite expensive compared to the price of the thyristor full bridge in Fig. 2.17.

In conclusion, the squirrel cage induction machine is advantageous over a separately excited DC machine in every aspect except the cost of power converter. But with the developments of the power electronics, the cost of the power electronics components, especially the price of microelectronics and power semiconductor, is dropping down rapidly, and the advantage of the induction machine will become dominant in the near future. Hence, in the recently installed variable-speed drive system, regardless of the size and the speed, the choice for the variable speed/variable torque is an AC machine whether it is a permanent magnet synchronous machine or an induction machine, not a DC machine.

2.16.2 Comparison of a Permanent Magnet AC Machine and a Separately Excited DC Machine

In addition to the merits of the squirrel cage rotor induction machine, the permanent magnet AC machine has no copper loss in the rotor circuit. And, the efficiency of the permanent magnet AC machine can be further enhanced even compared to that of the induction machine. The power density and torque density of the permanent magnet machine is the highest among DC machine and AC machine. Also, because the excitation flux is provided by the permanent magnet, the flux can be maintained as constant without any additional control means. Hence, the torque can be easily controlled by simply adjusting the torque component current of the machine, which is decided by the magnitude of the stator current and the relative position of the rotating MMF to the rotor position. The detailed description of the torque control of the permanent magnet AC machine is in Section 5.1.2. But, the permanent magnet, especially a neodium–iron–boron-based magnet, cannot be operated above 150°C. Hence, the operation of the machine at an elevated temperature environment is limited. Also, due to the centrifugal force applied to the magnet at the high speed, the surface mount permanent magnet motor cannot be operated in the extreme high speed without proper countermeasures. In the case of an interior permanent magnet (IPM) machine, the higher-speed operation is possible but the torque control of the machine and the optimal design of the machine are difficult due to the coupling of torque component and flux component currents. However, due to the merits such as higher torque density and accurate torque control, the application field of the permanent magnet AC machine is getting wider. In particular, as the control and design techniques of IPM machine is getting improved, the IPM machine is applied to the various field of the industry, especially automotive industry as a traction machine of the hybrid and electrical vehicle, where the flux weakening control is requisite.

2.17 VARIABLE-SPEED CONTROL OF INDUCTION MACHINE BASED ON STEADY-STATE CHARACTERISTICS

The control of the torque and speed of the induction machine inevitably results in transient states. However, if the variation of the torque and speed is slow enough, then some control methods based on the steady-state equivalent circuit developed in Section 2.10 can be used, though the control performances of these methods are limited. However, for variable-speed drive of pump and fan motors, the methods can be applied without troubles. Also, in the case of the synchronous machine a method using constant air gap flux control based on the steady-state circuit in Fig. 2.27 can be applied. The method has been used in the textile industry to run the multiple synchronous reluctance motors together. But, again, the performance of the control method is limited. The control methods of AC machines based on the transient state analysis are discussed in detail at Chapter 5.

2.17.1 Variable Speed Control of Induction Machine by Controlling Terminal Voltage [6]

The torque–speed curve of induction machine varies according to the terminal voltage. In Fig. 2.59, the torque–speed curves of NEMA B-type and NEMA D-type induction machines in the steady state are shown at various magnitudes of the stator voltage.

With decrease of the terminal voltage (stator voltage) of the machine, the magnitude of the air gap flux, $|\boldsymbol{\lambda}_m|$, decreases proportionally after neglecting the voltage drop due to the stator impedance. As $|\boldsymbol{\lambda}_m|$ decreases, pull-out torque decreases proportionally to the square of $|\boldsymbol{\lambda}_m|$ as seen by (2.65). The torque–speed curve of the standard general-purpose induction machine, the NEMA B-type machine, varies according to the terminal voltage. The operating point, which is the crossing point of torque–speed curve of the induction machine and that of the load, varies from "a" to "b" to "c" as shown in Fig. 2.59. But the speed variation itself due to the voltage variation is small. However, the speed variation of NEMA D-type machine by the same voltage variation is quite large and the operating point moves from "α" to "β" to "γ". But, the operating efficiency drops as slip increases, because an inequality in (2.81) holds regarding the efficiency, η, and the slip, S.

$$\eta < (1 - S) \tag{2.81}$$

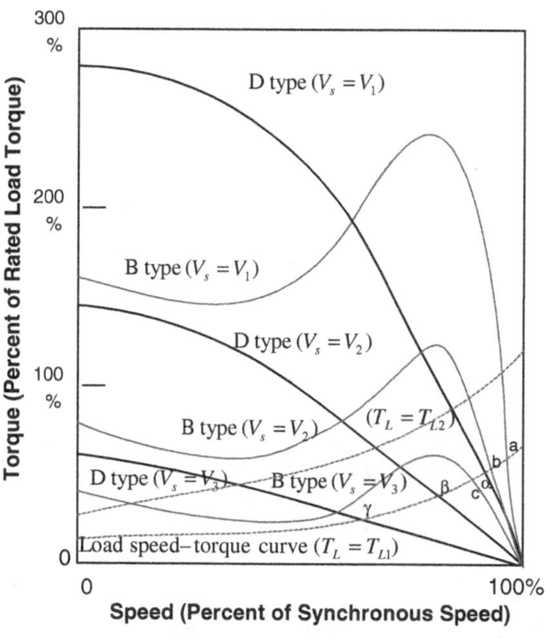

$V_1 > V_2 > V_3$, $T_{L1} < T_{L2}$

Figure 2.59 Variation of torque–speed curves of NEMA B-type and NEMA D-type induction machines with the variation of stator voltage.

At operating point, γ, the slip looks like 35% and the efficiency would be less than 65%. In conclusion, it can be said that the variable speed control of a NEMA D-type induction machine is possible by simply controlling the terminal voltage but at the cost of the efficiency. In addition to the demerit of the efficiency, if the load torque increases rapidly as shown in figure ($T_{L1} \rightarrow T_{L2}$), the load torque may be above the pull-out torque of the machine, the machine would be stalled, and the permanent damage may occur in the machine and the power converter to control the magnitude of the stator voltage. In the case of the synchronous machine, the speed is solely decided by the stator frequency, and the speed cannot be controlled by the variation of the terminal voltage.

2.17.2 Variable Speed Control of Induction Machine Based on Constant Air-Gap Flux ($\approx V/F$) Control [11]

As discussed in Section 2.10.2, if the terminal voltage and the frequency of the induction machine is controlled to keep the air gap flux, $|\boldsymbol{\lambda}_m|$, as constant up to the base speed, ω_b, and if above the base speed the frequency increases while the voltage is kept as constant, the capability of the induction machine can be exploited maximally. The block diagram of this control method is shown in Fig. 2.60. In

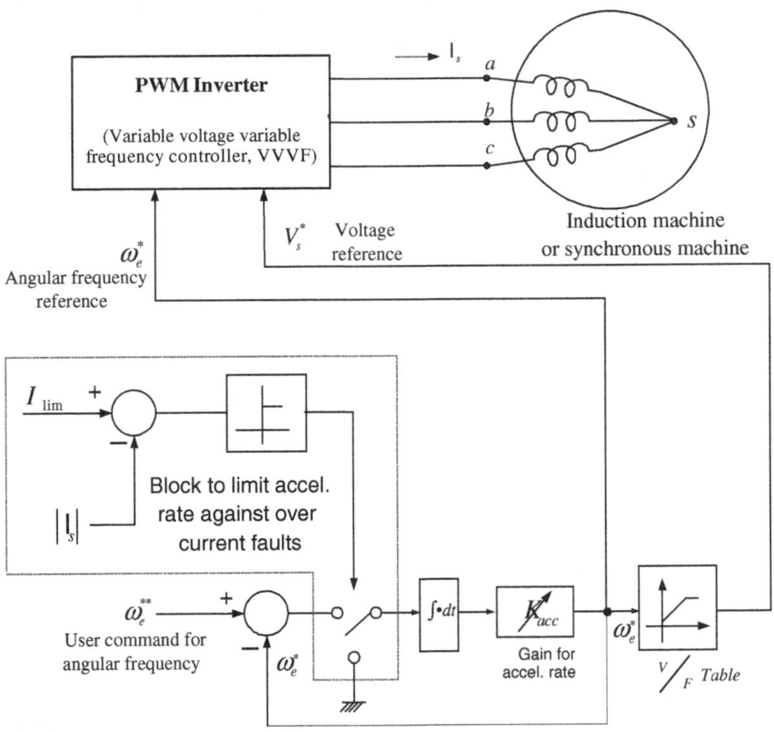

Figure 2.60 Block diagram of a variable-speed drive system based on constant air-gap flux ($\approx V/F$) of an AC machine.

the figure, the block to prevent over current faults by limiting the acceleration and deceleration rate is also included. The V/F table in the figure provides the magnitude of the stator voltage, $|V_s|$, according to the frequency, ω_e, to keep the air gap flux as constant. This constant air-gap flux control ($\approx V/F$) can be applicable to the variable-speed drive system of the synchronous machine, but the magnitude of the load angle, $|\delta|$, should be always kept less than 90° to prevent out of synchronization.

2.17.3 Variable Speed Control of Induction Machine Based on Actual Speed Feedback [12]

In the control method shown in Fig. 2.60, as the load torque increases, the slip increases. And even if the user command of operating frequency is fixed, the actual rotating speed of the machine decreases as the load torque increases, and accurate speed control is difficult. To overcome this problem, the actual speed can be fed back to the controller to keep the operating speed as constant regardless of the load torque. Because the slip angular frequency is almost proportional to the torque as seen in (2.63), the speed controller of the induction machine can be designed similarly to that of a DC machine. An example of the speed control method by the actual speed feedback is shown in Fig. 2.61 as a block diagram. The torque control method only by controlling the slip frequency in this figure is called a scalar control in contrast to the vector control discussed in Chapter 5.

However, the method in Fig. 2.61 is based on (2.63), which only holds in the steady state. Thus, the performance of the speed regulation is limited especially in the transient states. Furthermore, in flux weakening range, because (2.63) is the approximation of (2.62) under the assumption that $\omega_{sl}L_{lr}$ is much smaller than R_r, the error due to the approximation would be large as ω_e increases and $\omega_{sl}L_{lr}$ increases. In addition to this error, the variation of the magnitude of the air gap flux in the flux weakening region should be considered too. With these reasons, the accurate torque control in overall operating condition is extremely difficult with this scalar control method.

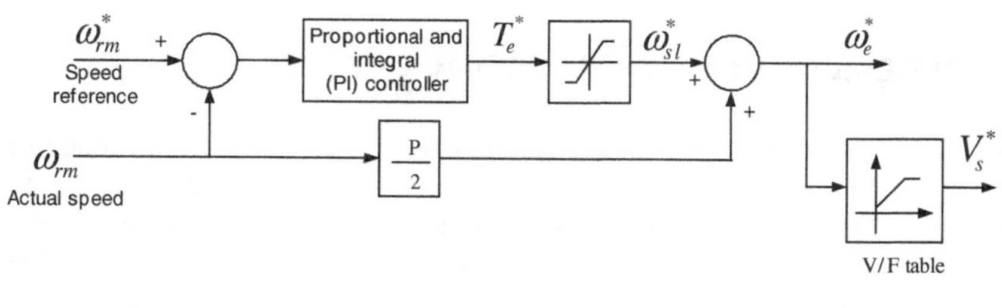

Figure 2.61 Control block diagram of a speed control system of the induction machine using actual speed feedback.

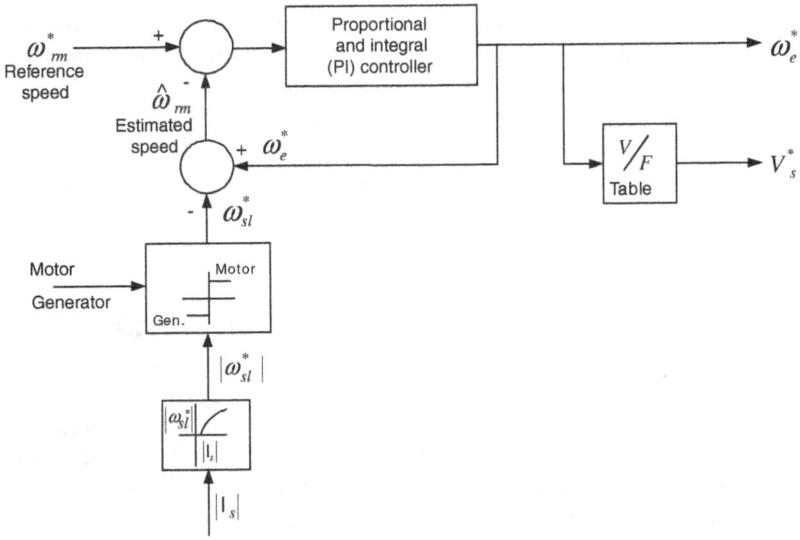

Figure 2.62 Block diagram of speed control of an induction machine using stator current magnitude feedback.

2.17.4 Enhancement of Constant Air-Gap Flux Control with Feedback of Magnitude of Stator Current

The method discussed in Section 2.17.3 has shortcomings such as the measurement of the actual speed, and the method in Section 2.17.2 has shortcomings such as the speed variation with the load torque. To overcome these shortcomings, the magnitude of the stator current, which can be easily measurable, is used to compensate the speed variation according to the load torque. Because the slip angular frequency is the function of the magnitude of the stator current under the constant air gap flux operation, the speed variation due to the load torque can be canceled out by increasing the stator frequency by the slip frequency as shown in Fig. 2.62. This simple speed sensorless control method is widely used in the field, where the moderate accuracy of the speed control is required without any measurement of the speed.

2.18 MODELING OF POWER CONVERTERS

The power converters based on power semiconductors are widely used alone or together with the electric machines to convert a form of the electrical energy to another form. The converter can be modeled as an electronics circuit using the simulation language such as SPICE. But, because of the fast enough response of the semiconductors compared to other parts of the system connected to the power converter, in order to understand the transient behavior of the over all system, the power converter may be modeled as a linear or a nonlinear gain element with some time delay. In this section, typical power converters such as a three-phase diode/thyristor rectifier, a PWM boost rectifier, two- and four-quadrant DC/DC converters, a PWM inverter, and a matrix converter are modeled to incorporate with the electric machines.

2.18.1 Three-Phase Diode/Thyristor Rectifier [13]

From a three-phase full-bridge-controlled rectifier shown in Fig. 2.63, by adjusting the gating angle, α, the variable DC voltage can be obtained. In the case of the diode rectifier, it is equivalent that the gating angle is zero at Fig. 2.63. And the output voltage of the diode rectifier is fixed by the input AC voltage. If the filtering inductance is large enough at the circuit in Fig. 2.63, the current through the inductor is continuous. And if the internal resistance of the inductor can be neglected, then the circuit can be represented as an equivalent circuit in Fig. 2.64, where the voltage drop at the resistance stands for the voltage drop due to overlap angle in the rectifier by the internal inductance of AC source, L_s. At the circuit in Fig. 2.64, the voltage drop due to L_s does not contribute to the loss of the system. It simply represents a drop characteristic of the output voltage according to the output current due to the internal inductance of AC source. In the figure, ω_e means the angular frequency of AC source. The DC output voltage in Fig. 2.63 is the cosine function of gating angle, α. Hence, there is nonlinearity between gating angle and output voltage. However, if the gating angle is obtained through the inverse cosine table as shown in Fig. 2.65, the output voltage of the rectifier would be proportional to the control command, V_c.

In this case the output voltage, V_d, can be represented in terms of the control command, V_c, and output current, I_d, as

$$V_d = V_c \frac{3\sqrt{2}}{\pi} V_{L-L} e^{-\frac{2\pi}{12\omega_e}s} - \frac{3\omega_e L_s}{\pi} I_d \tag{2.82}$$

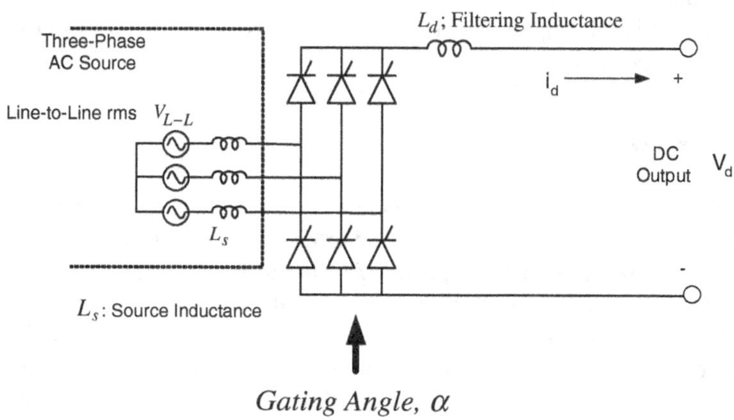

Figure 2.63 Three-phase-controlled rectifier based on thyristors.

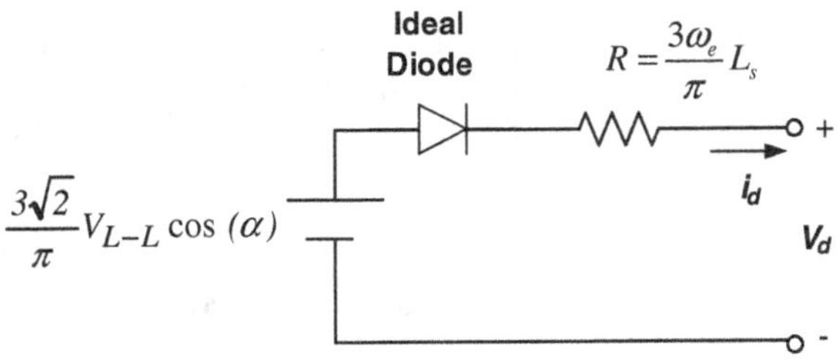

Figure 2.64 Equivalent circuit of a three-phase-controlled rectifier.

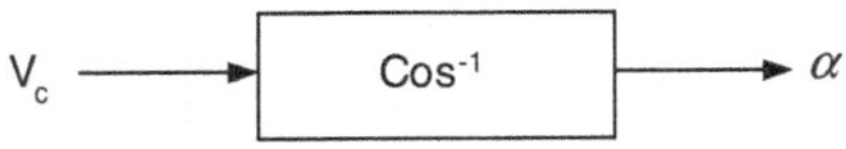

Figure 2.65 Gating angle generator to circumvent the nonlinearity.

where I_d is Laplace transformation of DC output current i_d. And $e^{-\frac{2\pi}{12\omega_e}s}$ represents the average time delay, which is one-twelfth of a period of AC source. The delay function can be approximated with the first-order low-pass filter, and (2.82) can be simplified as

$$V_d \approx V_c \frac{3\sqrt{2}}{\pi} V_{L-L} \frac{1}{1+T_d s} - \frac{3\omega_e L_s}{\pi} I_d \tag{2.83}$$

where $T_d = \frac{\pi}{6\omega_e}$ (seconds). In the case of a diode rectifier, where $\alpha = 0$, the output voltage can be expressed as

$$V_d = \frac{3\sqrt{2}}{\pi} V_{L-L} - \frac{3\omega_e L_s}{\pi} \langle i_d \rangle \tag{2.84}$$

where $\langle i_d \rangle$ is the average value of DC output current, i_d. However, if the current, i_d, is discontinuous, then the output voltage is described as a solution of nonlinear equations, and simple analytic expression of the output voltage is impossible. The current in the armature winding of DC machine driven by a three-phase-controlled rectifier can be described as the equations in Section 4.2.2.2 according to continuity or discontinuity of the current.

2.18.2 PWM Boost Rectifier

The PWM boost rectifier shown in Fig. 2.66 is widely used to get DC voltage output from three-phase AC utility source. The rectifier can provide boosted DC output, and DC output voltage is usually larger than the peak of AC line-to-line voltage. The input displacement power factor can be adjusted if needed. But, it is usually set as unity to minimize loss of the system. By using this PWM boost rectifier, total harmonic distortion (THD) of input AC current can be suppressed to meet the IEEE 519 standard. In some case, to cancel out the harmonics of the AC source actively, low-order harmonic current can be synthesized together with a fundamental frequency component of an AC source. Because of the bidirectional power flow capability, the

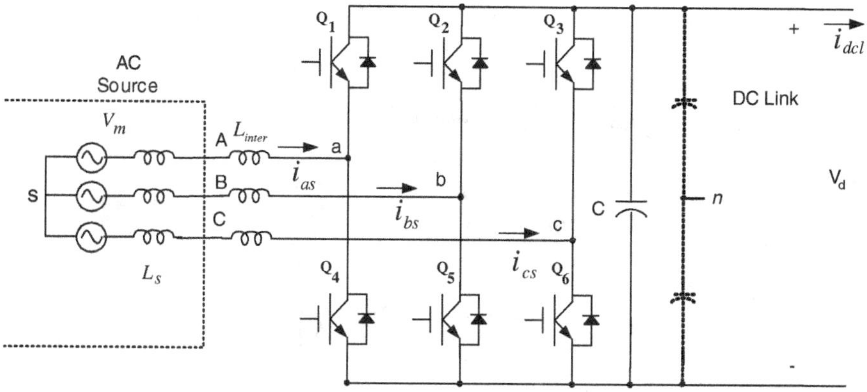

Figure 2.66 Power circuit of a PWM boost rectifier and an input AC source.

rectifier in Fig. 2.66 acts sometimes as an inverter, which converts the electric power from a DC source to an AC line. The frequency of an AC source may vary widely if the AC source voltage comes from a variable-speed engine generator. Moreover, DC output voltage would be varied if required. To keep the displacement power factor (DPF) as unity, DC link voltage should be larger than the peak of line-to-line voltage of the AC source, $\sqrt{3}V_m$, where V_m is the peak of the phase voltage of the AC source. If the AC voltage comes from an AC electric machine, there may be no need to install an interface inductor, L_{inter}, because of enough internal inductance of AC machine. However, if the boost rectifier is connected to a utility line, where the internal impedance is less than a few percentages of the rating of rectifier itself, there should be an interface inductor as shown in Fig. 2.66 to suppress the harmonic current to the utility line and to prevent the distortion of the voltage waveform of the utility line. If THD is fixed as a constant value, then the inductance of the interface inductor is inversely proportional to the switching frequency of the boost rectifier. In the PWM boost rectifier, if the DC link is considered as input and the AC line is considered as output, then from the equivalent circuit in Fig. 2.67

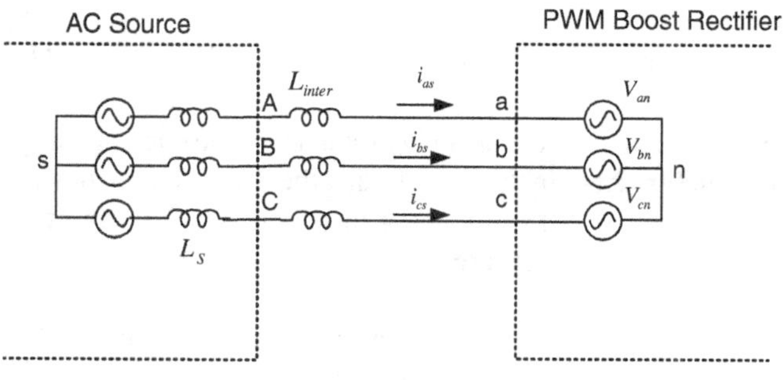

Figure 2.67 Equivalent circuit of a PWM boost rectifier.

the output AC voltage can be described as (2.85) in terms of DC link voltage, V_d.

$$
\begin{aligned}
V_{an} &= m\frac{V_d}{2}\sin(\omega_e t+\phi) \\
V_{bn} &= m\frac{V_d}{2}\sin\left(\omega_e t+\frac{2\pi}{3}+\phi\right) \\
V_{cn} &= m\frac{V_d}{2}\sin\left(\omega_e t-\frac{2\pi}{3}+\phi\right)
\end{aligned}
\tag{2.85}
$$

where m is modulation index defined as the ratio between the peak of the phase voltage, V_m, and a half of the DC link voltage, $V_d/2$, and $m \equiv \frac{V_m}{V_d/2}$. Also, n is the center point of the DC link as shown in Fig. 2.66 by a dashed line, and the point may be a conceptual one to define the pole voltages. And ϕ is the phase difference between AC source voltage and AC output voltage of the boost rectifier. The modulation index, m, varies from 0 to $4/\pi$ according to PWM. The modulation index, m, and ϕ can be changed by PWM at every half of the PWM period.

In the figure, if the AC line is a balanced three-phase source, then the voltage of the source can be described as

$$
\begin{aligned}
V_{As} &= V_m\sin(\omega_e t) \\
V_{Bs} &= V_m\sin\left(\omega_e t+\frac{2\pi}{3}\right) \\
V_{Cs} &= V_m\sin\left(\omega_e t-\frac{2\pi}{3}\right)
\end{aligned}
\tag{2.86}
$$

where ω_e is the angular frequency of AC source voltage. If the loss of the boost rectifier is neglected, then the relationship between the power from a AC source and that to a DC link voltage can be deduced as

$$
V_d\left(C\frac{dV_d}{dt}+i_{dcl}\right) = V_{an}i_{as}+V_{bn}i_{bs}+V_{cn}i_{cs} = V_{as}i_{as}+V_{bs}i_{bs}+V_{cs}i_{cs} \tag{2.87}
$$

where, C stands for the capacitance in the DC link and i_{dcl} stands for the current going out from the DC link as shown in Fig. 2.66. In the steady state, the displacement power factor at the AC source can be controlled from lagging to leading by adjusting the magnitude of the AC output voltage of the boost rectifier, V_{an}, V_{bn}, and V_{cn}, as shown in Fig. 2.68.

As seen in Fig. 2.68, at given output power, the magnitude of AC voltage of the boost rectifier, $|\mathbf{V}_n|$, in lagging power factor operating mode is the smallest among three operation modes, namely lagging, unity, and leading modes. Hence, even if

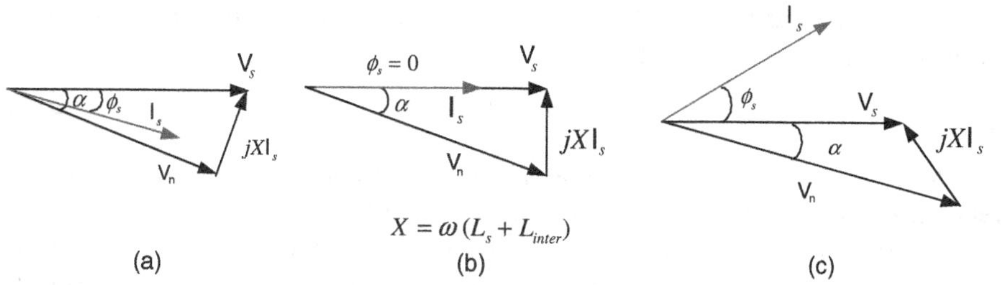

Figure 2.68 Phasor diagram of a PWM boost rectifier under the constant output power and constant AC source voltage at different displacement power factor ($\mathbf{V}_s$, $\mathbf{V}_n$, and $\mathbf{I}_s$ are phasors of AC source voltage excluding voltage drop by the internal inductance, phasors of PWM boost rectifier AC voltage, and AC source output current, respectively.) (a) Lagging power factor. (b) Unit power factor. (c) Leading power factor.

DC link voltage is smaller than the peak of line-to-line voltage of AC source, the PWM boost rectifier can still be operated by maintaining the displacement power factor as lagging.

2.18.3 Two-Quadrant Bidirectional DC/DC Converter

In Fig. 2.69, a circuit diagram of a two-quadrant DC/DC converter is shown, where a DC voltage source, V_s, can be connected to another DC voltage source, V_d, whose magnitude is larger than the magnitude of V_s. The electric power can be transferred bidirectionally, and both DC sources can act as a source or a load. If current flowing out from the smaller DC source, i_s, is continuous and the duty factor of a switch Q_1 is denoted as D, then the average voltage equation between two DC sources for a

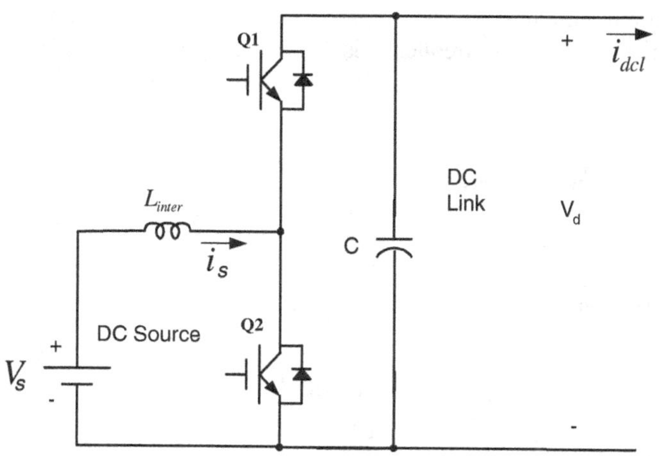

Figure 2.69 Power circuit of a bidirectional two-quadrant DC/DC power converter.

switching period can be derived as (2.88) under the assumption of the negligible losses of the converter and inductors.

$$V_s = V_d D \tag{2.88}$$

Also, the average currents in both DC sources for a switching period have the relationship shown in (2.89) under the assumption of the constant DC link voltage. The current ripples in both sources are inversely proportional to the product of the inductance of the interface inductor and the switching frequency. To reduce the ripples at given switching frequency, an interleaving technique can be used by connecting several bidirectional DC/DC converters in parallel and shifting the phase of switching period of each converter. For example, if three DC/DC converters are connected in parallel, then the phase of each converter should be shifted one-third of the switching period. In this way, the current ripple in both sources can be reduced remarkably. In particular, in the case of an interleaving operation of three DC/DC converters, if duty factor, D, equals one-third or two-thirds, the ripple can be eliminated perfectly in the steady state.

$$i_s D = i_{dcl} \tag{2.89}$$

2.18.4 Four-Quadrant DC/DC Converter

By the four-quadrant DC/DC converter shown in Fig. 2.18, where the DC link is obtained from an AC source through a diode rectifier, the voltage varying from $+V_d$ to $-V_d$ can be applied to the output, which is the armature circuit of DC machine. To control four switches, T_1, T'_1, T_2, and T'_2, for varying armature voltage, a control block diagram shown in Fig. 2.70 can be used [13].

If the reference voltage for the armature circuit, V_a^*, is given, the voltage is limited within the allowable control range, which is usually the minimum value of DC link voltage in Fig. 2.18, through the limiter. The output of the limiter, V_c^*, is compared with the triangular carrier wave, whose frequency is the switching frequency of DC/DC converter. Through the comparison, the switches are turned on or turned off as

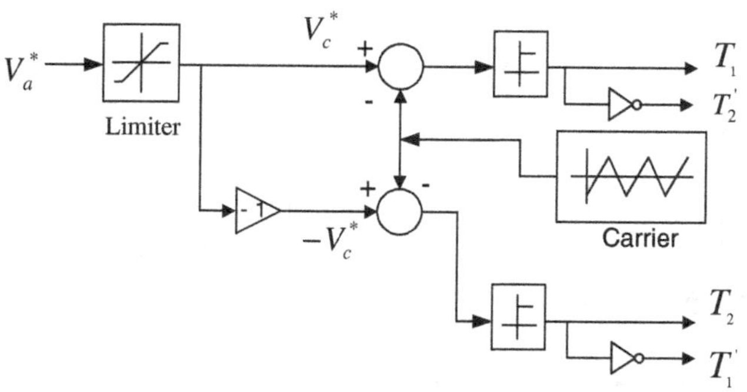

Figure 2.70 Control block diagram of pwm signals for a four-quadrant dc/dc converter.

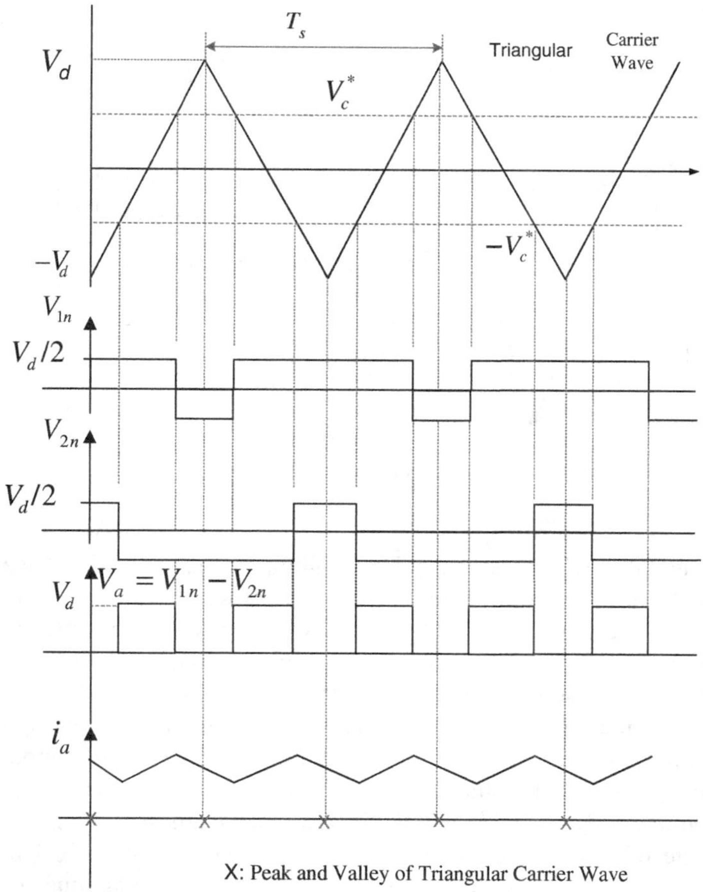

Figure 2.71 Pwm signals, voltages, and current of a four-quadrant DC/DC converter.

shown in Fig. 2.70. The waveform of signals in Fig. 2.70 and the voltages and current in Fig. 2.18 can be drawn as shown in Fig. 2.71.

As shown in Fig. 2.71, the frequency of current ripples to the armature winding is twice the switching frequency or twice of the carrier frequency. Hence, by increasing the switching frequency, the torque ripples and the acoustic noise due to the current ripples can be reduced. In particular, if the switching frequency is above 10 kHz, then audible noise from a DC machine can be perfectly eliminated.

2.18.5 Three-Phase PWM Inverter

As shown in Fig. 2.72, the circuit topology of a three-phase PWM inverter is the same as that of a three-phase PWM boost rectifier except the input and the output is

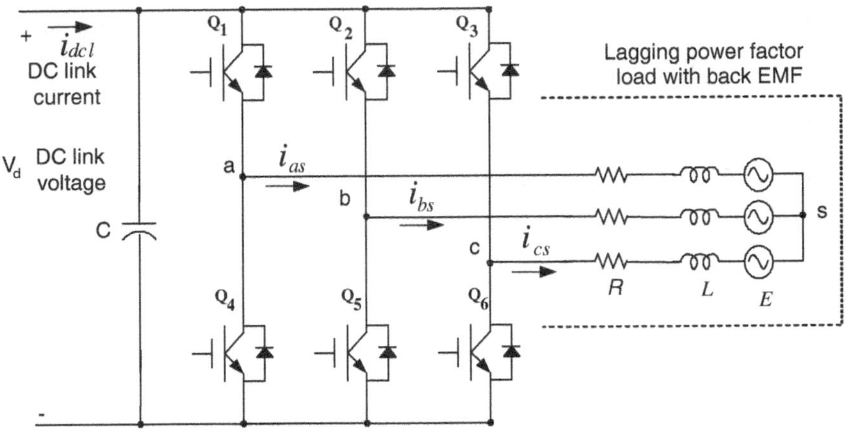

Figure 2.72 VVVF inverter with pulse width modulation (PWM).

reversed. The inverter takes DC voltage as an input and transforms it to AC voltage and outputs to AC load. As with the boost rectifier, the power can flow in both directions and DC power can be converted to AC or AC power can be converted to DC. The inverter synthesizes variable-voltage and variable-frequency (VVVF) AC voltage as the output. As shown in Fig. 2.72, if the power factor of an AC load of the inverter is lagging, by reducing low-order current harmonics through PWM, THD of output current can be minimized. The magnitude of AC output voltage is limited by DC input voltage. The maximum available peak phase voltage from DC voltage, V_d, is $2V_d/\pi$ and the inverter can synthesize AC phase voltage in the range of $0 \sim 2V_d/\pi$. In this case, in the range from $V_d/\sqrt{3}$ to $2V_d/\pi$, the control of low-order harmonics is getting difficult as the magnitude of phase voltage increases. In the extreme case, if phase voltage is $2V_d/\pi$, then the output voltage is a six-step waveform and the magnitude of the fifth harmonic is one-fifth that of fundamental wave, the seventh is one-seventh, the eleventh is one-eleventh, and so on. In the range from 0 to $V_d/\sqrt{3}$, if the higher-order harmonic components are neglected, output voltage can be expressed as (2.90). The modulation index, m, and angular frequency, ω, can be controlled at every half period of PWM switching frequency.

$$
\begin{aligned}
V_{as} &= m\frac{V_d}{2}\sin(\omega t) \\
V_{bs} &= m\frac{V_d}{2}\sin\left(\omega t + \frac{2\pi}{3}\right) \\
V_{cs} &= m\frac{V_d}{2}\sin\left(\omega t - \frac{2\pi}{3}\right)
\end{aligned}
\tag{2.90}
$$

If the loss in the inverter itself is neglected, then input and output power of the inverter would have the relationship shown in (2.91).

$$V_d\left(i_{dcl}-C\frac{dV_d}{dt}\right) = V_{as}i_{as}+V_{bs}i_{bs}+V_{cs}i_{cs} \tag{2.91}$$

2.18.6 Matrix Converter

The matrix converter is a power converter that transforms an arbitrary AC to another arbitrary AC without a DC link. The power can flow in both direction, and the four-quadrant operation in voltage and current plane is possible. The power factor of input and output can be independently regulated under the condition of the instantaneous power balance. In a switching period, which may be less than a couple hundred microseconds, one AC should work as a voltage source and the other AC should work as a current source. For convenience, AC that works as the voltage source is usually depicted as an input, while AC that works as the current source may be depicted as an output. If the utility line is connected to a matrix converter, then an L-C filter can be inserted between an AC input line. The circuit diagram of a matrix converter can be drawn as shown in Fig. 2.73, and an AC source with the filter can be approximated as a voltage source. And if an AC machine is connected to a matrix converter, the AC machine can be approximated as a current source due to its internal inductances. In Fig. 2.73, because the switches should block voltage bidirectionally and conduct current bidirectionally, a switch of the matrix converter is usually implemented by connecting two pairs of IGBT (insulated gate bipolar transistor) and diode as shown in Fig. 2.74.

I n the matrix converter, through PWM the segments of input line-to-line voltage are connected to the output while the segments of output current flow into the input line. To control input current and output voltage simultaneously with the regulation of

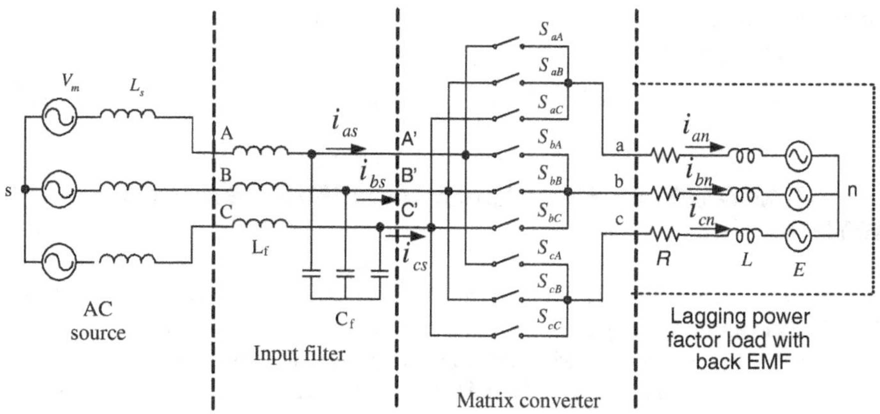

Figure 2.73 Power circuit of a matrix converter with an AC line and an input filter.

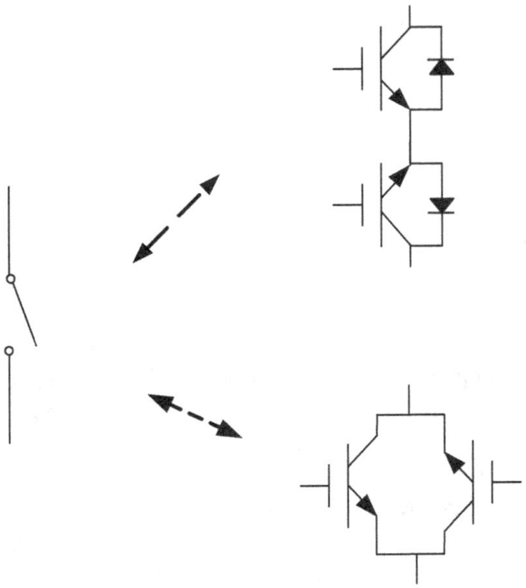

Figure 2.74 Two possible implementations of an AC switch with an IGBT and a diode.

the input power factor, which is usually unity, a proper PWM method should be used. Regarding the PWM method of the matrix converter, there are many studies based on the space vector concept or based on a carrier wave [14–17]. In the viewpoint of the implementation, the carrier-based method has some advantages over other methods. If the losses in the matrix converter itself and harmonics of the input and output are neglected, then power balance in (2.92) should be satisfied instantaneously. With the requirement of no low-order harmonics in output voltage and input current, the maximum magnitude of output phase voltage is limited within $\sqrt{3}/2$ of peak of input phase voltage under the assumption of unity power factor. If the displacement power factor of input, $\cos(\phi_s)$, is not unity, then the maximum magnitude is limited within $\frac{\sqrt{3}}{2}\cos(\phi_s)$. If some low-order harmonics are allowed in input and output of the matrix converter, even the larger output voltage than the input voltage is possible [16].

$$V_{A'n}i_{as} + V_{B'n}i_{bs} + V_{C'n}i_{cs} = V_{an}i_{an} + V_{bn}i_{bn} + V_{cn}i_{cn} \tag{2.92}$$

2.19 PARAMETER CONVERSION USING PER UNIT METHOD

The parameters in MKS units of the electric machinery including power transformers, DC and AC machines, and power converters do not have much meaning to compare the characteristics of the electric machinery in different ratings because of wide differences of the parameters not only due to the characteristics of the machinery itself but also due to the their different rated power, different rated voltage, different rated speed, and so on. For example, with the fact that the stator resistance of a 2-kW, 440-V,

DC machine is 6.28 Ω and that of a 110-kW, 220-V, DC machine is 0.1027 Ω, it cannot be said that the stator copper loss of the small machine is 60 times that of the large machine. Also in a system consisting of the multiple electric machines whose rated voltages and powers are different, absolute values of parameters in MKS units of the electric machines would be inconvenient to understand the effect of each machine or power converter on the overall system. To handle these problems, the parameters of electric machines and converters can be expressed on the basis of a specific power, voltage, and speed. The expressed parameters have no unit and they are represented as per unit value, which is the relative magnitude to the basis value. In the power system analysis, where several electric machines and power converters are interconnected, the apparent power in volt-amperes (VA) of the receiving terminal of the power system is usually used as the basis power of the system. And rms rated phase voltage of the receiving terminal is used as the basis voltage, V_B. In the case of electric machines, the rated output power (in watts) of the machine is usually used as the basis of power, P_B, and rms rated phase voltage is V_B. From P_B and V_B, the base current, I_B, can be calculated as (2.93), and the base impedance, Z_B, can be defined as (2.94).

$$I_B = \frac{P_B}{m \cdot V_B} \tag{2.93}$$

where m is the number of phase of input source. Also, for a DC machine or single-phase AC machine we have $m = 1$, and for a three-phase AC machine we have $m = 3$.

$$Z_B = \frac{V_B}{I_B} \tag{2.94}$$

With the base impedance, the armature winding resistance of DC machine can be represented as

$$R_{ap \cdot u} = \frac{R_a}{Z_B} \tag{2.95}$$

In the case of an AC machine, the rated angular frequency can be used as the base angular frequency, ω_B, and the reactance, ωL_s, of AC machine can be expressed in per unit as

$$X_{sp \cdot u} = \frac{\omega_B \cdot L_s}{Z_B} \tag{2.96}$$

And the base value of the torque of AC machine can be represented as

$$T_B = \frac{P_B}{\frac{2}{P} \cdot \omega_B} \tag{2.97}$$

where P is the number of pole of machine. And the torque, T_e, of an AC machine can be expressed in per unit as

$$T_{e p \cdot u} = \frac{T_e}{T_B} \tag{2.98}$$

PROBLEMS

1. A 100-Hp (74.6-kW) separately excited DC machine, which has the following parameters, is controlled by adjusting field current as shown in Fig. P2.1. The speed response to the step change of the armature voltage is in critical damping. The load inertia is 10 times that of the inertia of the machine itself. The friction can be neglected.

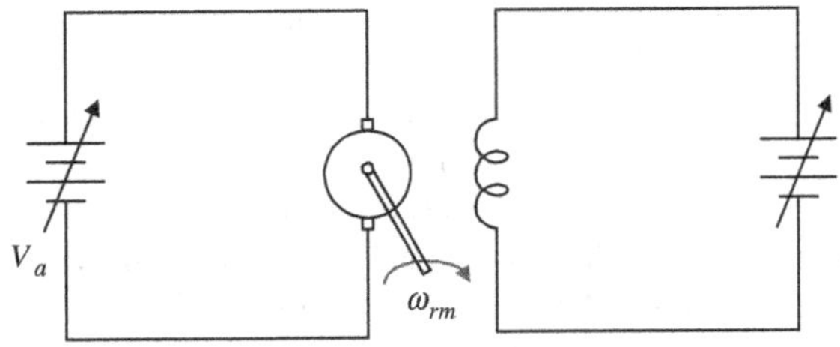

Figure P2.1 Field weakening control.

DC Machine Parameters

$R_a = 0.0144\,(\Omega)$, $L_a = 1.10\,(\text{mH})$, Rated armature current: 349 (A)

Rated speed: 1750 (r/min) Rated armature voltage: 230 (V)

$L_f = 10.77\,(\text{H})$, $R_f = 13.58\,(\Omega)$, Rated field current: 14.4 (A)

Inertia: $J_M = 1.82\,(\text{kg-m}^2)$

(1) To make the response be in the critical damping, what is the relative magnitude of the field flux to the rated field flux in percentage?

(2) With the flux set by the part 1, if the armature current and voltage are limited under the rated values, then calculate the maximum speed in revolutions per minute, the maximum torque in newton-meters, and the maximum output power in kilowatts, all in the steady state.

2. A 110-kW separately excited DC machine, which has following parameters, is running in a steady state with following operating condition. The load inertia is the same with the DC machine inertia itself. The friction can be neglected.

DC Machine Parameters

$R_a = 0.025\,(\Omega)$, $L_a = 1.557\,(\text{mH})$, Rated armature current: 274 (A)

Rated speed: 560 (r/min), Rated armature voltage: 440 (V)

$L_f = 10.77\,(\text{H})$, $R_f = 13.58\,(\Omega)$, Rated field current: 14.4 (A)

Inertia: $11.5\,(\text{kg-}m^2)$, Weight: 2000 (kg)

Operating Conditions

Armature voltage: $V_a = 220\,(\text{V})$, Field voltage: $V_f = 220\,(\text{V})$

Load torque: $T_L = 1800\,(\text{N-}m)$

(1) By small signal analysis, calculate the transfer function, $\Delta\omega_{rm}/\Delta V_f$, at the given steady-state operating point, where the speed of the machine, ω_{rm}, is expressed in revolutions per minute.

(2) Based on the result by part 1, plot the speed of the machine in revolutions per minute according to the time when the field voltage varies from 196 V to 186 V in step.

(3) By solving the nonlinear differential equations numerically with the computer simulation, plot the speed of the machine in r/min according to the time when the field voltage varies from 196 V to 186 V in step.

(4) Calculate how much the armature voltage should vary in step to get the same speed variation with the result in the part 3 at the constant 196-V field voltage, and plot the speed response with the calculated armature voltage variation by the computer simulation.

3. With a separately excited 2-kW DC machine, which has the following parameters, draw the capability curve in the first quadrant and the second quadrant of the torque–speed plane. Also in the plane, simultaneously plot the electric power to (from) the machine, along with the armature voltage, and the field current according to the torque. Because of the commutation problem, the field flux can be reduced down to one-third of the rated value. It is assumed that the field flux is proportional to the field current.

DC Machine Parameters

$R_a = 6.28\,(\Omega)$, $L_a = 53.6\,(\text{mH})$, $L_f = 221\,(\text{H})$, $R_f = 858\,(\Omega)$

Rated armature current: 5.7 (A), Rated speed: 2243 (r/min),

Rated field current: 0.225 (A)

Inertia: 0.0224 (kg-m^2), Rated armature voltage: 440 (V)

4. In Fig. P2.2, the balanced fifth harmonic currents flow in three-phase balanced windings, which are apart to each other by 120° spatially. Prove that the direction of the rotating MMF by the fifth harmonic currents is the reverse of the rotating MMF by the fundamental frequency current, and its speed is five times that of the MMF by the fundamental harmonics.

5. In the equivalent circuit of a round rotor synchronous machine, shown in Fig. 2.27, in the case where the speed of the machine is low and the magnitude of R_a almost equals that of X_s, represent the torque of the machine in terms of V_s, E_i, δ, R_a, X_s, P, and ω_e.

6. A 2250-Hp induction machine, which has the parameters listed below, is running in the steady state with the rated load at rated frequency. At the instant ($t = 0\,\text{s}$) when the "A" phase current is zero crossing from negative to positive, the input power source to the machine is disconnected and reconnected after 375 ms. The inertia of the machine and drive system can be assumed to be infinite, and the speed of the machine is constant during the transient of input power source. And it is assumed that the currents flowing through three-phase stator windings of the induction machine are immediately zero after disconnecting the input power source.

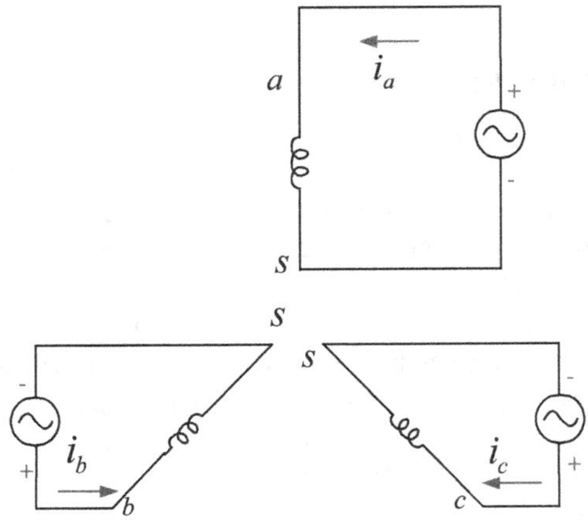

Figure P2.2 Balanced three-phase windings.

Rated power: 2250-Hp, Rated voltage (line-to-line rms): 2300V,

Rated speed: 1786 r/min, Rated frequency: 60Hz,

Number of pole: 4

$R_s = 0.029\ \Omega$, $R_r = 0.022\ \Omega$

$X_{ls} = 0.226\ \Omega$, $X_{lr} = 0.226\ \Omega$

$X_m = 13.04\ \Omega$

(1) Find the magnitude of the stator current, $|\mathbf{I}_s|$, and rotor current, $|\mathbf{I}_r|$, just before disconnection.

(2) For $0^+ \leq t < 375$ ms represent a line-to-line voltage, V_{ab}, in terms of the machine parameters and time, t.

(3) For $t \geq 375$ ms represent a line current, i_{as}, from the source to the machine in terms of the machine parameters and time, t. In this case, the stator and rotor resistance can be neglected, and they can be assumed to be $R_s = R_r = 0$. During the power source disconnection, the magnitude and phase of the input source voltage itself are well kept.

7. **(1)** Represent the parameters of the equivalent circuit of a 5.5-kW induction machine in per unit, which has following parameters in MKS units.

Rated power: 5.5 kW, Rated voltage (line-to-line rms): 440 V,

Rated speed: 1755 r/min, Rated frequency: 60 Hz,

Number of poles: 4

$R_s = 0.69\ \Omega$, $X_{ls} = X_{lr} = 2.24\ \Omega$

$X_m = 45.43\ \Omega$ $R_r = 0.451\ \Omega$

(2) Represent the parameters of the equivalent circuit of a 2250-Hp induction machine given in problem 6 in per unit.

(3) Represent the parameters of the equivalent circuit of a 22-kW induction machine in per unit, which has the following parameters in MKS units.

Rated power: 22 kW, Rated voltage (line-to-line rms): 220V,

Rated speed: 1755r/min Base frequence: 60 Hz,

Number of poles: 4, Rated current: 75A

$R_s = 0.044\,\Omega$, $R_r = 0.0252\,\Omega$, $L_{ls} = 0.55\,\text{mH}$, $L_{lr} = 0.47\,\text{mH}$,

$L_m = 12.90\,\text{mH}$

(4) Represent the parameters of the equivalent circuit of above three induction machine in per unit at the base values of line-to-line voltage, 380 Vrms, and apparent power, 400 KVA.

8. Find the parameters of the equivalent circuit of a 60-Hz, 100-Hp, 380-V, 6-pole standard general purpose induction machine in ohms, whose parameters are decided by (2.72)–(2.76).

9. A 2.7-kW, 220-V, 4-pole, 60-Hz induction machine, which has following equivalent circuit parameters in per unit, is running in the steady state at 1810 r/min with no load through a variable-voltage variable-frequency (VVVF) inverter. The input frequency of the stator voltage has been changed to 58 Hz in step. The speed of the machine is kept as constant because of the large inertia of the drive system and the machine is again running in the steady state with the 58-Hz source voltage. Regardless of the frequency, the magnitude of the stator voltage is always 220 V in line-to-line rms. The magnetic saturation of the machine can be ignored. The harmonics in the current and voltage to the machine can be ignored. Calculate the active power (watts) to the machine from the VVVF inverter.

$$R_s = 0.023(\text{P.U.}),\quad X_{ls} = 0.09(\text{P.U.}),\quad X_m = 1.5(\text{P.U.})$$
$$R_r = 0.03(\text{P.U.}),\quad X_{ls} = 0.11(\text{P.U.})$$

10. Draw the capability curve of the synchronous machine, which has following equivalent circuit parameters and limiting conditions, in the first quadrant of the torque–speed plane. The speed range is from 0 r/min to 6000 r/min.

$$3\text{ phases}, 4\text{ poles}, 60\text{ Hz}, 380\text{ V}, 200\text{ Hp},\quad X_s = 0.6\,\Omega, R_a = 0.02\,\Omega$$

Limiting conditions: Line-to-line voltage in rms: $|V_s| \le 380$ V

Line current in rms: $|\mathbf{I}_s| \le 240$A

In this problem, the magnitude of the field current should be equal to or less than the rated value, where the rated power, 200 Hp, occurs with the frequency, 60 Hz, voltage, 380 V, current 240 A. With the speed (r/min) as the horizontal axis, draw simultaneously the torque, the magnitude of the line-to-line stator voltage in rms, the magnitude of the line current in rms, the power factor at the terminal of the machine, and the mechanical power as the vertical axis.

11. A surface-mount permanent-magnet synchronous machine, which has the following ratings and parameters, is running as a motor in maximum torque per ampere operation mode at 800 r/min. The torque of the machine is 10 N-m, and the drive system is in the steady state.

5 Hp, 4 poles, 60 Hz, $R_s = 0.423\,\Omega$, $L_s = 4.96$ mH, $K_f \cdot I_f = 0.422$ V/(rad/s)

(1) Calculate the angular frequency of the stator voltage, ω_e(rad/s).
(2) Calculate the line-to-line voltage in rms and the line current in rms.
(3) Calculate the power factor at the terminal of the machine.
(4) Calculate the load angle δ.

12. **(1)** For the control system shown in Fig. 2.60, prepare the $^V/_F$ table of the figure in the frequency range from 0 to 360π (rad/s) for the induction machine, whose parameters are given as follow.

Rated power: 22 kW, Rated voltage (line-to-line rms): 220 V,

Rated speed: 1755 r/min,

Base frequency: 60 Hz, Number of poles: 4,

Inertia of the machine itself: 0.122 kg-m^2, Rated current : 75 A

$R_s = 0.044\,\Omega$, $R_r = 0.0252\,\Omega$, $L_{ls} = 0.55$ mH, $L_{lr} = 0.47$ mH,

$L_m = 12.90$ mH

(2) The machine in part 1 is driven by the control block diagram shown in Fig. 2.61. Plot the slip angular frequency, ω_{sl}^*, according to the torque, T_e^*, range from –200 N-m to 200 N-m.

(3) The machine in part 1 is driven by the control block diagram shown in Fig. 2.62. Prepare the table of $|\mathbf{I}_s|$ in terms of $|\omega_{sl}^*|$.

13. A 22-kW induction machine, whose parameters and limiting conditions are listed below, are driven by 280-Hz voltage source. The machine is generating the available maximum torque as a motor in the steady state keeping the limiting conditions.

Rated power: 22 kW, Rated voltage (line-to-line rms): 220 V,

Rated speed: 1755 r/min

Base frequency: 60Hz, Number of poles : 4

Inertia of the machine itself: 0.122 kg-m^2, Rated current: 75 A

Limiting conditions: Line-to-line voltage in rms: $|V_s| \le 220$ V,

Line current in rms: $\mathbf{I}_s \le 75$ A

$R_s = 0.044\,\Omega$, $R_r = 0.0252\,\Omega$, $L_{ls} = 0.55$ mH, $L_{lr} = 0.47$ mH,

$L_m = 12.90$ mH

(1) Calculate the slip angular frequency, ω_{sl} (rad/s).
(2) Calculate the torque (N-m) and mechanical output power (kW).

14. A 22-kW induction machine, whose parameters and limiting conditions are listed below, are controlled in V/F operation mode. The magnitude of the stator voltage is proportional to the source frequency up to 60 Hz, and above 60 Hz, the magnitude is kept constant in line-to-line 220 V rms.

Rated power: 22 kW, Rated voltage (line-to-line rms): 220 V,

Rated speed: 1755 r/min

Base frequency: 60 Hz, Number *of poles*: 4

Inertia of the machine itself: 0.122 kg-m^2, Rated current: 75 A

Limiting conditions: Line-to-line voltage in rms: $|V_s| \leq 220\,\text{V}$

Line *current in rms*: $|\mathbf{I}_s| \leq 75\text{A}$, Maximum speed: 6000 r/min

$R_s = 0.044\,\Omega$, $R_r = 0.0252\,\Omega$, $L_{ls} = 0.55\,\text{mH}$, $L_{lr} = 0.47\,\text{mH}$,

$L_m = 12.90\,\text{mH}$

(1) When the frequency changes from 5 Hz to 170 Hz in 15-Hz intervals, plot the steady-state torque–speed curves of the machine at each frequency in the torque–speed plane.

(2) Repeat part 1 when the ratio of the air-gap voltage in phase rms to the frequency of the machine is kept as $^{111}/_{60}$ up to 60 Hz, and above 60 Hz the air-gap voltage is constant at 111 V in phase voltage rms.

(3) Plot the capability curve of the machine operating according to the conditions in part 2 at the first quadrant of the torque–speed plane.

15. Plot the capability curve of the induction machine, whose parameters and limiting conditions are listed belows, in the format shown in Fig. 2.77.

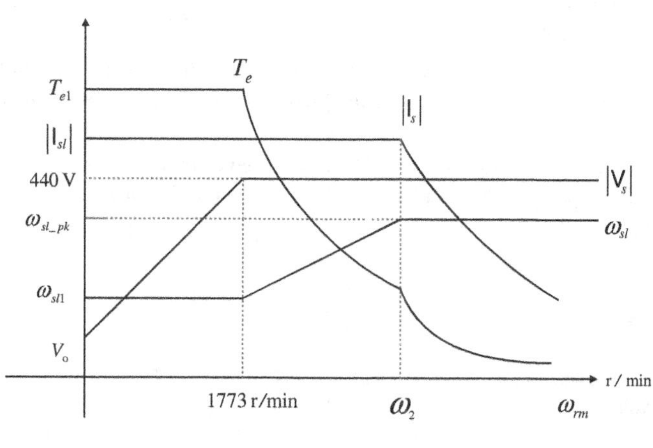

Figure P2.3 Capability curve of an induction machine.

Rated power: 185 kW, Rated voltage (line-to-line rms) : 440 V,

Rated speed: 1773 r/min,

Base frequency: 60 Hz, Number *of poles* : 4

$R_s = 0.014\,\Omega$, $L_{ls} = L_{lr} = 0.278\text{mH}$, $L_m = 15\text{ mH}$, $R_r = 0.014\,\Omega$

Limiting conditions: Line-to-line voltage in rms : $|V_s| \leq 440\text{ V}$

Line current in rms: $|\mathbf{I}_s| \leq$ Rated value, where the rated torque occurs at rated air-gap flux

Air-gap flux : $|\boldsymbol{\lambda}_m| \leq$ Rated value

(1) Calculate ω_{sl1}, where the rated torque occurs at the rated stator voltage and frequency, in electrical angular frequency, rad/s.

(2) Calculate the magnitude of line-to-line voltage in rms, V_0, where the rated torque occurs at zero running speed, which is a stalled condition.

(3) Calculate rated torque, T_{el}

(4) Calculate the magnitude of the stator current in rms, $\mathbf{I}_s$, where the rated torque occurs at the rated stator voltage and frequency.

(5) Calculate ω_{sl_pk}, where pull-out torque occurs.

(6) Calculate the angular frequency of the speed, where the flux weakening region 2 (characteristics region) starts.

16. The operating speed of a NEMA D-type induction machine, whose parameters are listed belows, is adjusted by controlling the terminal voltage of the machine.

Rated power : 5.5 kW, Rated voltage (line-to-line rms) : 440 V

Rated frequency : 60 Hz, *Number of poles* : 4

$R_s = 0.69\,\Omega$, $L_{ls} = L_{lr} = 5.9\text{ mH}$, $L_m = 0.12\text{ H}$, $R_r = 2.25\,\Omega$

The torque–speed curve of the load is given as $T_L = 30\left(\frac{\omega_{rm}}{1800}\right)^2$ [N-m], where ω_{rm} stands for the speed of the machine (=load) in r/min.

(1) When the line-to-line voltage of the stator terminal in rms is 440 V, find the operating speed in the steady state. Also find the efficiency of the machine at that operating point.

(2) When line-to-line voltage of the stator terminal in rms is reduced to 220 V, find the operating speed in the steady state. Also find the efficiency of the machine at that operating point.

REFERENCES

1. S. C. Hong et al., Microprocessor-based high-efficiency drive of a DC motor, *IEEE Trans. on Ind. Electron.*, Vol. 34, No. 4, November 1987, pp. 433–440.
2. D. Novotny et al., *Electromechanical Systems,* ECE411 Course Note, University of Wisconsin—Madison, 1986, Section 1.
3. P. C. Sen, *Thyristor DC Drives,* John Wiley & Sons, New York, 1981, Chapter 3.

4. T. Wildi, *Electrical Machines, Drives, and Power Systems*, 2nd edition, Prentice-Hall, Englewood Cliffs, NJ, 1991.
5. G. R. Slemon et al., *Electric Machines*, Addison-Wesley, Reading, MA, 1980.
6. D. Novotny et al., *Electromechanical Systems*, ECE411 Course Note, University of Wisconsin—Madison, 1986, Section 2.
7. P. L. Cochran, *Polyphase Induction Motors*, Marcel Dekker, New York, 1989.
8. C. G. Veinott, *Theory and Design of Small Induction Motors*, McGraw-Hill, New York, 1959.
9. H. C. J. de Jong, *AC Motor Design*, Hemisphere Publishing, New York, 1989.
10. K. K. Schwarz, *Design of Industrial Electric Motor Drives*, Butterworth-Heinemann, Oxford, UK, 1991.
11. D. W. Novotny et al., *Vector Control and Dynamics of AC Drives*, Oxford University Press, Oxford, UK, 1996.
12. J. M. D. Murphy, *Thyristor Control of AC Motors*, Pergamon Press, Elmsford, NY, 1972.
13. N. Mohan et al., *Power Electronics: Converters, Applications and Design*, John Wiley & Sons, New York, 1989.
14. L. Huber et al., Space vector modulator for forced commutated cycloconverters, *IEEE Trans. Ind. Appl.*, Vol. 31, No. 6, 2006, pp. 1234–1246.
15. D. Casadei et al., Matrix converter modulation strategies, *IEEE Trans. Ind. Electron.*, Vol. 49, No. 2, April, 2002, pp. 371–381.
16. Y. D. Yoon et al., Carrier based modulation technique for matrix converter, *IEEE Trans. Power Electron.*, Vol. 21, No. 6, 2006, pp. 1691–1703.
17. Y. D. Yoon et al., *Carrier Based Modulation Method for Matrix Converter with Input Power Factor Control and Unbalanced Input Voltage Conditions*, 22nd Applied Power Electronics Conference (APEC) Record, 2007, pp. 310–314.

Chapter 3

Reference Frame Transformation and Transient State Analysis of Three-Phase AC Machines

The inductances of all AC machines vary according to the rotor position. Because of that, the voltage equations of an AC machine are expressed as time-varying differential equations as long as the rotor of the machine rotates. The transformation of physical variables of an AC machine using reference frame theory could make the analysis be easy by transforming the time varying differential equations to the time-invariant differential equations. The electrical variables such as voltage, current, and flux in a, b, and c phases of a three-phase system can be transformed to the variables in d, q, and n (direct, quadrature, and neutral) orthogonal axes, where the magnetic couplings between axes are zero. Usually, the d-axis, which means the direct axis, is the axis where the main flux directs. And the q axis, which means the quadrature axis, lies 90° ahead of the d axis spatially with regard to the positive rotational direction of a rotating MMF. Also, the n axis, which means the neutral axis and sometimes called as zero sequence axis, is orthogonal to the d–q axes in three-dimensional space, and the n axis is perpendicular to the plane where the rotating MMF lies; hence the current or voltage at the n axis does not contribute to the rotating MMF and to the torque either, but only to losses. At the three-phase electric circuit including a three-phase AC machine, if the neutral point of a Y-connected three-phase circuit is not connected to a source and/or to other electric machines or power converters, and also if the impedance of each phase is the same and the instantaneous sum of the back EMF of all phases is zero, then there is no n-axis current and voltage component. Hence, in this case, there is no need to consider the n-axis components. Most normal AC machines are usually running in this condition when there is no fault internally at the

Control of Electric Machine Drive Systems, by Seung-Ki Sul

electric machine. And, the three-phase system can be easily represented only by d–q components. The d–q components, which are orthogonal to each other, can be expressed simply by a complex number, where real part stands for d-axis components and imaginary part stands for q-axis components.

This kind of transformation from the a, b, and c phases to the orthogonal axes can be done by complex vector algebra and also by using a matrix algebra [1–3]. For theoretical analysis and physical understanding, the transformation based on the complex vector would be easier compared to that on the matrix algebra. But for the computer simulation and programming of the real-time control software of the electric machine, the transformation with the matrix algebra is more convenient. In this chapter, for the convenience of mathematical presentation and for better physical understanding, the complex vector is used to present d–q axes components. The transformation of three-phase variables to d–q–n axes variables by the matrix algebra is described in Appendix B. In particular, when the instantaneous sum of physical variables of each phase is zero, which means no n-axis component, the representation by the complex vector is very convenient. In the viewpoint of the magnitude of variables at d–q–n axes compared to that at three phases, there are two methods of transformation. One is called the phase magnitude invariance method, where the magnitude of variables at each phase of the three-phase system is the same to that of d–q axes components in the balanced steady state. But the power and torque expressed in d–q–n axes should be multiplied by 3/2 to get the same torque and power expressed in terms of three phase variables [1, 4, 5]. The other one is called the power invariance method, where the magnitude of power or torque expressed in a three-phase system is the same as those in d–q–n axes. But in the power invariance method, the magnitude of variables at each d–q axis is $\sqrt{\frac{3}{2}}$ times that of variables at a three-phase system [6]. In this book, for the convenience of comparison of experimental results with calculated values, and also for easy conversion to three-phase variables to d–q variables, the phase magnitude invariance method is adopted.

3.1 COMPLEX VECTOR [1–3]

In the electromagnetic energy conversion, the d–q components are only contributed to the energy conversion but the components at the n axis only generate losses. So, if the electromechanical power conversion is only the concern, then the analysis of the electric machines can be done by only d–q components. In this case the complex vector can solely represent the three-phase electrical system by only two orthogonal components such as real and imaginary components of a complex vector. Here, the d component is represented by a real part, and the q component is represented by an imaginary part in a complex vector. This complex vector is called a complex space vector, or simply as a space vector. The definition of the space vector by three phase components is in (3.1):

$$\mathbf{f}_{abc} \equiv \tfrac{2}{3}\left(f_a + \mathbf{a}f_b + \mathbf{a}^2 f_c\right) \tag{3.1}$$

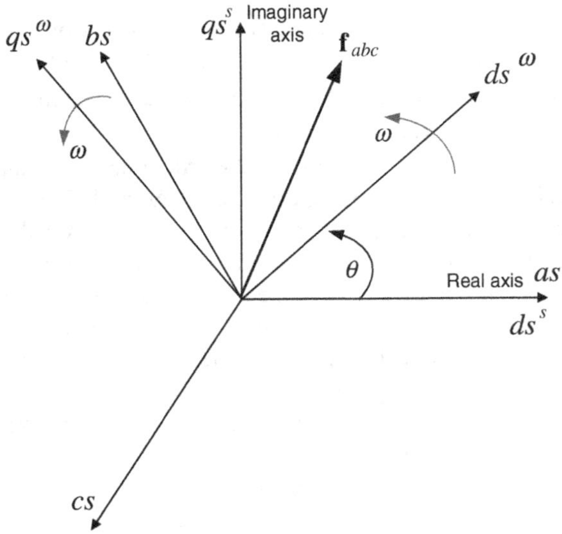

Figure 3.1 Relationship between different axis.

where $\mathbf{a} = e^{j\frac{2\pi}{3}}$ and f_a, f_b, and f_c are the variables at each a, b, and c phase, respectively. For example, if f_a, f_b, f_c is the flux linkage by each stator winding spatially 120° apart, then $\mathbf{f}_{abc}$ stands for the total flux linkage by the windings, which may rotate. In Fig. 3.1, the relationships between three-phase axes (as, bs, cs), d–q axes (ds^s and qs^s) which are in the stationary, and d–q axes (ds^{ω} and qs^{ω}) which are rotating in arbitrary speed, ω are shown graphically.

In the figure, the angle θ is defined as

$$\theta = \int_0^t \omega(\zeta)\, d\zeta + \theta(0) \tag{3.2}$$

where $\theta(0)$ is the angle between the d axis rotating with an arbitrary speed, ω (d^{ω}), and the stationary d axis ds^s (or usually a-phase axis of a three-phase system) at time point 0. Usually $\theta(0) = 0$.

In this text, f_{jk}^{i} means the following: f represents a specific physical variable such as voltage, current, or flux linkage. The superscript i represents the kinds of axis where $i = e$ stands for the axis rotating synchronously according to the rotating MMF ($\omega = \omega_e$) referred to as a synchronously rotating reference frame, $i = r$ for the axis rotating synchronously according to the rotor of the machine ($\omega = \omega_r$) referred to as a rotor reference frame, and $i = s$ for the stationary axis referred as a stationary reference frame. The subscript j represents the variables in d–q–n axes or a, b, and c phases. Here, $j = d$ stands for a d-axis variable, $j = q$ stands for a q-axis variable, $j = n$ stands for an n-axis (or zero sequence) variable, and $j = a$ stands for an a-phase variable among three phase variables. The subscript k represent where the variables exist. Here, $k = s$ stands for the variables in the stator, and $k = r$ stands for the variables in the rotor. As an example, i_{qs}^{e} means the stator current at the q axis rotating synchronously to the rotating MMF. In some countries, α–β axes are used to represent a stationary d–q axis, and a γ–δ axis is used to represent a synchronously rotating *d–q* axis. The equation to transform a space vector in three phases, $\mathbf{f}_{abc}$, to the space vector in *d–q–n* axes rotating with an arbitrary speed, ω can be derived as

$$\mathbf{f}^{\omega}_{dq} = \mathbf{f_{abc}} \cdot e^{-j\theta} = f^{\omega}_{d} + jf^{\omega}_{q}$$

$$f^{\omega}_{n} = \tfrac{1}{3}(f_a + f_b + f_c) \tag{3.3}$$

Some people, especially those working for the power system engineering field, use the different convention of the space vector such as $\mathbf{f}^{\omega}_{qd} = f^{\omega}_{q} - jf^{\omega}_{d}$ [4–6]. In this text, (3.3) is the definition of the space vector to transform three phase variables to *d–q* variables. By extending (3.3), the transformation to other reference frame can be easily deduced as

$$\mathbf{f}^{s}_{dq} = \mathbf{f_{abc}},\ \mathbf{f}^{e}_{dq} = \mathbf{f}^{s}_{dq} e^{-j\theta_e} \tag{3.4}$$

where $\theta_e = \int_0^t \omega_e(\zeta)\, d\zeta + \theta_e(0)$.

The instantaneous power represented with space vectors can be defined as

$$\text{Power} \equiv \tfrac{3}{2}\text{Re}\left(V_{abc} \cdot \mathrm{I}^{*}_{abc}\right) = \frac{3}{2}\text{Re}\left(V^{\omega}_{dq} \cdot \mathrm{I}^{*\omega}_{dq}\right) \tag{3.5}$$

where Re stands for the real part of the complex space vector, and $\mathbf{I}^{*}_{abc}$ stands for the conjugate of $\mathbf{I}_{abc}$.

The space vector is similar to phasor notation in expression and calculation of power. But the definition and application of the variable in space vector is totally different from that in phasor notation. The phasor is used to represent a sinusoidal electrical variable in the steady state, where the frequency of the variable would not vary, and the phase and magnitude of the variable are the only concern and are expressed based on the complex number; that is, $A\cos(\omega t + \phi) = \text{Re}\left[A\, e^{j(\omega t + \phi)}\right] = \text{Re}\left[A\, e^{j\omega t} \cdot e^{j\phi}\right]$, where $Ae^{j\phi}$ is $\alpha + j\beta$, and $Ae^{j\phi}$ is the phasor of $A\cos(\omega t + \phi)$.

Also, $Ae^{j\phi} = \alpha + j\beta = \sqrt{\alpha^2 + \beta^2}\angle \tan^{-1}\frac{\alpha}{\beta}$, where $\sqrt{\alpha^2 + \beta^2} = A$ is the magnitude of a sinusoidal variable, and $\tan^{-1}\frac{\beta}{\alpha}$ is the phase of the sinusoidal variable. In this case, α, β is not the function of time, but a fixed value. However, the imaginary and real part of a space vector is generally a function of time, and it can be differentiated or integrated with regard to time.

The variable at each of the three phases can be deduced as (3.6)–(3.8) from the space vector, $\mathbf{f}_{abc}$, where $\mathbf{a} = e^{j\frac{2\pi}{3}}$.

$$f_a = \text{Re}[\mathrm{f}_{abc}] + f^{\omega}_{n} \tag{3.6}$$

$$f_b = \text{Re}\left[\mathrm{a}^2\mathrm{f}_{abc}\right] + f^{\omega}_{n} \tag{3.7}$$

$$f_c = \text{Re}[\mathrm{a}\mathrm{f}_{abc}] + f^{\omega}_{n} \tag{3.8}$$

3.2 *d–q–n* MODELING OF AN INDUCTION MACHINE BASED ON COMPLEX SPACE VECTOR [1,4]

To derive voltage equations of a three-phase induction machine, it is assumed that the windings at the stator and rotor are symmetry, ideally distributed to generate sinusoidal air gap flux, the turn ratio between the stator winding and rotor winding is unity, and that there is no magnetic saturation and no eccentricity. If the turn ratio is not unity, which is the most of cases, the parameters of a rotor circuit can be referred to the stator circuit. And if the dynamics of the induction machine is observed and controlled in the stator side, which is all the cases of the squirrel cage induction machine, the turn ratio can be assumed as unity without loss of the generality. Under the above assumptions, the voltage equations described by three phase variables, which are time-varying differential equations, can be transformed to the equation described by *d–q–n* variables, which may be time-invariant differential equations.

3.2.1 Equivalent Circuit of an Induction Machine at *d–q–n* AXIS

In Fig. 3.2, the distributed winding of a stator $(as-as', bs-bs', cs-cs')$ and that of a rotor $(ar-ar', br-br', cr-cr')$ of a two-pole induction machine is modeled as a single-turn winding. In the figure, θ_r is the angle from the axis of an MMF by a-phase stator winding to the axis of an MMF by a-phase rotor winding, and ω_r is the time differentiation of θ_r in rad/s.

In Fig. 3.2, the pole number of the machine is two. If the number of pole is P, then the actual mechanical rotation angle, θ_{rm}, has the following relationship with the rotation angle, θ_r in electric angle:

$$\theta_{rm} = \theta_r/(P/2) \tag{3.9}$$

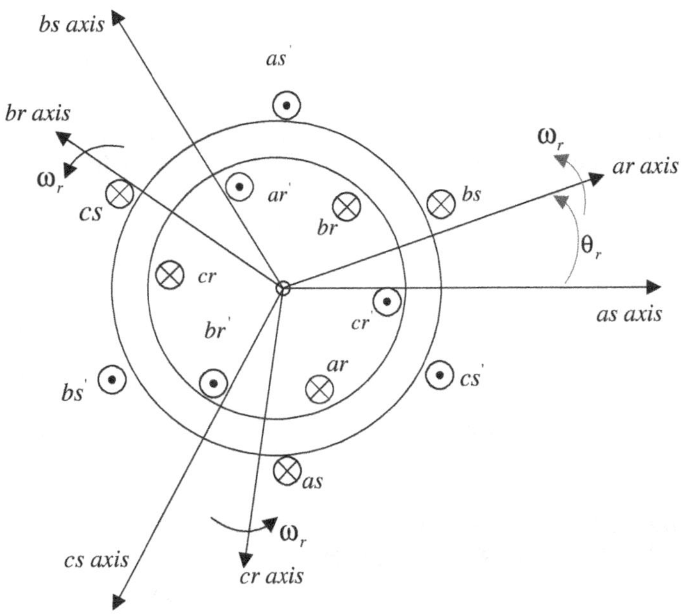

Figure 3.2 Winding model of an induction machine.

The stator and rotor voltage equations can be represented as (3.10) and (3.11) in vector form, respectively:

$$\mathbf{V}_{abcs} = R_s\mathbf{I}_{abcs} + p\boldsymbol{\lambda}_{abcs} \tag{3.10}$$

$$\mathbf{V}_{abcr} = R_r\mathbf{I}_{abcr} + p\boldsymbol{\lambda}_{abcr} \tag{3.11}$$

where p is a differential operator as $d \cdot /dt$, R_s stands for the stator winding resistance, and R_r stands for rotor winding resistance referred to the stator side. In the equations, each vector has the following elements:

$$\mathbf{V}_{abcs} = [V_{as} V_{bs} V_{cs}]^T \tag{3.12}$$

$$\mathbf{I}_{abcs} = [i_{as}\, i_{bs}\, i_{cs}]^T \tag{3.13}$$

$$\boldsymbol{\lambda}_{abcs} = [\lambda_{as}\lambda_{bs}\lambda_{cs}]^T \tag{3.14}$$

$$\mathbf{V}_{abcr} = [V_{ar} V_{br} V_{cr}]^T \tag{3.15}$$

$$\mathbf{I}_{abcr} = [i_{ar} i_{br} i_{cr}]^T \tag{3.16}$$

$$\boldsymbol{\lambda}_{abcr} = [\lambda_{ar}\lambda_{br}\lambda_{cr}]^T \tag{3.17}$$

where $[\ldots]^T$ stands for the transpose of a matrix or a vector, $[\ldots]$.

The flux linkages for the stator and rotor winding can be expressed as (3.18) in matrix form.

$$\begin{bmatrix} \boldsymbol{\lambda}_{abcs} \\ \boldsymbol{\lambda}_{abcr} \end{bmatrix} = \begin{bmatrix} \mathbf{L}_s & \mathbf{L}_{sr} \\ (\mathbf{L}_{sr})^T & \mathbf{L}_r \end{bmatrix} \begin{bmatrix} \mathbf{I}_{abcs} \\ \mathbf{I}_{abcr} \end{bmatrix} \tag{3.18}$$

where

$$\mathbf{L}_s = \begin{bmatrix} L_{ls} + L_{ms} & -\frac{1}{2}L_{ms} & -\frac{1}{2}L_{ms} \\ -\frac{1}{2}L_{ms} & L_{ls} + L_{ms} & -\frac{1}{2}L_{ms} \\ -\frac{1}{2}L_{ms} & -\frac{1}{2}L_{ms} & L_{ls} + L_{ms} \end{bmatrix} \tag{3.19}$$

$$\mathbf{L}_r = \begin{bmatrix} L_{lr} + L_{mr} & -\frac{1}{2}L_{mr} & -\frac{1}{2}L_{mr} \\ -\frac{1}{2}L_{mr} & L_{lr} + L_{mr} & -\frac{1}{2}L_{mr} \\ -\frac{1}{2}L_{mr} & -\frac{1}{2}L_{mr} & L_{lr} + L_{mr} \end{bmatrix} \tag{3.20}$$

$$\mathbf{L}_{sr} = L_{sr}\begin{bmatrix} \cos\theta_r & \cos\left(\theta_r+\frac{2\pi}{3}\right) & \cos\left(\theta_r-\frac{2\pi}{3}\right) \\ \cos\left(\theta_r-\frac{2\pi}{3}\right) & \cos\theta_r & \cos\left(\theta_r+\frac{2\pi}{3}\right) \\ \cos\left(\theta_r+\frac{2\pi}{3}\right) & \cos\left(\theta_r-\frac{2\pi}{3}\right) & \cos\theta_r \end{bmatrix} \tag{3.21}$$

In the above equations, L_{ms}stands for a mutual inductance, which is two times the mutual inductance between the stator windings, L_{mr} stands for a mutual inductance of the rotor, which is the two times of the mutual inductance between the rotor windings, L_{sr} stands for a mutual inductance between a-phase stator winding and a-phase rotor winding when θ_r is zero, stands L_{ls} for a leakage inductance of a stator winding, and L_{lr} stands for a leakage inductance of a rotor winding. As seen from (3.10)–(3.21), the dynamics of the induction machine are described by a time-varying differential equations. To solve these differential equations simultaneously regarding to the stator flux linkage and rotor flux linkage, the inverse matrix, which has time-varying elements, should be derived analytically, but it is very difficult to deduce. By transforming the three phase variables to d–q–n variables, the time-varying equations can be transformed to time-invariant equations.

Based on the relationship of axes shown in Fig. 3.1 and the winding model in Fig. 3.2, the stator variables at d–q–n axes which are rotating at arbitrary speed, ω can be obtained as (3.22) from (3.1) and (3.3). Similarly, the rotor variables at d–q–n axes can be obtained as (3.23).

$$\mathbf{f}^{w}_{dqs} = \mathbf{f}_{abcs}e^{-jo} \tag{3.22}$$

$$\mathbf{f}^{\omega}_{dqr} = \mathbf{f}_{abcr}e^{-j\beta}, \qquad \beta = \theta - \theta_r \tag{3.23}$$

where β is the angle from the axis of an MMF by a-phase rotor winding to the d axis that is rotating at arbitrary speed, ω. If d–q–n axes rotating synchronously ($\omega = \omega_e$) with the rotating MMF are chosen as a reference frame, then the stator variables in a three-phase system can be transformed to the variables at a synchronously rotating reference frame by (3.24) and the variables at a stationary reference frame can be transformed to the variables at a synchronously rotating reference frame by (3.25).

$$\mathbf{f}^{e}_{dqs} = \mathbf{f}_{abcs}e^{-j\omega_e t} \tag{3.24}$$

$$\mathrm{f}^{e}_{dqs} = \mathrm{f}^{s}_{dqs}e^{-j\theta_e}, \qquad \theta_e = \omega_e t \tag{3.25}$$

Also, similarly, the rotor variables in a three-phase system can be transformed to the variables at a synchronously rotating reference frame by (3.26) and the variables at a stationary reference frame can be transformed to the variables at a rotor reference frame by (3.27).

$$\mathbf{f}^{e}_{dqr} = \mathbf{f}_{abcr}e^{-j(\theta_e-\theta_r)} \tag{3.26}$$

$$\mathbf{f}^{r}_{dqr} = \mathbf{f}^{s}_{dqr}e^{-j\theta_r}, \qquad \theta_r = \omega_r t \tag{3.27}$$

where ω_r is the angular speed of the rotor expressed in the electrical angle and it is assumed that $\theta_e = \theta_r = 0$ at $t = 0$.

By (3.10) and (3.11) and the definition of the space vectors of three phase variables as (3.28)–(3.31), the stator and rotor voltage equations, (3.10) and (3.11), can be rewritten as (3.32), (3.33).

$$\mathbf{V}_{abcs} \equiv \tfrac{2}{3}(V_{as} + \mathbf{a}V_{bs} + \mathbf{a}^2V_{cs}) \tag{3.28}$$

$$\boldsymbol{\lambda}_{abcs} \equiv \tfrac{2}{3}(\lambda_{as} + \mathbf{a}\lambda_{bs} + \mathbf{a}^2\lambda_{cs}) \tag{3.29}$$

$$\mathbf{i}_{abcs} \equiv \tfrac{2}{3}(i_{as} + \mathbf{a}i_{bs} + \mathbf{a}^2 i_{cs}) \tag{3.30}$$

$$\mathbf{i}_{abcr} \equiv \tfrac{2}{3}(i_{ar} + \mathbf{a}i_{br} + \mathbf{a}^2 i_{cr}) \tag{3.31}$$

$$\mathbf{V}_{abcs} = R_s\mathbf{i}_{abcs} + \frac{d}{dt}\boldsymbol{\lambda}_{abcs} \tag{3.32}$$

$$\mathbf{V}_{abcr} = R_r\mathbf{i}_{abcr} + \frac{d}{dt}\boldsymbol{\lambda}_{abcr} \tag{3.33}$$

In (3.32), the complex space vector of the stator flux linkage, $\boldsymbol{\lambda}_{abcs}$, can be deduced as follows. From (3.18)–(3.21), the stator flux linkage can be represented as (3.34) by the product of the inductances and currents.

$$\begin{bmatrix}\lambda_{as}\\ \lambda_{bs}\\ \lambda_{cs}\end{bmatrix} = \begin{bmatrix} L_{ls}+L_{ms} & -\dfrac{L_{ms}}{2} & -\dfrac{L_{ms}}{2} \\ -\dfrac{L_{ms}}{2} & L_{ls}+L_{ms} & -\dfrac{L_{ms}}{2} \\ -\dfrac{L_{ms}}{2} & -\dfrac{L_{ms}}{2} & L_{ls}+L_{ms}\end{bmatrix}\begin{bmatrix}i_{as}\\ i_{bs}\\ i_{cs}\end{bmatrix} + L_{sr}\begin{bmatrix}\cos\theta_r & \cos\left(\theta_r + \dfrac{2\pi}{3}\right) & \cos\left(\theta_r - \dfrac{2\pi}{3}\right) \\ \cos\left(\theta_r - \dfrac{2\pi}{3}\right) & \cos\theta_r & \cos\left(\theta_r + \dfrac{2\pi}{3}\right) \\ \cos\left(\theta_r + \dfrac{2\pi}{3}\right) & \cos\left(\theta_r - \dfrac{2\pi}{3}\right) & \cos\theta_r\end{bmatrix}\begin{bmatrix}i_{ar}\\ i_{br}\\ i_{cr}\end{bmatrix} \tag{3.34}$$

In (3.34), by using the equalities such as $\cos\theta_r = \dfrac{e^{j\theta_r} + e^{-j\theta_r}}{2}$, $\cos\theta_r + \mathbf{a}\cos\left(\theta_r - \frac{2\pi}{3}\right) + \mathbf{a}^2\cos\left(\theta_r + \frac{2\pi}{3}\right) = \frac{3}{2}e^{j\theta_r}$, and $\mathbf{1} - \frac{\mathbf{a}}{\mathbf{2}} - \frac{\mathbf{a}^2}{\mathbf{2}} = \frac{3}{2}$, the complex space vector of the stator flux linkage can be presented as

$$\boldsymbol{\lambda}_{abcs} = \left(L_{ls} + \tfrac{3}{2}L_{ms}\right)\mathbf{i}_{abcs} + \tfrac{3}{2}L_{ms}\mathbf{i}_{abcr}e^{j\theta_r} \tag{3.35}$$

where $a = e^{j\frac{2}{3}\pi}$.

Similarly, the complex space vector of the rotor flux linkage, λ_{abcr}, can be presented as

$$\boldsymbol{\lambda}_{abcr} = \left(L_{lr} + \tfrac{3}{2}L_{ms}\right)\mathbf{i}_{abcr} + \tfrac{3}{2}L_{ms}\mathbf{i}_{abcs}e^{-j\theta_r} \tag{3.36}$$

In (3.36), if a mutual inductance, L_m, is defined as $L_m \equiv \frac{3}{2}L_{ms}$, then the voltage equations in (3.32) and (3.33) can be presented in terms of stator and rotor current with the space vector variables as (3.37) and (3.38):

$$\mathbf{V}_{abcs} = R_s\mathbf{i}_{abcs} + (L_{ls} + L_m)\frac{d\mathbf{i}_{abcs}}{dt} + L_m\frac{d}{dt}\left(\mathbf{i}_{abcr}e^{j\theta_r}\right) \tag{3.37}$$

$$\mathbf{V}_{abcr} = R_r\mathbf{i}_{abcr} + (L_{lr} + L_m)\frac{d\mathbf{i}_{abcr}}{dt} + L_m\frac{d}{dt}\left(\mathbf{i}_{abcs}e^{-j\theta_r}\right) \tag{3.38}$$

By using equalities $\mathbf{V}^{\omega}_{dqs} = \mathbf{V}_{abcs} \cdot e^{-j\theta}$, $\mathbf{i}^{\omega}_{dqr} = \mathbf{i}_{abcr} \cdot e^{-j(\theta-\theta_r)}$, the stator voltage equation (3.37) can be rewritten as (3.39) in terms of the space vector variables at *d–q–n* axes rotating in arbitrary speed, ω.

$$\begin{aligned}
\mathbf{V}^{\omega}_{dqs} &= \left[R_s e^{-j\theta}\,\mathbf{i}_{abcs} + (L_{ls} + L_m)e^{-j\theta}p\mathbf{i}_{abcs} + L_m e^{-j\theta}p(\mathbf{i}_{abcr}e^{j\theta_r})\right] \\
&= R_s\,\mathbf{i}^{\omega}_{dqs} + (L_{ls} + L_m)p\left[e^{-j\theta}\mathbf{i}_{abcs}\right] - (L_{ls} + L_m)\mathbf{i}_{abcs} \cdot p\left[e^{-j\theta}\right] \\
&\quad + L_m p\left(\mathbf{i}_{abcr}e^{-j(\theta-\theta_r)}\right) - L_m\left[pe^{-j\theta}\right] \cdot \mathbf{i}_{abcr}e^{j\theta_r} \\
&= R_s\mathbf{i}^{\omega}_{dqs} + (L_{ls} + L_m)p\mathbf{i}^{\omega}_{dqs} + j\omega(L_{ls} + L_m)\mathbf{i}^{\omega}_{dqs} + L_m p\mathbf{i}^{\omega}_{dqr} + j\omega L_m\mathbf{i}^{\omega}_{dqr} \\
&= R_s\mathbf{i}^{\omega}_{dqs} + (L_{ls} + L_m)p\mathbf{i}^{\omega}_{dqs} + L_m p\mathbf{i}^{\omega}_{dqr} + j\omega\left[(L_{ls} + L_m)\mathbf{i}^{\omega}_{dqs} + L_m\mathbf{i}^{\omega}_{dqr}\right]
\end{aligned} \tag{3.39}$$

Similarly, the rotor voltage equation (3.38) can be rewritten as (3.40) in terms of the space vector variables at *d–q–n* axes rotating in the arbitrary speed, ω:

$$\mathbf{V}^{\omega}_{dqr} = R_r\mathbf{i}^{\omega}_{dqr} + (L_{lr} + L_m)p\mathbf{i}^{\omega}_{dqr} + L_m p\mathbf{i}^{\omega}_{dqs} + j(\omega - \omega_r)\left[(L_{lr} + L_m)\mathbf{i}^{\omega}_{dqr} + L_m\mathbf{i}^{\omega}_{dqs}\right] \tag{3.40}$$

In (3.39) and (3.40), we have

$$L_s\mathbf{i}^{\omega}_{dqs} + L_m\mathbf{i}^{\omega}_{dqr} = \lambda^{\omega}_{dqs} \tag{3.41}$$

$$L_m\mathbf{i}^{\omega}_{dqs} + L_r\mathbf{i}^{\omega}_{dqr} = \lambda^{\omega}_{dqr} \tag{3.42}$$

where

$$L_s = L_m + L_{ls} \tag{3.43}$$

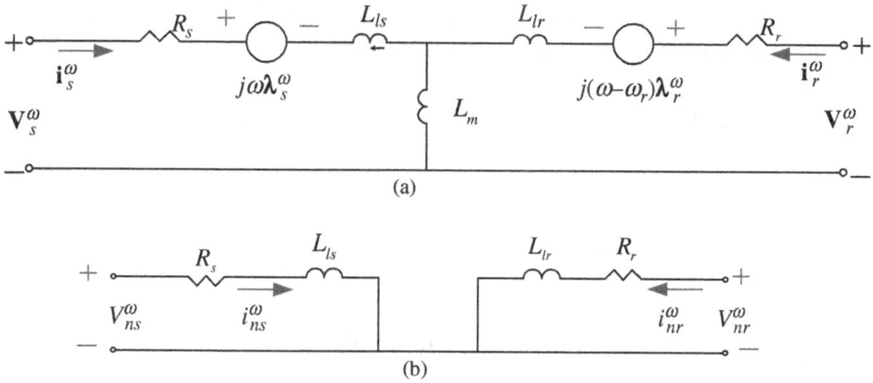

Figure 3.3 Equivalent circuit of an induction machine at *d–q–n* axes rotating in arbitrary speed, ω in terms of complex space vector variables. (a) equivalent circuit at *d–q* axes rotating in arbitrary speed, ω. (b) equivalent circuit at the *n* axis rotating in arbitrary speed, ω.

$$L_r = L_m + L_{lr} \tag{3.44}$$

If λ_{dqs}^ω, λ_{dqr}^ω, V_{dqs}^ω, V_{dqr}^ω, i_{dqs}^ω, and i_{dqr}^ω are simply expressed as $\boldsymbol{\lambda}_s^\omega$, $\boldsymbol{\lambda}_r^\omega$, V_s^ω, V_r^ω, i_s^ω, and i_r^ω, then a equivalent circuit shown in Fig. 3.3a is obtained at *d–q* axes rotating in the arbitrary speed, ω. The equivalent circuit at the *n* axis shown in Fig. 3.3b can be found by applying the definition of the neutral component expressed by three phase variables in (3.3) to the stator and rotor voltage equations, (3.10), and (3.11). As seen in the figure, the neutral component stator circuit and rotor circuit are decoupled to each other, and they are not involved to the electromechanical energy conversion, but generate only copper losses and some reactive voltages due to leakage inductances.

In the case of a squirrel cage rotor induction machine, the rotor is shorted by the end ring, and the rotor voltage, $\mathbf{V}_r^\omega$ and V_{nr}^ω, are always zero. Also, if the instantaneous sum of each phase stator current is zero and the stator impedance and the flux distribution are balanced, then there is no neutral component current, that is, $i_{ns}^\omega = 0$, $i_{nr}^\omega = 0$. So, there is no need to consider the *n*-axis equivalent circuit in this case. In the case of the most normal operation of a squirrel cage rotor induction machine, only *d–q* axes equivalent circuit is enough to evaluate the transient response of the induction machine.

3.2.2 Torque of the Induction Machine

The force or torque of an electric machine based on the electromagnetic energy conversion principle can be evaluated by differentiating the coenergy of the machine with regard to the displacement [7–9]. In another way, the force and torque can be evaluated from the equivalent circuit by using power balance [4]. In this text, the latter method is used to evaluate the torque of the induction machine. By (3.5), the input power to a three-phase circuit can be represented as

$$\begin{aligned}\mathrm{Re}\left[\mathbf{V}_{abcs}\mathbf{i}^{*}_{abcs}\right] &= \mathrm{Re}\left[\tfrac{2}{3}\left(V_{as} + \mathbf{a}V_{bs} + \mathbf{a}^2 V_{cs}\right)\cdot\tfrac{2}{3}\left(i_{as} + \mathbf{a}^2 i_{bs} + \mathbf{a}\, i_{cs}\right)\right] \\ &= \tfrac{4}{9}\Big\{V_{as}i_{as} + V_{bs}i_{bs} + V_{cs}i_{cs} - \tfrac{1}{2}\big[V_{as}(i_{bs} + i_{cs}) \\ &\quad + V_{bs}(i_{as} + i_{cs}) + V_{cs}(i_{as} + i_{bs})\big]\Big\}\end{aligned} \tag{3.45}$$

Under the assumption that $i_{as} + i_{bs} + i_{cs} = 0$, we obtain

$$\mathrm{Re}\left[\mathbf{V}_{abcs}\mathbf{i}^{*}_{abcs}\right] = \tfrac{2}{3}\left[V_{as}i_{as} + V_{bs}i_{bs} + V_{cs}i_{cs}\right] \tag{3.46}$$

By using (3.46), the input power to the induction machine (Fig. 3.4) can be represented as

$$\begin{aligned}P_{in} &= V_{as}i_{as} + V_{bs}i_{bs} + V_{cs}i_{cs} + i_{ar}V_{ar} + i_{br}V_{br} + i_{cr}V_{cr} \\ &= \tfrac{3}{2}\mathrm{Re}\left[\mathbf{V}_{abcs}\mathbf{i}^{*}_{abcs}\right] + \tfrac{3}{2}\mathrm{Re}\left[\mathbf{V}_{abcr}\mathbf{i}^{*}_{abcr}\right]\end{aligned} \tag{3.47}$$

Under the assumption of the balanced impedance and flux distribution of the machine and by equality in (3.48), (3.47) can be rewritten in *d–q* axis as (3.49).

$$\begin{aligned}&\mathbf{V}_{abcs}\mathbf{i}^{*}_{abcs} + \mathbf{V}_{abcr}\mathbf{i}^{*}_{abcr} \\ &= e^{j\theta}\mathbf{V}^{\omega}_{dqs}\, e^{-j\theta}\mathbf{i}^{\omega *}_{dqs} + e^{j(\theta-\theta_r)}\,\mathbf{V}^{\omega}_{dqr}e^{-j(\theta-\theta_r)}\mathbf{i}^{\omega *}_{dqr} \\ &= \mathbf{V}^{\omega}_{dqs}\mathbf{i}^{\omega *}_{dqs} + \mathbf{V}^{\omega}_{dqr}\mathbf{i}^{\omega *}_{dqr}\end{aligned} \tag{3.48}$$

Hence,

$$\begin{aligned}P_{in} &= \tfrac{3}{2}\mathrm{Re}\left[\mathbf{V}^{\omega}_{dqs}\mathbf{i}^{\omega *}_{dqs} + \mathbf{V}^{\omega}_{dqr}\mathbf{i}^{\omega *}_{dqr}\right] \\ &= \tfrac{3}{2}\left[V^{\omega}_{ds}i^{\omega}_{ds} + V^{\omega}_{qs}i^{\omega}_{qs} + V^{\omega}_{dr}i^{\omega}_{dr} + V^{\omega}_{qr}i^{\omega}_{qr}\right]\end{aligned} \tag{3.49}$$

In particular, in the case of the squirrel cage rotor machine, $V^{\omega}_{dr} = V^{\omega}_{qr} = 0$. Also,

$$P_{in} = \tfrac{3}{2}\left[V^{\omega}_{ds}i^{\omega}_{ds} + V^{\omega}_{qs}i^{\omega}_{qs}\right] \tag{3.50}$$

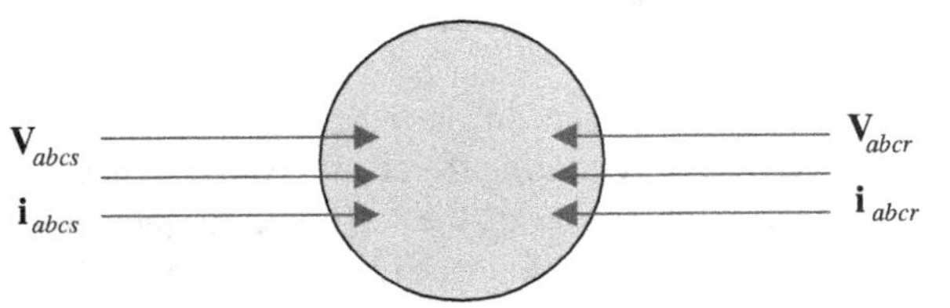

Figure 3.4 Input power of an induction machine.

By substituting the stator and rotor voltage equations into (3.49), the input power can be expressed in terms of the stator and rotor current as

$$\begin{aligned}
P_{in} &= \tfrac{3}{2}\mathrm{Re}\Big[\Big\{R_s\mathbf{i}^{\omega}_{dqs} + (L_{ls}+L_m)p\mathbf{i}^{\omega}_{dqs} + L_m p\mathbf{i}^{\omega}_{dqr} + j\omega\Big((L_{ls}+L_m)\mathbf{i}^{\omega}_{dqs} + L_m\mathbf{i}^{\omega}_{dqr}\Big)\Big\}\mathbf{i}^{\omega*}_{dqs} \\
&\quad + \Big\{\Big(R_r\mathbf{i}^{\omega}_{dqr} + (L_{lr}+L_m)p\mathbf{i}^{\omega}_{dqr} + L_m p\mathbf{i}^{\omega}_{dqs} + j(\omega-\omega_r)\Big((L_{lr}+L_m)\mathbf{i}^{\omega}_{dqr} + L_m\mathbf{i}^{\omega}_{dqs}\Big)\Big)\Big\}\mathbf{i}^{\omega*}_{dqr}\Big] \\
&= \tfrac{3}{2}R_s\left|\mathbf{i}^{\omega}_{dqs}\right|^2 + \tfrac{3}{2}R_r\left|\mathbf{i}^{\omega}_{dqr}\right|^2 + \tfrac{3}{2}p\left[\frac{L_{ls}}{2}\left|\mathbf{i}^{\omega}_{dqs}\right|^2 + \frac{L_{lr}}{2}\left|\mathbf{i}^{\omega}_{dqr}\right|^2 + \frac{L_m}{2}\left|\mathbf{i}^{\omega}_{dqs} + \mathbf{i}^{\omega}_{dqr}\right|^2\right] \\
&\quad + \tfrac{3}{2}\mathrm{Re}\Big[j\omega\Big\{(L_{ls}+L_m)\left|\mathbf{i}^{\omega}_{dqs}\right|^2 + L_m\mathbf{i}^{\omega}_{dqr}\mathbf{i}^{\omega*}_{dqs}\Big\} + j(\omega-\omega_r)\Big\{(L_{lr}+L_m)\left|\mathbf{i}^{\omega}_{dqr}\right|^2 + L_m\mathbf{i}^{\omega}_{dqs}\mathbf{i}^{\omega*}_{dqr}\Big\}\Big]
\end{aligned} \tag{3.51}$$

In the above equation, first and second terms on the right-hand side represent copper losses by the stator and rotor winding resistances, respectively. And the third term stands for the variation of the energy stored in the inductances regarding to the time, which is nothing to do with the electromechanical energy conversion. Hence, the last term may be expected as the mechanical output, which is the product of the rotating speed and the torque.

In the last term, $j\omega(L_{ls}+L_m)\left|\mathbf{i}^{\omega}_{dqs}\right|^2 + j(\omega-\omega_r)(L_{lr}+L_m)\left|\mathbf{i}^{\omega}_{dqr}\right|^2$ is pure imaginary number and the real part of that is zero. Hence, the mechanical output could be deduced as

$$P_m = \tfrac{3}{2}\mathrm{Re}\Big[j\omega L_m\Big(\mathbf{i}^{\omega}_{dqr}\mathbf{i}^{\omega*}_{dqs} + \mathbf{i}^{\omega}_{dqs}\mathbf{i}^{\omega*}_{dqr}\Big) - j\omega_r L_m\mathbf{i}^{\omega}_{dqs}\mathbf{i}^{\omega*}_{dqr}\Big] \tag{3.52}$$

However, if $A = a_r + ja_i, B = b_r + jb_i$, then $A \cdot B^* + A^* \cdot B = 2(a_r b_r + a_i b_i)$, which is a real number.

So, the first part of right-hand side of (3.52) cannot contribute to the mechanical output power. Finally, the mechanical output power can be deduced as

$$\begin{aligned}
P_m &= -\tfrac{3}{2}\mathrm{Re}\Big[j\omega_r L_m\mathbf{i}^{\omega}_{dqs}\mathbf{i}^{\omega*}_{dqr}\Big] = \tfrac{3}{2}\mathrm{Im}\Big[\omega_r L_m\mathbf{i}^{\omega}_{dqs}\mathbf{i}^{\omega*}_{dqr}\Big] \\
&= \tfrac{3}{2}\omega_r L_m \mathrm{Im}\Big[\Big(i^{\omega}_{ds} + ji^{\omega}_{qs}\Big)\Big(i^{\omega}_{dr} - ji^{\omega}_{qr}\Big)\Big] = \tfrac{3}{2}\omega_r L_m\Big(i^{\omega}_{qs}i^{\omega}_{dr} - i^{\omega}_{ds}i^{\omega}_{qr}\Big)
\end{aligned} \tag{3.53}$$

where Im stands for a imaginary part of a complex number. From the mechanical power in (3.53), the torque can be expressed as

$$T_e = \frac{P_m}{\left(\frac{\omega_r}{P/2}\right)} = \frac{3}{2}L_m\frac{P}{2}\Big(i^{\omega}_{qs}i^{\omega}_{dr} - i^{\omega}_{ds}i^{\omega}_{qr}\Big) \tag{3.54}$$

where P is the number of poles of the machine.

By using equalities such as $\boldsymbol{\lambda}^{\omega}_s = L_s\mathbf{i}^{\omega}_s + L_m\mathbf{i}^{\omega}_r$, $\boldsymbol{\lambda}^{\omega}_r = L_m\mathbf{i}^{\omega}_s + L_r\mathbf{i}^{\omega}_r$, and $\boldsymbol{\lambda}^{\omega}_m = L_m\mathbf{i}^{\omega}_s + L_m\mathbf{i}^{\omega}_r$, the torque,$T_e$, can be represented by the products of two variables among five variables such as $\boldsymbol{\lambda}^{\omega}_s$, $\boldsymbol{\lambda}^{\omega}_r$, $\boldsymbol{\lambda}^{\omega}_m$, $\mathbf{i}^{\omega}_s$, and $\mathbf{i}^{\omega}_r$. So, the torque can be expressed in 10 different ways. Equation (3.54) is the one way to express the torque among 10 ways.

3.3 *d–q–n* MODELING OF A SYNCHRONOUS MACHINE BASED ON COMPLEX SPACE VECTOR [4]

3.3.1 Equivalent Circuit of a Synchronous Machine at *d–q–n* AXIS

A three-phase synchronous machine can be generally modeled as shown in Fig. 3.5, where three stator (armature) windings on the stator, one field winding, and two damper windings on the rotor are depicted. The electric power for the electromechanical energy conversion is mainly provided by the armature winding, and the power of the armature winding is much larger than that of damper winding or field winding. Hence, the armature windings are usually on the stator, which is fixed, but the other two windings—field winding and damper winding—are on the rotor. Thus, the armature windings are usually the stator windings. The power for the field winding is supplied through the slip ring from the external DC source. Or it is supplied through a rotating rectifier connected to a permanent magnet-based AC generator. The damper winding is physically constructed with many bars as shown in Fig. 3.5 and they are shorted by the end connection like the cage rotor bars of the squirrel cage induction machine, and the damper winding has no output terminals. Physical damper bars in d or q axis can be modeled as multiple damper windings to consider several different time constants of the damper circuit due to the skin effect of the damper bars. However, in this chapter, two damper windings, which are namely d-axis damper and q-axis damper winding, are only considered. If it is needed, the multiple damper windings—some of which are in the d axis while others are in the q axis—can be considered. In conclusion, there are six windings in the synchronous machines in Fig. 3.5, which are, namely, a-phase winding, b-phase winding, c-phase winding in the

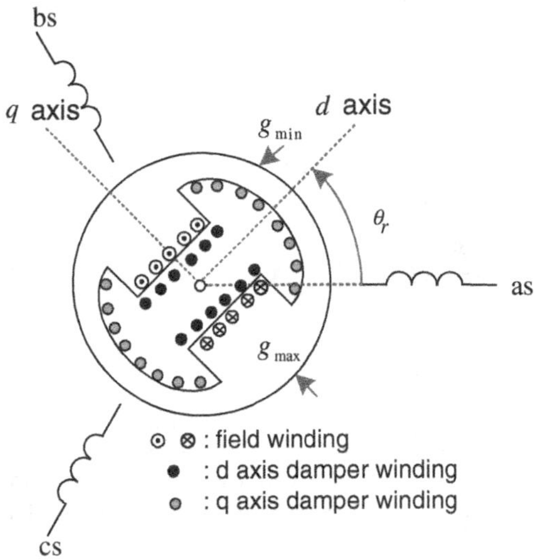

Figure 3.5 Modeling of a three phase synchronous machine.

stator, field winding and d axis damper winding which are in d axis of the rotor, and q-axis damper winding in the q axis of the rotor.

The voltage equation for a-, b-, and c-phase winding is in (3.55) represented in vector form.

$$\mathbf{V}_{abcs} = R_s\mathbf{I}_{abcs} + \frac{d}{dt}\boldsymbol{\lambda}_{abcs} \tag{3.55}$$

where R_s is the resistance of one of three phase stator winding.

Similarly, the voltage equation for rotor winding—namely field winding, d-axis damper winding, and q-axis damper winding—is in

$$\mathbf{V}_{fdqk} = \mathbf{R}_{fdqk}\mathbf{I}_{fdqk} + \frac{d}{dt}\boldsymbol{\lambda}_{fdqk} \tag{3.56}$$

In the above equations, we have

$$\mathbf{V}_{abcs} = \left[V_{as}\, V_{bs}\, V_{cs}\right]^T \tag{3.57}$$

$$\mathbf{I}_{abcs} = \left[i_{as}\, i_{bs}\, i_{cs}\right]^T \tag{3.58}$$

$$\lambda_{abcs} = \left[\lambda_{as}\, \lambda_{bs}\, \lambda_{cs}\right]^T \tag{3.59}$$

$$\mathbf{V}_{fdqk} = \left[V_{fk}\, V_{dk}\, V_{qk}\right]^T \tag{3.60}$$

$$\mathbf{I}_{fdqk} = \left[i_{fk}\, i_{dk}\, i_{qk}\right]^T \tag{3.61}$$

$$\lambda_{fdqk} = \left[\lambda_{fk}\, \lambda_{dk}\, \lambda_{qk}\right]^T \tag{3.62}$$

$$\mathbf{R}_{fdqk} = \begin{bmatrix} R_{fk} & 0 & 0 \\ 0 & R_{dk} & 0 \\ 0 & 0 & R_{qk} \end{bmatrix} \tag{3.63}$$

where V_{fk}, V_{dk}, and V_{qk} are the voltage to the field winding, d-axis damper winding, and q-axis damper winding, respectively. Because the damper windings are short-circuited, we have $V_{dk} = V_{qk} = 0$. And i_{fk}, i_{dk}, and i_{qk} are the current to the field winding, d-axis damper winding, and q-axis damper winding, respectively. If the speed of the rotor is exactly same as the rotating MMF by the stator windings, then there is no current in the damper windings as with the rotor current of the squirrel cage induction machine when slip $= 0$. In this case, $i_{dk} = i_{qk} = 0$. Also, R_{fk}, R_{dk}, and R_{qk} represent the winding resistance of the field winding, d-axis damper winding, and q-axis damper winding, respectively.

The flux linkages for the stator and rotor winding can be expressed as (3.64) in matrix form.

$$\begin{bmatrix} \boldsymbol{\lambda}_{abcs} \\ \boldsymbol{\lambda}_{fdqk} \end{bmatrix} = \begin{bmatrix} \mathbf{L}_s & \mathbf{L}_{sr} \\ (\mathbf{L}_{sr})^T & \mathbf{L}_r \end{bmatrix} \begin{bmatrix} \mathbf{I}_{abcs} \\ \mathbf{I}_{fdqk} \end{bmatrix} \tag{3.64}$$

where

$$\mathbf{L}_s = \begin{bmatrix} L_{as,as} & L_{bs,as} & L_{cs,as} \\ L_{as,bs} & L_{bs,bs} & L_{cs,bs} \\ L_{as,cs} & L_{bs,cs} & L_{cs,cs} \end{bmatrix} \tag{3.65}$$

$$\mathbf{L}_r = \begin{bmatrix} L_{fk,fk} & L_{dk,fk} & L_{qk,fk} \\ L_{fk,dk} & L_{dk,dk} & L_{qk,dk} \\ L_{fk,qk} & L_{dk,qk} & L_{qk,qk} \end{bmatrix} \tag{3.66}$$

$$\mathbf{L}_{sr} = \begin{bmatrix} L_{fk,as} & L_{dk,as} & L_{qk,as} \\ L_{fk,bs} & L_{dk,bs} & L_{qk,bs} \\ L_{fk,cs} & L_{dk,cs} & L_{qk,cs} \end{bmatrix} \tag{3.67}$$

In (3.65)–(3.67), $L_{x,y}$ stands for the inductance between x and y windings. Also, because the circuit is reciprocal, $L_{x,y} = L_{y,x}$.

The self- and mutual inductance of each winding can be evaluated as follows.

In the case of a round rotor synchronous machine, the length of the air gap between stator and rotor is independent of the position of the rotor, θ_r under the assumption of infinitesimal winding thickness and no slot on the stator and rotor. But in the case of a salient rotor synchronous machine, the length of the air gap varies from the maximum value, $g_{\max}$, to minimum value, $g_{\min}$, according to the rotor position. If the air gap length varies, then the self-inductance of each stator winding and the mutual inductance between the stator windings also vary. If the fundamental component in the spatial flux distribution is only considered after neglecting the harmonic components, then the self-inductance of a-phase winding can be modeled as (3.68) [4, 10].

$$L_{as,as} = L_{ls} + L_A + L_B \cos 2\theta_r \tag{3.68}$$

where θ_r is the angle from the axis of MMF by a-phase stator winding to the d axis of the rotor reference frame, which is defined as the position of the rotor. In (3.68), L_{ls} stands for the leakage inductance of the stator winding, L_A stands for the inductance independent on the rotation of the rotor, and L_B is the maximum value of the inductance varying with the rotation. L_A and L_B can be expressed as

$$L_A = \frac{K}{2} \cdot N_s^2 \left(\frac{1}{g_{\min}} + \frac{1}{g_{\max}} \right) \tag{3.69}$$

$$L_B = \frac{K}{2} \cdot N_s^2 \left(\frac{1}{g_{\min}} - \frac{1}{g_{\max}} \right) \tag{3.70}$$

where N_s stands for the effective number of turns of a stator (armature) winding, and K stands for a proportional constant set by $K = \mu_0 r l \cdot \frac{\pi}{4}$. Here, μ_0 is the permeability, r, is the radius of the rotor, and l is the effective axial stacking length of the rotor.

Hence, the maximum value of $L_{as,as}$ is $L_{ls}+K\cdot N_s^2\frac{1}{g_{\min}}$ at $\theta_r=0$, and the minimum value is $L_{ls}+K\cdot N_s^2\frac{1}{g_{\max}}$ at $\theta_r=\frac{\pi}{2}$.

Similarly, the self-inductance of the b and c phases can be represented as (3.71) and (3.72), respectively:

$$L_{bs,bs}=L_{ls}+L_A+L_B\cos\left(2\theta_r+\frac{2\pi}{3}\right) \tag{3.71}$$

$$L_{cs,cs}=L_{ls}+L_A+L_B\cos\left(2\theta_r-\frac{2\pi}{3}\right) \tag{3.72}$$

Also, the mutual inductance between stator windings can be represented as

$$L_{as,bs}=-\frac{1}{2}L_A+L_B\cos\left(2\theta_r-\frac{2\pi}{3}\right) \tag{3.73}$$

$$L_{as,cs}=-\frac{1}{2}L_A+L_B\cos\left(2\theta_r+\frac{2\pi}{3}\right) \tag{3.74}$$

$$L_{bs,cs}=-\frac{1}{2}L_A+L_B\cos 2\theta_r \tag{3.75}$$

The mutual inductance between the stator winding and the field winding can be represented as

$$L_{as,fk}=L_{sfk}\cos\theta_r \tag{3.76}$$

$$L_{bs,fk}=L_{sfk}\cos\left(\theta_r-\frac{2\pi}{3}\right) \tag{3.77}$$

$$L_{cs,fk}=L_{sfk}\cos\left(\theta_r+\frac{2\pi}{3}\right) \tag{3.78}$$

where L_{sfk} is the maximum value of the mutual inductance between the stator winding and the field winding and it can be represented as

$$L_{sfk}=KN_sN_{fk}\frac{1}{g_{\min}} \tag{3.79}$$

where N_{fk} is an effective number of turn of the field winding.

Similarly, the mutual inductance between the stator winding and the damper winding on the d axis can be represented as

$$L_{as,dk}=L_{sdk}\cos\theta_r \tag{3.80}$$

$$L_{bs,dk}=L_{sdk}\cos\left(\theta_r-\frac{2\pi}{3}\right) \tag{3.81}$$

$$L_{cs,dk} = L_{sdk} \cos\left(\theta_r + \frac{2\pi}{3}\right) \tag{3.82}$$

where L_{sdk} is the maximum value of the mutual inductance between the stator winding and the damper winding on the d axis and it can be represented as

$$L_{sdk} = KN_sN_{dk}\frac{1}{g_{\min}} \tag{3.83}$$

Here, N_{dk} is an effective number of turn of the damper winding.

Also, the mutual inductance between the stator winding and the damper winding on the q axis can be represented as

$$L_{as,qk} = -L_{sqk} \sin\theta_r \tag{3.84}$$

$$L_{bs,qk} = -L_{sqk} \sin\left(\theta_r - \frac{2\pi}{3}\right) \tag{3.85}$$

$$L_{cs,qk} = -L_{sqk} \sin\left(\theta_r + \frac{2\pi}{3}\right) \tag{3.86}$$

where L_{sqk} is the maximum value of the mutual inductance between the stator winding and the damper winding on the q axis and it can be represented as

$$L_{sqk} = KN_sN_{qk}\frac{1}{g_{\max}} \tag{3.87}$$

where N_{qk} is an effective number of turns of the damper winding on the q axis.

The inductances of rotor winding can be evaluated similarly as derived in (3.88)–(3.93). Because the d and q axes are perpendicular, the mutual inductance between the d-axis damper winding and the q-axis damper winding is null. And because of the same reason, the mutual inductance between field winding and the q-axis damper winding is also null.

$$L_{fk,fk} = KN_{fd}^2\frac{1}{g_{\min}} \tag{3.88}$$

$$L_{fk,dk} = KN_{dk}N_{fk}\frac{1}{g_{\min}} \tag{3.89}$$

$$L_{dk,dk} = KN_{dk}^2\frac{1}{g_{\min}} \tag{3.90}$$

$$L_{qk,qk} = KN_{qk}^2\frac{1}{g_{\max}} \tag{3.91}$$

$$L_{dk,qk} = 0 \tag{3.92}$$

$$L_{fk,qk} = 0 \tag{3.93}$$

Based on the inductances of each winding, the stator flux linkage can be represented as

$$\boldsymbol{\lambda}_{abcs} = \begin{bmatrix} L_{ls}+L_A+L_B\cos 2\theta_r & -\frac{1}{2}L_A+L_B\cos\left(2\theta_r-\frac{2\pi}{3}\right) & -\frac{1}{2}L_A+L_B\cos\left(2\theta_r+\frac{2\pi}{3}\right) \\ -\frac{1}{2}L_A+L_B\cos\left(2\theta_r-\frac{2\pi}{3}\right) & L_{ls}+L_A+L_B\cos\left(2\theta_r+\frac{2\pi}{3}\right) & -\frac{1}{2}L_A+L_B\cos 2\theta_r \\ -\frac{1}{2}L_A+L_B\cos\left(2\theta_r+\frac{2\pi}{3}\right) & -\frac{1}{2}L_A+L_B\cos 2\theta_r & L_{ls}+L_A+L_B\cos\left(2\theta_r-\frac{2\pi}{3}\right) \end{bmatrix}\mathbf{I}_{abcs}$$

$$+\begin{bmatrix} L_{sfk}\cos\theta_r & L_{sdk}\cos\theta_r & -L_{sqk}\sin\theta_r \\ L_{sfk}\cos\left(\theta_r-\frac{2\pi}{3}\right) & L_{sdk}\cos\left(\theta_r-\frac{2\pi}{3}\right) & -L_{sqk}\sin\left(\theta_r-\frac{2\pi}{3}\right) \\ L_{sfk}\cos\left(\theta_r+\frac{2\pi}{3}\right) & L_{sdk}\cos\left(\theta_r+\frac{2\pi}{3}\right) & -L_{sqk}\sin\left(\theta_r+\frac{2\pi}{3}\right) \end{bmatrix}\mathbf{I}_{fdqk} \tag{3.94}$$

Similarly, the rotor flux linkage can be represented as

$$\boldsymbol{\lambda}_{fdqk} = \begin{bmatrix} L_{sfk}\cos\theta_r & L_{sfk}\cos\left(\theta_r-\frac{2\pi}{3}\right) & L_{sfk}\cos\left(\theta_r+\frac{2\pi}{3}\right) \\ L_{sdk}\cos\theta_r & L_{sdk}\cos\left(\theta_r-\frac{2\pi}{3}\right) & L_{sdk}\cos\left(\theta_r+\frac{2\pi}{3}\right) \\ -L_{sqk}\sin\theta_r & -L_{sqk}\sin\left(\theta_r-\frac{2\pi}{3}\right) & -L_{sqk}\sin\left(\theta_r+\frac{2\pi}{3}\right) \end{bmatrix}\mathbf{I}_{abcs}$$

$$+\begin{bmatrix} L_{fk,fk} & L_{fk,dk} & 0 \\ L_{fk,dk} & L_{dk,dk} & 0 \\ 0 & 0 & L_{qk,qk} \end{bmatrix}\mathbf{I}_{fdqk} \tag{3.95}$$

By similar derivation process which was used to derive (3.35) from (3.34), the stator flux linkage can be deduced as (3.96) in terms of complex space vector variables and rotor currents.

$$\begin{aligned} \lambda_{abcs} &= \frac{2}{3}\left(\lambda_{as}+\mathbf{a}\lambda_{bs}+\mathbf{a}^2\lambda_{cs}\right) \\ &= \left(L_{ls}+\frac{3}{2}L_A\right)\mathbf{i}_{abcs}e^{j2\theta_r} \\ &\quad + L_{sfk}e^{j\theta_r}i_{fk}+L_{sdk}e^{j\theta_r}i_{dk}+L_{sqk}e^{j\left(\theta_r+\frac{\pi}{2}\right)}i_{qk} \end{aligned} \tag{3.96}$$

where $\mathbf{i}_{abcs}=\frac{2}{3}\left(i_{as}+\mathbf{a}i_{bs}+\mathbf{a}^2 i_{cs}\right)$.

Finally, the stator voltage equation with the complex space vector variables can be written as

$$\mathbf{V}_{abcs} = \frac{2}{3}(V_{as} + \mathbf{a}V_{bs} + \mathbf{a}^2 V_{cs}) = R_s \mathbf{i}_{abcs} + \frac{d}{dt}\boldsymbol{\lambda}_{abcs} \tag{3.97}$$

Equation (3.97) can be transformed to the equation expressed by the space vector variables on rotor reference frame, which is synchronously rotating to the rotor as follows:

After multiplying $e^{-j\theta_r}$ to both sides of (3.97), like

$$\mathbf{V}_{abcs}e^{-j\theta_r} = R_s \mathbf{i}_{abcs}e^{-j\theta_r} + e^{-j\theta_r}\frac{d}{dt}\boldsymbol{\lambda}_{abcs} \tag{3.98}$$

$$\boldsymbol{\lambda}_{abcs}e^{-j\theta_r} = \left(L_{ls} + \frac{3}{2}L_A\right)\mathbf{i}_{abcs}e^{-j\theta_r} + \frac{3}{2}L_B \mathbf{i}^*_{abcs}e^{j\theta_r} + L_{sfk}i_{fk} + L_{sdk}i_{dk} + L_{sqk}e^{j\frac{\pi}{2}}i_{qk} \tag{3.99}$$

$$\mathbf{V}^r_{dqs} = V^r_{ds} + jV^r_{qs} = \mathbf{V}_{abcs}e^{-j\theta_r} \tag{3.100}$$

By using (3.99), (3.100), and $\boldsymbol{\lambda}_{abcs} = \boldsymbol{\lambda}^r_{dqs}e^{j\theta_r}$, the voltage equation expressed by the space vector variables on rotor reference frame can be represented as

$$\begin{aligned}\mathbf{V}^r_{dqs} &= R_s \mathbf{i}^r_{dqs} + e^{-j\theta_r}\frac{d}{dt}\left(\boldsymbol{\lambda}^r_{dqs}e^{j\theta_r}\right)\\ &= R_s \mathbf{i}^r_{dqs} + e^{-j\theta_r}\cdot e^{j\theta_r}\cdot\frac{d}{dt}\boldsymbol{\lambda}^r_{dqs} + e^{-j\theta_r}\cdot\boldsymbol{\lambda}^r_{dqs}j\omega_r\cdot e^{j\theta_r}\\ &= R_s \mathbf{i}^r_{dqs} + \frac{d}{dt}\boldsymbol{\lambda}^r_{dqs} + (j\omega_r)\boldsymbol{\lambda}^r_{dqs}\end{aligned} \tag{3.101}$$

where ω_r is the angular speed of the rotor, which is the differentiation of the rotor angle, θ_r, with regard to the time.

The stator flux linkage equation in (3.99) can be rewritten in terms of the stator current, the damper current, and the field current in the space vector form as

$$\begin{aligned}\boldsymbol{\lambda}^r_{dqs} &= \left(L_{ls} + \frac{3}{2}L_A\right)\mathbf{i}^r_{dqs} + \frac{3}{2}L_B \mathbf{i}^{r*}_{dqs} + L_{sfk}i_{fk} + L_{sdk}i_{dk} + jL_{sqk}i_{qk}\\ &= \left(L_{ls} + \frac{L_{md}+L_{mq}}{2}\right)\mathbf{i}^r_{dqs} + \left(\frac{L_{md}-L_{mq}}{2}\right)\mathbf{i}^{r*}_{dqs} + L_{sfk}i_{fk} + L_{sdk}i_{dk} + jL_{sqk}i_{qk}\end{aligned} \tag{3.102}$$

where the *d–q* axis inductances, L_{md} and L_{mq}, are defined as (3.103) and (3.104), respectively:

$$L_{md} \equiv \frac{3}{2}(L_A + L_B) = \frac{3}{2}K\cdot N_s^2\frac{1}{g_{\min}} \tag{3.103}$$

$$L_{mq} \equiv \frac{3}{2}(L_A - L_B) = \frac{3}{2}K \cdot N_s^2 \frac{1}{g_{\max}} \tag{3.104}$$

In (3.102), $\mathbf{i}_{dqs}^{r*}$ means conjugate of $\mathbf{i}_{dqs}^{r}$.

The effective number of turns of each winding is very difficult to identify. Also, the currents in the damper windings cannot be measured because of its caged bar structure. Hence, without the loss of generality, the number of turns ratio between the damper winding and the stator winding can be set as 1.5 as (3.105) to consider the effects of two phases of the rotor, which has two windings, namely the *d*-axis damper and the *q*-axis damper, against the stator, which has three phase windings.

$$N_{dk} = N_{qk} = \frac{3}{2}N_s \tag{3.105}$$

Under the assumption of (3.105), the mutual inductances between rotor windings and the stator winding and self-inductance of rotor windings can be represented as

$$L_{sfk} = KN_sN_{fk}\frac{1}{g_{\min}} = \frac{2}{3}L_{md}\frac{N_{fk}}{N_s} \tag{3.106}$$

$$L_{sdk} = KN_sN_{dk}\frac{1}{g_{\min}} = L_{md} \tag{3.107}$$

$$L_{sqk} = KN_sN_{qk}\frac{1}{g_{\max}} = L_{mq} \tag{3.108}$$

$$L_{fk,fk} = KN_{fk}^2\frac{1}{g_{\min}} = \frac{2}{3}L_{md}\frac{N_{fk}^2}{N_s^2} \tag{3.109}$$

$$L_{fk,dk} = KN_{dk}N_{fk}\frac{1}{g_{\min}} = L_{md}\frac{N_{fk}}{N_s} \tag{3.110}$$

$$L_{dk,dk} = KN_{dk}^2\frac{1}{g_{\min}} = \frac{3}{2}L_{md} \tag{3.111}$$

$$L_{qk,qk} = KN_{qk}^2\frac{1}{g_{\max}} = \frac{3}{2}L_{mq} \tag{3.112}$$

Also, the stator flux linkage in terms of the space vector variables at the rotor reference frame with *d*–*q*-axis inductances can be expressed as (3.113) using (3.106)–(3.108).

$$\boldsymbol{\lambda}_{dqs}^{r} = \left(L_{ls} + \frac{L_{md}+L_{mq}}{2}\right)\mathbf{i}_{dqs}^{r} + \left(\frac{L_{md}-L_{mq}}{2}\right)\mathbf{i}_{dqs}^{r*} + L_{md}(i_{fk}' + i_{dk}) + jL_{mq}i_{qk} \tag{3.113}$$

where i_{fk}' is the field current referred to the stator side by considering the corresponding effective turn ratios such as (3.114).

$$i_{fk}' = \frac{2}{3}\frac{N_{fk}}{N_s}i_{fk} \tag{3.114}$$

Similarly, from (3.95) the flux linkage of rotor windings can be represented as (3.115)–(3.117), respectively, by using (3.109)–(3.112).

$$\lambda_{fk} = \left(L_{lfk} + \frac{2}{3}L_{md}\frac{N_{fk}^2}{N_s^2}\right)i_{fk} + L_{md}\frac{N_{fk}}{N_s}i_{dk} + \frac{2}{3}L_{md}\frac{N_{fk}}{N_s}\cdot\frac{3}{4}\left(\mathbf{i}_{dqs}^r + \mathbf{i}_{dqs}^{r*}\right) \tag{3.115}$$

$$\lambda_{dk} = \left(L_{ldk} + \frac{3}{2}L_{md}\right)i_{dk} + L_{md}\frac{N_{fk}}{N_s}i_{fk} + L_{md}\frac{3}{4}\left(\mathbf{i}_{dqs}^r + \mathbf{i}_{dqs}^{r*}\right) \tag{3.116}$$

$$\lambda_{qk} = \left(L_{lqk} + \frac{3}{2}L_{mq}\right)i_{qk} + jL_{mq}\cdot\frac{3}{4}\left(\mathbf{i}_{dqs}^r + \mathbf{i}_{dqs}^{r*}\right) \tag{3.117}$$

Because the current of the damper windings cannot be measured and the voltage to the damper windings are always null, the flux linkage, the leakage inductance, and the winding resistance of the damper windings can be adjusted as (3.118)–(3.123) to simplify voltage equations of the rotor windings.

$$\lambda'_{dk} = \tfrac{2}{3}\lambda_{dk} \tag{3.118}$$

$$\lambda'_{qk} = \tfrac{2}{3}\lambda_{qk} \tag{3.119}$$

$$L'_{ldk} = \tfrac{2}{3}L_{ldk} \tag{3.120}$$

$$L'_{lqk} = \tfrac{2}{3}L_{lqk} \tag{3.121}$$

$$R'_{dk} = \tfrac{2}{3}R_{dk} \tag{3.122}$$

$$R'_{qk} = \tfrac{2}{3}R_{qk} \tag{3.123}$$

Also, the flux linkage, the leakage inductance, and resistance of the field winding can be adjusted considering turn ratio between the stator and field winding as

$$\lambda'_{fk} = \frac{N_s}{N_{fk}}\lambda_{fk} \tag{3.124}$$

$$L'_{lfk} = \frac{3}{2}L_{lfk}\left(\frac{N_s}{N_{fk}}\right)^2 \tag{3.125}$$

$$R'_{fk} = \frac{3}{2}R_{fk}\left(\frac{N_s}{N_{fk}}\right)^2 \tag{3.126}$$

With the above adjustments, the flux linkage of the rotor windings can be represented as

$$\lambda'_{fk} = \left(L'_{lfk} + L_{md}\right)i'_{fk} + L_{md}i_{dk} + \frac{1}{2}L_{md}\left(\mathbf{i}_{dqs}^r + \left(\mathbf{i}_{dqs}^r\right)^*\right) \tag{3.127}$$

$$\lambda'_{dk} = \left(L'_{ldk} + L_{md}\right)i_{dk} + L_{md}i'_{fk} + \frac{1}{2}L_{md}\left(\mathbf{i}_{dqs}^r + \left(\mathbf{i}_{dqs}^r\right)^*\right) \tag{3.128}$$

$$\lambda'_{qk} = \left(L'_{lqk} + L_{mq}\right) i_{qk} + \frac{j}{2} L_{mq} \left(\left(\mathbf{i}^r_{dqs} \right)^* - \mathbf{i}^r_{dqs} \right) \tag{3.129}$$

Also, the voltage and the flux linkages can be expressed as (3.130)–(3.139) in terms of the stator and rotor currents, and the parameters of the synchronous machine can refer to the stator side.

$$V^r_{ds} = R_s i^r_{ds} + \frac{d}{dt}\lambda^r_{ds} - \omega_r \lambda^r_{qs} \tag{3.130}$$

$$V^r_{qs} = R_s i^r_{qs} + \frac{d}{dt}\lambda^r_{qs} + \omega_r \lambda^r_{ds} \tag{3.131}$$

$$V'_{fk} = R'_{fk} i'_{fk} + \frac{d}{dt}\lambda'_{fk} \tag{3.132}$$

$$V'_{dk} = R'_{dk} i_{dk} + \frac{d}{dt}\lambda'_{dk} \tag{3.133}$$

$$V'_{qk} = R'_{qk} i_{qk} + \frac{d}{dt}\lambda'_{qk} \tag{3.134}$$

where

$$\lambda^r_{ds} = L_{ls} i^r_{ds} + L_{md}\left(i^r_{ds} + i'_{fk} + i_{dk}\right) \tag{3.135}$$

$$\lambda^r_{qs} = L_{ls} i^r_{qs} + L_{mq}\left(i^r_{qs} + i_{qk}\right) \tag{3.136}$$

$$\lambda'_{fk} = L'_{lfk} i'_{fk} + L_{md}\left(i'_{fk} + i_{dk} + i^r_{ds}\right) \tag{3.137}$$

$$\lambda'_{dk} = L'_{ldk} i_{dk} + L_{md}\left(i'_{fk} + i_{dk} + i^r_{ds}\right) \tag{3.138}$$

$$\lambda'_{qk} = L'_{lqk} i_{qk} + L_{mq}\left(i_{qk} + i^r_{qs}\right) \tag{3.139}$$

$$V'_{fk} = \frac{N_s}{N_{fk}} V_{fk} \tag{3.140}$$

Based on the above equations, the equivalent circuit of the synchronous machine referred to the stator side at the rotor reference frame can be depicted as Fig. 3.6. Because the damper windings are usually shorted, $V'_{dk} = V'_{qk} = 0$.

Like the induction machine, the current and voltages on the circuit on the *n* axis do not contribute to the electromechanical energy conversion. Furthermore, in the case of the synchronous machine, because the rotor circuit consists of *d–q* windings, there is no *n*-axis component in the rotor inherently. Again, like the induction machine, if the stator winding impedance is the same in all three phases and the flux linkage of each winding is well-balanced, then there is no *n*-component current in the stator side either. Hence, there is no need to consider an *n*-axis equivalent circuit in this case. The current and voltage to the field winding can be measured in the actual

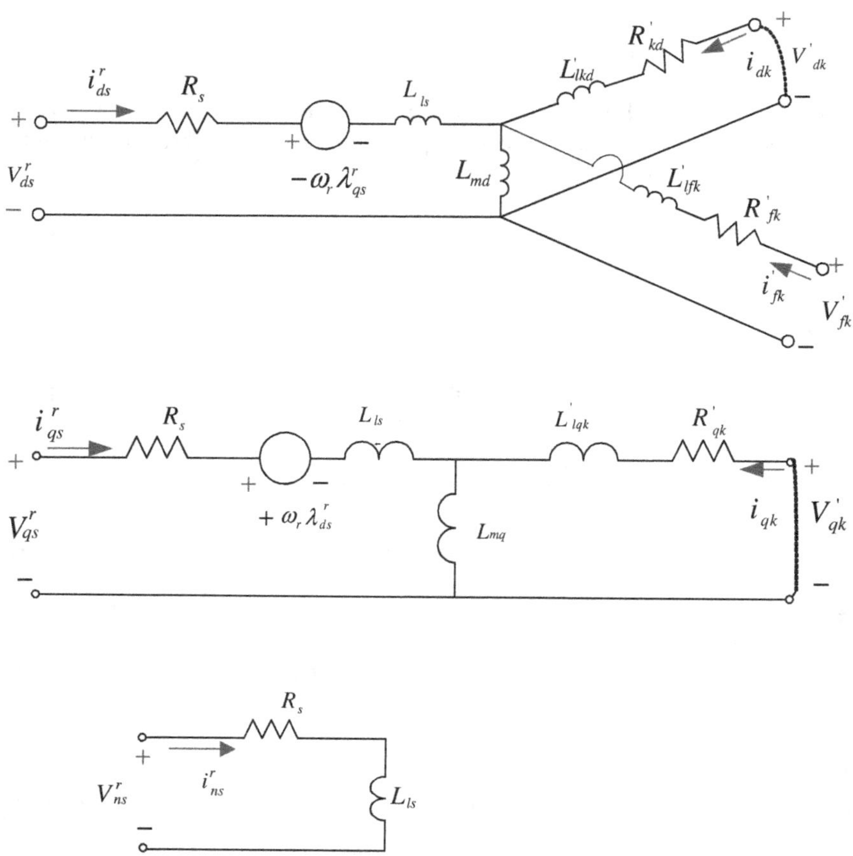

Figure 3.6 Equivalent circuit of synchronous machine at rotor reference *d–q–n* axis.

synchronous machine. To compare the measured values with the field current, i'_{fk}, and the field voltage, V'_{fk}, the turn ratio and the adjustment factors in (3.114) and (3.140) should be considered.

3.3.2 Torque of a Synchronous Machine

The torque of a synchronous machine can be evaluated by the power balance like the case of the induction machine. Under the assumption that $i^r_{ns} = 0$, the input power to the machine can be expressed as

$$P_{in} = \frac{3}{2}\left(V^r_{ds}i^r_{ds} + V^r_{qs}i^r_{qs} + V'_{fk}i'_{fk} + V'_{dk}i_{dk} + V'_{qk}i_{qk}\right) \tag{3.141}$$

In (3.30)–(3.134), the flux linkage of each winding expressed in (3.135)–(3.139) can be substituted and the voltage equation of the each winding can be represented as (3.142)–(3.146) in terms of the current of each winding and the parameters of the synchronous machine. These voltage equations can be substituted into (3.141). Also, the input power can be represented in the terms of stator and rotor current as (3.147).

$$V_{ds}^r = R_s i_{ds}^r - \omega_r \lambda_{qs}^r + p\left\{L_{ls} i_{ds}^r + L_{md}\left(i_{ds}^r + i_{dk} + i_{fk}'\right)\right\} \tag{3.142}$$

$$V_{qs}^r = R_s i_{qs}^r + \omega_r \lambda_{ds}^r + p\left\{L_{ls} i_{qs}^r + L_{mq}\left(i_{qs}^r + i_{kq}\right)\right\} \tag{3.143}$$

$$V_{fk}' = R_{fk}' i_{fk}' + p\left\{L_{lfk}' i_{fk}' + L_{md}\left(i_{ds}^r + i_{dk} + i_{fk}'\right)\right\} \tag{3.144}$$

$$V_{dk}' = R_{dk}' i_{dk} + p\left\{L_{ldk}' i_{dk} + L_{md}\left(i_{ds}^r + i_{dk} + i_{fk}'\right)\right\} \tag{3.145}$$

$$V_{qk}' = R_{qk}' i_{qk} + p\left\{L_{lqk}' i_{qk} + L_{mq}\left(i_{qs}^r + i_{qk}\right)\right\} \tag{3.146}$$

$$\begin{aligned} P_{in} = \frac{3}{2}\Bigg[& R_s\left(i_{ds}^{r2} + i_{qs}^{r2}\right) + R_{fk}' i_{fk}'^2 + R_{dk}' i_{dk}^2 + R_{qk}' i_{qk}^2 \\ & + p\Bigg\{\frac{1}{2}L_{ls}\left(i_{ds}^{r2} + i_{qs}^{r2}\right) + \frac{1}{2}L_{lfk}' i_{fk}'^2 + \frac{1}{2}L_{ldk}' i_{dk}^2 + \frac{1}{2}L_{lqk}' i_{qk}^2 \\ & + L_{md}\left(\frac{i_{ds}^{r2}}{2} + \frac{i_{fk}'^2}{2} + \frac{i_{dk}'^2}{2} + i_{ds}^r i_{dk} + i_{ds}^r i_{fk}' + i_{fk}' i_{dk}\right) \\ & + L_{mq}\left(\frac{i_{qs}^{r2}}{2} + \frac{i_{qk}^2}{2} + i_{qs}^r i_{qk}\right)\Bigg\} + \omega_r\left\{\lambda_{ds}^r \cdot i_{qs}^r - \lambda_{qs}^r \cdot i_{ds}^r\right\}\Bigg] \end{aligned} \tag{3.147}$$

Like the case of the induction machine in Section 3.2.2, the first part of the right-hand side of (3.147), $R_s\left(i_{ds}^{r2} + i_{qs}^{r2}\right) + R_{fk}' i_{fk}'^2 + R_{dk}' i_{dk}^2 + R_{qk}' i_{qk}^2$, represents the copper losses by the stator and rotor windings.

The next part,

$$\begin{aligned} & p\Bigg\{\frac{1}{2}L_{ls}\left(i_{ds}^{r2} + i_{qs}^{r2}\right) + \frac{1}{2}L_{lfk}' i_{fk}'^2 + \frac{1}{2}L_{ldk}' i_{dk}^2 + \frac{1}{2}L_{lqk}' i_{qk}^2 \\ & + L_{md}\left(\frac{i_{ds}^{r2}}{2} + \frac{i_{fk}'^2}{2} + \frac{i_{dk}'^2}{2} + i_{ds}^r i_{dk} + i_{ds}^r i_{fk}' + i_{fk}' i_{dk}\right) + L_{mq}\left(\frac{i_{qs}^{r2}}{2} + \frac{i_{qk}^2}{2} + i_{qs}^r i_{qk}\right)\Bigg\} \end{aligned}$$

is the variation of the magnetic energy stored in the inductors, which has nothing to do with the electromechanical energy conversion. And the last term, $\omega_r\left\{\lambda_{ds}^r \cdot i_{qs}^r - \lambda_{qs}^r \cdot i_{ds}^r\right\}$, is expected to be the mechanical output:

$$P_m = \frac{3}{2}\omega_r\left\{\lambda_{ds}^r i_{qs}^r - \lambda_{qs}^r i_{ds}^r\right\} \tag{3.148}$$

Then the torque can be calculated as $T_e = P_m / \left\{ \frac{\omega_r}{P/2} \right\}$ from (3.148), where P is the number of poles.

The flux linkages in (3.148) can be rewritten as

$$\lambda_{ds}^r = L_{md}\left(i_{ds}^r + i_{dk} + i_{fk}^{\prime}\right) + L_{ls} i_{ds}^r \tag{3.149}$$

$$\lambda_{qs}^r = L_{mq}\left(i_{qs}^r + i_{qk}\right) + L_{ls} i_{qs}^r \tag{3.150}$$

Also, the torque can be expressed as (3.151) in terms of the currents in the windings.

$$\begin{aligned} T_e &= \frac{3}{2}\frac{P}{2}\left[\left\{L_{ls} i_{ds}^r + L_{md}\left(i_{ds}^r + i_{fk}^{\prime} + i_{dk}\right)\right\} i_{qs}^r - \left\{L_{ls} i_{qs}^r + L_{mq}\left(i_{qs}^r + i_{qk}\right)\right\} i_{ds}^r\right] \\ &= \frac{3}{2}\frac{P}{2}\left[\left(L_{md} - L_{mq}\right) i_{ds}^r i_{qs}^r + L_{md} i_{fk}^{\prime} i_{qs}^r + L_{md} i_{dk} i_{qs}^r - L_{mq} i_{qk} i_{ds}^r\right] \end{aligned} \tag{3.151}$$

In the torque equation (3.151), the first term is the torque from the saliency of the rotor, which is the reluctance torque, the second term is the torque from the field current, which is the field torque, and the last two terms represent the torque from the damper winding currents, the damping torque. If the instantaneous speed of the rotating MMF, ω_e, and that of the rotor, ω_r, is the same as $\omega_e(t) = \omega_r(t)$, then there is no current in the damper windings as $i_{dk} = i_{qk} = 0$, and there is no damping torque. Otherwise, the damping torque occurs to damp out the speed difference between the rotating MMF by the stator winding and MMF by the rotor winding, whose rotating speed is the speed of the rotor. In the case of round rotor synchronous machine, because there is no saliency in the rotor like $L_{md} = L_{mq}$, the reluctance torque is not available. Based on the modeling of the synchronous machine in this section, the modeling of the machine that has multiple damper windings can be easily derived. Also, in the case of the permanent magnet synchronous machine, by omitting or simplifying some parts of the equivalent circuit in Fig. 3.6 and some terms in (3.151), the equivalent circuit and the torque equation of the permanent magnet synchronous machine can be easily derived as explained in Section 3.3.3.

3.3.3 Equivalent Circuit and Torque of a Permanent Magnet Synchronous Machine [11–13]

3.3.3.1 Surface-Mounted Permanent Magnet Synchronous Machine (SMPMSM)

The SMPMSM has the magnet on the rotor, whose cross-sectional view is shown in Fig. 3.7, and it is usually designed to generate sinusoidal back EMF. And there are no field winding and no damper windings either at the rotor. And the field flux is constant because of the permanent magnet. The effective air gap is larger compared to the

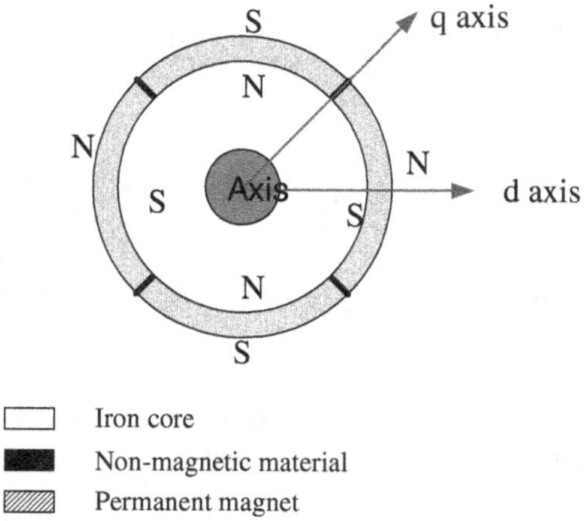

Figure 3.7 Cross-sectional view of the rotor of a four-pole surface-mounted permanent magnet synchronous motor (SMPMSM).

wound rotor synchronous machine because the relative permeability of the magnet is almost unity like air. By using these observations, the equivalent circuit in Fig. 3.6 can be modified as Fig. 3.8 by omitting damper winding circuits and by replacing the field winding circuit by a constant current source. However, there is no modification in the n-axis circuit, and it is not repeated in Fig. 3.8.

The mutual inductance on d–q axes are the same as $L_{md} = L_{mq} = L_m$, and the value is much smaller compared to the value of a wound rotor synchronous machine due to the increased effective air gap [13]. The typical impedance of the mutual inductance of a general-purpose SMPMSM is usually less than 0.2 per unit, such as $X_m = \omega_e L_m \leq 0.2$ per unit.

The field flux, λ_f, by the magnet can be expressed as $\lambda_f = i_f \cdot L_m$ based on the constant current, i_f, from the current source. Hence, the torque equation (3.151) can be

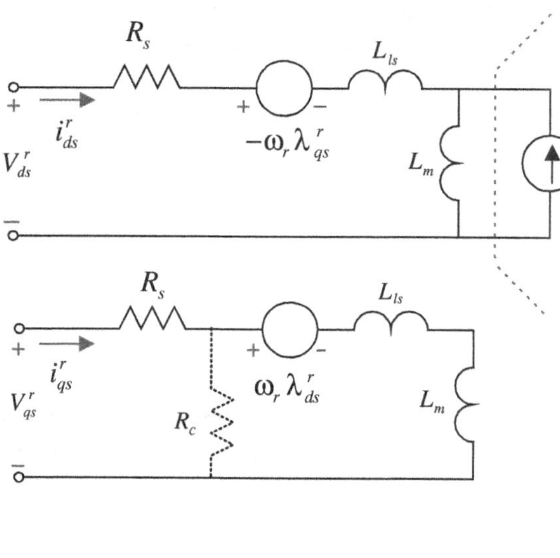

Figure 3.8 Equivalent circuit of a surface-mounted permanent magnet synchronous machine at rotor reference d–q axis.

modified as (3.152), where there is no reluctance torque and damping torque. The stator voltage equations in rotor reference frame are in (3.153) and (3.154).

$$T_e = \frac{3}{2}\frac{P}{2} L_m i_f \cdot i^r_{qs} = \frac{3}{2}\frac{P}{2} \lambda_f i^r_{qs} \tag{3.152}$$

$$V^r_{ds} = R_s i^r_{ds} - \omega_r \lambda^r_{qs} + p\left(L_s i^r_{ds} + \lambda_f\right) \tag{3.153}$$

$$V^r_{qs} = R_s i^r_{qs} + \omega_r \lambda^r_{ds} + p\left(L_s i^r_{qs}\right) \tag{3.154}$$

where $L_s = L_{ls} + L_m$ and $\lambda^r_{qs} = L_s i^r_{qs}$, $\lambda^r_{ds} = L_s i^r_{ds} + \lambda_f$.

It can be seen from (3.152) that the torque of SMPMSM is directly proportional to q-axis current, i^r_{qs}, because of the constant λ_f by the permanent magnet. Hence, the torque can be instantaneously controlled by adjusting i^r_{qs}. In the case of a general-purpose SMPMSM the magnitude of the current, i_f, is usually much larger than that of the rated current of the machine, and the flux weakening control by negative d-axis current, i^r_{ds}, is very limited. Recently, the neodymium–iron–boron magnet is widely used because of its higher energy density as the permanent magnet. However, the neodymium–iron–boron magnet is conductive itself, and an eddy current loss occurs through the magnet in the presence of the variation of flux linkage to the magnet. In a small machine, less than a few hundred watts, the eddy current loss may be negligible, but in the higher-power machine the eddy current loss, especially at higher operating speed, cannot be ignored. The eddy current loss can be modeled as R_c in the circuit in Fig. 3.8. The value of R_c would vary according to the operating speed.

3.3.3.2 Interior Permanent Magnet Synchronous Machine (IPMSM)

The IPMSM has the magnet in the rotor, whose cross-sectional view is in Fig. 3.9, and it is also designed to generate sinusoidal back EMF like SMPMSM. And there are no field windings and no damper windings like SMPMSM. But due to the machining process of the rotor surface, regardless of the lamination of the silicon steel sheet, there may be some conductive path in the surface of the rotor. Hence, some damping torque may appear from the conductive path, but the time constants given by

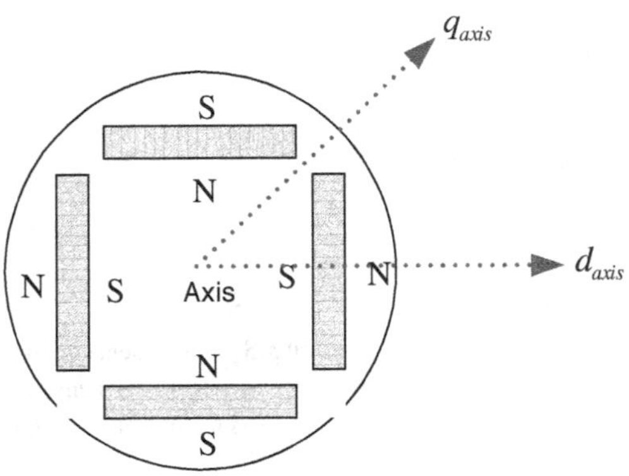

Figure 3.9 Cross-sectional view of the rotor of a four-pole interior permanent magnet machine (IPMSM).

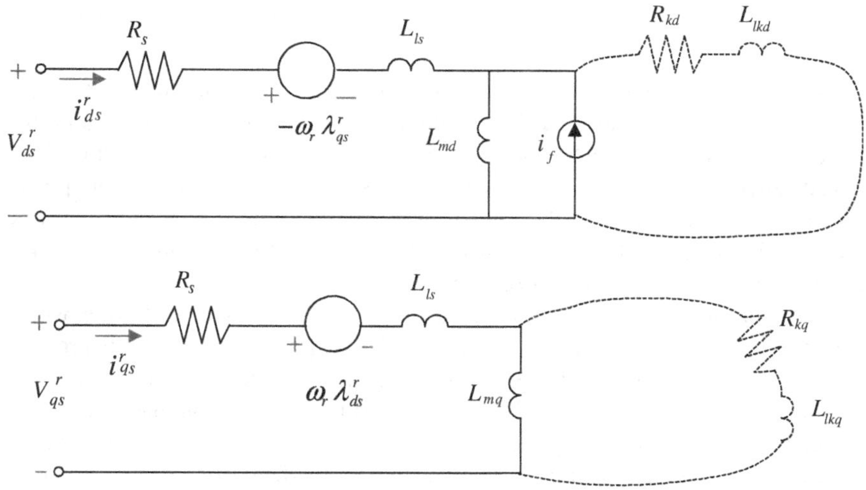

Figure 3.10 Equivalent circuit of an interior permanent magnet synchronous machine at rotor reference *d–q* axis. (Dotted parts are for consideration of pseudo-damper circuits by the conductivity of rotor outer surface.)

$\frac{L_{ldk}+L_{md}}{R_{dk}}$, $\frac{L_{lqk}+L_{mq}}{R_{qk}}$ in Fig. 3.10 are quite small because of the large resistance of the path and the current in the path is small and disappears rapidly. Also like an SMPMSM, the neodymium–iron–boron magnet of an IMPSM is conductive itself, and an eddy current loss occurs through the magnet in the presence of the variation of flux linkage to the magnet. But unlike an SMPMSM, the variation of the flux linkage to the magnet is smaller due to varied structure of the magnet. However, though the eddy current problem of an IPMSM is less than that of an SMPMSM, the problem is still there and the eddy current loss should be considered especially in the case of a high-speed and high-power IPMSM.

As mentioned above, because the time constant of the damping circuit is too small, the damping torque can be neglected in the dynamic performance evaluation of an IPMSM in most cases. The equivalent circuit in Fig. 3.10 is similar to the salient wound rotor synchronous machine. However, unlike the salient wound rotor synchronous machine, it should be noted that the mutual inductance on the *d* axis is smaller than that on the *q* axis because the relative permeability of the magnet is unity like air and the reluctance at the *d* axis is larger than that of the *q* axis. The torque of an IPMSM can be described in (3.155), and the stator voltage equations are (3.156) and (3.157).

$$T_e = \frac{3}{2}\frac{P}{2}\left\{\left(L_{md}-L_{mq}\right)i^r_{ds}i^r_{qs} + \lambda_f i^r_{qs}\right\} = \frac{3}{2}\frac{P}{2}\left\{\left(L_d-L_q\right)i^r_{ds}i^r_{qs} + \lambda_f i^r_{qs}\right\} \tag{3.155}$$

$$V^r_{ds} = R_s i^r_{ds} - \omega_r \lambda^r_{qs} + p\left(L_d i^r_{ds} + \lambda_f\right) \tag{3.156}$$

$$V^r_{qs} = R_s i^r_{qs} + \omega_r \lambda^r_{ds} + p\left(L_q i^r_{qs}\right) \tag{3.157}$$

where $\lambda_f = i_f \cdot L_{md}$, $L_d = L_{ls} + L_{md}$, $L_q = L_{ls} + L_{mq}$, $\lambda^r_{qs} = L_q i^r_{qs}$, and $\lambda^r_{ds} = L_d i^r_{ds} + \lambda_f$.

Like the conventional salient rotor synchronous machine, the reluctance torque contributes to the torque of an IPMSM, but unlike the wound rotor salient machine, the positive reluctance torque can be obtained with negative d-axis current under positive q-axis current because $L_d < L_q$. With the proper design of an IPMSM, constant power speed range (CPSR), which is the region where the constant output power can be maintained by flux weakening control, can be extended to infinity under the assumption of no limitation of speed from the mechanical reasons. In addition, by negative d-axis current, i^r_{ds}, for flux weakening, the positive reluctance torque, which is the contrast to the case of the conventional wound rotor salient synchronous machine, can be obtained and the torque capability of the machine in the flux weakening region can be enhanced [14]. In contrast to the induction machine, where the air-gap flux variation is delayed by the rotor time constant, L_r/R_r, an IPMSM has no rotor time constant and the flux is decided directly by the stator current i^r_{ds} and i^r_{qs}, and λ_f by the magnet without any time delay. Hence, highly dynamic flux weakening control is possible. However, as mentioned in Section 2.9.2, L_{mq} decreases as the magnitude of i^r_{qs} increases because of the magnetic saturation of the core. Also, L_{mq} varies according to i^r_{ds} too—that is, the so-called cross-saturation. And the control of the torque is getting difficult as the magnitude of the current increases because of these nonlinearities of the parameters of an IPMSM. The detailed description of the torque control of an IPMSM is given in Section 5.1.3.

3.3.4 Synchronous Reluctance Machine (SynRM) [15,16]

The synchronous reluctance machine (SynRM) operates only based on the reluctance difference of the rotor, and there is no permanent magnet and no winding on the rotor. The structure of the stator is almost same to other machines, but that of the rotor is quite simple. The cross-sectional view of a typical a four pole synchronous reluctance machine is shown in Fig. 3.11. The equivalent circuit can be depicted as Fig. 3.12 by modifying the circuit in Fig. 3.6. The typical power range of a SynRM is less than several tens of kilowatts, and the saliency ratio defined as L_{md}/L_{mq} can reach up to around 10. With this saliency ratio the torque density of a SynRM can be comparable to the torque density of the induction machine. Because of the large reactive power

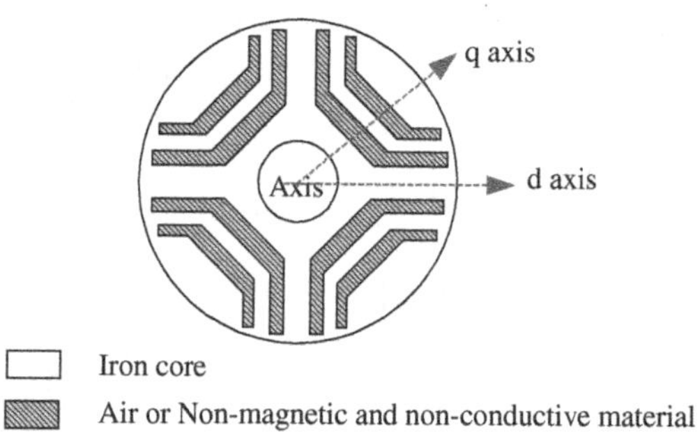

Figure 3.11 Cross-sectional view of the rotor of a four-pole synchronous reluctance machine (SynRM).

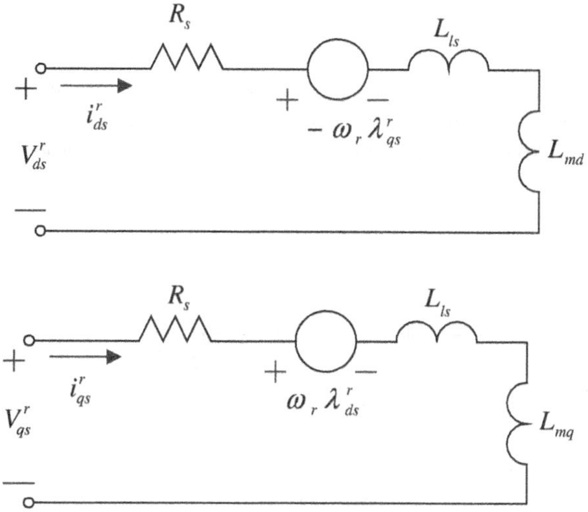

Figure 3.12 Equivalent circuit of a synchronous reluctance machine at rotor reference *d–q* axis.

due to L_{mq}, the power factor at the machine terminal is lower than that at any other machines. Also, the size of the power converter to drive a SynRM should be larger than that of the converter to drive other machines.

From the equivalent circuit and (3.151), the torque of a SynRM can be represented as

$$T_e = \frac{3}{2}\frac{P}{2}\left\{\left(L_{md}-L_{mq}\right)i^r_{ds}i^r_{qs}\right\} = \frac{3}{2}\frac{P}{2}\left\{\left(L_d-L_q\right)i^r_{ds}i^r_{qs}\right\} \tag{3.158}$$

Because of no magnet, there is no field torque unlike the case of an IPMSM.

The voltage equations, given in (3.159) and (3.160), are similar to that of an IPMSM, but there are no terms with λ_f because of no magnet in the rotor.

$$V^r_{ds} = R_s i^r_{ds} - \omega_r \lambda^r_{qs} + p(L_d i^r_{ds}) \tag{3.159}$$

$$V^r_{qs} = R_s i^r_{qs} + \omega_r \lambda^r_{ds} + p\left(L_q i^r_{qs}\right) \tag{3.160}$$

where $L_d = L_{ls} + L_{md}$, $L_q = L_{ls} + L_{mq}$, $\lambda^r_{qs} = L_q i^r_{qs}$, and $\lambda^r_{ds} = L_d i^r_{ds}$.

In this machine, because the air gap flux is solely decided by the stator current in the rotor reference current, CPSR has not been limited by electrical constraints but is only limited by mechanical reasons such as centrifugal forces and bearing problems. And flux weakening control performance is superior to any other machines [16]. But, like an IPMSM, the *d* inductance, L_{md}, varies by several hundred percentages according to the magnitude of i^r_{ds}, and it also varies with i^r_{qs}. And, the control task is complicated. Moreover, the saliency ratio decreases rapidly as the magnitude of the current increases due to the magnetic saturation, and torque density drops down severely in a higher torque region. Because the air gap must be small enough to get a higher saliency ratio, the higher-power machine such as one over several hundred kilowatts is inherently difficult to design.

PROBLEMS

1. Complete equivalent circuit shown in Fig. P3.1, which refers to the primary side, of a three-phase transformer at rotating *d–q–n* reference axes with an arbitrary speed, ω. The transformer windings and connection to the three-phase source and load are shown in Fig. P 3.2.

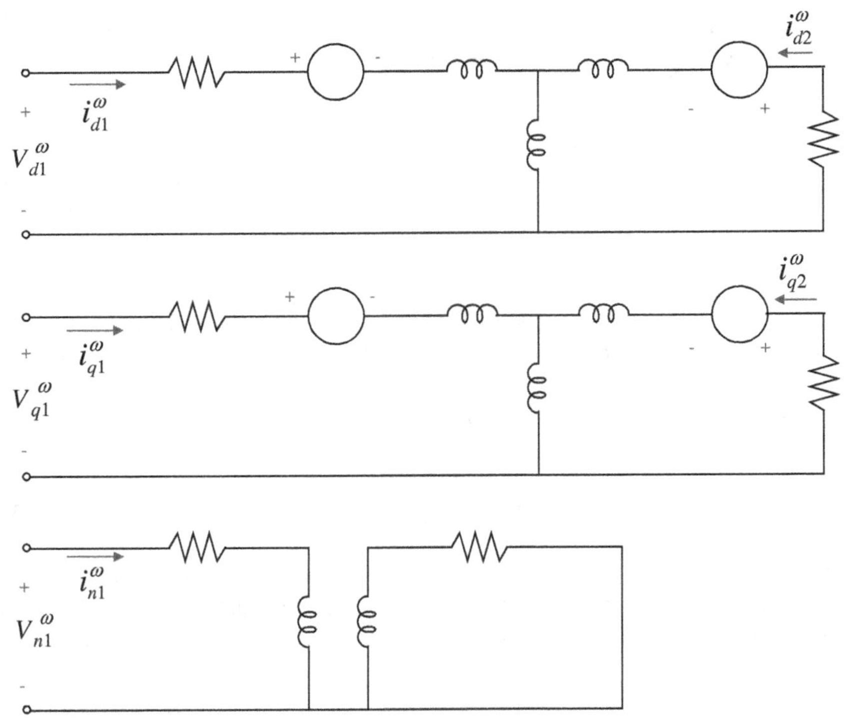

Figure P3.1 *d–q–n* equivalent circuit of a three-phase transformer.

The windings of the transformer are balanced and the parameters of the transformer are as follows.

Number of turns of each primary winding: N_1

Number of turns of each secondary winding: N_2

Resistance of each primary winding: R_1

Resistance of each secondary winding: R_2

Leakage inductance of each primary winding: L_{l1}

Leakage inductance of each secondary winding: L_{l2}

Mutual inductance between AA′ winding and aa′ winding: L_m

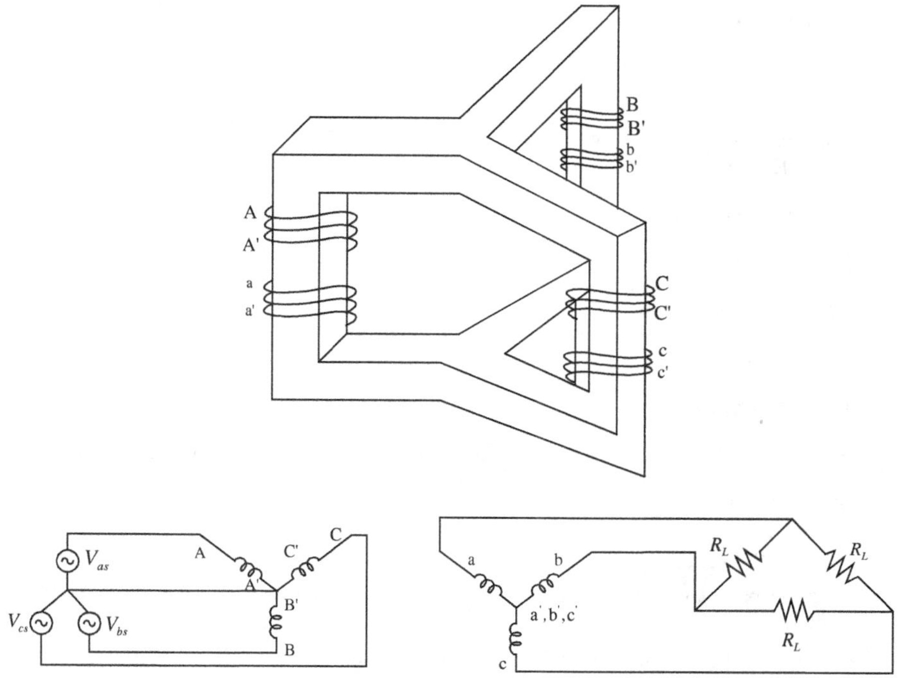

Figure P3.2 Windings and connection of a three-phase transformer.

2. From stationary reference frame d–q–n equivalent circuit of a balanced three-phase induction machine, which is the circuit in Fig. 3.3 with $\omega = 0$, deduce the steady state per phase equivalent shown in Fig. 2.45.

3. Simulate the starting characteristics of the induction machines, whose parameters are as follows, by using digital computer. At $t = 0$ s, a-phase voltage is crossing zero with a positive slope. The balanced three-phase voltage with the rated voltage and rated frequency of each induction machine is applied to the machine. The load torque and the inertia of load can be neglected. For each machine, plot a-phase voltage (V), a-phase current (A), torque of the machine (N-m), and speed (r/min) in the vertical axis simultaneously until the starting of the machine finished and the machine is in the steady state. Here, the steady state means less than 3% variation of the magnitude of the current between consecutive cycles. In the plot, horizontal axis is time from zero until the end of the starting. Also, for a starting interval, the plot torque–speed curve in the torque–speed plane, where the vertical axis is torque (N-m) and the horizontal axis is speed (r/min) [7].

Induction Machine 1

Rated power: 3 Hp, Rated voltage in line-to-line rms: 220 V, Rated speed: 1710 r/min
Rated frequency: 60 Hz, Number of poles: 4, Inertia of the machine: 0.089 kg-m^2
$R_s = 0.435\,\Omega$, $R_r = 0.816\,\Omega$
$X_{ls} = 0.754\,\Omega$, $X_{lr} = 0.754\,\Omega$
$X_m = 26.13\,\Omega$

Induction Machine 2

Rated power: 22 kW, Rated voltage in line-to-line rms: 220 V, Rated speed: 1765 r/min
Rated frequency: 60 Hz, Number of poles: 4, Inertia of the machine: 0.122 kg-m^2
$R_s = 0.044\,\Omega$, $R_r = 0.0252\,\Omega$
$L_{ls} = 0.55$ mH, $L_{lr} = 0.47$ mH
$L_m = 12.90$ mH

Induction Machine 3

Rated power: 185 kW, Rated voltage in line-to-line rms, 440 V
Rated frequency: 60 Hz, Number of poles: 4, Inertia of the machine: 4 kg-m^2
$R_s = 0.013\,\Omega$, $R_r = 0.075\,\Omega$
$L_{ls} = 0.1$ mH, $L_{lr} = 0.1$ mH
$L_m = 12$ mH

Induction Machine 4

Rated power: 2250 Hp, Rated voltage in line-to-line rms: 2300 V, Rated speed: 1786 r/min
Rated frequency: 60 Hz, Number of poles: 4, Inertia of the machine: 63.87 kg-m^2
$R_s = 0.029\,\Omega$, $R_r = 0.022\,\Omega$
$X_{ls} = 0.226\,\Omega$, $X_{lr} = 0.226\,\Omega$
$X_m = 13.04\,\Omega$

4. Using the parameters of the induction machine 2 in problem 3, answer the following questions.

(1) During the starting interval found from problem 3, plot V^s_{ds}, V^s_{qs} (V), i^s_{ds}, i^s_{qs} (A), $\lambda^s_{ds}, \lambda^s_{qs}, \lambda^s_{dr}, \lambda^s_{qr}$ (Wb-t) regarding time, t. Plot torque (N-m) and input power (W) using rotor flux, $\lambda^s_{dr}, \lambda^s_{qr}$, and stator current, i^s_{ds}, i^s_{qs}, in the stationary reference d–q frame regarding time, t.

(2) During the starting interval, plot V^e_{ds}, V^e_{qs} (V), i^e_{ds}, i^e_{qs} (A), $\lambda^e_{ds}, \lambda^e_{qs}, \lambda^e_{dr}, \lambda^e_{qr}$ (Wb-t) regarding time, t. Plot torque (N-m) and input power (W) using rotor flux, $\lambda^e_{dr}, \lambda^e_{qr}$, and stator current, i^e_{ds}, i^e_{qs}, in the synchronous reference d–q frame regarding time, t.

(3) During the starting interval, plot V^r_{ds}, V^r_{qs} (V), i^r_{ds}, i^r_{qs} (A), $\lambda^r_{ds}, \lambda^r_{qs}, \lambda^r_{dr}, \lambda^r_{qr}$ (Wb-t) regarding time, t. Plot torque (N-m) and input power (W) using rotor flux, $\lambda^r_{dr}, \lambda^r_{qr}$, and stator current, i^r_{ds}, i^r_{qs}, in the rotor reference d–q frame regarding time, t.

(4) During the starting interval, plot $V^\omega_{ds}, V^\omega_{qs}$ (V), $i^\omega_{ds}, i^\omega_{qs}$ (A), $\lambda^\omega_{ds}, \lambda^\omega_{qs}, \lambda^\omega_{dr}, \lambda^\omega_{qr}$ (Wb-t) regarding time, t. Plot torque (N-m) and input power (W) using rotor flux, $\lambda^\omega_{dr}, \lambda^\omega_{qr}$, and stator current, $i^\omega_{ds}, i^\omega_{qs}$, in the rotating reference d–q frame, whose speed is ω and given as follows, regarding time, t. In the starting interval, during the first 1/3 of the interval the speed of the d–q axis, ω equals zero, during next 1/3 we have $\omega = \omega_e$, and during the last 1/3 we have $\omega = -\omega_e$.

5. As shown in Fig. P3.3, there is a synchronous reluctance machine, whose rotor shape is ellipse. And there are no windings on the rotor. The stator windings are the balanced three-phase windings. The self- and mutual inductances of each stator winding are as follows.

$$L_{as,as} = L_{ls} + L_A + L_B \cos 2\theta_r$$

$$L_{bs,bs} = L_{ls} + L_A + L_B \cos 2\left(\theta_r - \frac{2\pi}{3}\right)$$

$$L_{cs,cs} = L_{ls} + L_A + L_B \cos 2\left(\theta_r + \frac{2\pi}{3}\right)$$

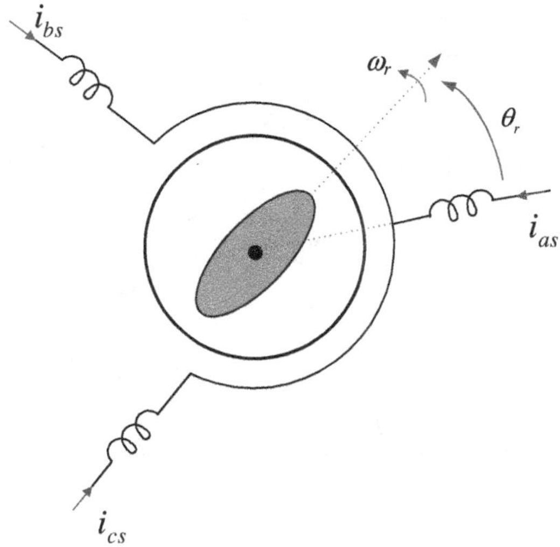

Figure P3.3 Synchronous reluctance machine (SynRM).

$$L_{as,bs} = -\frac{1}{2}L_A + L_B\cos 2\left(\theta_r - \frac{\pi}{3}\right)$$

$$L_{as,cs} = -\frac{1}{2}L_A + L_B\cos 2\left(\theta_r + \frac{\pi}{3}\right)$$

$$L_{bs,cs} = -\frac{1}{2}L_A + L_B\cos 2\theta_r$$

(1) If the stator flux linkage is represented as

$$\begin{bmatrix}\lambda_{as}\\ \lambda_{bs}\\ \lambda_{cs}\end{bmatrix} = \mathbf{L}\begin{bmatrix}i_{as}\\ i_{bs}\\ i_{cs}\end{bmatrix}$$

find inductance matrix, **L**.

(2) Using $\mathbf{f}_{abcs} = \frac{2}{3}(f_{as} + \mathbf{a}f_{bs} + \mathbf{a}^2 f_{cs})$, show that

$$\mathbf{V}_{abcs} = R_s\mathbf{i}_{abcs} + \frac{d}{dt}\left[\left(L_{ls} + \frac{3}{2}L_A\right)\mathbf{i}_{abcs} + \frac{3}{2}L_B e^{j2\theta_r}\mathbf{i}^*_{abcs}\right]$$

where R_s stands for the winding resistance of each phase and $\mathbf{i}^*_{abcs}$ is the conjugate of $\mathbf{i}_{abcs}$.

(3) Derive the following voltage equation of the SynRM at a rotor reference d–q frame from the voltage equation in part 2, where rotor angle is given by $\theta_r = \omega_r t$.

$$\mathbf{V}^r_{dqs} = R_s\mathbf{i}^r_{dqs} + \left(L_{ls} + \tfrac{3}{2}L_A\right)(p + j\omega_r)\mathbf{i}^r_{dqs} + \tfrac{3}{2}L_B(p + j\omega_r)\mathbf{i}^{r*}_{dqs}$$

where p is a differential operator such as d/dt.

(4) After finding input power, and prove that the torque of the SynRM is given by $T_e = \frac{9}{2}\cdot L_B\left(i^r_{qs}\cdot i^r_{ds}\right)$.

6. The self-inductance of a stator winding of a salient electric machine, shown in Fig. P3.4, is given by $L(\theta) = (L_A + L_B \cos 2\theta_r)e^{-i}$. Answer the following questions [9].
   (1) Derive the field energy of the machine.
   (2) Derive the coenergy of the machine.
   (3) Derive the torque equation.

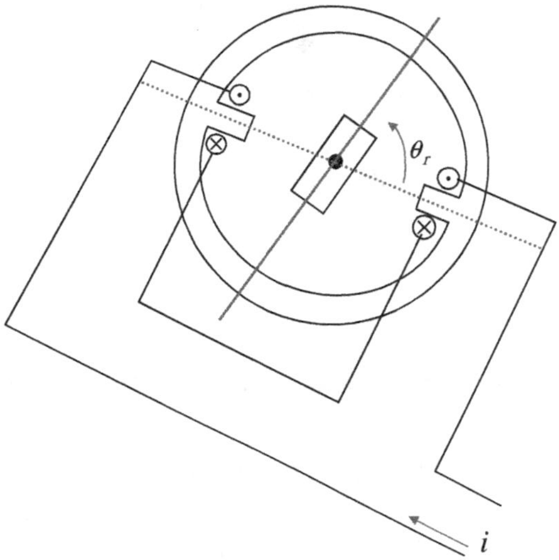

Figure P3.4 Salient electric machine.

7. As shown in Fig. P3.5, a-phase terminal of a three-phase induction machine is opened, and between other two terminals a constant current source, I_{dc}, is connected. When the machine is run by an external prime mover at constant speed, ω_{rm}(rad/s), in the mechanical angle, represent the following variables in terms of the machine parameters, R_s, R_r, L_m, L_{ls}, L_{lr}, number of poles, P in the steady state.
   (1) i^s_{ds}, i^s_{qs}
   (2) V^s_{ds}, V^s_{qs}
   (3) i^s_{dr}, i^s_{qr}
   (4) Torque of the machine

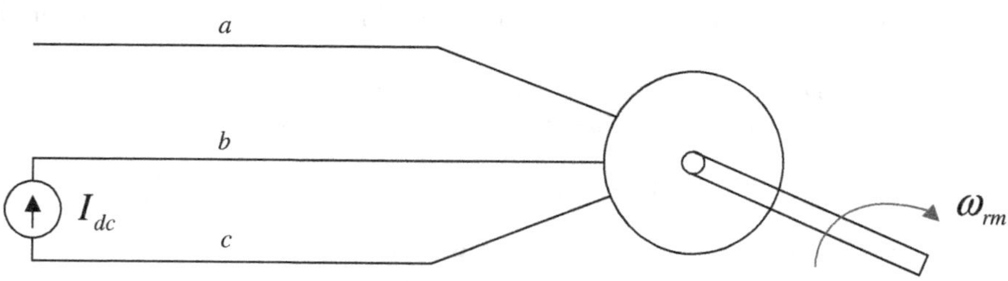

Figure P3.5 Single-phase excitation of a three-phase induction machine.

8. At a rotating reference d–q frame with arbitrary speed ω, the torque of the induction machine can be represented by the following equation.

$$T_e = \frac{3}{2}\frac{P}{2} L_m \mathrm{Im}\left[i_{dqs}^{\omega} \cdot i_{dqr}^{\omega^*}\right]$$

By using the above equation and the equivalent circuit represented with space vector variables in Fig. 3.3, derive the following torque equation.

$$T_e = \frac{3}{2}\frac{P}{2}\frac{L_m}{\left(1-\frac{L_m^2}{L_r L_s}\right)L_r L_s}\mathrm{Im}\left[\lambda_{dqs}^{\omega} \cdot \lambda_{dqr}^{\omega^*}\right]$$

where Im[•] means an imaginary part of [•].

9. Derive the reactive power to the machine at the stator terminal of an induction machine using rotor flux, $\boldsymbol{\lambda}_{dqr}^e = \lambda_{dr}^e + j\lambda_{qr}^e$, stator current, $\mathbf{i}_{dqs}^e = i_{ds}^e + ji_{qs}^e$, and parameters of the induction machine through the following procedures. Here, there is no neutral component voltage, and the impedances of the machine are balanced.

Hint: The reactive power to a three-phase system is expressed as $Q_{in} = -\frac{3}{2}\mathrm{Im}\left[\mathbf{V}_{abcs}\mathbf{i}_{abcs}^*\right]$, where $\mathbf{i}_{abcs}^*$ is the conjugate of $\mathbf{i}_{abcs}$.

(1) Express $\mathbf{V}_{abcs}, \mathbf{i}_{abcs}$ in terms of $\mathbf{V}_{dqs}^e, \mathbf{i}_{dqs}^e$.

(2) Express $\mathbf{V}_{dqs}^e$ in terms of $\boldsymbol{\lambda}_{dqr}^{\mathbf{e}}$, $\mathbf{i}_{dqs}^{\mathbf{e}}$ and differential operator, $p = d/dt$.

(3) Finally, express Q_{in} in terms of $i_{ds}^e, i_{qs}^e, \lambda_{dr}^e, \lambda_{qr}^e$ and differential operator, $p = d/dt$.

10. An SMPMSM, which has the following parameters, is driven by an inverter. At $t = 0$ s, it is decelerated from 2000 r/min with –1 N-m torque. If $i_{ds}^r = 0$ A and the friction and load torque is negligible, answer following questions. At $t = 0$ s, $\theta_r = 0$.

SMPMSM Parameters:

$L_m = 2$ mH, $L_{ls} = 0.35$ mH, $R_s = 0.65\ \Omega$, Rated Power: 200 W, 4-pole,
$\lambda_f = L_m i_f = 0.061$ Wb–t, Sum of machine and load inertia: $J_{M+L} = 1.29 \times 10^{-4}$ kg-m^2

(1) Represent i_{qs}^r(A) in terms of the machine parameters and time, t.

(2) Represent mechanical output of an SMPMSM, P_m(W), in terms of the machine parameters and time, t.

(3) Represent V_{qs}^r(V) in terms of the machine parameters and time, t.

(4) Represent electrical input to the machine, P_{in}(W), in terms of the machine parameters and time, t.

(5) Show that the summation of, P_m, and copper loss equal to the input power, P_{in}, instantaneously. Here, the copper loss means the power dissipated by the resistance of the windings.

11. A SynRM, whose parameters are listed below, is controlled as $i_{ds}^r = i_{qs}^r$. The machine is in the steady state with constant load torque, 15 N-m at constant speed, $\omega_{rm} = 1800$ r/min. At $t = 0$ s, the angle of the rotor expressed in mechanical angle, θ_{rm}, is zero. It is assumed that all parameters are constant regardless of operating conditions. Answer the following questions.

$R_s = 0.232\ \Omega$, $L_d = 32$ mH, $L_q = 2.54$ mH
Sum of inertias of load and motor itself: $J_{M+L} = 0.024$ kg-m^2
4-pole, Rated frequency: 60 Hz,
Rated voltage in line-to-line rms: 170 V,

Rated output power: 5 Hp
$L_d = L_{ls} + L_{md}$, $L_q = L_{ls} + L_{mq}$

(1) Represent i^r_{qs}(A) in terms of the machine parameters and time, t.
(2) Represent V^r_{ds}(V) in terms of the machine parameters and time, t.
(3) Represent V^r_{qs}(V) in terms of the machine parameters and time, t.
(4) Find the power factor at the machine terminal.

12. With an SMPMSM whose equivalent circuit is in Fig. P3.6, the terminals are shorted, which means $V^r_{ds} = 0$, $V^r_{qs} = 0$, and the rotor of the machine is driven by an external prime mover in constant speed, ω_{rm}(rad/s), in the mechanical angle. The machine and drive system are in the steady state.

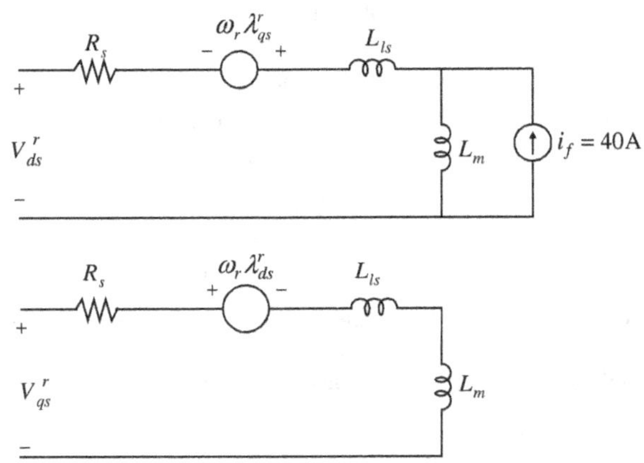

Figure P3.6 Equivalent d–q circuit of SMPMSM.

Find the torque of the prime mover to run the machine and the a-phase rms current to the machine in terms of ω_{rm} in rad/s.

$R_s = 0.22\,\Omega$
$L_m = 2\,\text{mH}$
$L_{ls} = 0.2\,\text{mH}$
8-pole
Rated phase voltage in rms: 87 V
Rated line current in rms: 8.3 A

13. An IPMSM, whose voltage equations are represented by (3.156) and (3.157), is rotating in the steady state. By using the following calculation, what parameter of the machine can be obtained?

$$\frac{\text{Re}(V^s_{dqs} \cdot \lambda^{s*}_{dqs})}{\text{Re}(i^s_{dqs} \cdot \lambda^{s*}_{dqs})}$$

where Re[•] means real part of [•].

14. Derive (3.113) from (3.96).

REFERENCES

1. I. Bolea et al., *Electric Drives,* CRC Press, Boca Raton, FL, 1998, Chapter 8.
2. P. Vas, *Electrical Machines and Drives,* Oxford University Press, Oxford, UK, 1992, Chapter 2.
3. W. Leonhard, *Control of Electrical Drives,* 2nd edition, Springer, Berlin, 1996, Chapter 10.
4. D. W. Novotny et al., *Vector Control and Dynamics of AC Drives,* Oxford University Press, Oxford, UK, 1996, Chapter 2.
5. P. C. Krause, *Analysis of Electric Machinery,* McGraw-Hill, New York, 1987, Chapter 3.
6. A. R. Bergen, *Power System Analysis,* Prentice-Hall, Englewood Cliffs, NJ, 1986, Chapter 10.
7. P. C. Krause, *Analysis of Electric Machinery,* McGraw-Hill, New York, 1987, Chapter 4.
8. H. H. Woodson, *Electromechanical Dynamics,* John Wiley & Sons, New York, 1968, Chapter 3.
9. P. C. Krause et al., *Electromechanical Motion Devices,* McGraw-Hill, New York, 1989, Chapter 2.
10. I. Bolea et al., *Electric Drives,* CRC Press, Boca Raton, FL, 1998, Chapter 10.
11. H. W. Beaty et al., *Electric Motor Handbook,* McGraw-Hill, New York, 1998, Chapter 6.
12. I. Bolea et al., *Vector Control of AC Drives,* CRC Press, Boca Raton, 1992, Chapter 7.
13. W. Leonhard, *Control of Electrical Drives,* 2nd edition, Springer, Berlin, 1996.
14. T. M. Jahns, Flux-weakening regime operation of an interior permanent-magnet motor drive, *IEEE Trans. Ind. Appl.*, Vol. IA-23, July/August 1996, pp. 681–689.
15. A. Vagati et al., *Synchronous Reluctance Motors and Drives—A New Alternative*, IEEE Industry Application Society 1994 Annual Meeting Tutorial Note Book, IEEE, New York, 1994.
16. S. J. Kang et al., *Highly dynamic torque control of synchronous reluctance motor, IEEE Trans. Power Elec.* Vol. 13, No. 4, July 1998, pp. 793–798.

Chapter 4

Design of Regulators for Electric Machines and Power Converters

To meet control objectives of electric machines and power converters through electrical and/or electromechanical energy conversion, the input and output variables of the electric machine and the power converter should be in the acceptable ranges, and the variables should be regulated instantaneously or in average sense. Usually, the electrical variables under control are current, voltage, and power. The flux linkage can be estimated based on the measured variables, and the flux linkage can also be regulated based on the estimated value. The mechanical variables controlled are usually torque, acceleration, speed, and position (angle). Through the regulation of these electric and mechanical variables, the electric machines and converters can accomplish the desired functions. In Fig. 4.1, a block diagram of a typical electric power control system and a typical electromechanical system are shown. In Fig. 4.1a, a system provides electric power to the load from a particular power source, where in some cases the load can supply power to the source. The source may be either DC or AC whose frequency might be variable. By measuring the voltage and current of input and/or output instantaneously, the control function can be accomplished. In Fig. 4.1b, a system controlling an electric machine through a power converter is shown, where the machine can operate as a motor or as a generator. Hence, the flow of the energy from the machine to the source via the converter may be bidirectional. If the converter cannot accept the regenerated power, then the power should be dissipated by other means such as a braking chopper and resistor. In the control of the electric machine, usually the current to the machine is measured as well as the voltage and current to the converter. In addition,

Control of Electric Machine Drive Systems, by Seung-Ki Sul

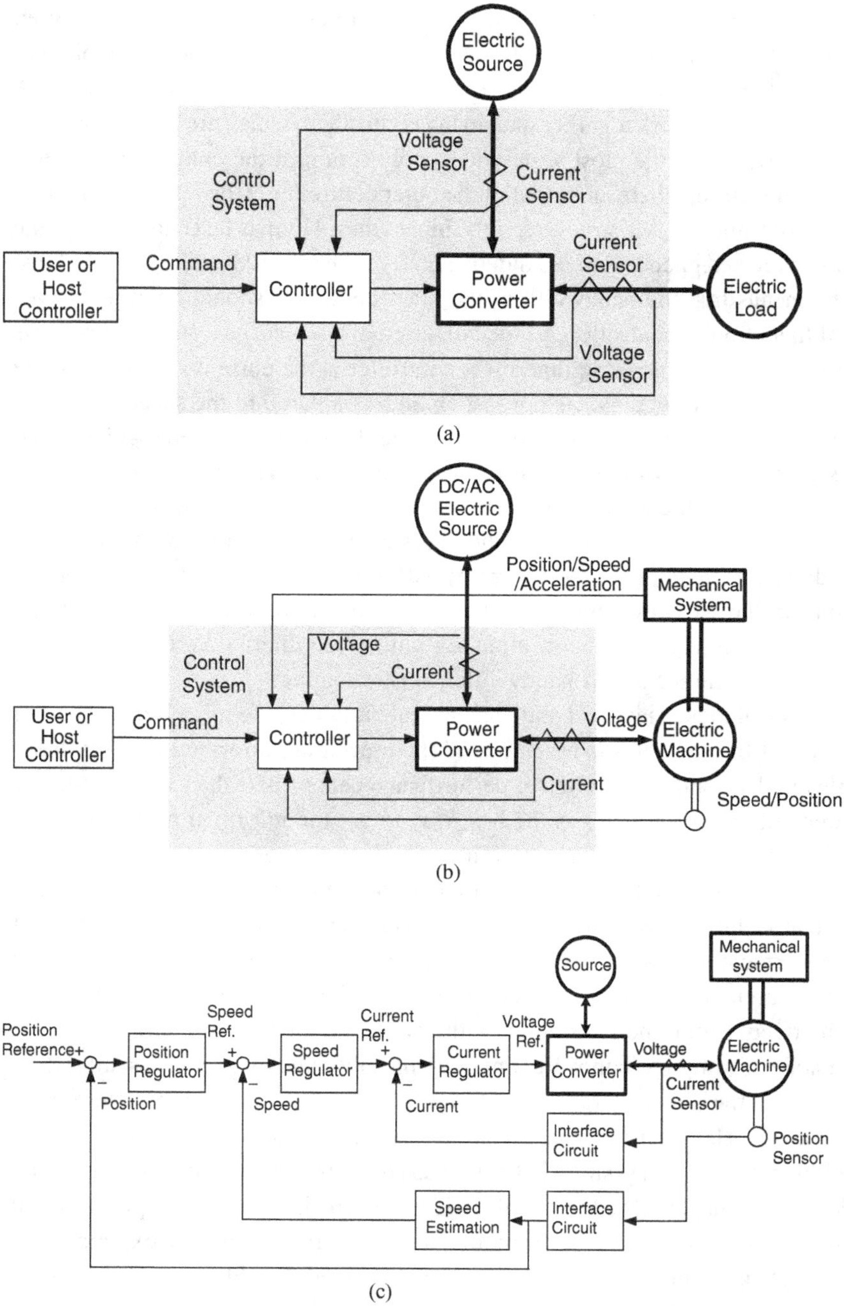

Figure 4.1 Block diagrams of typical control systems of electric machine and power converter. (a) Electric power control system. (b) Electric machine control system. (c) A typical cascaded control system.

the position, speed, and acceleration are measured or estimated from the measured variables. Also, the measured and estimated electrical and mechanical variables are fed back to the control loop to enhance the performance of the control system. A control system may work together with many controllers, which are interconnected. In an industrial field, the most widely used connection of the controllers is the cascade connection, where an output of the outer controller is the reference input to the inner controller. In this structure, if the inner control loop is fast enough, then the controllers are designed and tuned independently with other controllers. Here, "fast enough" means that the bandwidth of the inner controller should be at least five times of that of outer controller. A block diagram of a typical cascaded controller is shown in Fig. 4.1c, where the innermost controller is the current regulator. In the figure, if the current regulator is not fast enough compared to the speed regulator, there may be interference between the current regulator and the speed regulator, and the design and tuning of the regulators are complicated. The definition of the bandwidth, ω_{bw}, is the angular frequency of the input signal in rad/s where the magnitude of the response of the controlled system is less than the magnitude of the sinusoidally varying input of the system by 3 dB—that is, $1/\sqrt{2}$ of the magnitude of the input. In the measurement of bandwidth, the inner controllers should not be saturated. If there is saturation in input, output, or the internal variables of any controller, then the measured bandwidth has no meaning.

To control these electrical and mechanical variables, the measurement and/or estimation of the variables is essential. Without accurate information about the variables under control, the control performance cannot be enhanced. The direct measurement of the variables is the best way to get the information. However, in many cases the direct measurement of the variables is impossible or too expensive. In these cases, the variables can be estimated through the state equation of the dynamic model of the system under control. This kind of the estimation instead of the measurement may reveal robustness to the external noise even compared to the direct measurement. Also, in the measurement and/or the estimation, the linearity and time delay should be considered. Otherwise, the inherently stable system or system stabilized by the controllers may be unstable unexpectedly. Usually, the delay reduces the phase margin of the controlled system and let the system be unstable easily. Hence, in many cases the delay would be the main reason to limit the bandwidth of the control system. In the industrial fields, many control strategies and theories have been developed and applied. However, in this chapter, the proportional and integral (PI) controllers based on the classic control theory are explained, and some design guides of the controllers are described. In addition to the classical PI controller, the observers based on state equation of the dynamic system under control are exploited to enhance the control performance. Also, to improve the performance, the so-called "active damping," which is a kind of state feedback control, is incorporated in the control loop.

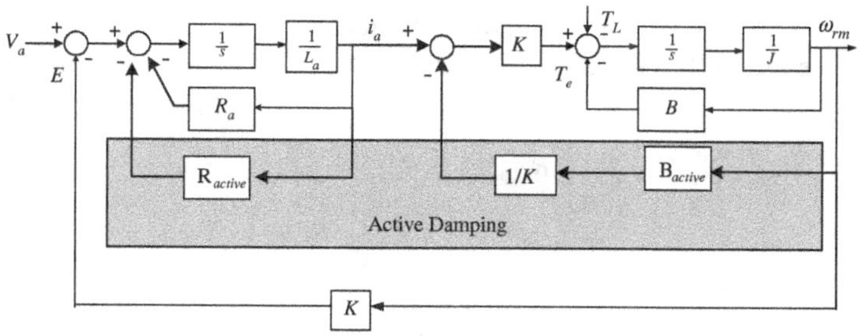

Figure 4.2 Control block diagram of separately excited DC machine with active damping.

4.1 ACTIVE DAMPING [1]

As mentioned in Section 2.3, in the control of a separately excited DC machine, by increasing the armature resistance and/or the friction coefficient, the damping of the system increases and the speed response to the voltage step input would be less oscillatory. However, the friction and resistance degrade the efficiency of the control system. In contrast to those passive damping components, by adding the active damping by feedback control as shown in Fig. 4.2, the oscillatory, underdamped responses can be changed to overdamped responses.

In Fig. 4.2, by adding R_{active} together with R_a, the armature resistance has been increased by R_{active}, and the eigenvalues of the system have been changed. Furthermore, if R_{active} is much larger thanR_a, the controller with the active damping would be robust to the variation of armature resistance, R_a. The same principles can be applied to the case of friction, and the system friction, B, can be manipulated by active damping, B_{active}. As seen from this example, based on the control block diagram and physical understanding of the system, the damping terms can be actively inserted using feedback control to enhance the dynamic performance of the control system. This kind of manipulation of the eigenvalues of the system by feedback is referred to as the state feedback control in the modern control theory.

In Fig. 4.2, the transfer function between armature voltage to armature current is in (4.1):

$$\frac{i_a}{V_a} = \frac{1}{sL_a + (R_a + R_{\text{active}})} \tag{4.1}$$

If R_{active} is set to be much larger than the magnitude of sL_a within the control bandwidth, ω_{bw}, then (4.1) can be approximated as

$$\frac{i_a}{V_a} \approx \frac{1}{R_a + R_{\text{active}}} \tag{4.2}$$

By using the fact that the torque of the separately excited DC machine is proportional to the armature current ($T_e = Ki_a$), the voltage, V_a, can be set as

$V_a = (T_e^*/K) \cdot (R_a + R_{\text{active}})$ to get the commanded torque, T_e^*, where K is the torque constant. With this setup, (4.2) can be rewritten as

$$\frac{i_a}{V_a} = \frac{i_a}{(T_e^*/K)(R_a + R_{\text{active}})} \approx \frac{1}{R_a + R_{\text{active}}} \tag{4.3}$$

Hence,

$$\frac{T_e}{T_e^*} = \frac{Ki_a}{T_e^*} \approx \frac{(R_a + R_{\text{active}})}{(R_a + R_{\text{active}})} = 1 \tag{4.4}$$

As seen from (4.4), if the active damping, R_{active}, is large enough, the torque can be controlled instantaneously within the control bandwidth. However, due to the time delay and parameter errors, there is limitation of the magnitude of R_{active} to keep the stability of the system. In real implementation of the active damping, the integral term is added to improve the control performance of torque. This kind of design of the active damping with the integral term is described in Section 4.2.3 as a current regulator.

4.2 CURRENT REGULATOR

The driving force, thrust, or torque, which comes from electromagnetic energy conversion, can be expressed as the function of the current flowing through the energy converter. Hence, the measurement of the current in the energy converter, which is usually electric machine or power converter based on power electronics, is essential to control the thrust or torque of the electric machine. Furthermore, if the machine is driven by a power converter, the accurate and instantaneous measurement of the current is also important not only for better control performance such as higher efficiency and wider control bandwidth but also for the protection of the machine and the power converter itself from the overload and the short circuit.

4.2.1 Measurement of Current

As mentioned before, to get better control performance of physical variables, the measurement of the physical variables should be better. In particular, the current regulation loop is usually the innermost loop, and its bandwidth is a key factor to increase the bandwidths of other outer control loops such speed and position loops. To measure the current flowing to the machine or to the power converter, the physical connection between sensors and the conductors where the current flows is inevitable whether the connection is direct contact, magnetic coupling, or optical coupling. Thus, the measurement may be sensitive to the measurement noise from high dv/dt or high di/dt of the conductor due to the switching of the power converter. Measuring the current immune to the noise without time delay is quite a difficult task. In this section, several methods of the current measurement used widely in the industry are introduced.

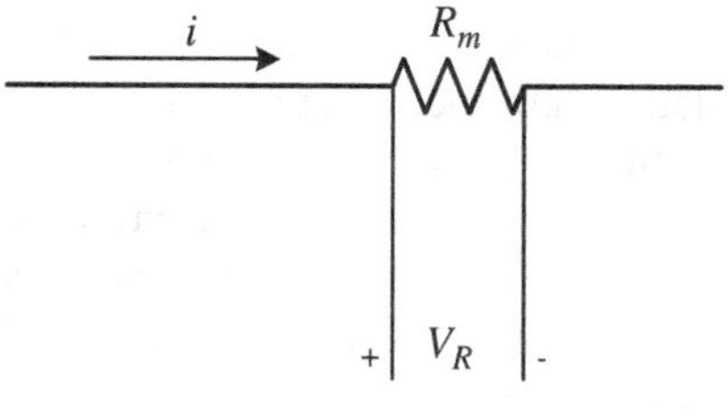

Figure 4.3 Measurement of current using a resistor.

4.2.1.1 Measurement of the Current by a Resistor

This method is the most simple and easy when the magnitude of the current is small. As shown in Fig. 4.3, the current, i, can be measured by sensing the voltage drop of the resistor, R_m, inserted in series to the circuit, where the current should be measured. Because of the ohmic loss given by iV_R, the magnitude of the current should be limited. Also, to reduce the loss, the resistance of the resistor should be minimized, but in this case the magnitude of the sensed voltage decreases together and the signal-to-noise ratio is getting worse. Furthermore, if the current varies rapidly, because of the stray inductance of the resistor itself the measurement would be inaccurate. In addition, the measured signal is not galvanically isolated, and the power converter and the control system are connected electrically to the circuit where the current is flowing. This may result in severe common mode noise problems and safety issues. For the isolation, the optical isolator or high-impedance differential amplifier may be used after the measurement of the voltage of the resistor, but this isolation stage may incur additional cost and time delay. Regardless of these shortcomings, due to the simplicity, this method has been used for the measurement of a current that is less than hundreds of amperes, where a few milli-ohms of accurate noninductive measurement resistor can be used, especially for home appliances and also for the low-cost general-purpose variable-speed drive system.

4.2.1.2 Measurement of the Current by a Current Transformer (CT)

This method exploits the principles of the transformer as shown in Fig. 4.4. The AC current in the primary winding, which is usually single-turn, induces the flux variation in the secondary winding, where the current proportional to the primary current flows to cancel the flux variation by the primary current. Through the measurement of the secondary current by a resistor, R_m, the primary current can be measured easily. The voltage across the resistor is proportional to the current in the primary side. While this

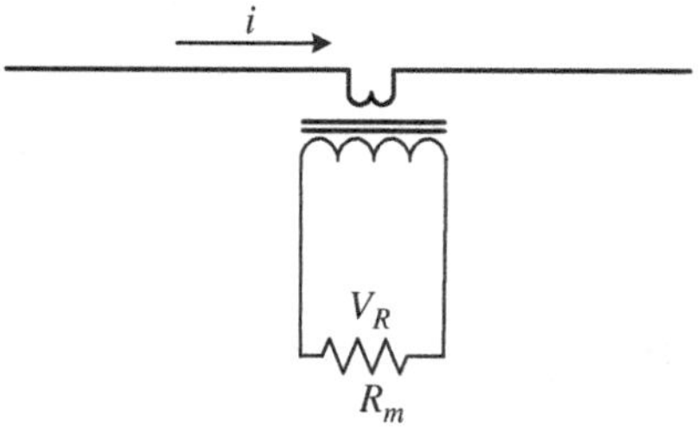

Figure 4.4 Measurement of current using a current transformer (CT).

method is simple and easy to implement, especially in the case of the measurement of large AC current, DC current cannot be measured inherently, because DC current does not incur any flux variation. Hence, this method cannot be applied to the measurement of the current of DC machine and that of AC machine when the frequency of current of AC machine is down to zero, which means DC current. But as a special case, as introduced in Section 2.4.1, if a DC machine is driven by a static Ward–Leonard system, the armature current, i_a, can be reconstructed by AC source current according to the gating sequences. While the measurement bandwidth of the current transformer (CT) can be extended up to several megahertz, that of the conventional CT based on a simple iron core is extended up to 1 kHz. And, the current regulation bandwidth using the conventional CT can reach up to 100 Hz. Furthermore, due to the magnetic saturation, if the magnitude of the current in the primary side is over the range, the linearity of the measurement may be endangered.

4.2.1.3 Measurement of the Current by a Hall Effect Sensor

When the current flows in a conductor, there is the magnetic field around conductor. The magnitude of the field is proportional to the current, and there is no time delay between the current and the magnetic field. Hence, by measuring the magnetic field, the current can be measured indirectly. To measure the magnetic field, a Hall effect sensor can be used. The operating principle of a Hall effect sensor is shown in Fig. 4.5. If the current, i, flows in a semiconductor and the magnetic flux, B, is applied perpendicularly to the direction of the current flow, then the voltage difference, the so-called "Hall voltage," V_{Hall}, occurs between two sides of a semiconductor. The direction of the voltage is normal to the direction of the current flow and also normal to the magnetic flux. The magnitude of the voltage is proportional to the magnitude of the flux under the constant current. Hence, basically, the Hall sensor measures the magnetic flux passing through the sensor. To measure the current flowing through the conductor based on this principle, the setup shown in Fig. 4.6 can be arranged. In the figure, if the current flows through the conductor located in the center of a soft ferromagnetic material core such as a ferrite core, the magnetic flux flows through the core. The flux is proportional to the current with no time delay. A Hall effect sensor driven by a constant current source is inserted in the core to the normal direction of the flow of the flux, and the sensor provides a voltage signal proportional to the current of the conductor due to the Hall effect. Because of temperature dependency and nonlinearity of the Hall effect sensor,

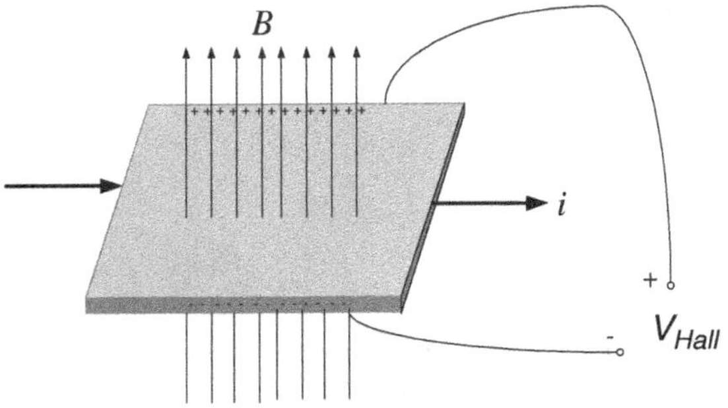

Figure 4.5 Voltage by the Hall effect.

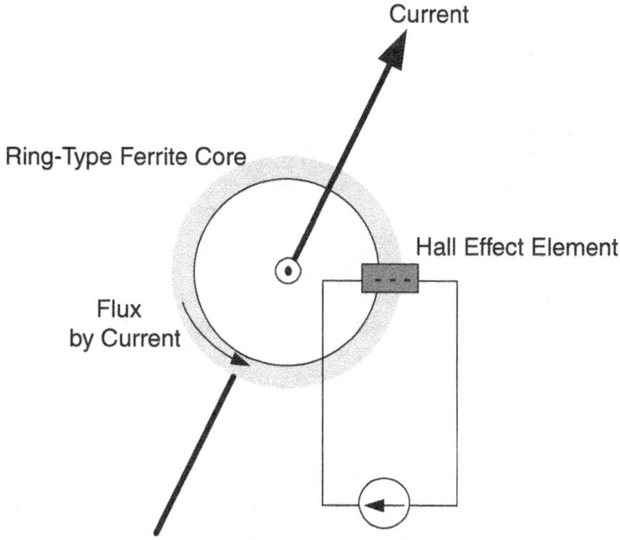

Figure 4.6 Current measurement by a Hall effect sensor.

the measurement by a simple Hall effect sensor would not be accurate in wide operating temperature and current range. Hence, an improved technique to overcome the above-mentioned problems is widely used, where the flux by the small current in another winding wound around the core in Fig. 4.6 cancels out the flux of the current of the conductor in the center of the core to make the flux in the core be null by feedback control. This closed-loop sensor is reasonably accurate, and the measurement bandwidth can be extended up to several hundred kilohertz.

Recently, a surface-mount chip type Hall effect current sensor is available, and the cost and size decreases and the application area of the sensor increases.

Also, the giant magneto-resistive (GMR) sensor, which varies the resistance according to the magnetic flux, is also used to measure the magnetic field of the current of the conductor for the measurement of the current. But, still there are some problems such as temperature dependency and nonlinearity. However, with the development microelectronics and GMR sensor itself, the problems would be overcome soon. Already several commercial current sensors based on GMR are available on the market, and galvanically isolated current measurement with a GMR effect would be more cost effective and accurate than any others in the near future. Also, there is the possibility that the size of the sensor would be shrunk very much.

4.2.2 Current Regulator for Three-Phase-Controlled Rectifier

4.2.2.1 Proportional and Integral (PI) Regulator

For the regulation of the current of the three-phase-controlled rectifier shown in Fig. 2.17 and Fig. 2.63, a proportional and integral (PI) regulator as shown in Fig. 4.7

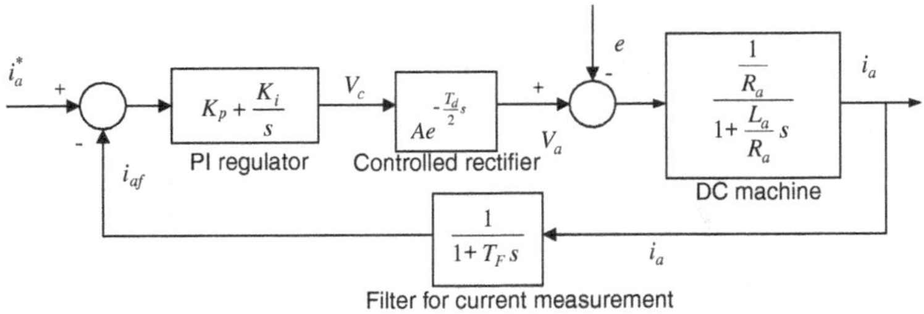

Figure 4.7 Control block diagram of current control system of a separately excited DC machine driven by a three-phase-controlled rectifier with proportional and integral (PI) current regulator.

is used widely because of its simplicity and relatively easy gain tuning. In Fig. 4.7, a control block diagram of separately excited DC machine driven by a three-phase-controlled rectifier with a PI current regulator is shown. In the figure, *e* stands for back EMF of the machine. In the diagram, it is assumed that the current of the machine is continuous and the nonlinearity between control voltage, V_c, and gating angle, α, is resolved by using an inverse cosine function as mentioned in Section 2.18.1. Also, the source voltage drop due to the overlapping angle from the source inductance, shown in Fig. 2.64 and in (2.82), has been neglected. Hence, the voltage gain of the controlled rectifier in Fig. 4.7 can be represented as (4.5), which is deduced from (2.82) and the time delay as (4.6), which is one-twelfth of a period of input source voltage as mentioned in Section 2.18.1.

$$A = \frac{3\sqrt{2}}{\pi} V_{L-L} \tag{4.5}$$

$$T_d = \frac{2\pi}{12\omega_e} \tag{4.6}$$

where ω_e is the angular frequency of the source voltage.

In Fig. 4.7, the armature current, i_a, includes the ripple current, whose frequency is the six times of the source frequency. Also, to get average value of the armature current, the ripple current in the measured current should be filtered out. In the designing of PI regulator, the filter is assumed as the first-order low-pass filter, whose time constant is T_F, and the cutoff angular frequency is $1/T_F$. The cutoff frequency can be set the same as the source frequency, which is one-sixth of the ripple frequency. The time delay due to switching action of the thyristors in the controlled rectifier can be modeled as the first-order delay as derived in (2.83), and then the control block diagram in Fig. 4.7 can be simplified as shown in Fig. 4.8.

The loop gain includes three time lagging elements, namely the delay due to the rectifier itself, $T_d/2$, the delay due to the filter, T_F, and DC machine itself, $T_a = L_a/R_a$,

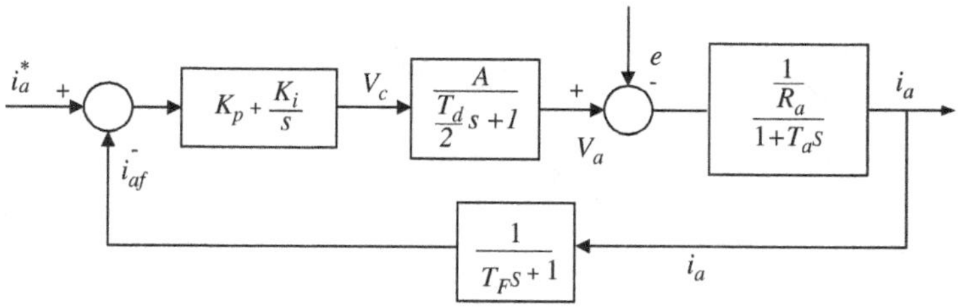

Figure 4.8 Simplified control block diagram.

which is the armature circuit time constant. The first and the second delay can be added and approximated as one delay:

$$\begin{aligned} G_{\text{open-}loop} &= \frac{K_p s + K_i}{s} \cdot \frac{A}{\frac{T_d}{2}s+1} \cdot \frac{\frac{1}{R_a}}{1+T_a s} \cdot \frac{1}{1+T_F s} \\ &\approx \frac{K_p s + K_i}{s} \cdot \frac{1}{R_a}\frac{A}{1+T_a s} \cdot \frac{1}{1+T_e s} \end{aligned} \tag{4.7}$$

where $T_e = T_d/2 + T_F$.

In the case of a large DC machine drive system over a several-hundred-kilowatt drive system, where three-phase controlled rectifiers are used, usually $T_a \gg T_e$, and the term $1 + T_a s$ in (4.7) can be approximated as $T_a s$. And, the open-loop gain can be simplified as

$$G_{\text{open-loop}} \approx \frac{AK_p\left(s + \frac{1}{T_i}\right)}{s^2 R_a T_a} \cdot \frac{1}{1+T_e s} \tag{4.8}$$

where T_i is the integral time constant of PI regulator, which is set by the two gains of the regulator like K_p/K_i. There are several gain tuning methods for this system. But, the symmetric optimum method is the one used widely [2, 3]. The integral time constant, T_i, is decided by the system damping factor, ζ. To get the maximum phase margin, ζ can be set as $1/\sqrt{2}$. Then $T_i = T_e \cdot (1+2\zeta)^2$ and $K_p = \frac{1}{A} \cdot \frac{T_a}{(1+2\zeta)T_e}$. In this closed-loop system, the bandwidth is around $1/15.1\ T_e$. With this tuning procedure, in the case of the drive system with 60-Hz source frequency, the bandwidth is approximately 16 (rad/s). Through this kind of design procedures and gain tuning, the current control system with the controlled rectifier can provide reasonable performance, but still there are several shortcomings as follows. The closed-loop bandwidth is limited to less than a few tens of rad/s, and the current response to the step input of the current command reveals overshoot due to the zero of PI regulator. Furthermore, if the armature current is discontinuous, then the transfer function of the rectifier in (2.83) would not hold and the overall transfer function is totally different from the function in (4.7) and the control performance degrades conspicuously.

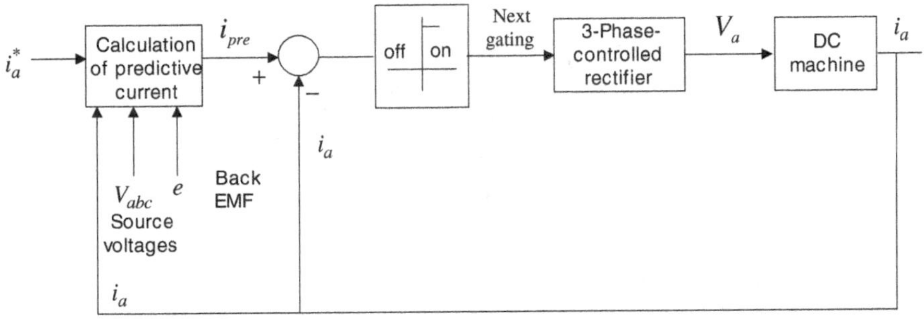

Figure 4.9 Control block diagram of a predictive current regulator for a three-phase controlled rectifier driving DC machine.

4.2.2.2 Predictive Current Regulator

To solve the problems of a PI regulator, as mentioned in Section 4.2.2.1, the powerful computing capability of a digital signal processor (DSP), which is recently available in a cost effective manner, can be applied. By emulating the whole control system including DC machine and the rectifier, the gating time point to regulate the armature current according to the commanded current, i_a^*, can be directly obtained based on the predicted armature current. Hence, by using this type of regulator, referred to as a predictive current regulator, the armature current can track the commanded one without any control delay [4]. The control block diagram of the predictive regulator is shown in Fig. 4.9.

The equivalent circuit of the system including AC source, DC machine, and thyristor switches is shown in Fig. 4.10, when two thyristors of the three-phase-controlled rectifier is conducting. In the circuit, the conduction voltage drop of the thyristor is neglected and V_s stands for a source line-to-line voltage to which the conducting two thyristors are connected. According to the conduction of thyristors,

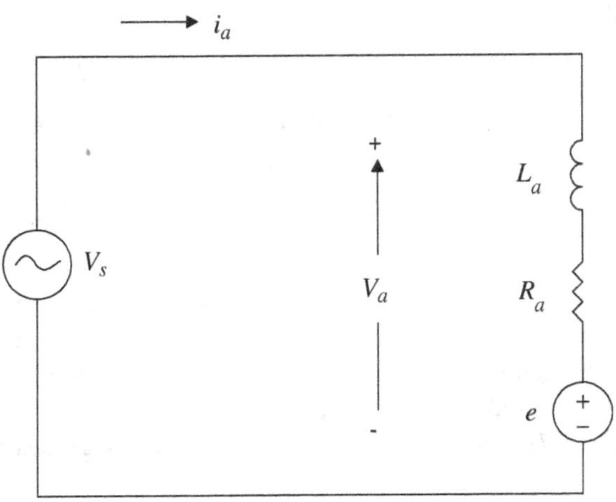

Figure 4.10 Equivalent circuit of a three-phase-controlled rectifier drive system.

this voltage varies sequentially and is one of among six line-to-line voltages, namely, $V_{ac}, V_{bc}, V_{ba}, V_{ca}, V_{cb}, V_{ab}$.

The voltage loop equation of the circuit can be described as (4.9), and the solution of the equation under the assumption of constant back EMF, e, during one-sixth of the source voltage period can be found as (4.10). The assumption holds if the speed of the DC machine is almost constant over one-sixth of the period of the source voltage:

$$L_a \frac{di_a}{dt} + R_a i_a + e = V_s \tag{4.9}$$

$$i_a = I_1 \cos\left(\omega_e t - \frac{\pi}{6} - \phi\right) + I_2 e^{(-\rho\omega_e t)} + I_3 \tag{4.10}$$

where ω_e is the angular frequency of the source voltage, $\phi = \tan^{-1}(\omega_e L_a / R_a)$, $\rho = R_a/\omega_e L_a$, $I_1 = \sqrt{2}V_{\ell-\ell rms}/\sqrt{R_a^2 + (\omega_e L_a)^2}$, and $I_3 = -e/R_a$.

In (4.10) the first term in the right-hand side is the steady-state solution by AC source, and the second, transient solution by AC source, and third, the steady-state solution by back EMF. The first and the third term are independent of the gating angle, α. Thus, to regulate the current by gating angle, the magnitude of the second term should be controlled. According to the continuous or discontinuous conduction of the current, the magnitude of the second term is given by (4.11), or (4.12), respectively.

$$I_2 = \left[\left\{i_a^* - I_3 - \frac{3}{\pi} I_1 \cos(\alpha - \phi)\right\} \cdot \frac{\pi}{3}\rho \cdot e^{\rho\left(\alpha + \frac{\pi}{6}\right)}\right] \frac{1}{2\sinh\left(\rho\frac{\pi}{6}\right)} \tag{4.11}$$

where $\alpha = \cos^{-1}\left(\dfrac{R_a i_a^* + e}{\sqrt{2}V_{\ell-\ell rms}} \cdot \dfrac{\pi}{3}\right)$.

$$\begin{aligned} I_2 &= \left[-I_3 - I_1 \cos\left(\alpha - \frac{\pi}{6} - \phi\right)\right] e^{\rho\alpha} \\ &= \left[-I_3 - I_1 \cos\left(\beta - \frac{\pi}{6} - \phi\right)\right] e^{\rho\beta} \end{aligned} \tag{4.12}$$

where α stands for the firing angle of the thyristor and β stands for the extinction angle. In the case of discontinuous conduction, by using the constraint that the average voltage across the inductor, L_a, equals zero, α, β, and I_2 can be calculated by simultaneously solving (4.12) and (4.13). And based on the calculated α, β, and I_2, the instantaneous armature current can be predicted from (4.10).

$$\sqrt{2}V_{\ell-\ell rms}\left(\sin\left(\beta - \frac{\pi}{6}\right) - \sin\left(\alpha - \frac{\pi}{6}\right)\right) + e\left(\frac{\pi}{3} + \alpha - \beta\right) - \frac{\pi}{3}\left(R_a i_a^* + e\right) = 0 \tag{4.13}$$

The control sequence is as follows. First, at every sampling interval, which should be much less than a period of the source voltage—for example, less than two-hundredths of the period—the voltage, V_s, back EMF, e, calculated by the product of the speed of the machine and back EMF constant, and the instantaneous armature current, i_a, are sampled. Next, the armature current under the assumption that the next thyristor turns on—that is, the predictive current, i_{pre}—is calculated by (4.10). Finally, if the predictive current is equal to or larger than the sampled armature current, then the next thyristor turns on at that sampling instant. By this method, the armature current can be regulated as fast as possible regardless of continuous or discontinuous conduction of the armature current. The bandwidth of the regulator can be extended up to a few hundred rad/s, which is more than 10 times the bandwidth of the PI regulator in Section 4.2.2.1. In Fig. 4.11, the waveforms of the commanded current and actual current are displayed.

The shortcomings of this regulator are (a) the computational burden due to the real-time emulation and (b) the sensitivity to the machine parameters such as armature resistance, inductance, and back EMF constant. With parameter errors, the current regulation may have the steady-state error. This problem can be resolved by augmenting a simple PI regulator in addition to the predictive regulator.

4.2.3 Current Regulator for a DC Machine Driven by a PWM Chopper

For the torque control of a DC machine whose power rating is less than several tens of kilowatts, the four-quadrant DC/DC power converter described in Section 2.4.2 is widely used. In Section 2.18.4, the operating principle of the power converter has been briefly described. At this power converter or at the bidirectional chopper in Section 2.18.3, the switching frequency, $f_s = 1/T_s$, of the power semiconductors in the power converter is usually much larger than the bandwidth frequency of the current regulator, $f_{bw} = \omega_{bw}/2\pi$ such as $f_s > 10f_{bw}$. In this case the voltage applied to the electric machine can be assumed as the commanded voltage itself without any time delay, and the power converter can be modeled as a voltage amplifier without any internal dynamics and as a simple amplifier with constant gain k. At given constant switching frequency, the available maximum current regulation bandwidth depends on the current sampling method and control period as shown in Table 4.1 in Section 4.2.3.2.

4.2.3.1 Proportional and Integral (PI) Current Regulator [5,6]

If the inertia of the machine and load is large enough to consider back EMF as constant in the design of the current regulator, then the machine can be modeled as simple R–L load, $R_a - L_a$, with constant disturbance voltage, e, as shown in Fig. 4.12. If the back EMF voltage can be estimated from the speed of the machine and back EMF constant, then by feed-forward compensation the disturbance by back EMF can be rejected perfectly as the dotted part in Fig. 4.12. In addition to this rejection of the disturbance voltage, by canceling the pole of the system, which comes from $R_a - L_a$ with a zero of

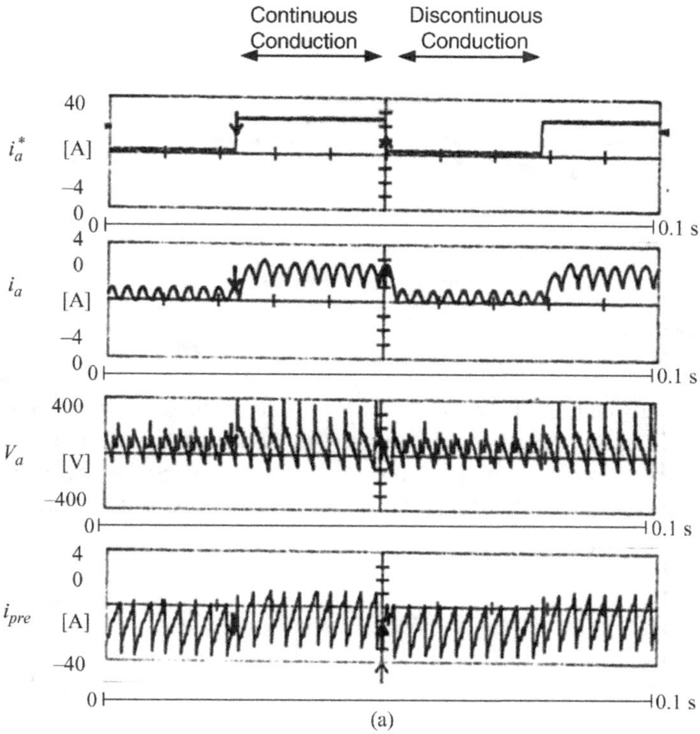

(a)

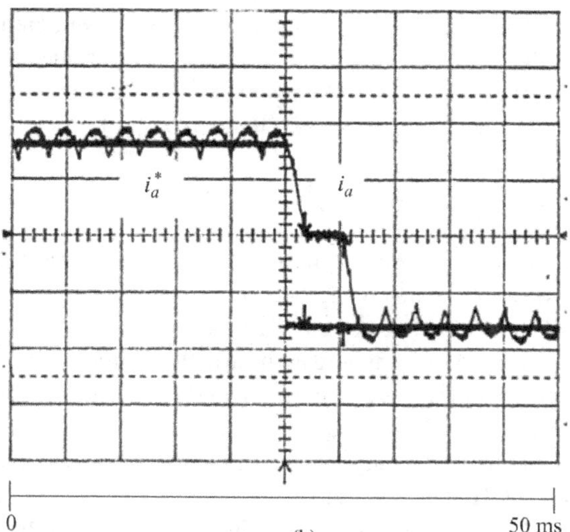

(b)

Figure 4.11 Regulation of the armature current of a three-phase-controlled rectifier feeding dc machine by a predictive current regulator. (a) Continuous and discontinuous conduction. (b) Polarity changing of the commanded current (20 A/div.).

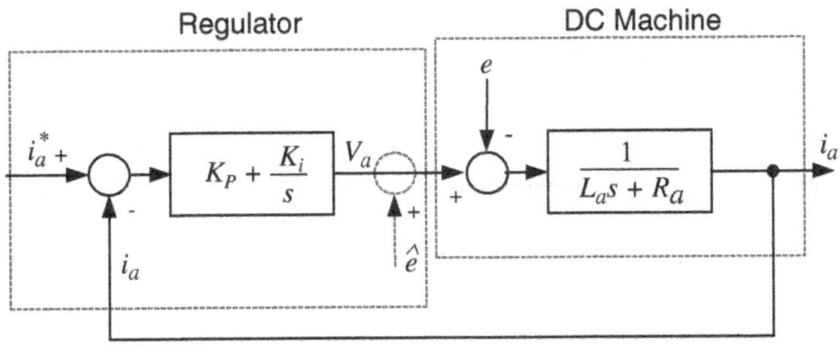

Figure 4.12 Proportional and Integral (PI) regulator.

PI regulator, the transfer function of the closed-loop system can be set as that of the first-order low-pass filter. This kind of gain tuning method is referred to as the technical optimum. Under this gain tuning method, the proportional and integral gain can be set as (4.14), and then the closed-loop transfer function of the current regulator is deduced like (4.15).

$$\begin{aligned} K_p &= L_a\omega_c \\ K_i &= R_a\omega_c \end{aligned} \tag{4.14}$$

$$\frac{i_a(s)}{i_a^*(s)} = \frac{\omega_c}{s+\omega_c} = \frac{1}{T_c s+1}, \qquad T_c = \frac{1}{\omega_c} \tag{4.15}$$

As seen in (4.15), the bandwidth of the regulator, ω_{bw}, is set as ω_c. Hence, after deciding the bandwidth, the gains are calculated directly from the parameters of a DC machine. With this gain setting, from (4.15) the response of the current regulator is well-defined without any overshoot and no steady-state error.

4.2.3.2 Implementation Issues [6]

Due to the instantaneous voltage difference between the voltage applied to DC machine, V_a, and back EMF, e, of the machine, there are inevitable ripple currents in the armature current, i_a. If the switching period, T_s, of a DC/DC converter, the so-called PWM chopper, is small enough compared to the armature circuit time constant, $T_a = L_a/R_a$, then the current variation can be approximated as a linear function of the instantaneous voltage difference. In this case, the armature current has ripple components whose fundamental frequency is two times the switching frequency, $f_s = 1/T_s$, as shown in Fig. 2.71. For accurate regulation of the average armature current in a switching period, the ripple components should be removed. In addition, because the armature voltage is synthesized by pulse width modulation (PWM) of the converter, the commanded armature voltage can be updated by PWM at best twice per switching period. Due to this inherent limitation in the implementation of the current regulator, there are inevitable time delays even if the PI regulator itself is designed in the analog form. But, nowadays, most of the regulators are designed in digital form, and there is additional delay due to the execution of the digital algorithm of the current regulator.

If the ripple components of the armature current are filtered out by the second-order low-pass filter, there is additional time delay. As an example, if the cutoff frequency of the second-order filter to remove the ripples is $f_s/5$, there is time delay between the actual current and the filtered current by $1.25T_s$. The average value of the armature current in a sampling period can be obtained by averaging the current as in (4.16). Here, the averaging interval can be a half of a switching period, which gives the minimum time delay, $\frac{1}{2} \cdot \frac{1}{2}\ T_s$, to get the average value.

$$\langle i_a \rangle = \frac{1}{\frac{T_s}{2}} \int_{(n-1)\frac{T_s}{2}}^{n\frac{T_s}{2}} i_a(\tau)dt \tag{4.16}$$

In another way, without low-pass filtering or averaging, the average value of the current in a switching period can be obtained by exploiting the feature of PWM. If the current is sampled at the peak or the valley of the carrier wave of PWM as shown in Fig. 2.71, then the average current can be obtained under the assumption of the linear variation of the current in the period. Though this method has shortcomings such that the measurement is sensitive to noise because of no low-pass filtering, the average current can be obtained without any time delay, and it has the potential to give the highest regulation bandwidth compared to other methods. Further issues with this method such as adding low-pass filtering to improve the noise immunity is discussed in Section 7.2.1.

The delays, mentioned before, limit the bandwidth of the current regulation. If the delay is reduced by other means such as prediction of the current or sampling twice per switching period, the bandwidth can be increased. To eliminate the digital delay for the execution of the regulator algorithm, the current at the next sampling point, $i_a(n+1)$, can be predicted based on the present commanded voltage, $V_a^*(n)$, estimated back EMF, $\hat{e}(n)$, estimated parameters such as $\hat{R}_a$ and $\hat{L}_a$, and the sampled current, $i_a(n)$, as in (4.17). In this prediction, if there are errors in the parameters, the sampled current, and estimated back EMF, then there are some current ripples due to these errors even in the steady state.

$$i_a(n+1) = i_a(n) + \frac{T_s}{2}\left[V_a^*(n) - \hat{e}(n) - \hat{R}_a \cdot i_a(n)\right]\frac{1}{\hat{L}_a} \tag{4.17}$$

The maximum bandwidths of the current regulator based on the technical optimum method with back EMF compensation are given in Table 4.1 according to different current measurement methods and different PWM update intervals. With the developments of the microelectronics technology, the current sampling and the execution of the regulator algorithm can be done in a few microseconds or even in a several hundred nanoseconds. In this case, by eliminating the execution delay of the regulator itself in digital, the bandwidth of the current regulator can reach up to $2\pi \cdot \frac{f_s}{3.5}$ regardless of the parameter and estimation errors [7].

Table 4.1 Maximum Bandwidth of Digital PI Current Regulator with a Four-Quadrant Chopper

Current Measurement Method	Low-Pass Filtering $f_{cut} = f_s / 5$	Sampling at Peak or Valley of Carrier Waveform	Sampling at Peak and Valley of Carrier Waveform	Sampling at Peak and Valley of Carrier and Prediction of Current
Command voltage updating interval	Once per T_s synchronized to the current sampling	Once per T_s synchronized to the current sampling	Twice per T_s synchronized to the current sampling	Twice per T_s synchronized to the current sampling
Total time delay	$2.75\ T_s$	$1.5\ T_s$	$0.75\ T_s$	$0.25\ T_s$
Maximum bandwidth, ω_{cc}(rad/s)	$2\pi \cdot \frac{f_s}{40}$	$2\pi \cdot \frac{f_s}{21}$	$2\pi \cdot \frac{f_s}{10}$	$2\pi \cdot \frac{f_s}{3.5}$

[a]Here, $T_s = \frac{1}{f_s}$, and f_s is the switching frequency of the four-quadrant chopper.

4.2.4 Anti-Wind-Up [8]

All physical variables in the control system are bounded by physical limitations. As shown in Fig. 4.13, the output of the current regulator, which is the input to the power amplifier such as a three-phase-controlled rectifier or a four-quadrant DC/DC power converter, decides the terminal voltage of DC machine, V_a. In this case the magnitude of the voltage is limited by input source voltage of the amplifier whatever the output of the regulator itself is. If the terminal voltage is saturated by the physical limitation but if the output of the regulator is not bounded at that physical limitation, then the control system including a DC machine is out of control of the closed loop, where the regulator output has nothing to do with the input to the physical system under control. In this case the response to the command of the control system is sluggish and oscillatory and is uncontrollable by the regulator. To prevent this situation, the output of the regulator can be simply bounded by a limiter as shown in Fig. 2.70. However, if the integrator terms are included in the output of the regulator, then the output of the integrator itself, $\int_0^t e(\tau)d\tau$ or $\sum_{k=0}^{n} e(k)$, is not bounded by the limiter and it would be wound up above the limited value of the regulator though the output of the regulator is

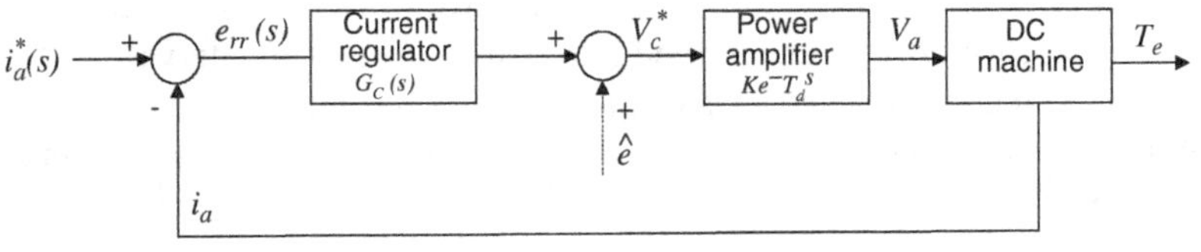

Figure 4.13 Control system including regulator, power amplifier, and DC machine.

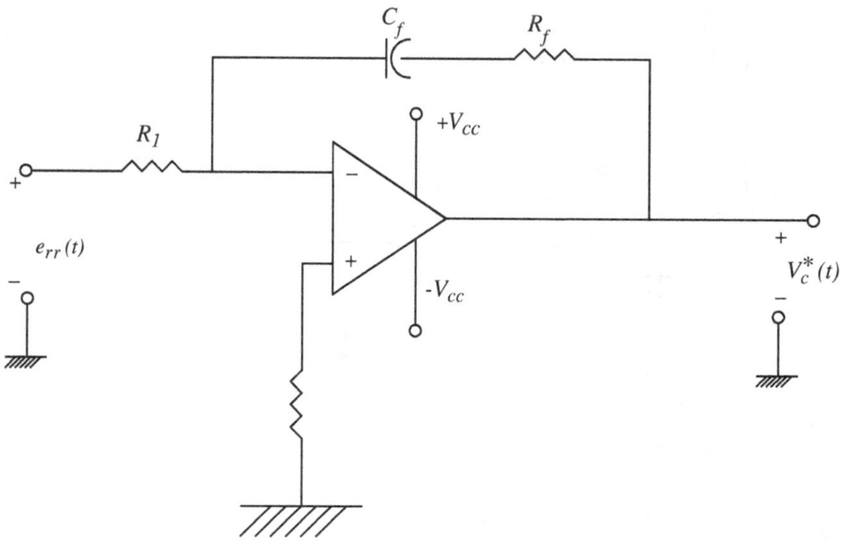

Figure 4.14 PI regulator implemented by an operational amplifier.

limited. This phenomenon is called "wind-up." If the wind-up of the integrator of the regulator occurs, then the input to the integrator, which is the regulation error, should have a sign opposite to that of the output of the integrator for a while to clear the wind-up. In this reason, the response would have a large overshoot or undershoot because of the integrator wind-up. In the worst case the control system may be unstable. In a conventional PI regulator implemented with operational amplifier (OP Amp), the output of OP Amp including the integrator itself is inherently limited by the supply voltage such as $+V_{cc}$ or $-V_{cc}$ as shown in Fig. 4.14. To limit the output voltage of the integrator further, back-to-back connected Zener diodes can be used as shown in Fig. 4.15.

$$V_c^*(t) = -\frac{R_f}{R_1}e_{rr}(t) - \frac{1}{C_f R_1}\int_0^t e_{rr}(\tau)d\tau \tag{4.18}$$

In the case of digital signal processor (DSP) using fixed-point operation, the variables can be limited by a proper saturation function and also by scaling of the variables. However, in the case of DSP using floating point operation or in the case of computer simulation, the output of the integrator should be properly bounded with consideration of the physical limitation of the variables. The limitation of the integrator output, which is called an "anti wind-up," can be done in several ways as shown in Fig. 4.16. The method in Fig. 4.16a, which is simple to implement digitally, has shortcomings such that the limiting value of the output of the integrator does not match to the limiting value of the output of the regulator. And the setting of the limiting value of the integrator is not easy. The method in Fig. 4.16b, which is

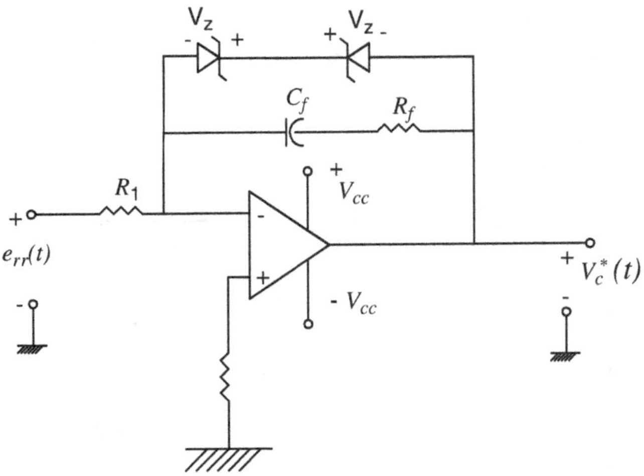

Figure 4.15 PI regulator implemented by an operational amplifier with Zener diodes to limit output of the integrator.

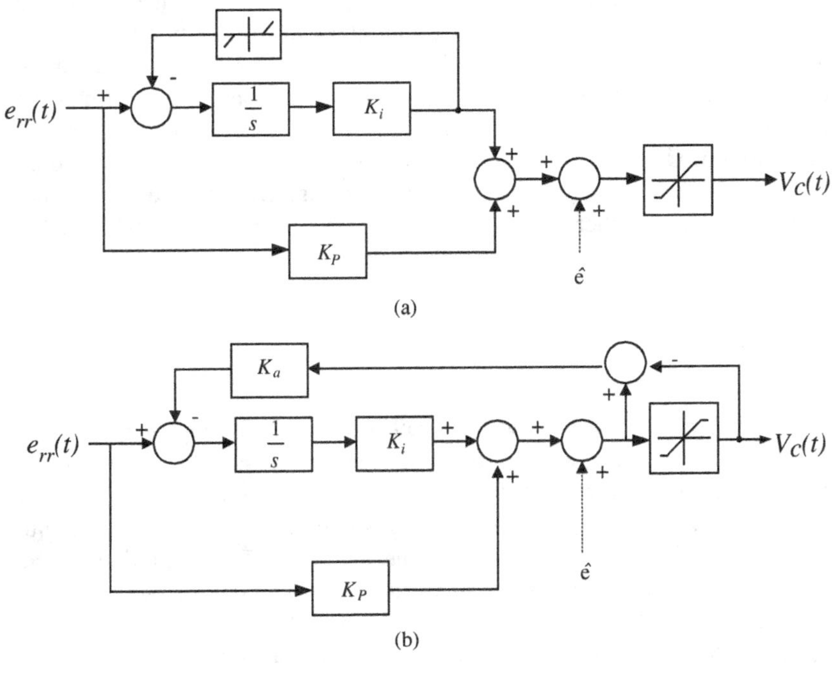

Figure 4.16 Implementation of anti-wind-up. (a) Limited integral method. (b) Anti-wind-up controller.

called the back calculation method, is easy to tune and gives the reasonable performance in most of application fields. In all implementations, the back EMF can be compensated in feed-forward manner to improve the control performance.

In Fig. 4.16b, K_a is usually set as $1/K_p$. However, in general, such a setting gives a reasonable performance, but the anti-wind-up gain, K_a, can be tuned to get better performance in the range of $1/3 \sim 3$ of K_p.

Other methods such as turning-on or turning-off of the integrator of the PI regulator according to the operating condition can be used [9]. Because the main reason to use the integrator in the regulator is to null the steady-state error, the integrator may be turned off whenever the error is large. If the error is small and the output of the regulator is within the physical bound, then the integrator may be turned on to reduce the steady-state error.

The above-mentioned anti-wind-up functions should be incorporated not only in a PI regulator but also in any regulator that has integral terms.

4.2.5 AC Current Regulator

Like a DC machine, for the high-performance control of an AC machine the regulation of the current is essential. Several methods for the current regulation of AC machine have been developed, such as hysteresis regulator, predictive regulator, dead beat regulator, and so on. In this section a PI regulator mentioned in Section 4.2.3 is extended and modified for the regulation of current of an AC machine [10]. The extended PI regulator reveals reasonable dynamic performance together with easy gain tuning and robustness to the parameter variation of AC machine and the power converters.

4.2.5.1 Current Regulator for a Balanced Three-Phase Circuit

A balanced three-phase electric circuit consisting of a resistor, an inductor, and an EMF in series per phase can be depicted as in Fig. 4.17. Here, it is assumed that the instantaneous sum of three EMF voltages is zero: $e_{as} + e_{bs} + e_{cs} = 0$. And EMF is sinusoidal and its angular frequency is ω_e.

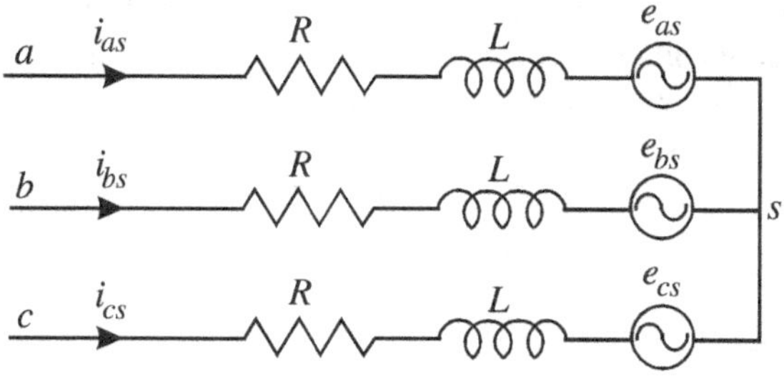

Figure 4.17 Balanced three-phase R–L–EMF circuit.

The above circuit can be described by

$$\begin{aligned} V_{as} &= Ri_{as} + L\frac{di_{as}}{dt} + e_{as} \\ V_{bs} &= Ri_{bs} + L\frac{di_{bs}}{dt} + e_{bs} \\ V_{cs} &= Ri_{cs} + L\frac{di_{cs}}{dt} + e_{cs} \end{aligned} \tag{4.19}$$

Equation (4.19) can be rewritten in a stationary d–q reference frame as (4.20). Because of the balanced impedance and EMF, there is no n-term voltage and current in d–q–n axes.

$$\begin{aligned} V_{ds}^{s} &= Ri_{ds}^{s} + L\frac{di_{ds}^{s}}{dt} + e_{ds}^{s} \\ V_{qs}^{s} &= Ri_{qs}^{s} + L\frac{di_{qs}^{s}}{dt} + e_{qs}^{s} \end{aligned} \tag{4.20}$$

The equation can be further expressed in synchronously rotating reference frame, whose rotating speed is ω_e, as in (4.21):

$$\begin{aligned} V_{ds}^{e} &= Ri_{ds}^{e} + L\frac{di_{ds}^{e}}{dt} - \omega_e L i_{qs}^{e} + e_{ds}^{e} \\ V_{qs}^{e} &= Ri_{qs}^{e} + L\frac{di_{qs}^{e}}{dt} + \omega_e L i_{ds}^{e} + e_{qs}^{e} \end{aligned} \tag{4.21}$$

If all electrical variables of the circuit in Fig. 4.17 vary sinusoidally at angular frequency, ω_e, then the variables in (4.21) are strictly DC quantities. Hence, the design methodology of a PI regulator for the DC machine introduced in Section 4.2.3 can be applied to the design of an AC current regulator. In Fig. 4.18, a block diagram of a typical AC current PI regulator on the synchronous d–q reference frame, which is extended from a PI regulator of DC current, is depicted including anti-wind-up function due to the limitation of the output voltage of the inverter.

Because of the d–q axis, two sets of PI regulators are used for a three-phase circuit. All gain-tuning, including anti-wind-up gain and back EMF feed-forwarding at each axis, is the same as in the case of PI regulator in Section 4.2.3. The only difference is that the feed-forwarding voltage includes not only back EMF, e_{ds}^{e} and e_{qs}^{e}, but also coupling terms due to the rotation of the reference frame such as $-\omega_e \hat{L} i_{qs}^{e}$ in the d axis and $+\omega_e \hat{L} i_{ds}^{e}$ in the q axis as in (4.22):

$$\begin{aligned} V_{ds\_ff}^{e} &= -\omega_e \hat{L} i_{qs}^{e} + \hat{e}_{ds}^{e} \\ V_{qs\_ff}^{e} &= \omega_e \hat{L} i_{ds}^{e} + \hat{e}_{qs}^{e} \end{aligned} \tag{4.22}$$

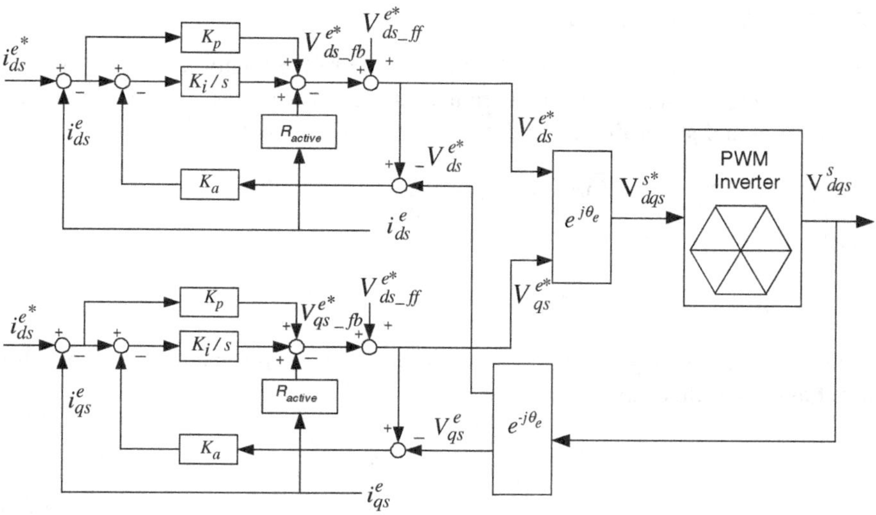

Figure 4.18 PI current regulator for a balanced three-phase AC circuit.

In addition, the output voltage of each axis PI regulator is limited not in d or q axis separately but in d–q axis simultaneously. The voltages are limited within the hexagon at a stationary d–q voltage reference plane. Hence, the voltages after limitation at each axis are coupled with each other. If the output voltage, in space vector form, of the regulator, V_{dqs}^{s*}, is out of the hexagon boundary, then the voltage should be limited by an overmodulation scheme, and after overmodulation the voltage would lie at the boundary of the hexagon—that is, V_{dqs}^{s}. Based on this voltage, the anti-wind-up controller works [11].

Like PI regulator in Section 4.2.3, if the bandwidth of the regulator is set as ω_{bw}, then the gains of PI regulator can be easily calculated. For further improvement of the robustness to the parameter variations, the active damping term can be added and the stator resistance can be increased by R_{active}. In this case the gains of the regulator are given as

$$
\begin{aligned}
K_p &= L\omega_{bw} \\
K_i &= (R + R_{\text{active}})\omega_{bw} \\
K_a &= \frac{1}{K_p}
\end{aligned}
\tag{4.23}
$$

Because of the simplicity and no steady-state error, the PI regulator on the synchronous reference frame with feed-forwarding terms is widely used in the industrial field to drive a high-performance AC machine. The gain-tuning for each type of AC machine can be derived as follows from the modeling of AC machine discussed in Sections 3.2 and 3.3. Also, the bandwidth of the regulator according to the current sampling methods and PWM update intervals is exactly same as in the case of the DC machine mentioned in Section 4.2.3.1 and Table 4.1.

4.2.5.2 Induction Machine

To derive the input voltage equation for the design of PI regulator mentioned before, the stator voltage equation of the induction machine can be represented in terms of the stator current and stator flux linkage at a synchronous reference frame as

$$
\begin{aligned}
V_{ds}^e &= R_s i_{ds}^e + \frac{d\lambda_{ds}^e}{dt} - \omega_e \lambda_{qs}^e \\
V_{qs}^e &= R_s i_{qs}^e + \frac{d\lambda_{qs}^e}{dt} + \omega_e \lambda_{ds}^e
\end{aligned}
\tag{4.24}
$$

Similarly, the voltage equation of the rotor can be represented in terms of rotor current and rotor flux linkage as

$$
\begin{aligned}
0 &= R_r i_{dr}^e + \frac{d\lambda_{dr}^e}{dt} - \omega_{sl} \lambda_{qr}^e \\
0 &= R_r i_{qr}^e + \frac{d\lambda_{qr}^e}{dt} + \omega_{sl} \lambda_{dr}^e
\end{aligned}
\tag{4.25}
$$

After rewriting the stator and rotor flux linkage in terms of stator and rotor current as (4.26) and (4.27), the stator flux linkage can be represented in terms of stator current and rotor flux linkage as

$$
\begin{aligned}
\lambda_{ds}^e &= L_s i_{ds}^e + L_m i_{dr}^e \\
\lambda_{qs}^e &= L_s i_{qs}^e + L_m i_{qr}^e
\end{aligned}
\tag{4.26}
$$

$$
\begin{aligned}
\lambda_{dr}^e &= L_m i_{ds}^e + L_r i_{dr}^e \\
\lambda_{qr}^e &= L_m i_{qs}^e + L_r i_{qr}^e
\end{aligned}
\tag{4.27}
$$

$$
\begin{aligned}
\lambda_{ds}^e &= \sigma L_s i_{ds}^e + \frac{L_m}{L_r} \lambda_{dr}^e \\
\lambda_{qs}^e &= \sigma L_s i_{qs}^e + \frac{L_m}{L_r} \lambda_{qr}^e
\end{aligned}
\tag{4.28}
$$

By using (4.27), the rotor current in (4.25) can be eliminated and the rotor flux linkage can be represented in terms of stator current and slip angular frequency, $(\omega_e - \omega_r)$ as shown in (4.29).

$$
\begin{aligned}
\frac{d\lambda_{dr}^e}{dt} &= R_r \frac{L_m}{L_r} i_{ds}^e - \frac{R_r}{L_r} \lambda_{dr}^e + (\omega_e - \omega_r) \lambda_{qr}^e \\
\frac{d\lambda_{qr}^e}{dt} &= R_r \frac{L_m}{L_r} i_{qs}^e - \frac{R_r}{L_r} \lambda_{qr}^e - (\omega_e - \omega_r) \lambda_{dr}^e
\end{aligned}
\tag{4.29}
$$

By substituting (4.28) and (4.29) into (4.24), finally, the stator voltage equation in terms of stator current and rotor flux linkage can be derived as

$$
\begin{aligned}
V_{ds}^{e} &= \left(R_s + R_r\frac{L_m^2}{L_r^2}\right) i_{ds}^{e} + \sigma L_s\frac{di_{ds}^{e}}{dt} - \omega_e\sigma L_s i_{qs}^{e} - R_r\frac{L_m}{L_r^2}\lambda_{dr}^{e} - \omega_r\frac{L_m}{L_r}\lambda_{qr}^{e} \\
V_{qs}^{e} &= \left(R_s + R_r\frac{L_m^2}{L_r^2}\right) i_{qs}^{e} + \sigma L_s\frac{di_{qs}^{e}}{dt} + \omega_e\sigma L_s i_{ds}^{e} + \omega_r\frac{L_m}{L_r}\lambda_{dr}^{e} - R_r\frac{L_m}{L_r^2}\lambda_{qr}^{e}
\end{aligned}
\tag{4.30}
$$

When the induction machine is controlled with the rotor-flux-oriented vector control method described in Section 5.2, $\lambda_{qr}^{e} = 0$. Hence, (4.30) can be simplified as

$$
\begin{aligned}
V_{ds}^{e} &= \left(R_s + R_r\frac{L_m^2}{L_r^2}\right) i_{ds}^{e} + \sigma L_s\frac{di_{ds}^{e}}{dt} - \omega_e\sigma L_s i_{qs}^{e} - R_r\frac{L_m}{L_r^2}\lambda_{dr}^{e} \\
V_{qs}^{e} &= \left(R_s + R_r\frac{L_m^2}{L_r^2}\right) i_{qs}^{e} + \sigma L_s\frac{di_{qs}^{e}}{dt} + \omega_e\sigma L_s i_{ds}^{e} + \omega_r\frac{L_m}{L_r}\lambda_{dr}^{e}
\end{aligned}
\tag{4.31}
$$

By comparing (4.31) and (4.21), the parameters for gain-tuning and feed-forwarding voltage terms can be deduced as given in Table 4.2.

4.2.5.3 Synchronous Machine

The synchronous machine, including the synchronous reluctance machine (SynRM), can be generally described as (4.32) at the rotor reference frame from Fig. 3.6. Again, by comparing (4.32) and (4.21), the parameters for gain-tuning and feed-forwarding voltage terms for the synchronous machine can be deduced as given in Table 4.2.

$$
\begin{aligned}
V_{ds}^{r} &= R_s i_{ds}^{r} + L_d\frac{di_{ds}^{r}}{dt} - \omega_r L_q i_{qs}^{r} \\
V_{qs}^{r} &= R_s i_{qs}^{r} + L_q\frac{di_{qs}^{r}}{dt} + \omega_r L_d i_{ds}^{r} + \omega_r\lambda_f
\end{aligned}
\tag{4.32}
$$

Table 4.2 Circuit Parameters

Balance Three-Phase Circuit in (4.21)	Induction Machine	Synchronous Machine
Resistance R	$R_s + R_r\left(\frac{L_m}{L_r}\right)^2$	R_s
Inductance L	σL_s	L_d, L_q
Angular frequency ω	$\omega_e(=\omega_r + \omega_{sl})$	ω_r
Feed-forwarding term in d axis, $-\omega_e L i_{qs}^{e} + e_{ds}^{e}$	$-\omega_e\sigma L_s i_{qs}^{e} - R_r\frac{L_m}{L_r^2}\lambda_{dr}^{e}$	$-\omega_r L_q i_{qs}^{r}$
Feed-forwarding term in q axis, $\omega_e L i_{ds}^{e} + e_{qs}^{e}$	$\omega_e\sigma L_s i_{ds}^{e} + \omega_r\frac{L_m}{L_r}\lambda_{dr}^{e}$	$\omega_r L_d i_{ds}^{r} + \omega_r\lambda_f$

where λ_f stands for the flux linkage by the magnet or by the field winding, and $L_d = L_{ls} + L_{md}$ and $L_q = L_{ls} + L_{mq}$. In the case of the wound rotor synchronous machine we have $\lambda_f = L_{md} i'_{fk}$, and in the case of SynRM we have $\lambda_f = 0$. For the case of an SMPMSM or round rotor synchronous machine, $L_d = L_q$.

4.2.5.4 Complex Vector Current Regulator

The above-mentioned PI regulator at synchronous reference frame decouples the coupling terms due to the rotation of axis by the feed-forwarding terms, $-\omega_e \hat{L} i^e_{qs}$ and $\omega_e \widehat{L} i^e_{ds}$. If the inductance varies according to the operating conditions as mentioned in Section 2.9 and Section 2.12, the decoupling would be imperfect and the control performance degrades rapidly. However, after understanding that the coupling terms due to parameter errors are covered by the integral term of a PI regulator, as shown in Fig. 4.19 the decoupling term can be added to the integral term instead of a feed-forwarding term. Such a current regulator is called a complex vector current regulator because the transfer function analysis of the regulator is done based on the complex frequency response function (complex FRF), where input and output of the transfer function are described as complex numbers [12]. The complex vector current regulator for the balanced three-phase circuit can be depicted as Fig. 4.19, where the estimated coupling terms are the input of the integrators of the regulator for decoupling. By using these inputs, the coupling terms due to rotation of the reference frame can be canceled out. Against the wind-up of the integrator, anti wind-up controller is also incorporated as shown in Fig. 4.19, where anti-wind-up controller includes a cross-coupling term that is not in Fig. 4.18 [13]. If the parameters of the regulator have no errors, then the transfer functions of both regulators in Figs. 4.18 and Fig. 4.19 are the same and the closed-loop transfer function of the current regulation

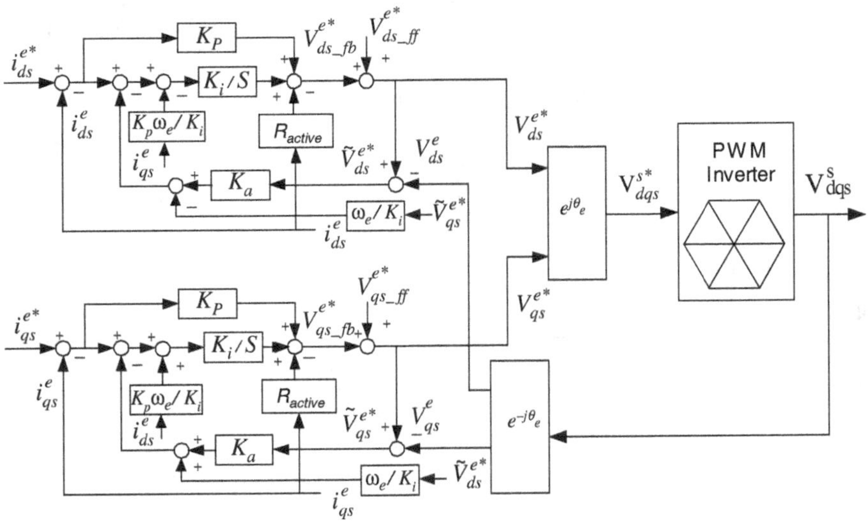

Figure 4.19 Complex vector current regulator for a three-phase AC circuit.

loop is described as the first-order low-pass filter, where the cutoff frequency of the filter is ω_{bw}. However, if the inductances vary according to the operating condition, which is the case in most drive systems, the regulator in Fig. 4.19 reveals a better decoupling effect and a better current regulation performance compared to the regulator in Fig. 4.18. In particular, in an IPMSM drive system, where the q-axis inductance varies by several hundred percentages, the performance improvement by the complex vector current regulator is eminent [14].

In the figure, $\tilde{V}_{ds}^{e*}$ stands for the difference between the commanded voltage of the regulator and the limited voltage by the hexagon boundary in the d axis, and $\tilde{V}_{qs}^{e*}$ stands for that in the q axis.

4.3 SPEED REGULATOR

The torque of an electric machine occurs through the interaction between the flux linkage and the current. Usually, the flux linkage is kept as constant by the permanent magnet or constant field current. Also, the electric machine can be modeled as a converter that converts the current to the torque or to the thrust force. The rotating (moving) speed of the electric machine is decided by the torque (force), friction, the load torque (force), and the inertia (mass) of the rotating (moving) part of the system under control. Hence, if the load torque, inertia, and friction are given, the speed can be regulated by the torque of the machine. So, to control the speed in the speed regulation loop of an electric machine drive system, the torque of the machine should be controlled, and the output of the speed regulator is naturally torque. In this section the design of the speed regulation loop is discussed.

4.3.1 Measurement of Speed/Position of Rotor of an Electric Machine

To regulate the rotor speed, the speed should be measured or estimated. In this section, several methods to measure the position and/or the speed of the rotor are introduced. In the case of the linear electric machine, the position and/or speed also can be measured similarly.

4.3.1.1 Tacho-Generator

A tacho-generator has been used to measure the rotational speed of the rotor. The output voltage of the tacho-generator is proportional to the speed because it is a small DC machine excited by a permanent magnet. The output of the tacho-generators is usually several volts/(1000 r/min), and it includes ripple components due to the limited number of poles of DC machine. The output signal of tacho-generator is given in analog form as voltage, and it is sensitive to measurement noise. Furthermore, at the constant speed the output voltage varies according to the temperature due to the magnet characteristics. Because of the above shortcomings, the accuracy of speed regulation using tacho-generator is limited, and the speed regulation accuracy is usually at best 0.1% of the rated speed in overall operating temperature. Regardless of the shortcomings, the tacho-generator is still used in many simple speed regulation systems due to its merits such as simplicity, mechanical robustness, and direct measurement of the speed without any signal processing.

4.3.1.2 Encoder [15]

To measure the speed or the position, the digital encoder is widely used, which generates pulse train whose frequency is exactly proportional to the speed. The encoder can be classified as the incremental type, where the pulse corresponds to the relative position of the rotor, and the absolute type, where the pulse corresponds to the absolute position of the rotor. For the control of synchronous machine, where the absolute angle of rotor is essential for control, the absolute encoder has been used. However, with the development of techniques of control of the electric machine, the incremental encoders with several auxiliary pulses are getting popularity for the control of the synchronous machine in the industry because of its cost effectiveness. In particular, for the control of the DC machine or the induction machine, where the absolute position has no meaning for speed regulation, the incremental encoders are used without exception.

The incremental-type encoder can be classified according to the operating principles as the magnetic type and the optical type. The maximum number of *pulses-p*er *r*evolution (P_{PR}) of the magnetic-type encoder is usually smaller than that of the optical-type encoder, but the magnetic-type encoder is robust to environmental conditions and is cost effective. Hence, the magnetic encoder is usually used for (a) traction application when the operating environment is hostile and (b) home appliances, where the cost is the first concern. However, in this section the optical encoder, which is the most widely used for high-performance speed control, is introduced in detail. In the control viewpoint, the difference between two types of encoder is only *pulse p*er *r*evolution, and all control concepts developed here can be applied to the speed control system with any type of encoder.

As shown in Fig. 4.20, regardless of absolute or incremental type, the optical encoder consists of a light-emitting diode (LED), a condensing lens, a rotating disk with slits, a light receiver that is usually a photosensitive semiconductor such as a

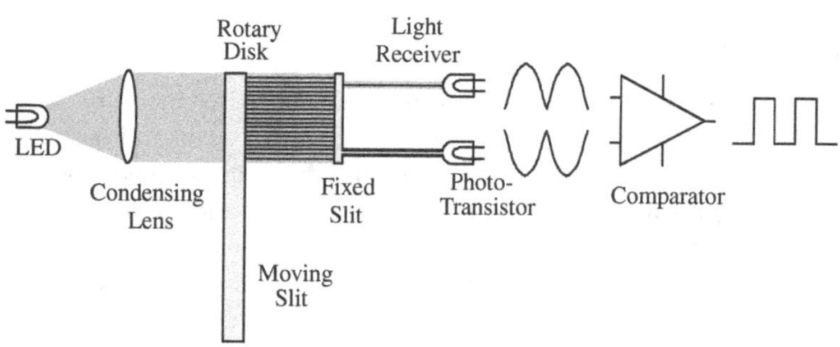

Figure 4.20 Basic configuration of optical encoder.

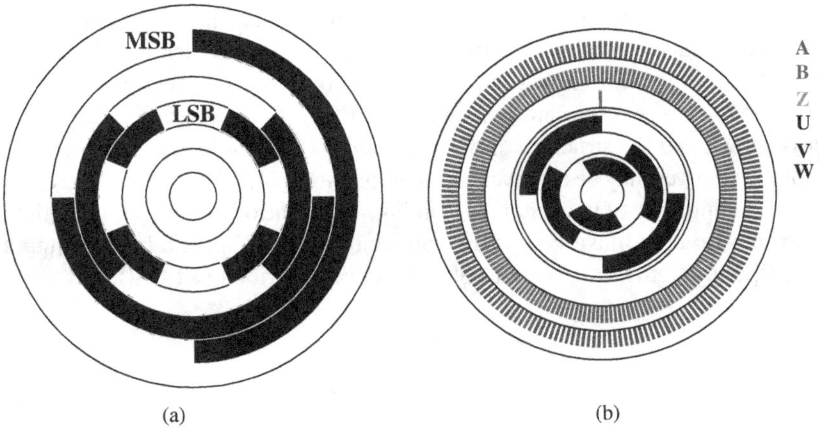

Figure 4.21 Slit of optical encoders. (a) Four-bit absolute encoder. (b) Incremental encoder.

phototransistor, and microelectronics to process the output of the phototransistor. The pulse from the encoder comes in differential output for noise immunity. In Fig. 4.21a, the slit for a 4-bit Gray code absolute encoder is shown, and Fig. 4.21b shows the slit for the incremental encoder, which generates not only incremental "A" pulse and "B" pulse, but also some auxiliary pulses such as an index pulse, "Z," and "U," "V," and "W" pulses for the absolute position of the rotor.

While the absolute encoder provides the rotor position in forms of digital bits, the incremental encoder provides usually two sets of pulse train called "A" pulse and "B" pulse, where the phase of the "A" pulse train is leading that of the "B" pulse at positive rotational direction. In the incremental encoder, the relative position of the rotor can be measured by accumulating the pulses. Usually, the incremental encoders provide a "Z" pulse, which is an index pulse occurring once per revolution at a certain position of the rotor. And by counting the number of pulses of the "A" or "B" pulse train from the "Z" pulse, the absolute position can be identified after the "Z" 'pulse comes out. However, before the "Z" pulse, the absolute position cannot be identified. To solve this problem, "U," "V," and "W," pulses are sometimes provided, which are the pulses with duty factor 50%, and the phase of each pulse train is 120° offset to each other. Hence, the absolute position of the rotor can be identified within ±30° accuracy. In most industry applications, the incremental encoders with less than 8192 P_{PR} are used; but for highly dynamic and very accurate speed and position control, $2^{23}P_{PR}$ is already available. Such a high P_{PR} is usually obtained by interpolation of several thousand P_{PR} signals based on digital signal processing. By using this high P_{PR} encoder, even at less than a few r/min speed, several thousand rad/s of speed regulation bandwidth is possible [16].

4.3.1.3 Resolver

Resolver is a type of synchronous reluctance machine, where the reluctance is the function of the rotor position. The resolver usually has three sets of winding as shown in Fig. 4.22. One is used as an excitation source driven by constant voltage sinusoidal AC source. Also, others are used to detect the rotor position, and each winding is 90° apart spatially to other windings. From the voltage of two sets of the winding, the absolute position of the rotor can be identified. In Fig. 4.22a, the outer shape of a resolver is shown and the structure of the rotor and stator is in Fig. 4.22b. The pole number of the resolver may be any even number, but in Fig. 4.22b a two-pole resolver is depicted for simplicity of understanding. In Fig. 4.22c, the output signals according to the rotor position are illustrated. If the voltage in (4.33) is applied between input terminals, R_1 and R_2, then the output voltage can be described as (4.34).

$$V_{R1-R2} = E \sin \omega t \tag{4.33}$$

$$\begin{aligned} V_{S1-S3} &= KE \sin \omega t \cos(PP\,\theta_r) \\ V_{S2-S4} &= KE \sin \omega t \sin(PP\,\theta_r) \end{aligned} \tag{4.34}$$

where K is the turn ratio between input and output windings, and PP is the number of the pole pair of the resolver, which is half the number of poles. And PP of the resolver should be matched to the number of pole pairs of the electric machine.

As seen from (4.34), the output voltages include the information regarding the rotor position, and by demodulating two output voltages the absolute position can be measured. Because the structure of the resolver is the same as the reluctance machine, the resolver is robust to mechanical stress and can be adapted to the hostile environment. Hence, it is widely used as a position sensor of the electric machine for vehicle application such as a hybrid or electric car. Together with the resolver, the resolver-to-digital converter (RDC) can provide the rotor position and speed directly in digital format. The resolver has many merits; however, because the resolution of the position is quite poor compared to the optical encoder, the application of the resolver to the high-performance speed regulation system is limited.

4.3.2 Estimation of Speed with Incremental Encoder

After getting position information through the resolver or the encoder, for speed regulation or for the feed-forward compensation of the coupling voltage in current regulator mentioned in Sections 4.2.3 and 4.2.5, the speed should be identified. While RDC can provide the speed information directly in the case of the resolver, in the case of an incremental encoder the speed should be estimated from the pulse train. In this section, the methods to calculate the speed from the pulse train are introduced, and a method to estimate the speed based on the observer theory is discussed.

4.3.2.1 Multiplication of Pulse per Revolution (P_{PR})

With more P_{PR}, the speed can be regulated in a higher control bandwidth at lower speed [16]. However, as P_{PR} increases, the cost does and, furthermore, the higher number of P_{PR} encoder is sensitive to mechanical shock. P_{PR} can be increased by

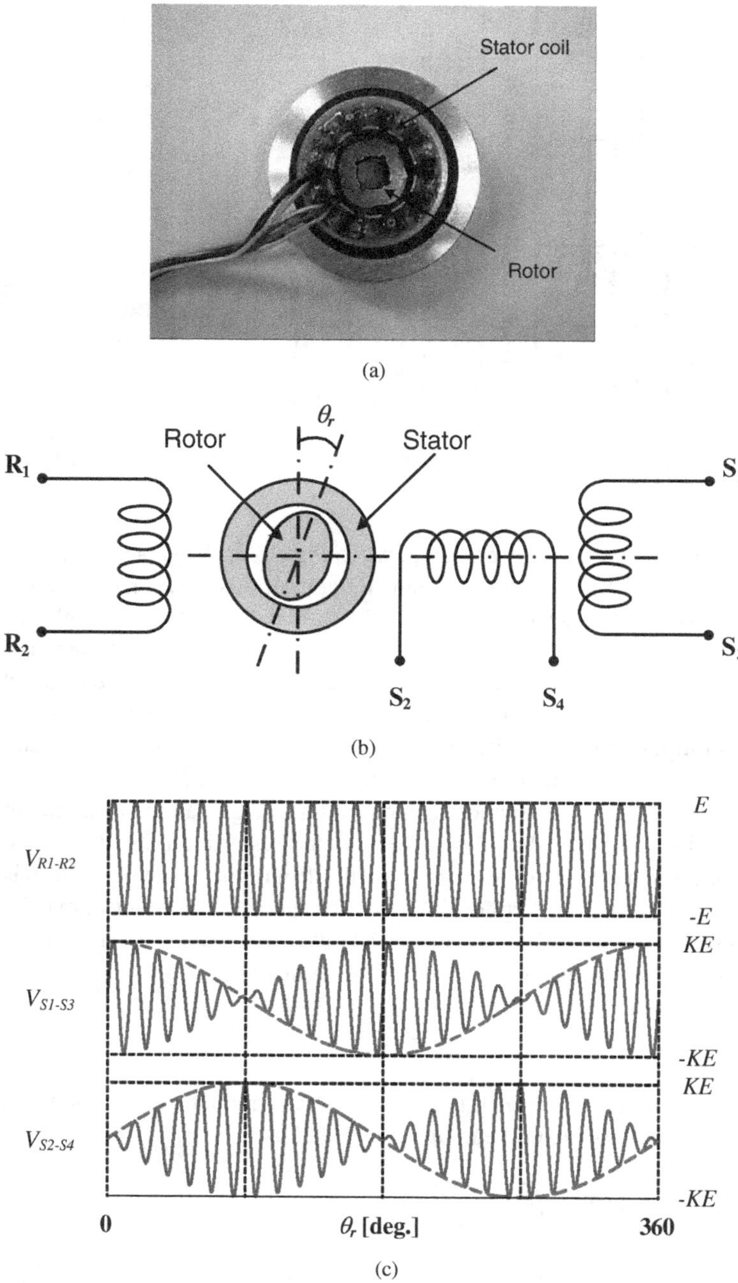

Figure 4.22 Operating principles of a resolver. (a) Outer shape of four-pole resolver. (b) Conceptual diagram of a two-pole resolver. (c) Excitation voltage and output voltage according to the rotor position.

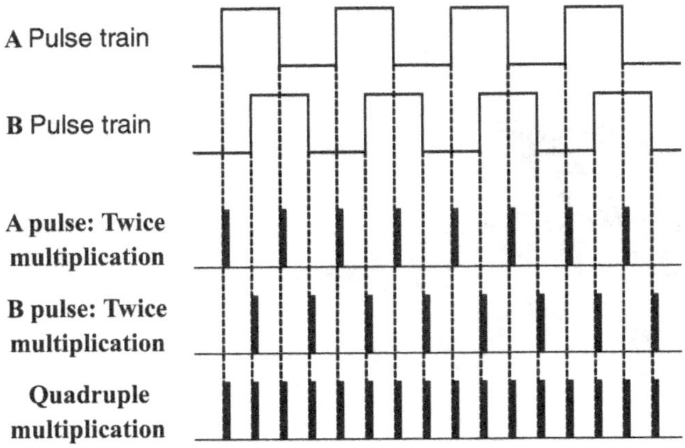

Figure 4.23 Multiplication of P_{PR} by rising and falling edges of pulses.

signal processing of "A" and "B" pulses. As shown in Fig. 4.23, by using falling edge and rising edge of a pulse train between "A" and "B" pulse trains, P_{PR} can be doubled easily. If both edges of both pulse trains are used, P_{PR} can be quadrupled. Though P_{PR} is increased, because of the errors in duty factor of each pulse train and errors in the phase difference between the pulse trains, the time interval between each multiplied pulses may not be constant even at constant rotational speed as shown in Fig. 4.24. So, sometimes, the multiplication of P_{PR} by the above signal processing makes the speed regulation performance worse than the case without multiplication.

The measured value from the encoder is the rotational angle, and in order to get the speed information the angle should be differentiated regarding time. The angle represented by the pulse train is discrete nature, and the differentiation of the angle has multiple options. According to the options, the performance of speed control may be different. In this section, several differentiation methods, which are widely used in the industry, are discussed with their merits and shortcomings [17].

4.3.2.2 "M" Method

This method is counting the number of pulses for a fixed time interval as shown in Fig. 4.25. The speed by this method can be represented as

$$N_M = \frac{m\alpha}{T_{sp}} = \frac{60m}{P_{PR}T_{sp}} \qquad \text{(r/min)} \tag{4.35}$$

where α stands for the rotational angle of the rotor in mechanical degree between two adjacent pulses, and T_{sp} stands for the sampling period of the speed.

The "M" method has merits such that the implementation is simple and that the measurement interval, which is the speed sampling period, is constant at any speed. But the speed error, which is expressed by $\frac{60}{P_{PR}T_{sp}}$ (r/min), is constant regardless of the speed, and at lower speed the relative accuracy and resolution of the speed measure-

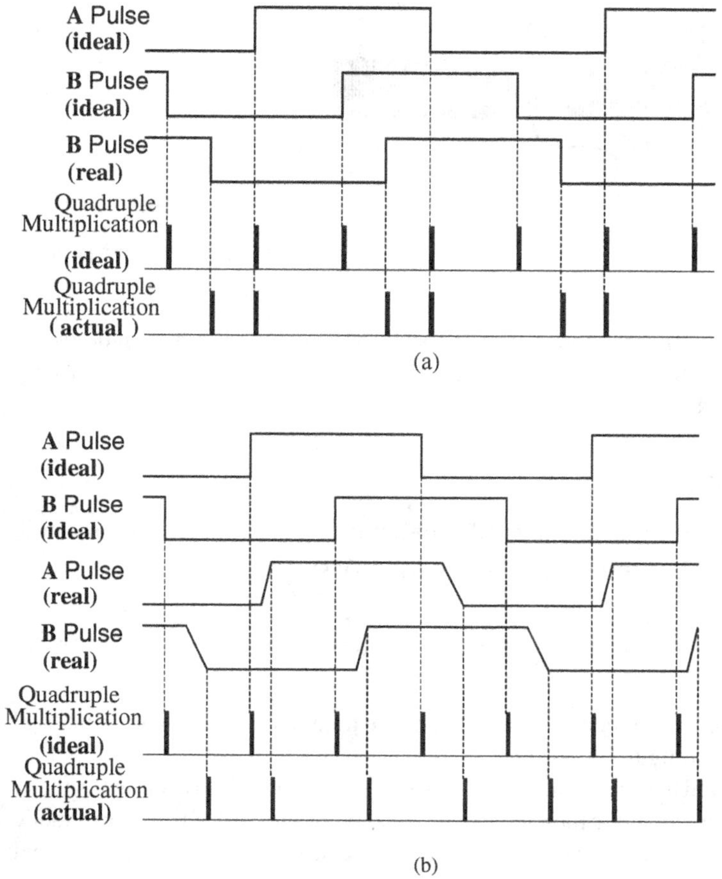

Figure 4.24 Problems of multiplication of P_{PR} by signal processing. (a) Problem due to phase errors between "A" and "B" pulse trains. (b) Problems due to nonideal duty factors.

ment is getting poorer. With a conventional several thousand P_{PR} encoder, the speed measurement error would be several tens of r/min with the several hundred micro-seconds sampling interval. And the speed control performance may be unacceptable for most servo applications. However, if the extremely high P_{PR} encoder is used such

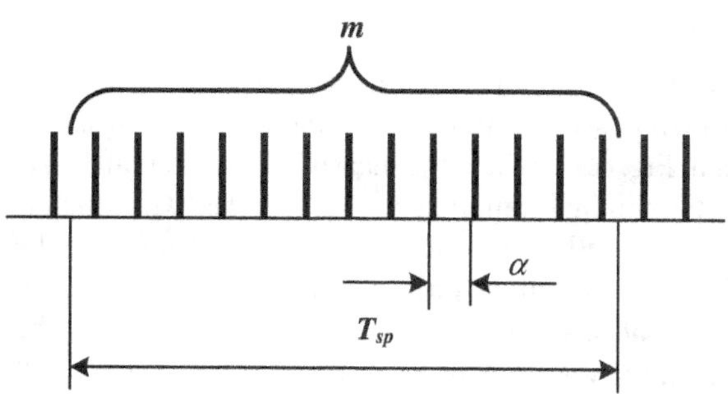

Figure 4.25 "M" method.

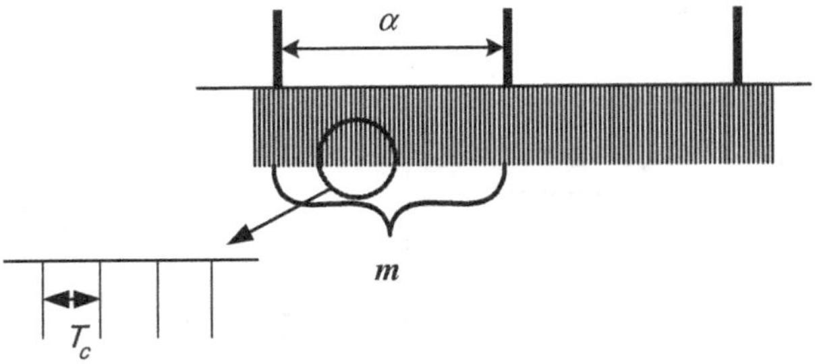

Figure 4.26 "T" method.

as $2^{23}P_{PR}$, by calculating the speed even at less than a few r/min with a hundreds-of-microseconds sampling interval, the speed regulation bandwidth can be extended up to several thousand rad/s.

4.3.2.3 "T" Method

This method measures the time interval between two adjacent pulses as shown in Fig. 4.26. The speed by this method can be represented as

$$N_T = \frac{\alpha f_c}{m} = \frac{60 f_c}{m P_{PR}} \quad \text{(r/min)} \tag{4.36}$$

where f_c stands for the frequency of the clock pulse train to measure the time interval and $f_c = 1/T_c$ in the figure.

The "T" method has merits such that the speed can be measured accurately at low speed and that time delay for the measurement is negligible. But it needs arithmetic division to calculate speed, which is usually a burden to a low-cost digital signal processor. Also, because the sampling interval varies according to the speed, though the delay of the measurement itself is negligible, the calculated speed information may have additional time delay with the constant sampling frequency digital speed regulation loop due to the asynchronization between the execution of the regulation algorithm and the sampling of the speed. Also, to get higher resolution at extremely low speed, where α would be very large, the number of clock pulse would be too large to handle.

4.3.2.4 M/T Method

In the industry, a method mixed with "T" and "M" methods is used widely [17]. This method gets the merits of both methods while minimizing the shortcomings. The principle of the method is shown in Fig. 4.27, where the measurement interval, T_D, is synchronized to the pulse right after nominal sampling period, T_{sp}. And T_D, which is the speed sampling interval, varies slightly larger than T_{sp} by ΔT at most of the operating speeds as shown in Fig. 4.27. The slight variation of the sampling interval may not be a problem in the digital control system. The effect of the variation may be almost the same effect with the slight variation of integrator gain if a digital PI

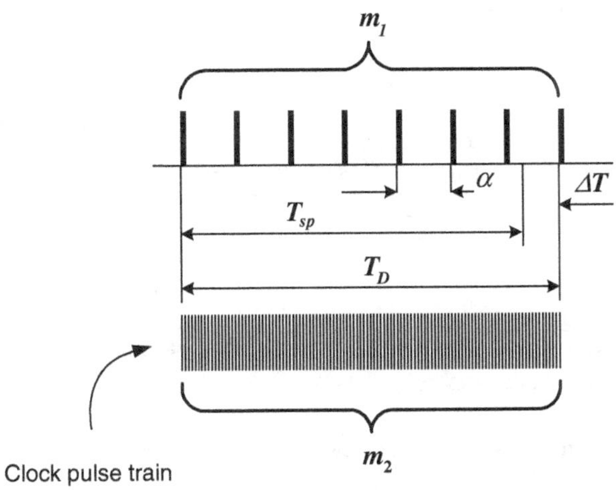

Figure 4.27 M/T method.

regulator is used for the speed regulation. The speed by this method can be represented as (4.37). As seen from (4.37), m_1 and m_2 are measured discrete integer values, and based on those values the speed can be calculated. The M/T method combines the T method, which reveals accurate speed measurement at lower speed, and the M method, which reveals accurate speed measurement at higher speed. And it reveals reasonably accurate speed measurement performance overall speed range. But, at extremely low speed, it operates like the T method, whose shortcoming is that variable sampling time occurs. And eventually, ΔT may be larger than the nominal sampling interval, T_{sp}; below a certain speed, the speed regulation bandwidth would be reduced remarkably [18].

$$N_{M/T} = \frac{m_1\,\alpha}{P_{PR}\left(T_{sp} + \Delta T\right)} = \frac{60 f_c m_1}{m_2 P_{PR}} \qquad (\mathrm{r/min}) \tag{4.37}$$

The speed measurement resolution, $Q_n(\mathrm{r/min})$, and the actual speed sampling interval, T_D, according to the speed can be represented as (4.38) and (4.39), respectively.

$$Q_N = \frac{60 f_c m_1}{P_{PR}}\left(\frac{1}{m_2 - 1} - \frac{1}{m_2}\right) = \frac{60 f_c m_1}{m_2(m_2 - 1)P_{PR}} \tag{4.38}$$

$$\begin{cases} T_D = T_{sp} & (\Delta T = 0) \\ T_D = \left\{\left[\mathrm{Integer}\left(\dfrac{P_{PR} N_{M/T} T_{sp}}{60}\right)\right] + 1\right\} \dfrac{60}{P_{PR} N_{M/T}} & (\Delta T \neq 0) \end{cases} \tag{4.39}$$

where $\mathrm{Integer}(x)$ means integer part of the number x, and $N_{M/T}$ represents speed of the rotor in r/min.

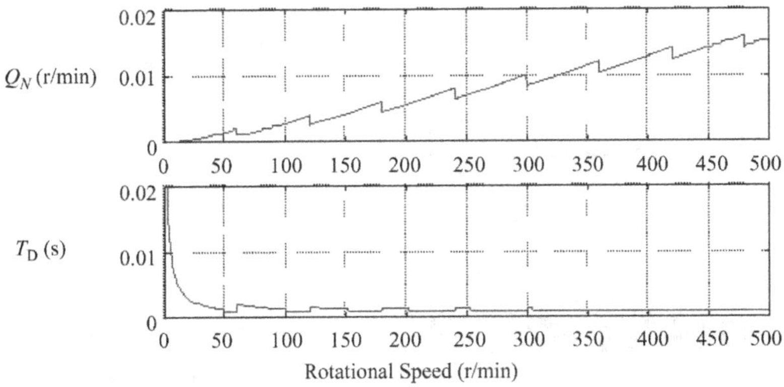

Figure 4.28 Resolution and detection time of speed measurement by the M/T method.

The above equations are plotted in the case of $P_{PR} = 1000$, $f_c = 30\,\text{MHz}$, and $T_c = 1$ ms from 3 r/min to 500 r/min in Fig. 4.28. As seen from the figure and (4.38), the resolution is getting worse as the speed increases, but the relative resolution defined as $Q_N/N_{M/T}$ is almost constant regardless of the speed. To increase the resolution, m_2 should be increased. But increasing m_2 above the resolution of encoder pulse train itself has no meaning in improving the resolution. The speed detection time, which is the actual speed sampling interval, is getting longer as the speed decreases, but above 100 r/min, it is almost constant as the nominal sampling interval as 1 ms in this example. At the extremely low speed, as shown in Fig. 4.29, the detection time, T_D, would be larger, the time delay in the speed measurement increases, and the delay may limit the speed regulation bandwidth.

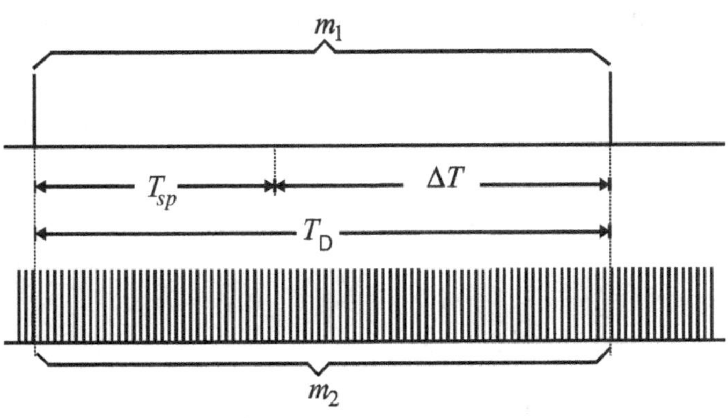

Figure 4.29 Time delay of the speed measurement by the M/T method at extremely low speed.

4.3.3 Estimation of Speed by a State Observer

In the design of a control system of the electric machines and power converters, some state variables are fed back for the closed-loop control to improve the control performance. However, in many cases the state variables are not measurable or the measurement of the variables is costly. In this section, a method to estimate the states in a time-invariant linear system is described based on the observer theory. In particular, a method to estimate the speed of the electric machine from the measured rotor angle by an encoder is discussed.

4.3.3.1 Full-Order Observer [19]

The state variable of a time-invariant linear system can be estimated directly from the state equation itself in (4.40) if the state is observable.

$$\dot{\hat{\mathbf{x}}} = \mathbf{A}\hat{x} + \mathbf{Bu} \tag{4.40}$$

where vector $\hat{\mathbf{x}}$ stands for the estimated state of the state variable vector $\mathbf{x}$, $\mathbf{A}$ stands for the system matrix, $\mathbf{B}$ stands for the input matrix or vector, and u stands for the input vector. If the initial value of the state vector, $\mathbf{x}(0)$, is known accurately and matrices $\mathbf{A}$ and '$\mathbf{B}$ are known correctly, $\hat{\mathbf{x}}$ can be calculated by (4.40). But, usually the initial value of the state is unknown or even if it is known, matrices $\mathbf{A}$ and $\mathbf{B}$ may have errors. Then the estimated state, $\hat{\mathbf{x}}$, would deviate from the real state, $\mathbf{x}$. By defining state error, $\tilde{\mathbf{x}}$, as (4.41), the error dynamics of the state variable can be described as (4.42):

$$\tilde{\mathbf{x}} = \mathbf{x} - \hat{\mathbf{x}} \tag{4.41}$$

$$\dot{\tilde{\mathbf{x}}} = \mathbf{A}\tilde{\mathbf{x}}, \qquad \tilde{\mathbf{x}}(0) = \mathbf{x}(0) - \hat{\mathbf{x}}(0) \tag{4.42}$$

If the system is stable, which means that all eigenvalues of the system lie in the left half-plane of Laplace domain in the case of a continuous system, then the error, $\tilde{\mathbf{x}}$, would converge to zero. The convergence rate is decided by the magnitude of the eigenvalues. As the magnitude decreases, the convergence rate decreases. To improve the convergence rate and the accuracy of the estimation, a part of the state error, $\tilde{\mathbf{x}}$, or the function of state error, $\tilde{\mathbf{y}}$, can be fed back to the system as shown in Fig. 4.30, which

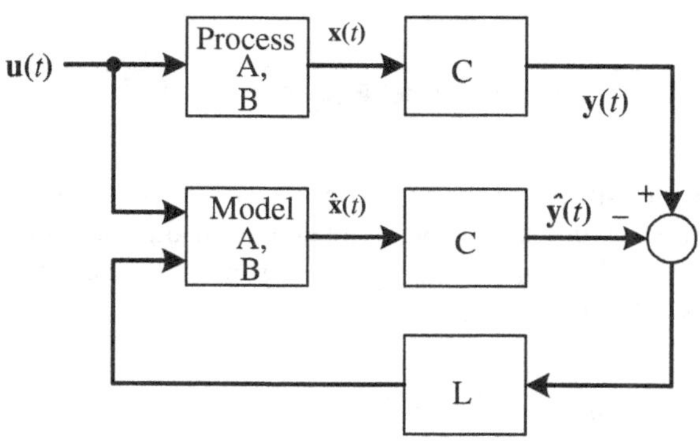

Figure 4.30 Closed-loop observer.

can be represented as (4.43). Here, $\mathbf{C}$ is the output matrix, which generates the measurable output, $\mathbf{y}$, of the system from the state, $\mathbf{x}$.

$$\dot{\hat{\mathbf{x}}} = \mathbf{A}\hat{\mathbf{x}} + \mathbf{B}\mathbf{u} + \mathbf{L}(\mathbf{y} - \mathbf{C}\hat{\mathbf{x}}) \tag{4.43}$$

where $\mathbf{L}$ is the proportional gain matrix, which is set to get the satisfactory performance of estimation of the state. The error dynamic of (4.42) can be modified to (4.44) with the closed-loop estimation process in Fig. 4.30.

$$\dot{\tilde{\mathbf{x}}} = (\mathbf{A} - \mathbf{L}\mathbf{C})\tilde{\mathbf{x}} + \mathbf{G}_1\mathbf{w} \tag{4.44}$$

where w stands for a vector representing disturbance of the system, and G_1 stands for the matrix to model the effect of the disturbance to the state. The characteristic equation (4.44) is deduced as

$$\det[s\mathbf{I} - (\mathbf{A} - \mathbf{L}\mathbf{C})] = 0 \tag{4.45}$$

where $\det[\bullet]$ means the determinant of matrix $[\bullet]$.

If $\mathbf{L}$ is set as the system $\mathbf{A}$-$\mathbf{LC}$ to be stable and if all matrices $\mathbf{A}$, $\mathbf{B}$, and $\mathbf{C}$ have no errors, then $\tilde{\mathbf{x}}$ would converge to zero in the case of no disturbance like $\mathbf{w} = 0$. Furthermore, by adjusting the gain matrix $\mathbf{L}$, regardless of the initial value, $\hat{\mathbf{x}}(0)$, $\tilde{\mathbf{x}}$ converges to zero and the convergence rate can be improved compared to the rate by (4.42). However, even if $\mathbf{A}$, $\mathbf{B}$, and $\mathbf{C}$ have reasonably small errors and if there are some disturbance like $\mathbf{w} \neq 0$, by adjusting $\mathbf{L}$, $\tilde{\mathbf{x}}$ can converge approximately to zero in the practical sense. If the roots of the characteristic equation, (4.45), are set as (4.46), then the equation can be represented as (4.47) in terms of the roots.

$$s_i = \beta_1, \beta_2, \ldots, \beta_n \tag{4.46}$$

$$\alpha_e(s) = (s - \beta_1)(s - \beta_2)(s - \beta_n) = 0 \tag{4.47}$$

By comparing the coefficient of equations (4.47) and (4.45), the proportional gain $\mathbf{L}$ can be calculated.

4.3.3.2 Speed Estimation from the Measured Encoder Angle by a Full-Order Observer [20]

To apply the above-mentioned state estimation method to an electric machine control system, where the speed of the machine should be estimated from the measured encoder angle, from a control block diagram in Fig. 4.31 and from the linear time-invariant motion equation (4.48), the state equation can be derived as (4.49).

$$T_e = J\frac{d\omega_{rm}}{dt} + B\omega_{rm} + T_L \tag{4.48}$$

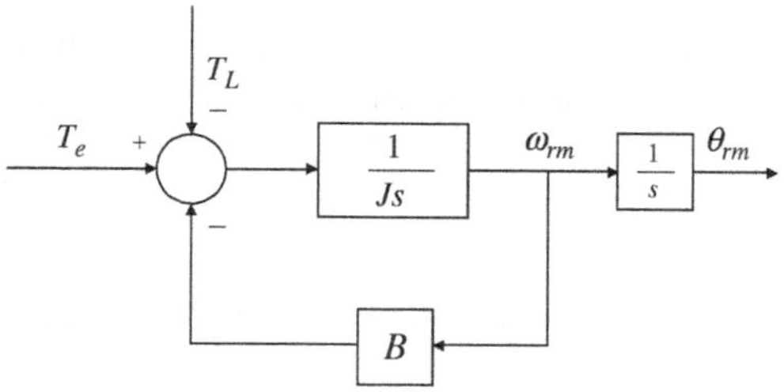

Figure 4.31 Control block diagram of a simple mechancical system.

$$\frac{d}{dt}\begin{bmatrix}\theta_{rm}\\ \omega_{rm}\\ T_L\end{bmatrix}=\begin{bmatrix}0 & 1 & 0\\ 0 & -\frac{B}{J} & -\frac{1}{J}\\ 0 & 0 & 0\end{bmatrix}\begin{bmatrix}\theta_{rm}\\ \omega_{rm}\\ T_L\end{bmatrix}+\begin{bmatrix}0\\ \frac{1}{J}\\ 0\end{bmatrix}T_e \tag{4.49}$$

where the state of the system is given by $\mathbf{x}=\begin{bmatrix}\theta\\ \omega\\ T_L\end{bmatrix}$ and the load torque, T_L, is augumented as a state under the assumption that $dT_L/dt=0$. The assumption is valid as long as the load torque varies much slowly compared to the variation of other state variables such as speed and angle. From (4.49), the observer equation such as (4.43) can be described as (4.50), which can be depicted as a block diagram shown in Fig. 4.32.

$$\frac{d}{dt}\begin{bmatrix}\hat{\theta}_{rm}\\ \hat{\omega}_{rm}\\ \hat{T}_L\end{bmatrix}=\begin{bmatrix}0 & 1 & 0\\ 0 & -\frac{\hat{B}}{\hat{J}} & -\frac{1}{\hat{J}}\\ 0 & 0 & 0\end{bmatrix}\begin{bmatrix}\hat{\theta}_{rm}\\ \hat{\omega}_{rm}\\ \hat{T}_L\end{bmatrix}+\begin{bmatrix}0\\ \frac{1}{\hat{J}}\\ 0\end{bmatrix}T_e+\begin{bmatrix}l_1\\ l_2\\ l_3\end{bmatrix}\left(\theta_{rm}-[1\ \ 0\ \ 0]\begin{bmatrix}\hat{\theta}_{rm}\\ \hat{\omega}_{rm}\\ \hat{T}_L\end{bmatrix}\right) \tag{4.50}$$

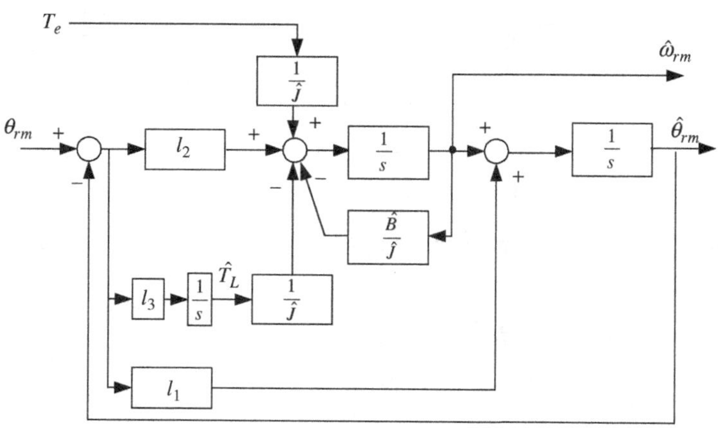

Figure 4.32 Block diagram of a speed observer.

The gain matrix, **L**, should be set to guarantee the stability of the system in Fig. 4.32 and to meet the reasonable speed estimation performance not only in the steady state but also in the transient state.

The characteristic equation of (4.50) can be deduced as

$$\det\left[s\mathbf{I}-(\mathbf{A}-\mathbf{LC})\right] = s^3 + \frac{l_1 J + B}{J}s^2 + \frac{l_2 J + l_1 B}{J}s - \frac{l_3}{J} = 0 \tag{4.51}$$

Equation (4.47) can be rewritten as

$$\begin{aligned} \alpha_e(s) &= (s-\beta_1)(s-\beta_2)(s-\beta_3) \\ &= s^3-(\beta_1+\beta_2+\beta_3)s^2+(\beta_1\beta_2+\beta_2\beta_3+\beta_3\beta_1)s-\beta_1\beta_2\beta_3 = 0 \end{aligned} \tag{4.52}$$

By comparing (4.51) and (4.52), the elements of **L** can be evaluated as

$$\begin{aligned} l_1 &= -(\beta_1+\beta_2+\beta_3)-\frac{B}{J} \\ l_2 &= (\beta_1\beta_2+\beta_2\beta_3+\beta_3\beta_1)-\frac{B}{J}l_1 \\ &= (\beta_1\beta_2+\beta_2\beta_3+\beta_3\beta_1)+(\beta_1+\beta_2+\beta_3)\frac{B}{J}+\left(\frac{B}{J}\right)^2 \\ l_3 &= \beta_1\beta_2\beta_3 J \end{aligned} \tag{4.53}$$

If the roots of (4.52) are set as triple roots as β, then (4.53) can be simplified as

$$\begin{aligned} l_1 &= -3\beta-\frac{B}{J} \\ l_2 &= 3\beta^2-\frac{B}{J}l_1 \\ &= 3\beta^2+3\beta\frac{B}{J}+\left(\frac{B}{J}\right)^2 \\ l_3 &= \beta^3 J \end{aligned} \tag{4.54}$$

If the friction of the system is neglected as $B = 0$, then (4.54) can be further simplified as

$$\begin{aligned} l_1 &= -3\beta \\ l_2 &= 3\beta^2 \\ l_3 &= \beta^3 J \end{aligned} \tag{4.55}$$

The observer bandwidth decided by the root, β and parameter errors, should be large enough compared to the speed regulation bandwidth such as at least three times the speed regulation bandwidth. The bandwidth of an observer can be evaluated by calculating the magnitude of the transfer functions in (4.58) and (4.59) in the next section. If β is too small, then the convergence is too slow. But, if β is too large, the system may be unstable due to measurement noise, time delays in the signal processing, and errors in system modeling matrices, **A**, **B**, and **C**. Instead of triple roots, the roots can be set as the characteristic poles of a third-order Butterworth low-pass filter, where the cutoff frequency of the filter decides the convergence rate. As the cutoff frequency increases, the convergence rate increases but the stability margin decreases. In another way, the roots may be set according to the optimal control theory, which optimizes a performance index.

4.3.3.3 Physical Understanding of a Full-Order Observer

Equation (4.50) can be rewritten as (4.56), which can be depicted as Fig. 4.33 in a control block diagram. From the figure, it can be seen that the speed observer is a kind of closed-loop regulator, which regulates the estimated angle, $\hat{\theta}_{rm}$, according to the reference input, θ_{rm}. The error between the reference input and the actual value is the input to a proportional, integral, and differential (PID) regulator. Setting the observer gain matrix, **L**, is equivalent to setting the gains of PID regulator. To circumvent differential operation (dotted part in Fig. 4.33), the differential term is added not to the torque junction but to the speed junction, where its dimension is rad/s as shown in Fig. 4.33. Then the block diagram of the PID regulator is exactly the same as the diagram in Fig. 4.32.

$$\dot{\hat{x}} = \boldsymbol{A}\hat{x} + \boldsymbol{B}u + \boldsymbol{L}(\mathrm{y} - \boldsymbol{C}\hat{x})$$

$$= \begin{bmatrix} 0 & 1 & 0 \\ 0 & -\dfrac{\hat{B}}{\hat{J}} & -\dfrac{1}{\hat{J}} \\ 0 & 0 & 0 \end{bmatrix} \begin{bmatrix} \hat{\theta}_{rm} \\ \hat{\omega}_{rm} \\ \hat{T}_L \end{bmatrix} + \begin{bmatrix} 0 \\ -\dfrac{1}{\hat{J}} \\ 0 \end{bmatrix} T_e + \begin{bmatrix} l_1 \\ l_2 \\ l_3 \end{bmatrix} \left(\theta_{rm} - \begin{bmatrix} 1 & 0 & 0 \end{bmatrix} \begin{bmatrix} \hat{\theta}_{rm} \\ \hat{\omega}_{rm} \\ \hat{T}_L \end{bmatrix} \right)$$

$$\begin{bmatrix} \dot{\hat{\theta}}_{rm} \\ \dot{\hat{\omega}}_{rm} \\ \dot{\hat{T}}_L \end{bmatrix} = \begin{bmatrix} \hat{\omega}_{rm} + l_1 \left(\theta_{rm} - \hat{\theta}_{rm} \right) \\ -\hat{B}\hat{J}\hat{\omega}_{rm} - \dfrac{1}{\hat{J}} \hat{T}_L + \dfrac{1}{\hat{J}} T_e + l_2 \left(\theta_{rm} - \hat{\theta}_{rm} \right) \\ l_3 \left(\theta_{rm} - \hat{\theta}_{rm} \right) \end{bmatrix} \tag{4.56}$$

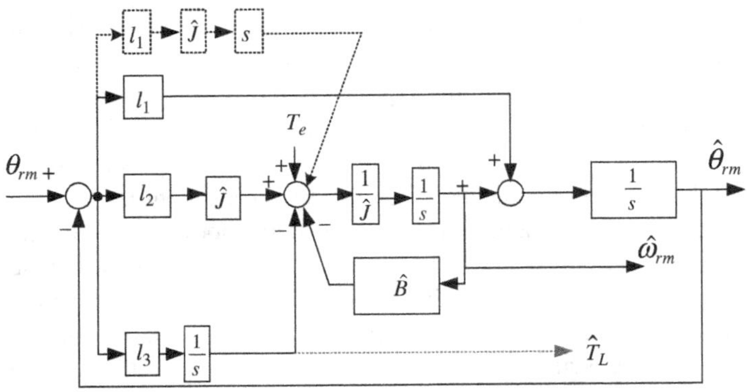

Figure 4.33 Physical understanding of an observer as a closed-loop regulator.

In (4.56) and Fig. 4.33, l_3 stands for integral gain, l_2 stands for proportional gain, and l_1 stands for differential gain. Figure 4.33 can be redrawn as Fig. 4.34 to emphasize the regulator aspect of the observer.

The gains of PID regulator can be set through the analysis of the transfer function. By substituting $T_e = J\frac{d\omega_{rm}}{dt} + B\omega_{rm}$, $\omega_{rm} = \frac{d\theta_{rm}}{dt}$, and $T_e = (Js^2 + Bs)\theta_{rm}$ into (4.56), the transfer function between the estimated angle and measured angle can be deduced as

$$\frac{\hat{\theta}_{rm}}{\theta_{rm}} = \frac{Js^3 + (B + \hat{J}K_1)s^2 + (\hat{B}K_1 + K_2)s + K_3}{\hat{J}s^3 + (\hat{B} + \hat{J}K_1)s^2 + (\hat{B}K_1 + K_2)s + K_3} \tag{4.57}$$

From (4.57), it can be said that at lower-frequency regions (considering $s = j\omega$), where the magnitude of the sum of the constant and the first order of s term, $(\hat{B}K_1 + K_2)s + K_3$, is much larger than that of the sum of the third and the second order of s term, $Js^3 + (B + \hat{J}K_1)s^2$, the estimated angle would well track the measured angle regardless of the parameter errors. Hence, even if there are some mismatches between actual inertia and friction coefficient, J and B, and nominal ones, $\hat{J}$ and $\hat{B}$, the estimated angle matches well to the real measured angle as long as

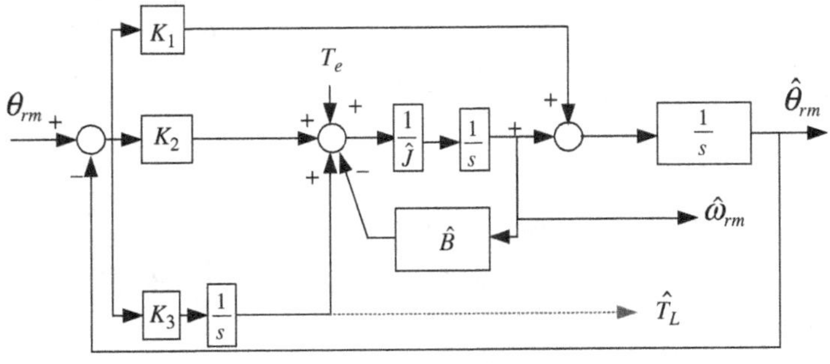

Figure 4.34 Representation of an observer as a closed-loop PID regulator. $K_1 = l_1$, $K_2 = \hat{J}l_2$, $K_3 = -l_3$.

the variation of angle is slow enough to guarantee $|(\hat{B}K_1 + K_2)j\omega + K_3| \gg |J(j\omega)^3 + (B + \hat{J}K_1)(j\omega)^2|$. In addition, if $\hat{J}K_1$ is much larger than the friction coefficient, B, the second-order terms in the numerator and denominator of the transfer function would be almost the same and that the regulation performance—that is, estimation performance—would be improved. Because the parameter errors in J and B are affected only in a higher-frequency region, if the bandwidth of the control system given by the transfer function is set high enough compared to the speed regulation bandwidth, the estimated speed, $\hat{\omega}_{rm}$, from the observer can be used as a speed feedback information without the direct measurement of the speed [21]. The transfer function between estimated speed and actual speed can be derived as (4.58) from (4.57) and Fig. 4.34.

$$\frac{\hat{\omega}_{rm}}{\omega_{rm}} = \frac{Js^3 + (B + JK_1)s^2 + (BK_1 + K_2)s + K_3}{\hat{J}s^3 + (\hat{B} + \hat{J}K_1)s^2 + (\hat{B}K_1 + K_2)s + K_3} \tag{4.58}$$

Comparing to the transfer function in (4.57), it can be seen that the estimated speed in (4.58) is more sensitive to the parameter errors, where the first-order terms in the denominator and numerator do not match in the case of error in B. However, if the estimated speed is obtained from the differentiated state of the estimated angle as shown in Fig. 4.35, the transfer function between estimated speed and actual speed is exactly same as that in (4.57). And the estimation of the speed is rather robust to the error in B. But in this case, the angle error directly affects the estimated speed, and the estimated speed would be sensitive to the measurement noise that is included in the measured angle, θ_{rm}.

$$\frac{\hat{\theta}_{rm}}{\theta_{rm}} = \frac{\hat{\omega}_{rm}}{\omega_{rm}} = \frac{Js^3 + (B + \hat{J}K_1)s^2 + (\hat{B}K_1 + K_2)s + K_3}{\hat{J}s^3 + (\hat{B} + \hat{J}K_1)s^2 + (\hat{B}K_1 + K_2)s + K_3} \tag{4.59}$$

Another way to set the gains of the observer represented as a PID regulator is to tune the gains as PID regulator in a general closed-loop control system. At the first step, in Fig. 4.35, input, θ_{rm}, is given by a function generator as a square wave, whose

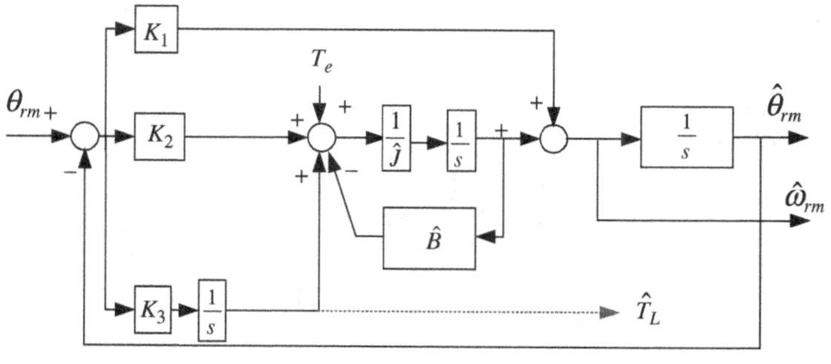

Figure 4.35 Modified speed observer robust to parameter error but sensitive to measurement noise.

frequency is around 10% of the desired bandwidth of the observer, and the magnitude should be small enough not to saturate any state variable and internal variable in the closed loop but large enough to increase signal-to-noise ratio. By observing the output, $\hat{\theta}_{rm}$, the gains can be adjusted as follows. In the tuning process, the feed-forwarding term, T_e, is kept as zero. And at the second step, all PID gains are set as zero and then increase the proportional gain, K_2, until the output, $\hat{\theta}_{rm}$, gives 10% overshoot to the input. If the system is unstable without the differential gain in tuning of K_2, then set K_1 as a minimum value to make the system stable and tune K_2 as mentioned before with this minimum K_1. After fixing K_2 as tuned, next, increases differential gain, K_1, until the overshoot disappears. After setting K_1 and K_2, increase the integral gain, K_3, again until the output, $\hat{\theta}_{rm}$, gives 10% overshoot to the input. After setting all gains, the control loops including the mechanical system in Fig. 4.31 are closed, and the gains may be fine-tuned with the measured angle and with the feed-forwarding term [22].

4.3.3.4 Disturbance Observer [23]

If the bandwidth of the speed observer is high enough, as shown in Fig. 4.35, the disturbance torque, T_L, can be estimated as a byproduct of the speed observer. By feed-forwarding this estimated disturbance torque, $\hat{T}_L$, to the speed regulation loop as shown in Fig. 4.36, the disturbance rejection performance of the speed regulation loop can be enhanced.

Because the disturbance torque obtained from the speed observer comes from the integral term of the observer, the response may be sluggish. To enhance the performance of the estimation of the disturbance torque, a separately designed disturbance torque observer can be augmented to the system as shown in Fig. 4.37. In this configuration, the input to the disturbance observer is the measured encoder angle like input to the speed observer but the torque, which is the feed-forwarding term to the observer, is different in each observer. While the torque in the speed observer is the reference torque, T_e^*, to the mechanical system, that in the disturbance observer is the sum of the reference torque, T_e^*, and the compensated disturbance torque, $\hat{T}_L$, that is, T_e^{**}. The gains of the disturbance observer, K_{1L}, K_{2L}, and K_{3L}, can be tuned separately with the gains of the speed observer to get the best performance in the disturbance estimation.

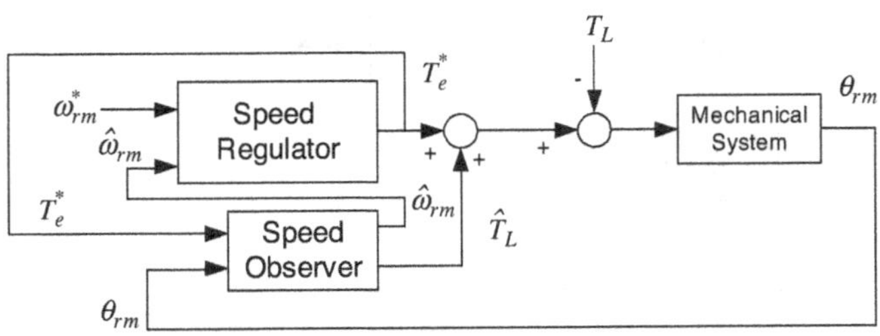

Figure 4.36 Feed-forwarding of estimated disturbance torque.

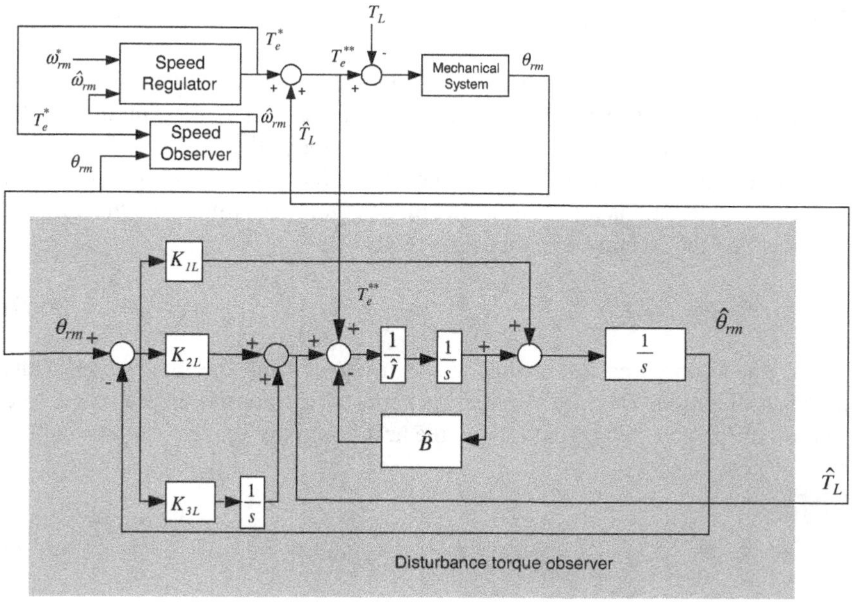

Figure 4.37 Separately designed disturbance observer.

4.3.3.5 Implementation of Observer in Discrete Time Domain [24]

The observers in the continuous time domain shown in Fig. 4.35 can be implemented in a discrete time domain as Fig. 4.38. In Fig. 4.38, the three integrators in Fig. 4.35 are implemented differently in a discrete time domain to reduce the discretization error. In

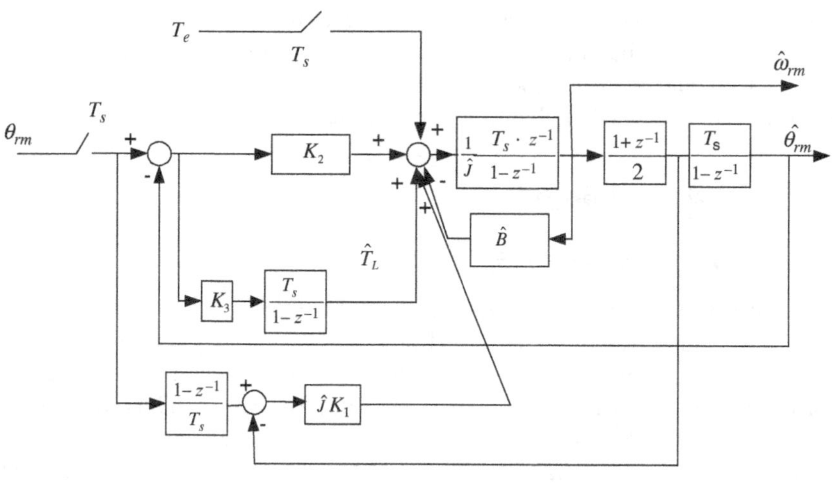

Figure 4.38 Implementation of speed observer in discrete time domain.

the figure, T_s stands for the sampling period of the observer and z stands for the z-transform operator.

The error in the estimated angle varies in impulse manner at the sampling instant, and the backward difference method in (4.60) is the best one to minimize the discretization error. In the backward difference method, the input to the present sampling time point is used as the input for the integration. So, the impulse response in the discrete time domain is the same as that in the continuous time domain. And, the method is called the impulse invariance method.

$$\frac{1}{s} \rightarrow \frac{T_s z}{z-1} = \frac{T_s}{1-z^{-1}} \tag{4.60}$$

In the speed observer, the estimated speed comes from the integration of the estimated acceleration. The fastest dynamics of the acceleration is the step type. And the forward difference method in (4.61) is the best one to minimize the discretization error.

$$\frac{1}{s} \rightarrow \frac{T_s}{z-1} = \frac{T_s z^{-1}}{1-z^{-1}} \tag{4.61}$$

To calculate the estimated angle, the input to the integrator is the estimated speed. The fastest dynamics of the speed is ramp type, because the acceleration is assumed as step function. And the bilinear transformation method, the so-called Tustin's method, in (4.62) is the best one to minimize the discretization error.

$$\frac{1}{s} \rightarrow \frac{T_s}{2}\frac{z+1}{z-1} = \frac{T_s}{2}\frac{1+z^{-1}}{1-z^{-1}} \tag{4.62}$$

In Fig. 4.35, the differential operation in the continuous time domain, where the operation is very sensitive to the measurement noise, is circumvented. But in discrete time domain the difference can be used to implement the differential operation, and the difference term can be implemented as Fig. 4.38 directly. The difference of the actual speed itself and estimated speed in discrete time domain can be obtained from the measured and estimated angle through the digital operation, $\frac{1-z^{-1}}{T_s}$.

4.3.4 PI/IP Speed Regulator

4.3.4.1 PI Speed Regulator [25]

The design of a proportional and integral (PI) regulator, which is the most widely used regulator in the industry, can be described in a speed control system as shown in Fig. 4.39, where a mechanical system is driven by an electric machine, whose torque is controlled instantaneously. In the figure, the torque due to the friction is considered as a part of load torque, and the current regulator is modeled as the first-order low-pass filter as mentioned in Section 4.2.3.1. In the case of a DC machine, it is assumed that the torque is directly proportional to the armature current neglecting the armature

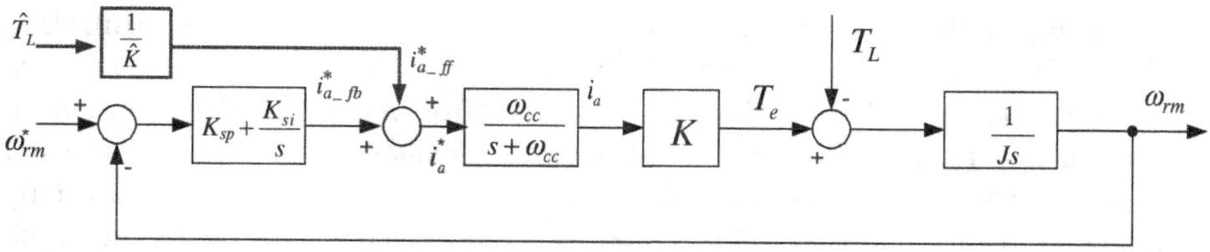

Figure 4.39 Proportional in integral (PI) speed regulator in a speed control loop.

reaction and other stray effects. In the case of an AC machine, it can also be assumed that the torque is directly proportional to the torque component current by the vector control, which is described in Chapter 5. In the figure, ω_{sc} stands for the control bandwidth of the speed regulator and ω_{cc} stands for that of the current regulator.

The transfer function of PI regulator can be described as

$$G_s(s) = K_{sp} + \frac{K_{si}}{s} \tag{4.63}$$

where K_{sp} and K_{si} are the proportional and integral gain of the regulator, respectively. If the torque constant, K, and the system inertia, J, is known accurately, then the open-loop transfer function, $G_{sc}(s)$, of the system in Fig. 4.39 can be derived as

$$G_{sc}(s) = \left(K_{sp} + \frac{K_{si}}{s}\right) \cdot \frac{\omega_{cc}}{s+\omega_{cc}} \cdot \frac{K}{Js} \tag{4.64}$$

The Bode plot of $G_{sc}(s)$ in magnitude can be approximately drawn as the plot in Fig. 4.40.

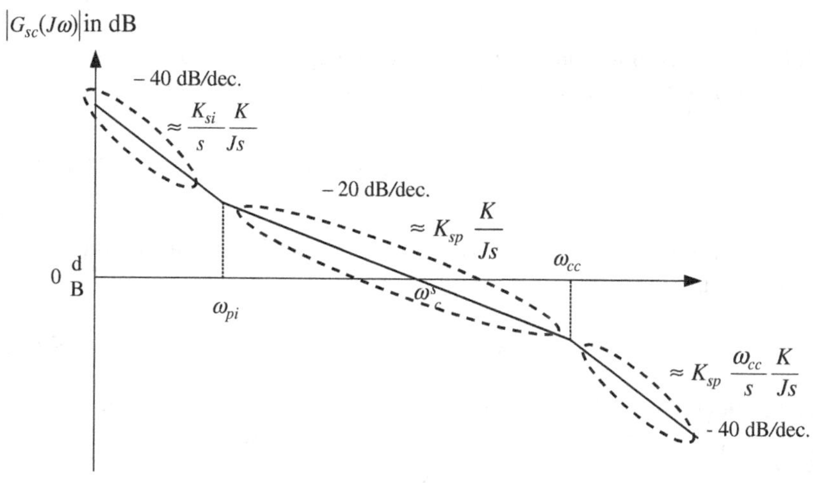

Figure 4.40 Open-loop frequency response of a speed control system with a PI regulator.

In Fig. 4.40, there are three distinct regions, where the slopes are -40 dB/dec., -20 dB/dec., and -40 dB/dec., respectively, from the left-hand side of the figure. The first part from the left-hand side of the plot comes from the inertia of the mechanical load and integral gain of the speed regulator. In this frequency region, the proportional gain can be neglected compared to the integral gain. And the current regulation loop can be set as unity gain like (4.65). The second part comes from the inertia of the mechanical load and the proportional gain of the speed regulator. In this frequency region, the integral gain can be neglected compared to the proportional gain. The second part starts from the region where the frequency is above ω_{pi} defined by (4.66). The last part comes from the current regulator and the load. In this frequency region, the speed regulator looks like only a proportional regulator while the current regulation loop looks like only an integral term. This region starts when the operating frequency is over the bandwidth of the current regulation loop, ω_{cc}.

As mentioned, the current regulation loop can be modeled as unity gain below ω_{cc} as

$$G_{cc}(s) = \frac{\omega_{cc}}{s+\omega_{cc}} \cong 1 \tag{4.65}$$

The cutoff frequency of a PI regulator, ω_{pi}, is defined as

$$\omega_{pi} = \frac{K_{si}}{K_{sp}} \tag{4.66}$$

If ω_{pi} is set as a fraction of ω_{sc}, the PI regulator can be approximated as (4.67) at the vicinity of ω_{sc} in the frequency region.

$$G_s(s) = K_{sp} + \frac{K_{si}}{s} \approx K_{sp} \quad \text{at } s \approx j\omega_{sc} \tag{4.67}$$

With these gain settings and approximation, the open-loop transfer function of the speed control system can be simplified as (4.68) at the vicinity of ω_{sc}.

$$G_{sc}(s) \approx K_{sp} \cdot \frac{K}{Js} \qquad \text{at } s \approx j\omega_{sc} \tag{4.68}$$

Hence, the frequency, where $|G_{sc}(j\omega_{sc})| = 1$, is the bandwidth of the speed control loop. So, the proportinal gain, K_{sp}, is deduced as

$$K_{sp} = \frac{J\omega_{sc}}{K} \tag{4.69}$$

And, the integral gain, K_{si}, can be set from the condition that the cutoff frequency, ω_{pi}, of PI regulator should be a fraction of the bandwidth, ω_{sc}, as

$$\omega_{pi} \leq \frac{\omega_{sc}}{5} \tag{4.70}$$

Also, K_{si} can be set as

$$K_{si} = K_{sp}\omega_{pi} \cong K_{sp}\frac{\omega_{sc}}{5} \tag{4.71}$$

If $\omega_{pi} = \omega_{sc}/5$, the damping coefficient, ζ, of the speed control system is designed as $\sqrt{5}/2$. The response of the system would be overdamped. But for the step input as the speed command, the response would be oscillatory due to the zero in the transfer function, which is derived in (4.72). This oscillatry response may be a problem in some applications, and it can be solved by the integral and proportional (IP) regulator discussed in the next section.

$$\frac{\omega_{rm}}{\omega_{rm}^{*}} = \frac{KK_{sp}s + KK_{si}}{Js^2 + KK_{sp}s + K\cdot K_{si}} = \frac{\frac{KK_{sp}s}{J} + \frac{KK_{si}}{J}}{s^2 + \frac{K\cdot K_{sp}}{J}s + \frac{K\cdot K_{s}}{J}} \tag{4.72}$$

4.3.4.2 Integral and Proportional (IP) Regulator

In Fig. 4.41, a speed control system employing an IP regulator is shown, where the proportional gain is applied not to the speed error but to the speed itself.

The transfer function of the control system employing IP regulator can be derived as (4.73), where there is no zero in the transfer function. And the response is solely decided by the damping coefficient set by the gains of the regulator.

$$\frac{\omega_{rm}}{\omega_{rm}^{*}} = \frac{\frac{K\,K_{si}}{J}}{s^2 + \frac{K\cdot K_{sp}}{J}s + \frac{K\cdot K_{si}}{J}} \tag{4.73}$$

The transfer function between the speed and the disturbance torque is derived in (4.74). The transfer function of an IP regulator is identical to that of a PI regulator as

$$\frac{\omega_{rm}}{T_L} = -\frac{s}{Js^2 + KK_{sp}s + KK_{si}} \tag{4.74}$$

This control system with an IP regulator can also be understood in the concept of the active damping in Section 4.1 as follows. Instead of the PI regulator in Fig. 4.39, an integral regulator is used. Hence, the output of the regulator to the step input command is smooth. While an active damping like $K_{sp}K$ is added to the system, the damping torque acts like an artificial friction torque. And this damping term improves the stability of the system.

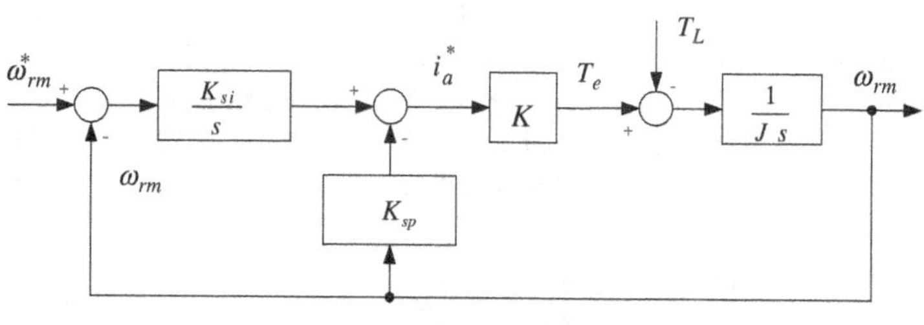

Figure 4.41 Speed control system employing integral and proportional (IP) regulator.

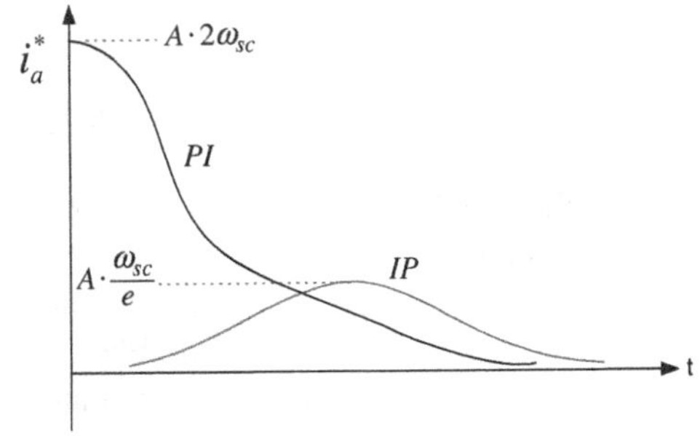

Figure 4.42 Output of IP and PI regulators with the same damping coefficient ($\zeta = 1$) and the same bandwidth at the same step input speed command.

In the case of an IP regulator, at the step input of the speed command the output of the regulator varies slowly and its magnitude is smaller than the magnitude of PI regulator at the same step input. If the outputs of both regulators are limited to the same value by physical constraints as explained in Section 4.2.4, then compared to the bandwidth of a PI regulator the bandwidth of an IP regulator can be extended without the saturation of the regulator output. In Fig. 4.42, there is a comparison of outputs of both regulators with the same damping coefficient ($\zeta = 1$) and the same bandwidth, ω_{sc}, at the same step input speed command. As seen from the figure, the peak of the output of IP regulator is much smaller than that of a PI regulator. Under the condtion that the peaks of both regulators are limited to the same value, the bandwidth of an IP regulator can be extended to $2e(\approx 5.44)$ times that of a PI regulator.

4.3.4.3 Blending of PI Regulator and IP Regulator: Two-Degree-of-Freedom Regulator

The block diagram of the speed control system employing the IP regulator in Fig. 4.41 can be redrawn as Fig. 4.43. It can be seen from the figure that the control system with the IP regulator is equivalent to the control system with the PI regulator after filtering the input command, ω_{rm}^*, by the first-order low-pass filter. This observation can be further extended to optimize the response to the input command and simultaneously to the disturbance torque as shown in Fig. 4.44, where the response to the input command can be optimized by input filter, $G_1(s)$, and the response to the disturbance torque can be optimized by the regulator, $G_2(s)$. In general, this control concept is called a two-degree-of-freedom controller [26].

As a simple example of the two-degree-of-freedom controller, an IP regulator and a PI regulator can be blended to improve the performance of the input command tracking and simultaneously to improve that of the disturbance rejection. Figure 4.45 shows a block diagram of the speed control system, where both IP and PI regulators are incorporated together with a blending factor, α, which is between zero and unity. In this case the transfer function to the disturbance torque is the same one as in (4.74) regardless of the factor α. But that to the input speed command is dependent on the factor α as in (4.75).

$$\frac{\omega_{rm}(s)}{\omega_{rm}^*(s)} = \frac{K(\alpha K_{sp}s + K_{si})}{Js^2 + KK_{sp}s + KK_{si}} \tag{4.75}$$

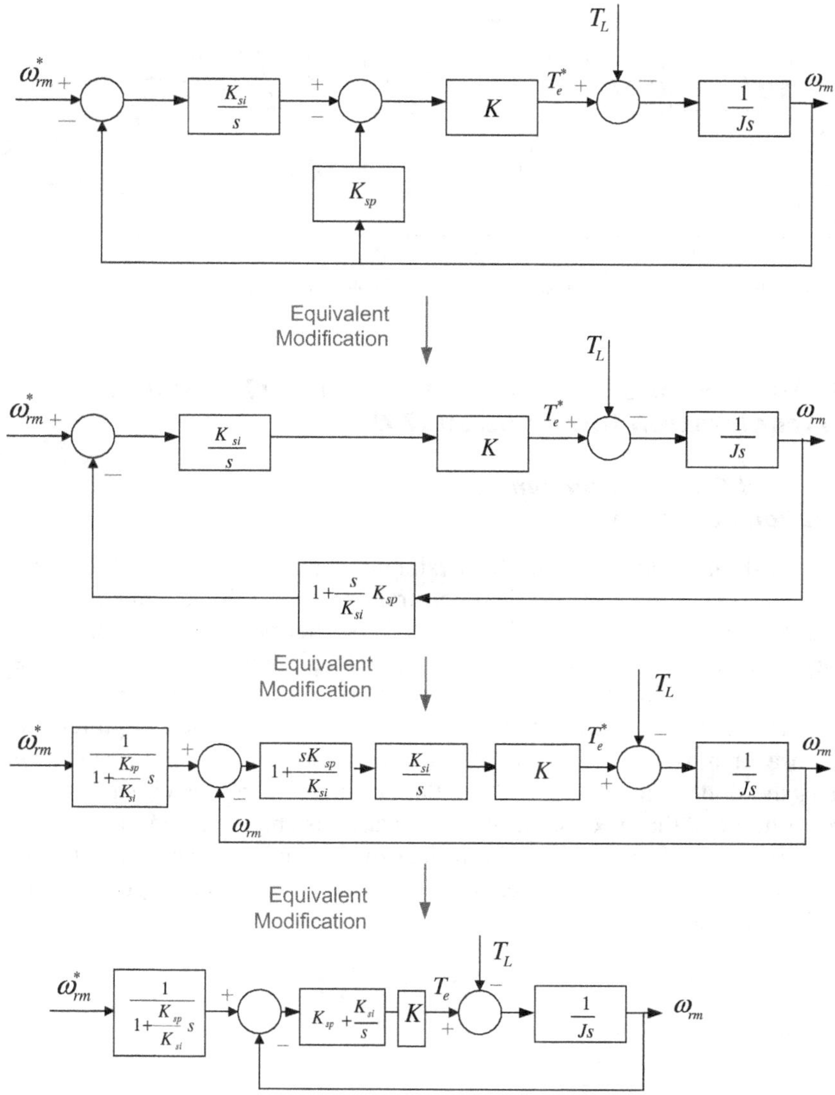

Figure 4.43 Equivalent modification of control block diagram of a speed control system with an IP regulator.

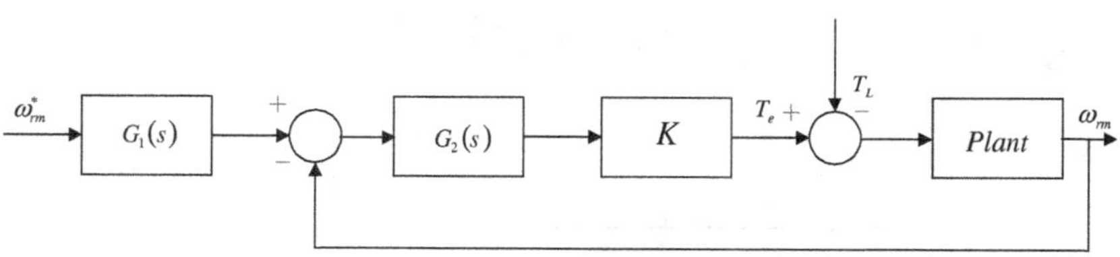

Figure 4.44 Two-degree-of-freedom controller for a speed control system.

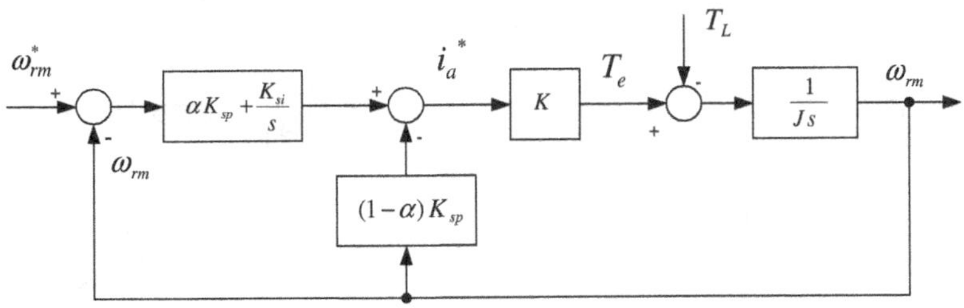

Figure 4.45 Blending of PI and IP regulators for a speed control system.

4.3.5 Enhancement of Speed Control Performance with Acceleration Information [27]

4.3.5.1 Feed-Forward Compensation of Acceleration Reference

As mentioned in Section 4.2.4, the available maximum torque of electric machine is always limited by physical constraints. Hence, the maximum acceleration and deceleration are also limited. In most motion control systems, the trajectory of the motion comes from the multiple integration of the jerk reference as discussed in Section 1.2.6. If the acceleration reference information is available from either the integration of the jerk reference (which is preferable in the sense of noise immunity) or the differentiation of the speed reference (which is the dotted line in Fig. 4.46), the acceleration reference can be fed forward to improve the speed control performance as shown in Fig. 4.46. With this acceleration feed-forwarding, the torque of the machine can adapt to the speed command even before speed error occurs, whereas the speed regulator only works after the speed error occurs.

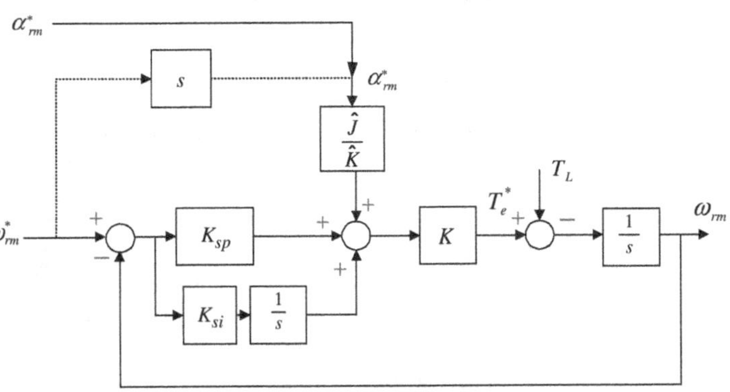

Figure 4.46 Speed control system with acceleration feed-forwarding term.

With acceleration feed-forwarding, the transfer function between the speed command and speed response of the system in Fig. 4.46 is modified from (4.72) to (4.76).

$$\frac{\omega_{rm}}{\omega_{rm}^*} = \frac{-\frac{\hat{J}}{\hat{K}}\frac{K}{J}s^2 + K_{sp}\frac{K}{J}s + K_{si}\frac{K}{J}}{s^2 + K_{sp}\frac{K}{J}s + K_{si}\frac{K}{J}} \tag{4.76}$$

As seen from (4.76), if there are no errors in torque constant, K, and the inertia, J, then $\hat{J}/\hat{K} = J/K$. Hence, the transfer function can be unity, which means that the response follows its command perfectly. In the practical sense, even if there are some errors in the parameters, the speed control performance regarding to the input command can be improved remarkably. This acceleration feed-forwarding is especially helpful to improve the speed control performance of the system where the speed control bandwidth is limited due to physical reasons such as the mechanical resonance, measurement delay, and noise.

4.3.5.2 Acceleration Feedback

At the transfer function between the speed response and the disturbance torque in (4.74), as the magnitude of the transfer function decreases, the control system reveals better disturbance rejection performance. By taking the magnitude of the inverse of (4.74) after substituting Laplace operator s with $j\omega$, and if it is defined as the dynamic stiffness, then it is similar to the stiffness of the materials, which represents the robustness of the material against the external stress. The dynamic stiffness, now a function of ω, which is in (4.77), means the robustness of the control system against the disturbance torque.

$$\left|\frac{T_L}{\omega_{rm}}\right| = J\frac{\left(j\omega^2 + K \cdot K_{sp} j\omega + K \cdot K_{si}\right)}{|j\omega|} \tag{4.77}$$

As seen from (4.77), at lower ω region, where the disturbance varies slowly, the dynamic stiffness is dominated by the integral gain of the speed regulator. At the constant disturbance, which means that $\omega = 0$, the disturbance can be rejected perfectly, and the constant disturbance torque cannot affect to the speed response in the steady state, as long as the output of the speed regulator is not saturated. In contrast to that, if the disturbance varies rapidly, which means that ω is large, the stiffness is dominated only by the inertia of the system. If the inertia is increased, the rejection to the fast varying disturbance can be enhanced. Increasing the inertia means physically heavier and more bulky system. Furthermore, the increased inertia degrades the acceleration performance and the efficiency of the drive system, because at decelerating the energy stored in the inertia should be discharged to somewhere. However, if the acceleration can be measured or estimated and if it is fed back to the control system as shown in Fig. 4.47, the equivalent inertia in the control loop could be increased without increasing the inertia physically.

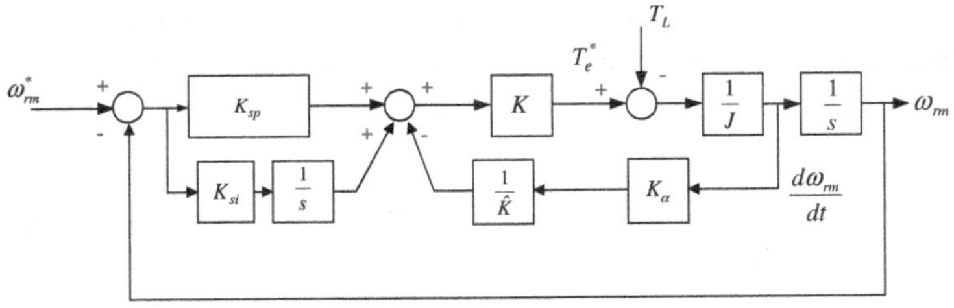

Figure 4.47 Speed control system with acceleration feedback.

With the acceleration feedback the dynamic stiffness of the system in Fig. 4.47 can be derived as

$$\frac{T_L}{\omega_{rm}} = \frac{\left| J\left(1 + \frac{K_\alpha}{J} \cdot \frac{K}{\hat{K}}\right) \cdot (j\omega)^2 + K \cdot K_{sp} j\omega + K \cdot K_{si} \right|}{|j\omega|} \tag{4.78}$$

As seen from (4.78), if $K/\hat{K}$ is 1, which means no error in torque constant, then the equivalent inertia can be increased by feedback gain of the acceleration feedback loop, K_α. By increasing the gain, the dynamic stiffness at the higher-frequency (ω) region, where the disturbance varies rapidly, can be increased as desired. Thanks to the recent development of micro-electromechanical system (MEMS) technology, the low-cost acceleration sensors are available and the dynamic bandwidth of the sensor is also being improved [28]. By employing such sensors in the speed control system, the disturbance rejection performance can be enhanced remarkably [29]. Because the acceleration feedback is mainly effective at the higher-ω region, the bandwidth of the sensor should be large enough compared to the frequency region where the disturbance would be rejected. Instead of the direct measurement of the acceleration with the sensors, the acceleration can be estimated from the measured encoder angle by the acceleration observer as shown in Fig. 4.48 like the speed observer in Fig. 4.35. In Fig. 4.48, the bandwidth of the acceleration observer should be large enough to reject the disturbance torque at the higher-ω region. Also, because the estimated acceleration is basically calculated by double differentiation of the measured angle, if the number of pulses of the encoder per revolution, P_{PR}, is not large enough, the estimated acceleration could not improve the performance of the disturbance rejection at all. With higher P_{PR} encoders such as above 2^{20} pulses per revolution, the estimated acceleration can suppress the disturbance effectively at several-hundred-hertz speed regulation bandwidth.

4.3.6 Speed Regulator with Anti-Wind-Up Controller

As mentioned in Sections 2.2, 2.8, and 2.15, the torque of the electric machine at a certain speed is limited by the capability curve. The output of the speed regulator, which is basically torque command to the electric machine, should also be limited properly. By this limitation the speed regulation performance and disturbance rejection performance may be degraded in the case of the rapid speed command change and large disturbance torque. This degradation of the performance can be

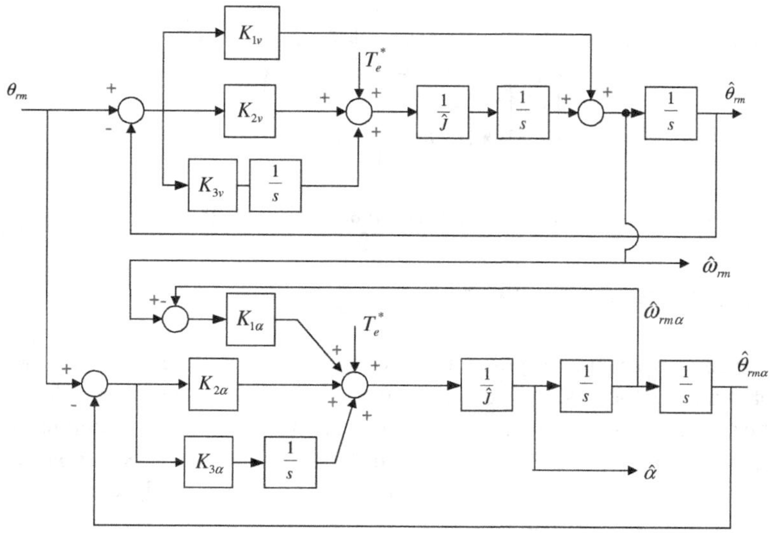

Figure 4.48 Acceleration observer implemented together with a speed observer [30].

reduced by employing the anti-wind-up controller as mentioned in Section 4.2.4. The gain of the anti-wind-up controller for the speed regulator can be set as the inverse of the proportional gain of PI regulator like the anti-wind-up gain of PI current regulator. The gain can be fine-tuned from the value set by the inverse of the proportional gain of the speed regulator. The overall control block diagram incorporating anti-wind-up controller, acceleration reference feed-forwarding, and acceleration feedback with a PI speed regulator is shown in Fig. 4.49.

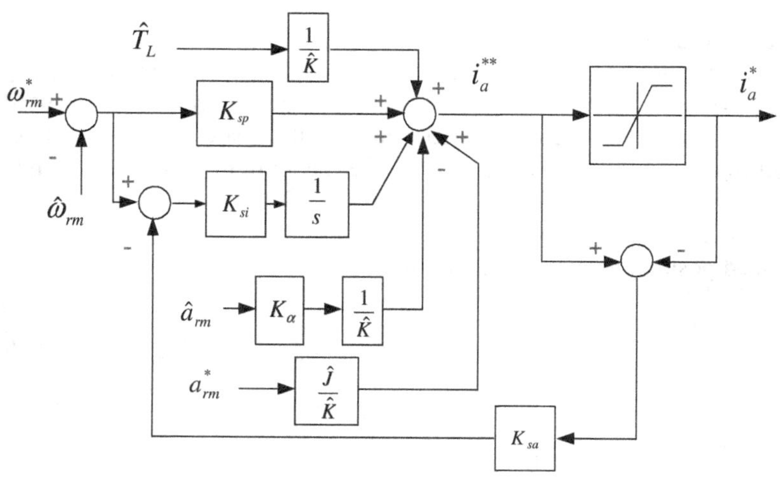

Figure 4.49 Overall control block diagram incorporating anti-wind-up controller, acceleration reference feed-forwarding, and acceleration feedback with PI speed regulator.

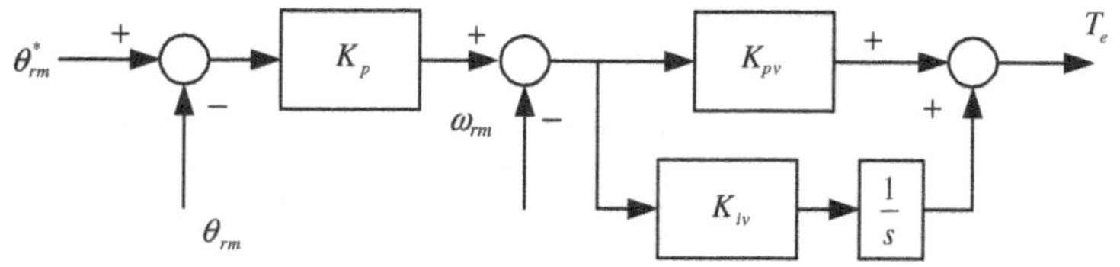

Figure 4.50 Cascaded connection of proportional–proportional integral (P–PI) regulator.

4.4 POSITION REGULATOR

In some motion control system, the position of the rotor or mover is the concern. Also, for the robot manipulator, because the trajectory of the end effector should be controlled, the position of the rotor of each axis of the servo motor of the manipulator should be regulated. Many types of position regulator have been developed and used in the industry, but in this chapter a type of position controller, widely used in the motion control industry, is introduced.

4.4.1 Proportional–Proportional and Integral (P–PI) Regulator [31]

In Fig. 4.50, a proportional(P) position regulator and a PI speed regulator are connected in a cascaded form. If PI speed regulator is designed as Section 4.3.3.1, then the gains of the speed regulator can be set as (4.79).

$$\begin{aligned} K_{pv} &= \hat{J} \cdot \omega_{sc} \\ K_{iv} &= \tfrac{1}{5} K_{pv} \cdot \omega_{sc} \end{aligned} \tag{4.79}$$

where $\hat{J}$ is the estimated system inertia.

Though the transfer function between the commanded speed and actual speed by the gains in (4.79) can be represented as (4.72), within the bandwidth of the speed regulator, the transfer function can be approximated as the first-order low-pass filter whose cutoff frequency is the bandwidth of the speed control loop, ω_{sc}, as shown in Fig. 4.51.

With the approximation of the speed control loop, the transfer function between the commanded position and the actual position can be represented as

$$\frac{\theta_{rm}}{\theta^*_{rm}} = \frac{K_p \frac{\omega_{sc}}{s+\omega_{sc}} \frac{1}{s}}{1 + K_p \frac{\omega_{sc}}{s+\omega_{sc}} \frac{1}{s}} = \frac{K_p \omega_{sc}}{s^2 + \omega_{sc} s + K_p \omega_{sc}} = \frac{{\omega_n}^2}{s^2 + 2\zeta \omega_n s + {\omega_n}^2} \tag{4.80}$$

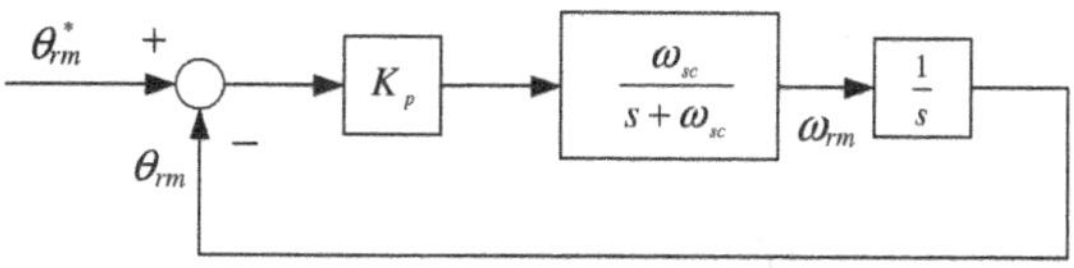

Figure 4.51 Control block diagram of position control loop with an inner speed control loop.

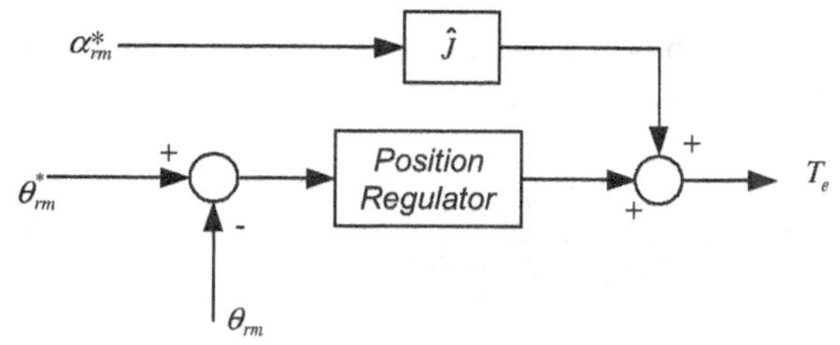

Figure 4.52 Acceleration feed-forwarding to a position regulator.

The overall transfer function is the form of the second-order low-pass filter. Because at the position response the overshoot is usually not allowed, the damping coefficient, ζ, should be over unity. If ζ is set as 1.5, then the proportional gain of the position regulator and the natural undamped frequency, ω_n, can be derived as

$$K_p = \frac{\omega_{sc}}{9}, \qquad \omega_n = \frac{\omega_{sc}}{3} \tag{4.81}$$

where ω_{sc} is the bandwidth of the inner speed control loop.

If the gain is set as (4.81), the bandwidth of the position control loop, ω_{pc}, is $0.374\omega_n$, which is approximately one-eighth of the bandwidth of the speed control loop as

$$\omega_{pc} = 0.374\omega_n = \omega_{sc}/8 \tag{4.82}$$

4.4.2 Feed-Forwarding of Speed Reference and Acceleration Reference

At the position control loop in Fig. 4.50, the acceleration reference can be fed forward to enhance the performance of the position control as shown in Fig. 4.52.

In the control loop at Fig. 4.52, the speed reference also can be fed forward to a PI speed regulator as shown in Fig. 4.53. In the figure, the speed reference and position reference are obtained by the successive integral operation of acceleration reference or one more integration of jerk reference.

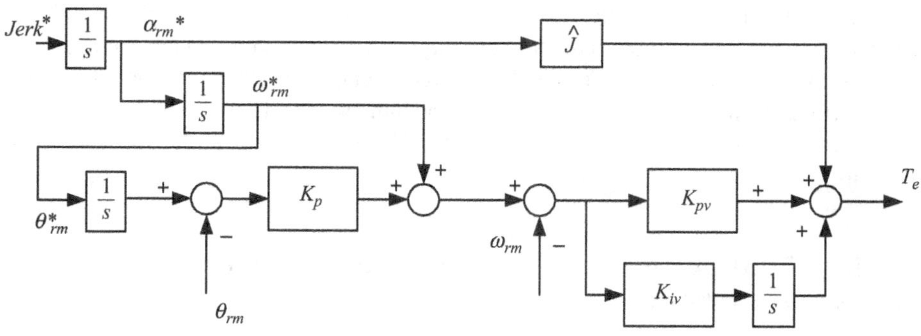

Figure 4.53 Overall block diagram of P–PI position regulator with acceleration and speed reference feed-forwarding terms.

The transfer function of the control block diagram in Fig. 4.53 can be derived as (4.83) with the assumption of a typical mechanical system like $T_e = Js^2\theta_{rm}$.

$$\frac{\theta_{rm}}{\theta^*_{rm}} = \frac{\hat{J}s^3 + K_{pv}s^2 + (K_pK_{pv} + K_{iv})s + K_pK_{iv}}{Js^3 + K_{pv}s^2 + (K_pK_{pv} + K_{iv})s + K_pK_{iv}} \tag{4.83}$$

As seen from (4.83), if the inertia is correctly estimated like $\hat{J} = J$, then the transfer function would be unity. Even if there is error in the estimation of the inertia, the position tracks the commanded position up to certain frequency region, where the magnitude of the third-order terms in the transfer function can be neglected compared to other terms. That can be easily understood by the observation that the numerator and denominator of the transfer function is identical up to second-order terms regarding $s(=j\omega)$. Hence, even if the error exists, the error only affects to the position control performance at the higher-frequency region, where the third-order terms in the transfer function dominate. With this feed-forwarding control, the position control performance can be improved remarkably even though the bandwidth of the position control loop is limited by several physical reasons such as mechanical resonance, measurement delay, and noise.

4.5 DETECTION OF PHASE ANGLE OF AC VOLTAGE [32]

In the control of power converters connected to an AC source such as a controlled rectifier, a PWM boost rectifier, and a matrix converter, the instantaneous AC source angle is the critical information for the control of the power converters. In an ideal three-phase AC source where each phase voltage is shifted by 120° without any harmonics, its magnitude is the same, and the frequency of the voltage is fixed, the instantaneous phase angle of three-phase voltage can be identified by simply monitoring the zero crossing point of a line-to-line voltage of AC input. However, in the case of the practical three-phase AC source, the frequency, the phase difference, the magnitude, and the harmonics are all time-varying [33]. Such a phenomenon would be getting severe with wide penetration of nonlinear elements like switching power converters, saturable reactors, discharge lamps, and so on. Furthermore, with errors of the phase angle of the AC source voltage in the operation of power converters, the harmonics in the AC source become severe. In this section, a method to detect an instantaneous phase angle from a practical three-phase AC source is discussed.

4.5.1 Detection of Phase Angle on Synchronous Reference Frame

If the three-phase voltage source is an ideal one with angular frequency, ω_e, then the three-phase voltage can be represented as

$$
\begin{aligned}
e_a &= -V_m \sin\omega_e t \\
e_b &= -V_m \sin\left(\omega_e t - \frac{2\pi}{3}\right) \\
e_c &= -V_m \sin\left(\omega_e t + \frac{2\pi}{3}\right)
\end{aligned}
\tag{4.84}
$$

where V_m is the peak of the phase voltage. The stationary and synchronous reference frame d–q voltage can be deduced as (4.85) and (4.86), respectively, from the definition of d–q variables in Section 3.1.

$$
\begin{aligned}
e_d^s &= -V_m \sin(\omega_e t) \\
e_q^s &= V_m \cos(\omega_e t)
\end{aligned}
\tag{4.85}
$$

$$
\begin{aligned}
e_d^e &= 0 \\
e_q^e &= V_m
\end{aligned}
\tag{4.86}
$$

where the instantaneous phase angle is given by $\theta_e = \omega_e t$. In this ideal case, the line-to-line voltage, phase voltage, and d–q voltage at stationary and synchronous reference frame with the phase angle can be depicted as shown in Fig. 4.54. In the

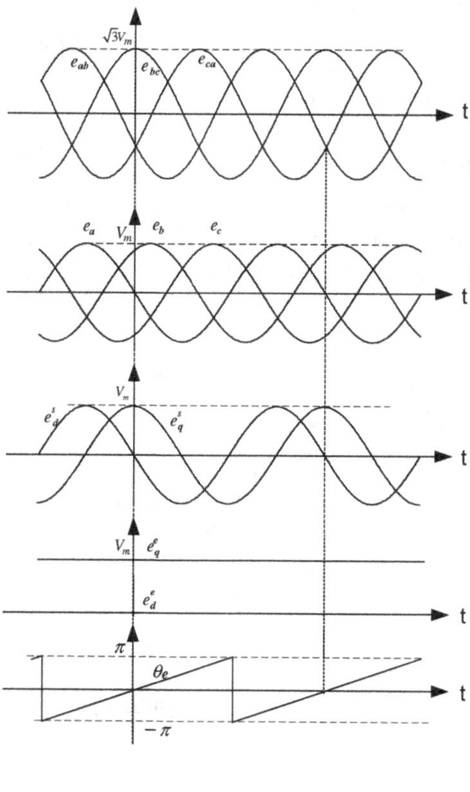

Figure 4.54 AC source voltages and phase angle.

figure, the instant when a-phase voltage crosses the zero voltage line from positive voltage to negative voltage is defined as the zero of the phase angle. Then, the instantaneous angle can be calculated as

$$\theta_e = \tan^{-1}\left(\frac{-e_d^s}{e_q^s}\right) \tag{4.87}$$

However, if in a practical AC source voltage, especially with some high-order harmonics and/or measurement noise, the angle from (4.87) varies suddenly and the control system would be unstable.

As seen from Fig. 4.55, d-axis voltage at synchronous reference frame is zero when the estimated phase angle, $\hat{\theta}_e$, coincides with the real phase angle in the case of Fig. 4.55a. But if the estimated angle is ahead of the real angle in the case of Fig. 4.55b, the d-axis voltage is positive. And if the angle is behind of the real angle in the case of Fig. 4.55c, the voltage is negative. In the figure, $\vec{\mathbf{e}}$ means the complex space vector of AC source voltage. By exploiting the relation of the d-axis voltage to the phase angle error, which is defined as the difference between real and estimated angle, the phase angle can be tracked with simple proportional and integral (PI) regulator as shown in Fig. 4.56. By using a PI regulator, thanks to the inherent filtering effects of the regulator, the tracking is robust to the high-order harmonics and measurement noises.

The d-axis voltage, which is the input to PI regulator, can be deduced as

$$e_d^e = V_m \sin(\hat{\theta}_e - \theta_e) \tag{4.88}$$

If angle error is small, then (4.88) can be approximated as

$$e_d^e \approx V_m(\hat{\theta}_e - \theta_e) \tag{4.89}$$

In Fig. 4.56, ω_N is the nominal angular frequency, and LPF block stands for low-pass filter to suppress the ripple and the noise of the calculated d-axis voltage. If the angle error is small enough and the time constant of the low-pass filter is

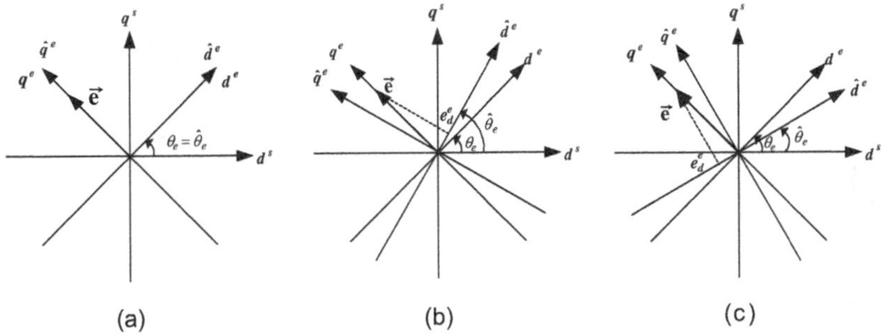

Figure 4.55 Phase angle error and d-axis voltage at synchronous reference frame. (a) $\theta_e = \hat{\theta}_e$, (b) $\theta_e < \hat{\theta}_e$, (c) $\theta_e > \hat{\theta}_e$

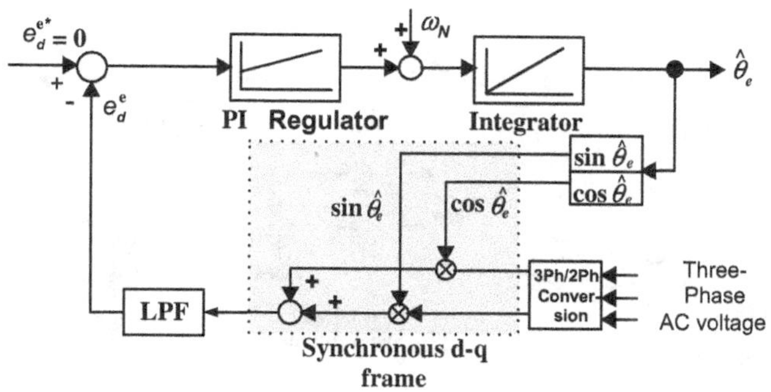

Figure 4.56 Control block diagram of phase angle detector.

much smaller that that of the PI regulator, the transfer function between real and estimated angle can be derived as (4.90) after neglecting dynamics of the low-pass filter.

$$\hat{\theta}_e = \left(K_p + \frac{K_i}{s}\right)\frac{1}{s}(-e_d^e) = \left(K_p + \frac{K_i}{s}\right)\frac{1}{s}V_m(\theta_e - \hat{\theta}_e)$$
$$\frac{\hat{\theta}_e}{\theta_e} = \frac{K_p V_m s + K_i V_m}{s^2 + K_p V_m s + K_i V_m} \quad (4.90)$$

To detect the phase angle, the method shown in Fig. 4.56 is simple, stable, and robust to measurement noise. However, if there is low-order harmonics and/or unbalance in source AC voltage, then the angle also includes the harmonics. Also, the current reference or voltage reference based on the angle from block diagram in Fig. 4.56 for the control of power converters may incur low-order harmonic currents into AC lines, whereas, in the control block diagram at Fig. 4.56, high-order harmonics in AC source voltage cannot affect the detection of the phase angle due to the inherent filtering effect of PI regulator.

4.5.2 Detection of Phase Angle Using Positive Sequence Voltage on Synchronous Reference Frame [34]

In the control block diagram shown in Fig. 4.56, the positive sequence voltage can be extracted from the unbalanced three-phase voltage including low-order harmonics. If the positive sequence voltage is used for the phase detection, the problems due to voltage with unbalance and/or low-order harmonics could be tolerated. In Fig. 4.57, a control block diagram for detecting a phase angle based on the positive sequence voltage is shown. In the figure, the detector can be decomposed into two parts. One is the part to extract the positive sequence voltage from the AC source voltage, and the other is the part to detect the phase angle using a low-pass filter and PI regulator on synchronous reference frame. The reason why the positive sequence voltage is used to

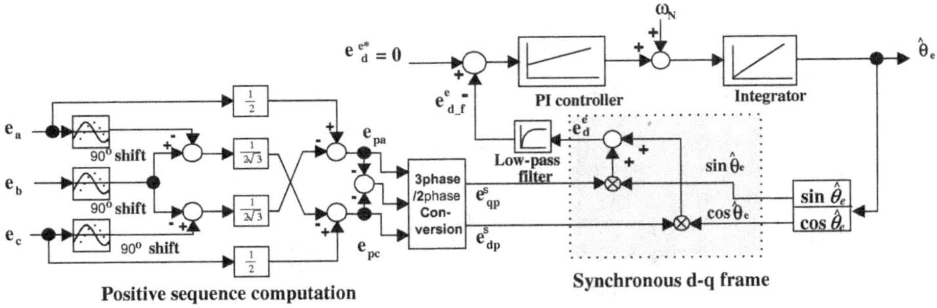

Figure 4.57 Control block diagram of a phase angle detector using a positive sequence voltage.

detect the phase angle is that the positive sequence voltage of the fundamental frequency component of three-phase voltage is always balanced regardless of imbalance and harmonics in the source voltage. By detecting the phase angle with this method, the second-order harmonic frequency, which comes from the negative sequence voltage, does not appear in the detected phase angle.

The positive sequence voltage can be obtained from three-phase voltages by

$$\begin{bmatrix} E_{pa} \\ E_{pb} \\ E_{pc} \end{bmatrix} = \frac{1}{3}\begin{bmatrix} \mathbf{1} & \mathbf{a} & \mathbf{a}^2 \\ \mathbf{a}^2 & \mathbf{1} & \mathbf{a} \\ \mathbf{a} & \mathbf{a}^2 & \mathbf{1} \end{bmatrix}\begin{bmatrix} e_a \\ e_b \\ e_c \end{bmatrix} = \begin{bmatrix} \frac{1}{2}e_a - \frac{1}{2\sqrt{3}j}(e_b - e_c) \\ -(E_{pa} + E_{pc}) \\ \frac{1}{2}e_c - \frac{1}{2\sqrt{3}j}(e_a - e_b) \end{bmatrix} \tag{4.91}$$

where $\mathbf{a} = e^{j\frac{2\pi}{3}}$.

To devise (4.91), the phase delay by 90° is implemented by the second-order all-pass filter whose transfer function is given by

$$\frac{s^2 - \sqrt{2}\omega_o s + \omega_o^2}{s^2 + \sqrt{2}\omega_o s + \omega_o^2} \tag{4.92}$$

where ω_o is $\frac{1+\sqrt{3}}{\sqrt{2}}\omega_N \approx 1.9319\omega_N$. At the nominal angular frequency of the AC source, ω_N, the phase delay by 90° occurs with this all pass filter.

To suppress the effects due to the high-order harmonics in AC source voltage, a low-pass filter is used in Fig. 4.57. Compared to the low-pass filter in Fig. 4.57, where the cutoff frequency of the filter is quite high only to cut measurement noise or high-order harmonics, the cutoff frequency of the filter in Fig. 4.56 should be low enough to cut off fifth and seventh harmonics of the AC source voltage. Even if the cutoff frequency of the filter in Fig. 4.56 is set low enough to eliminate low-order harmonics, the second-order harmonics due to the input voltage imbalance cannot be filtered out. To cut off the second-order harmonics, the cutoff frequency would be too low and the response of the phase detection loop would be too sluggish to detect the sudden variation of the phase angle resulting from AC source faults or from rapid speed variation of engine generator if the AC source comes from an engine generator set. With regard to the detection loop in Fig. 4.57, because there is no imbalance in the positive sequence voltage and, hence, no second-order harmonics, the cutoff frequency can be set low enough to cut low-order harmonics like fifth and seventh but high enough to detect sudden variation of the phase angle.

4.6 VOLTAGE REGULATOR

4.6.1 Voltage Regulator for DC Link of PWM Boost Rectifier

4.6.1.1 Modeling of Control System

The voltage equation of PWM boost rectifier shown in Fig. 2.66 can be derived as follows:

$$
\begin{aligned}
V_{As} &= L_{inter}\frac{di_a}{dt} + V_{as} \\
V_{Bs} &= L_{inter}\frac{di_b}{dt} + V_{bs} \\
V_{Cs} &= L_{inter}\frac{di_c}{dt} + V_{cs}
\end{aligned}
\tag{4.93}
$$

Equation (4.93) can be rewritten as (4.94) in a synchronously rotating d–q reference frame by the transformation method in Section 3.1:

$$
\begin{aligned}
e_d^e &= -\omega_e L_{inter} i_q^e + L_{inter}\frac{di_d^e}{dt} + e_{ds}^e \\
e_q^e &= \omega_e L_{inter} i_d^e + L_{inter}\frac{di_q^e}{dt} + e_{qs}^e
\end{aligned}
\tag{4.94}
$$

where e_d^e and e_q^e stand for the AC source voltage, e_{ds}^e and e_{qs}^e stand for the voltage by PWM boost rectifier, and i_d^e and i_q^e stand for the current flowing into PWM boost converter at the synchronous reference frame. If the phase angle of the AC source is set as shown in Fig. 4.54, the power from AC source to the boost converter can be described as

$$P_{in} = V_{as}i_{as} + V_{bs}i_{bs} + V_{cs}i_{cs} = \tfrac{3}{2}\left(e_q^s i_q^s + e_d^s i_d^s\right) = \tfrac{3}{2}\left(e_q^e i_q^e + e_d^e i_d^e\right) = \tfrac{3}{2}V_m i_q^e \tag{4.95}$$

As seen from (4.95), the power is dependent only on the q-axis current and the power has nothing to do with the d-axis current. Hence, the q-axis current at the synchronous reference frame is called the active power current, and it is similar to the torque component current of a vector-controlled drive system in Chapter 5. Also, the d-axis current is related to the reactive power of the system, and it is similar to the flux component current of a vector-controlled drive system. If the AC source voltage and current are pure sinusoidal, then the power factor can be represented as (4.96). For unity power factor operation, which guarantees minimum loss, the d-axis current should be null.

$$PF = \frac{\mathbf{e}_{dq}^{e} \cdot \mathbf{i}_{dq}^{e}}{\left|\mathbf{e}_{dq}^{e}\right|\left|\mathbf{i}_{dq}^{e}\right|} = \frac{i_q^e}{\sqrt{i_d^{e2} + i_q^{e2}}} \tag{4.96}$$

where $\mathbf{e}_{dq}^{e} = e_d^e + je_q^e$, $\mathbf{i}_{dq}^{e} = i_d^e + ji_q^e$, and $\mathbf{e}_{dq}^{e} \cdot \mathbf{i}_{dq}^{e}$ is the inner product of the source voltage space vector and the line current space vector. $\left|\mathbf{e}_{dq}^{e}\right|\left|\mathbf{i}_{dq}^{e}\right|$ represents the product of the magnitudes of the source voltage vector and the line current vector.

4.6.1.2 DC Link Voltage Regulator

As seen from (2.87), DC link voltage, V_d, increases when the active power from the AC source is larger than the power consumed by the load—that is, $P_{\text{out}} = i_{dcl} V_d$—and vice versa. Hence, DC link voltage can be regulated by adjusting active power to a DC link from an AC source. Hence, the q-axis current reference can be composed of the sum of a feed-forward term, which corresponds to the estimated load power, and a term to regulate DC link voltage as (4.97). As a feedback controller to regulate DC link voltage, an integral and proportional (IP) regulator can be used to prevent overshoot of DC link voltage even with the step change of DC link voltage reference. The block diagram of an IP DC link voltage regulator including the feed-forward term is shown in Fig. 4.58, where the d-axis current reference is set as zero to keep power factor as unity.

$$i_q^{e*} = -K_p V_d + K_i \int (V_d^* - V_d)\, dt + \hat{P}_{out} / \left(\tfrac{3}{2} V_m\right) \tag{4.97}$$

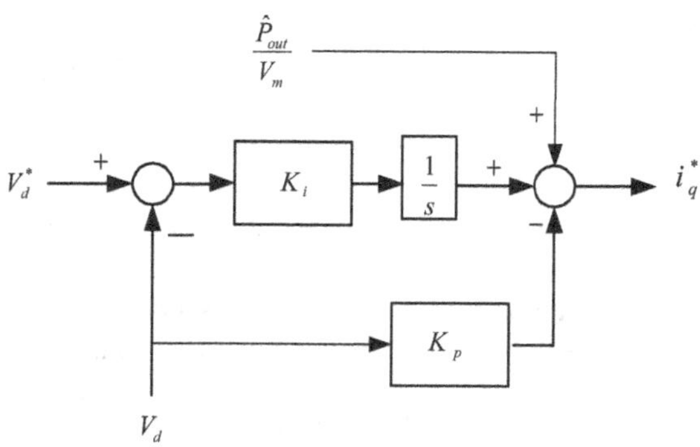

Figure 4.58 Control block diagram of a DC link voltage regulator for a PWM boost rectifier.

The differential equation to represent the DC link voltage can be represented as

$$\frac{C}{2}\frac{dV_d^2}{dt} = P_{\text{in}} - P_{\text{out}} \tag{4.98}$$

where C is the total capacitance in a DC link.

If the feed-forward compensation for load power is perfect, then DC link voltage can be described by (4.99) in terms of error of DC link voltage, $(V_d^* - V_d)$, and gains of IP regulator.

$$\frac{C}{2}\frac{dV_d^2}{dt} = \frac{3}{2}V_m\left(-K_pV_d + K_i\int(V_d^* - V_d)\,dt\right) \tag{4.99}$$

Equation (4.99) is a nonlinear differential equation, and it can be linearized at an operating point using small signal analysis. At that operating point, V_{d0}, DC link voltage can be represented as (4.100) by the small signal analysis.

$$V_d^2 \approx V_{d0}^2 + 2V_{d0}(V_d - V_{d0}) \tag{4.100}$$

Equation (4.100) can be substituted into (4.99), and then a linerized differential equation at an operating point can be derived as follows:

$$CV_{d0}\frac{dV_d}{dt} = \frac{3}{2}V_m\left(-K_pV_d + K_i\int(V_d^* - V_d)\,dt\right) \tag{4.101}$$

From (4.101), a transfer function between reference DC link voltage and actual DC link voltage can be deduced as follows:

$$\frac{V_d(s)}{V_d^*(s)} = \frac{\frac{\frac{3}{2}V_mK_i}{CV_{d0}}}{s^2 + \frac{\frac{3}{2}V_mK_p}{CV_{d0}}s + \frac{\frac{3}{2}V_mK_i}{CV_{d0}}} = \frac{\omega_n^2}{s^2 + 2\zeta\omega_n s + \omega_n^2} \tag{4.102}$$

where ω_n stands for the natural undamped frequency and ζ stands for a damping coefficient. The gains of an IP regulator can be calculated through the frequency domain analysis. Under the assumption of well-regulated DC link voltage, the operating point can be set as its reference. Then the gains can be represented as follows:

$$\begin{aligned} K_p &= 2\zeta\omega_n\frac{CV_d^*}{\frac{3}{2}V_m} \\ K_i &= \omega_n^2\frac{CV_d^*}{\frac{3}{2}V_m} \end{aligned} \tag{4.103}$$

With these gains, the control bandwidth of a DC link voltage regulation loop, ω_{bw}, is given by

$$\omega_{bw} = \omega_n[(1-2\zeta^2)+\sqrt{4\zeta^4-4\zeta^2+2}]^{1/2} \quad (4.104)$$

The modeling errors from the assumption of small signals at the operating point may affect the setting of gains of a PI regulator. However, if DC link voltage variation is controlled within 10% of its nominal value, the error due to small signal linearization can be neglected. To prevent overshoot of DC link voltage, the damping factor, ζ, should be above unity. If ζ is set as 1.2, then ω_{bw} is half of ω_n by (4.104). For power factor control, which is given by (4.96), a d-axis current reference can be set as

$$i_d^{e*} = i_q^{e*}\frac{\sqrt{1-PF^{*2}}}{PF^*} \quad (4.105)$$

For unity power factor control, the d-axis current should be regulated as null—that is, maximum power per ampere operation. If a PWM boost rectifier has to be operated as a leading power factor to an AC source, the d-axis current should be regulated as a positive value. In general, the AC distribution network, as well as the power factor at the AC source with industrial loads such as motors and lightings, is usually lagging. And, the PWM boost rectifier would be operated as leading power factor to compensate the lagging power of other loads. However, if the DC link voltage of PWM boost converter is not large enough or smaller compared to the peak of the AC source voltage, the PWM boost rectifier should be controlled in lagging power factor by regulating d-axis current as a negative value to ensure the current control voltage margin. In these cases of power factor control, the current rating of PWM boost converter increases due to the d-axis current.

PROBLEMS

1. As shown in Fig. P.4.1, DC machine is connected to a roll with a long shaft. The axis has limited stiffness. The torque of DC machine is given by $T_e = T\cdot\sin(\omega t)$ N-m. The friction torque and load torque are negligible. And the inertia of the long shaft can be neglected.

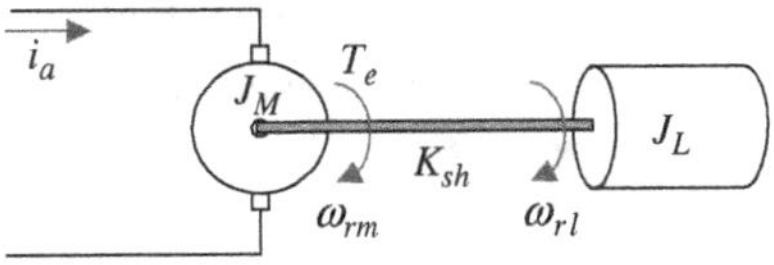

Figure P4.1 Drive system with mechanical resonance.

Inertia of load : $J_L = 0.5(\text{kg}\cdot\text{m}^2)$

Inertia of rotor of electric machine : $J_M = 0.1(\text{kg}\cdot\text{m}^2)$

Stiffness of the shaft : $K_{sh} = 700(\text{N-}m/\text{rad})$

(1) Find the transfer function regarding ω_{rl}/T_e.

(2) Find the angular frequency of torque, ω, where the magnitude of ω_{rl} is maximum.

(3) Design a control system employing active damping to suppress the mechanical resonance due to the limited stiffness of the shaft. And show the response, ω_{rl}, with the sinusoidally varying torque as $T_e = T \cdot \sin(\omega t)$, where the angular frequency, ω, is the value calculated in part 2. In this controller design, the torque applied to the shaft is not measurable and the angle of the load, $\theta_{rl} = \int \omega_{rl}\, dt$, and the angle of the machine, $\theta_{rm} = \int \omega_{rm}\, dt$, can be measured ideally. The torque of DC machine can be ideally estimated by the torque constant and current of machine.

2. For a 3.08-kW permanent magnet DC servo motor with the following ratings and parameters, answer the following questions.

$$R_a = 0.26\ (\Omega), \qquad L_a = 1.7(\text{mH})$$

Rated armature current : 24.9(A)

Rated speed : 3000(r/min)

Rated armature voltage : 139(V)

Inertia of the rotor of the motor : 0.00252(kg.m^2)

(1) After finding electric time constant, T_a, and electromechanical time constant, T_m, specify the speed response to the step change of the armature voltage among under damped, critically damped, or overdamped. It is assumed that the magnitude of the step is small enough not to saturate any part of the drive system. And the motor is not connected to any load.

(2) Now, a load is connected to the motor. The inertia of the load is the same as that of the motor itself. The stiffness coefficient of the connection shaft between load and the rotor of the motor can be assumed as infinity. Hence, the rotor and the load can be assumed as a single body. There is no load torque and the friction coefficient, B, equals 0.005 (N-m/(rad/s)). At time point, 0 s, the armature voltage, whose magnitude is 50 V, is applied to the motor in the step function. Plot speed (r/min), armature current, i_a(A), torque, T_e(N-m) for 0–0.2 s. Also plot T_e(N-m) and speed (r/min) in the torque speed plane, where the vertical axis is torque.

(3) Under the same operating conditions as in part 2, a current regulator as shown in Fig. P.4.2 is inserted into the control loop.

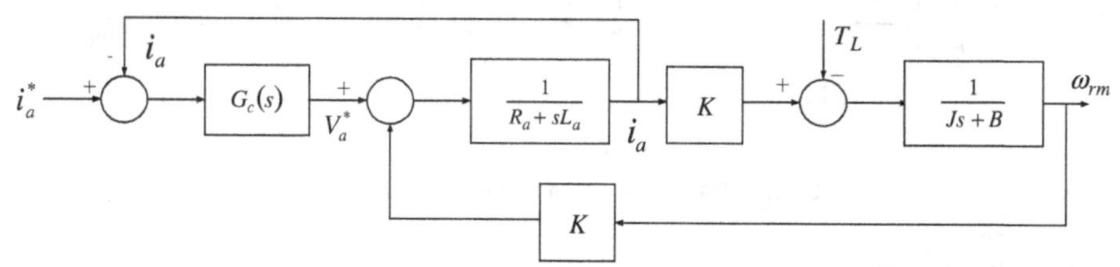

Figure P4.2 Insertion of a current regulator.

In the figure, the current regulator is a simple proportional gain like $G_c(s) = K_p$. When all poles of the transfer function, ω_{rm}/i_a^*, lie at the real axis of Laplace domain and the largest magnitude of pole is 10 times of the magnitude of the smallest pole, calculate

K_p. In this gain setting, at time point, 0 s, the armature current reference, i_a^*, increased from zero to 50 A in the step function. Plot speed (r/min), armature current, i_a(A), torque, T_e(N-m) for 0–0.2 s. Also, plot T_e(N-m) and speed (r/min) in the torque-speed plane.

(4) The gain of the current regulator is changed to $K_p = 50$. Repeat part 3. In this problem, there is no limitation in the armature voltage, V_a^*.

(5) Repeat part 4 with the limitation of the armature voltage, V_a^*, within ± 150 V.

(6) Repeat part 5 with feed-forward compensation of back EMF as shown in Fig. 4.12.

(7) There is a time delay by 200μs in the processing of the armature voltage as shown in Fig. P.4.3. The armature current reference, i_a^*, increases from zero to 50 A in the step function at time point 0 s. The armature voltage is limited as part 5, and the back EMF is compensated as part 6. For three different regulator gains as $K_p = 1$, 10, and 50, plot speed (r/min), armature current, i_a(A), armature current reference, i_a^*(A) for 0–0.2 s for each gain.

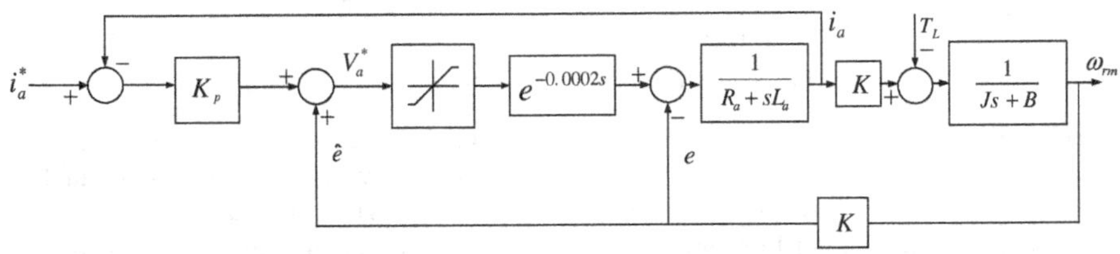

Figure P4.3 Current control system with time delay.

3. The servo motor in problem 2 is driven with the same conditions as in problem 2, part 2. The current regulator is proportional and integral (PI) type as shown in Fig. P.4.4. The back EMF is compensated by 95% of the actual back EMF in feed-forward manner—that is, $\hat{e} = 0.95e$. And the estimated back EMF constant, the torque constant, is 95% of the actual value as $\hat{K} = 0.95K$. The DC link voltage of the four-quadrant DC/DC converter, shown in Fig. 2.18, is 300 V and the voltage reference, V_a^*, is limited within ± 250 V. The switching frequency of the converter, which is the frequency of the carrier wave, is 5 kHz.

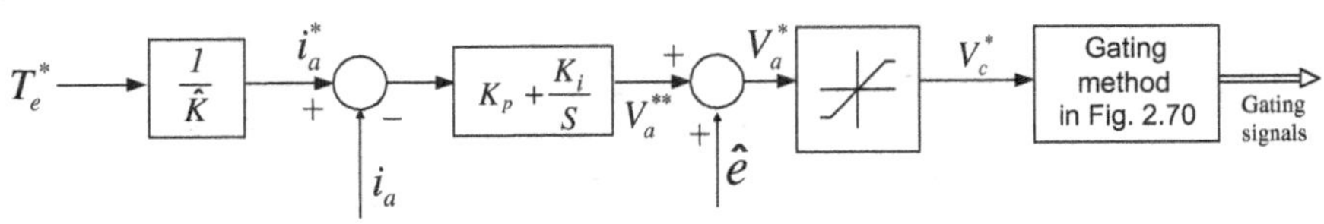

Figure P4.4 PI current regulator.

(1) All measurements and controls except PWM voltage synthesis is done in the analog domain. When the reference torque varies as follows, answer the following questions.

0–0.02 s, $T_e^* = 30$ N-m
0.02–0.03 s, $T_e^* = -30$ N-m
0.03 s ~ 0.04 s, $T_e^* = 2\sin(2000\pi t)$N-m

(A) Though the voltage is limited as $-250\text{ V} \leq V_c^* \leq 250\text{ V}$ by a limiter in Fig. P.4.4, there is no anti-wind-up controller. The control bandwidth of the current regulation loop is 1 kHz. Calculate the gains of the PI regulator by (4.14). Plot the following responses for 0–0.04 s.

(a) i_a and i_a^*

(b) T_e and T_e^*

(c) $\omega_{rm}(\text{r/min})$

(d) V_a^*

(e) For 0.039–0.04 s, plot V_c^*(output of the limiter) and the actual PWM voltage waveform to the motor.

(B) With anti wind-up controller, repeat part A. The gain of the anti-wind-up controller is set as $K_a = 1/K_p$.

(2) Now, all controls are done digitally in the discrete time domain. For the control algorithm execution, one sampling interval is required. The output of the regulator is updated at the next sampling point after finishing the execution of the algorithm. The sampling frequency is the same with the switching frequency. To measure the average current, the second-order Butterworth low-pass filter is used. The cutoff frequency of the filter is 1 kHz, and the filter is implemented in the analog domain. To achieve maximum current regulation bandwidth, decide the gains of PI regulator. Also, find the bandwidth of the current regulation loop. At this gain setting, plot the following responses.

(a) i_a and i_a^* for 0–0.04 s.

(b) T_e and T_e^* for 0–0.04 s

(c) $\omega_{rm}(\text{r/min})$ for 0–0.04 s

(d) V_a^* for 0–0.04 s

(e) For 0.039–0.04 s, plot V_c^*(output of the limiter) and the actual PWM voltage waveform to the motor.

(f) For 0.039–0.04 s, plot i_a^* and the filtered armature current, i_{af}, by the second-order low-pass filter.

(3) In part 2, the current is sampled at the peak of the carrier wave, and the second-order low-pass filter has been removed. To achieve maximum current regulation bandwidth, decide the gains of PI regulator. Also, find the bandwidth of the current regulation loop. At this gain setting, plot the following responses.

(a) i_a and i_a^* for 0–0.04 s.

(b) T_e and T_e^* for 0–0.04 s

(c) $\omega_{rm}(\text{r/min})$ for 0–0.04 s

(d) V_a^* for 0–0.04 s

(e) For 0.039–0.04 s, plot V_c^*(output of the limiter) and the actual PWM voltage waveform to the motor.

(f) For 0.039–0.04 s, plot i_a^* and the sampled armature current, $i_a(n)$, at the peak of the carrier wave.

(4) The sampling frequency has been doubled and the the sampling interval is a half of the case of part 3, and it is twice the switching frequency. And all control algorithms are executed within one sampling interval. The output of the regulator is updated at the next sampling point after finishing the execution of the algorithm. The sampling point is peak and valley of the carrier wave. To achieve maximum current regulation bandwidth, decide the gains of PI regulator. Also, find the bandwidth of the current regulation loop. At this gain setting, plot the following responses.

(a) i_a and i_a^* for 0–0.04 s
(b) T_e and T_e^* for 0–0.04 s
(c) $\omega_{rm}(\mathrm{r/min})$ for 0–0.04 s
(d) V_a^* for 0–0.04 s
(e) For 0.039–0.04 s, plot V_c^*(output of the limiter) and the actual PWM voltage waveform to the motor.
(f) For 0.039–0.04 s, plot i_a^* and the sampled armature current, $i_a(n)$, at the peak and the valley of the carrier wave.

(5) With the operating condition at part 4, to eliminate the time delay due to the algorithm execution, the current is predicted based on the sampled current. The back EMF information for the prediction is 95% of the actual value and the inductance for the prediction is 90% of the real value. Also, the voltage drop on the armature resistance is neglected in the prediction. To achieve maximum current regulation bandwidth, decide the gains of PI regulator. And, find the bandwidth of the current regulation loop. At this gain setting, plot the following responses.

(a) i_a and i_a^* for 0–0.04 s
(b) T_e and T_e^* for 0–0.04 s
(c) $\omega_{rm}(\mathrm{r/min})$ for 0–0.04 s
(d) V_a^* for 0–0.04 s
(e) For 0.039–0.04 s, plot V_c^*(output of the limiter) and the actual PWM voltage waveform to the motor.
(f) For 0.039–0.04 s, plot i_a^*, the sampled armature current, $i_a(n)$, at the peak and the valley of the carrier wave, and predicted current, $i_a(n+1)$.

4. The servo motor in problem 2 is driven with the same load condition as in problem 2, part 2, and with the current control system in problem 3, part 4. The speed command and load torque are given as follows. Answer the following questions.

$0 \le t < 0.005\ \mathrm{s}$	$\omega_{rm}^* = \dfrac{200}{0.005} t\ \mathrm{r/min}$
$0.005\ \mathrm{s} \le t < 0.025\ \mathrm{s}$	$\omega_{rm}^* = 200\ \mathrm{r/min}$
$0.025\ \mathrm{s} \le t < 0.055\ \mathrm{s}$	$\omega_{rm}^* = 200 + 20\sin(400\pi t)\ \mathrm{r/min}$
$0.055\ \mathrm{s} \le t$	$\omega_{rm}^* = 200\ \mathrm{r/min}$
$0 \le t < 0.065\ \mathrm{s}$	$T_L = 0\ \mathrm{N\text{-}m}$
$0.065\ \mathrm{s} \le t < 0.08\ \mathrm{s}$	$T_L = 10\ \mathrm{N\text{-}m}$
$0.08\ \mathrm{s} \le t$	$T_L = 0\ \mathrm{N\text{-}m}$

The current control loop can be assumed as the first-order low-pass filter, whose cutoff frequency is the bandwidth of the current control loop decided in problem 3, part 4. The output of the speed regulator, which is the current reference, i_a^*, is limited within ± 75 A, and the anti-wind-up controller is included. The estimated torque constant is 95% of the actual torque constant as $\hat{K} = 0.95K$.

(1) The speed regulator, which is PI type, is implemented in the analog domain. The control bandwidth of the speed control loop, f_{sc}, is 200 Hz and the damping factor is $\sqrt{5}/2$.
If the measured speed is identical to the real speed, plot the following responses.
(A) 0–0.09 s: ω_{rm}, ω^*_{rm}, i^*_a
(B) 0.045–0.055 s: ω_{rm}, ω^*_{rm}, i^*_a
(C) 0.065–0.08 s:T_e, T_L, ω_{rm}, i^*_a

(2) If the speed sampling is done at every 200 μs, plot the following responses. The speed regulator is implemented digitally in the discrete time domain.
(A) 0–0.09 s: ω_{rm}, ω^*_{rm}, i^*_a
(B) 0.045–0.055 s: ω_{rm}, ω^*_{rm}, i^*_a
(C) 0.065–0.08 s: T_e, T_L, ω_{rm}, i^*_a

(3) If the speed sampling is done at every 2 ms, plot the following responses. The speed regulator is implemented digitally in the discrete time domain.
(A) 0–0.09 s: ω_{rm}, ω^*_{rm}, i^*_a
(B) 0.045–0.055 s: ω_{rm}, ω^*_{rm}, i^*_a
(C) 0.065–0.08 s: T_e, T_L, ω_{rm}, i^*_a

(4) The speed is measured by the following encoder. The number of pulses per revolution, P_{PR}, is 8192 and the measured angle has white noise, whose magnitude is 4% of $2\pi/8192$ in root mean square; the sampling frequency of the noise is $100\mu s$. Also, the speed is calculated by the "M" method and its sampling time, T_{sp}, is 1 ms. Repeat part 1, but the speed regulator algorithm is implemented digitally and executed at every 1 ms just after the speed is calculated. It is assumed that time for the speed calculation by the "M" method and for the control algorithm execution is negligible.

(5) Now, the speed calculation method is changed from the "M" method to the "M/T" method. Repeat part 4. The speed calculation and the execution of the algorithm is at every T_D, and T_{sp} is set as 1 ms, and time for the speed calculation by the "M/T" method and for the control algorithm execution is negligible. And clock frequency for the "M/T" method is high enough.

(6) Design the speed observer, whose bandwidth is 500 Hz. The measured angle, which is the input to the observer, is ideal. The estimated parameters are listed below. Repeat part 1. The observer and the regulator are designed in the analog domain with following parameter errors, and the time can be measured continuously.

$$\hat{B} = 0.2B \quad \text{and} \quad \hat{J} = 0.7J$$

(7) If the angle is measured by the encoder in part 4 at every 100 μs, repeat part 6. The observer and regulator are designed digitally in the discrete time domain, and algorithm execution time for the observer and speed regulator can be neglected. The sampling frequency of the observer is 10 kHz, and that of the speed regulator is 1 ms.

(8) With the operating condition in part 7, the acceleration reference is fed forward. Repeat part 7. The estimated inertia for the acceleration reference feed-forwarding is 70% of the actual inertia. The acceleration reference can be obtained from the given speed reference by ideal mathematical differential operation.

(9) In addition to of the operating condition and the controller in part 8, the acceleration feedback is added to enhance the robustness to the disturbance torque. Now, because of the acceleration feedback the system works as like the total inertia increased twofold. At this condition set the feedback gain K_α. And repeat part 8. In this problem, the measurement of the acceleration is ideal.

(10) By the method in the part 7, the angle from the encoder is measured. Implement the acceleration observer and speed observer together. And set the gains in both observers to achieve reasonable speed control performance. In this gain setting, repeat part 9 based on the estimated acceleration.

(11) The encoder is changed. The number of pulses per revolution, $P_{PR,}$ is 2^{22} and the measured angle has white noise, whose magnitude is 4% of $2\pi/2^{22}$ in root mean square; the sampling frequency of the noise is 100 μs. The speed is calculated by the "M" method and its sampling time, T_{sp}, is 100 μs. Repeat part 4.

(12) By using the encoder in part 11, the angle is measured at every 100 μs. Repeat part 10.

5. The PI speed regulator in problem 4 is changed to IP regulator. All operating conditions of the speed control system are the same as those of problem 4. The speed regulator is implemented in the analog domain. The damping factor of the IP regulator is $\sqrt{5}/2$. It can be assumed that the measured speed is identical to the real speed. For 10-r/min step input to the speed control loop, both the PI and the IP regulators reveal the same peak of the current reference. Find the gains of the IP regulator and the bandwidth of speed regulation loop with the IP regulator. In that gain setting, plot the following responses.

(A) 0–0.09 s; ω_{rm}, ω_{rm}^*, i_a^*

(B) 0.045–0.055 s; ω_{rm}, ω_{rm}^*, i_a^*

(C) 0.065–0.08 s; T_e, T_L, ω_{rm}, i_a^*

6. As shown in Fig. P.4.5, there is a X–Y table driven by two X-axis linear motors and one Y-axis linear motor.

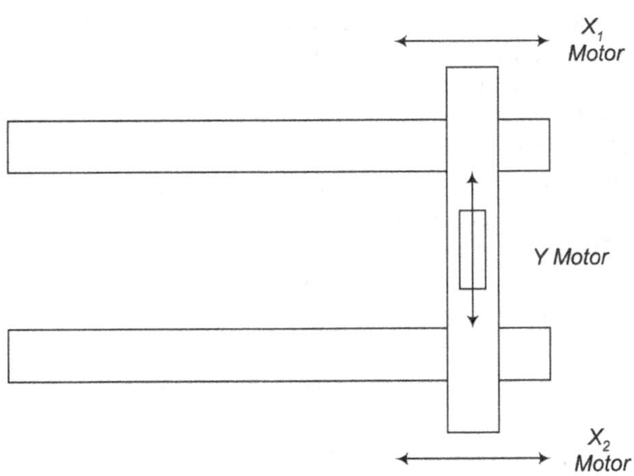

Figure P4.5 X–Y table driven by three linear motors.

By adjusting the thrust forces of X_1 and X_2motors, namely F_{X1} and F_{X2}, the position and the speed of the Y axis is controlled. According to the mover position on the Y axis, the equivalent mass to X_1 and X_2 motors, namely M_1 and M_2, vary from 5 kg to 15 kg. For the worst-case consideration, assume $M_1 = 5\,\text{kg}$, $M_2 = 15\,\text{kg}$. The block diagram of the mechanical system is shown in Fig. P4.6, where the coupling coefficient, K_{coup}, is $10^6\,\text{N/m}$. To control the mechanical system, design the speed control system of the X–Y table as shown in Fig. P4.7, where the speed regulator is P type. Together with the P regulator, we add the decoupling controller, the active damping term, and the acceleration reference feed-forward term to the output of the regulator. The P gains of the regulator of both X_1 and X_2 motors, namely K_{P1} and K_{P2}, are the same as K_P. Based on the block diagram of the mechanical system shown in Fig. P4.6, answer the following questions.

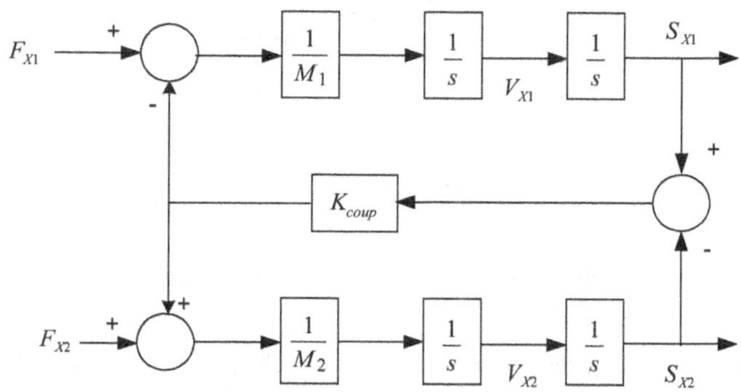

Figure P4.6 Block diagram of mechanical system of X axis of X–Y table.

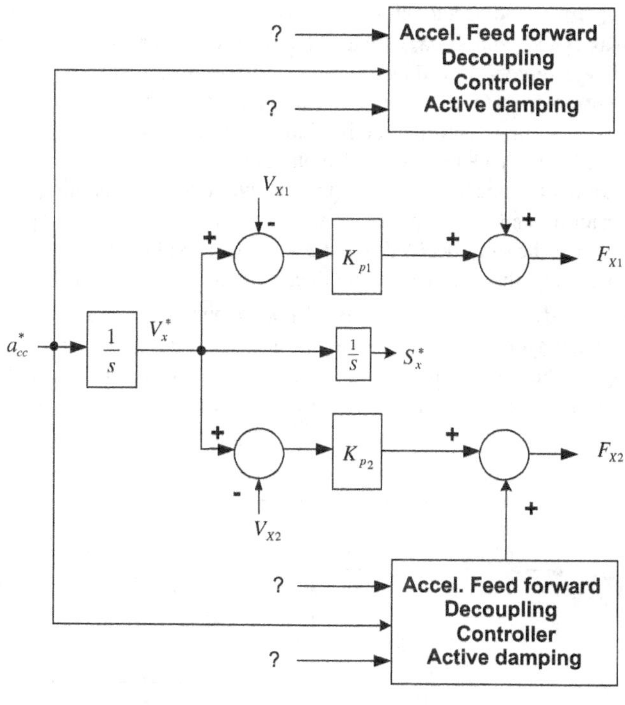

Figure P4.7 Block diagram of a controller.

(1) Calculate the moving distance if the movers of X_1, and X_2 motors move exactly according to the acceleration reference in Fig. P4.8.

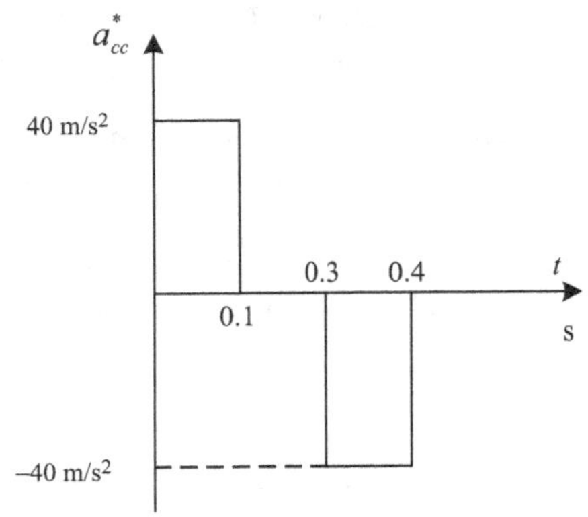

Figure P4.8 Acceleration reference.

(2) When the coupling coefficient is neglected like $K_{coup} = 0$, set the P gain of the speed regulator if the bandwidth of the transfer function of V_{x1}/V_X^* is 50 Hz. In the calculation of the gain, the mass of each mover is the same as 10 kg.

(3) If S_{x1}, S_{x2}, V_{x1}, and V_{x2} are ideally measurable, design the decoupling controller, active damping terms, and acceleration feed-forward terms to suppress the position errors defined as $S_x^* - S_{x1}$ and $S_x^* - S_{x2}$ are less than 10 μm in over all moving range of movers of the linear motors according to the acceleration reference in Fig. P.4.8. Plot the position errors from 0 s to 0.4 s with the acceleration reference in Fig. P4.8 in the case with the decoupling controller, active damping terms, and acceleration feed-forward terms together with a P regulator. Also plot the position error from 0 s to 0.4 s with the acceleration reference in Fig. P4.8 in the case without the decoupling controller, active damping terms, and acceleration feed-forward terms, with only a P regulator. For these plots, P gain is the value decided at part 2, and the mass of mover of X_1 is 5 kg and that of X_2 is 15 kg. The thrust forces of the linear motors are limited within $\pm$700 N.

(4) Design a speed observer that estimates V_{x1} based on S_{x1}, F_{x1} as shown in Fig. P4.9. Plot the estimated speed, $\hat{V}_{x1}$, and actual speed, V_{x1}. In the design of the observer, the coupling coefficient is assumed as zero and the mass of each mover is 10 kg. All poles of the observer lie at –150 Hz of real axis identically.

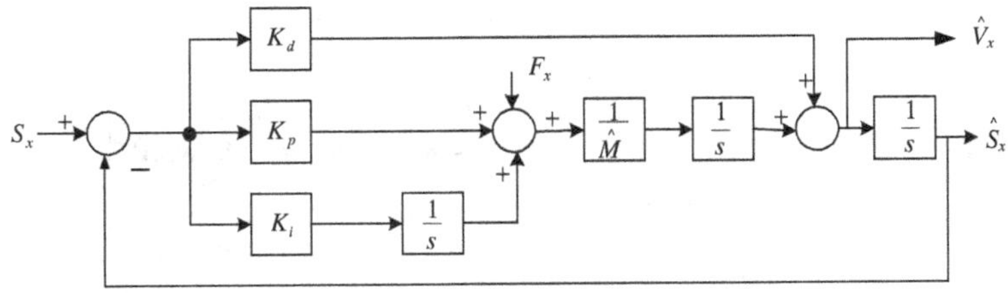

Figure P4.9 Speed observer.

By using observer designed in part 4, close the speed control loop with the estimated speed by the designed observer, and repeat part 3.

7. With the system shown in Fig. P4.1, the position of the load is controlled. The parameters of the load, the electric machine, and the connecting shaft are the same as those of problem 1. The final position is 30 revolutions of the load, and overshoot in the response is not allowed. The magnitude of jerk, acceleration, and speed is limited as follows: $100 \times 2\pi\ \text{rad/s}^3$, $20 \times 2\pi\ \text{rad/s}^2$, and $20 \times 2\pi\ \text{rad/s}$, respectively.

(1) Keeping the limitations, plot the jerk, acceleration, speed, and position (angle of load) references simultaneously with regard to time, t, which minimize the tracking time as short as possible. In the plot, the horizontal axis is time and its range is from 0–3 s. And the vertical axes are for jerk, acceleration, speed, and position reference in -1000 to $1000\ \text{rad/s}^3$, -200 to $200\ \text{rad/s}^2$, -200 to $200\ \text{rad/s}$, and -100 to 100 rad, respectively.

(2) If the angle and the speed are measured ideally, design a P–PI-type position regulator in Fig. 4.50, where the bandwidth of the speed control loop is 16 Hz and that of the position loop is 2 Hz. In the design of the regulator, the inertia of the electric machine and load can be assumed to be a single inertia, which means that there is no connection shaft and that two inertias are tied together. In the design of a P–PI regulator, the current regulator can be assumed to be an ideal one. In the computer simulation, the current regulator is modeled as the first-order filter as $\omega_c/s + \omega_c$, where $\omega_c = 200 \times 2\pi\ \text{rad/s}$. The maximum magnitude of the current is limited within 50 A and the torque (back EMF) constant is 2 N-m/A. In the design of the controller, the estimated torque constant is 90% of the real value as $\hat{K} = 0.9K$. Also, the estimated inertia is 90% of the actual value of the total inertia.

(3) By controlling the system using the regulator designed in part 2, plot the response of jerk, jerk reference, acceleration, acceleration reference, speed, speed reference, and position, position reference. In this control system, there is no load torque. In the plot, the horizontal axis is time and its range is from 0 s to 3 s. And the vertical axis is for jerk, acceleration, speed, and position in $-1000\ \text{rad/s}^3$ to $1000\ \text{rad/s}^3$, $-200\ \text{rad/s}^2$ to $200\ \text{rad/s}^2$, $-200\ \text{rad/s}$ to $200\ \text{rad/s}$, and -100 rad to 100 rad, respectively. Also calculate the value, which is the time integral of the square of the position error from time $t = 0$ s to time $t = 3$ s.

(4) If the speed of the load and the electric machine can be measured ideally, add an active damping term to the control system designed in part 2, and enhance the responses in part 3. Plot the responses as in part 3 with the active damping term. Also calculate the value, which is the time integral of the square of the position error from time $t = 0$ s to time $t = 3$ s. In the designing of active damping term, the current should not be oscillatory continuously.

(5) In addition to the control system designed in part 4, the speed and acceleration command feed-forward terms are added to enhance the command following performance. In the design of the controller, the estimated torque constant is 90% of the real value as $\hat{K} = 0.9K$. Also, the estimated inertia is 90% of the actual value of the total inertia. Plot the responses as in part 3 with the feed-forward terms. Also calculate the value, which is the time integral of the square of the position error from time $t = 0$ s to time $t = 3$ s.

(6) After finishing the position control with the control method in part 5, the load torque, $T_L = 60 \sin(20{*}\pi{*}t)$ N-m, is applied from time point $t = 3$ s. After 3 s, the position reference is constant. Plot the acceleration, speed, position, load torque, T_L, and the machine torque, T_e. In the plot, the horizontal axis is time and its range is from 3 s to 4 s. And the vertical axis is for acceleration, speed, position, and torque in $-200\,\mathrm{rad}/s^2$ to $200\,\mathrm{rad}/s^2$, $-200\,\mathrm{rad}/s$ to $200\,\mathrm{rad}/s$, -100 rad to 100 rad, and -120 Nm to 120 N-m, respectively.

(7) To enhance disturbance rejection performance, the acceleration of the load is fed back to the control system. The acceleration is measured ideally but with time delay by 10 ms like $\hat{\alpha}(s) = \alpha(s)*e^{-.01\,s}$. By adding the acceleration feedback term to the control system designed in part 5, plot the responses as plots in part 6.

REFERENCES

1. M. A. Valenzuela et al., Start-up and commissioning procedure for electronically line-shafted paper machine drive, *IEEE Trans. Ind. Appl.*, Vol. 38, July/August 2002, pp. 966–973.
2. T. Krishnan et al., Speed control of DC motor using thyristor dual converter, *IEEE Trans. Ind. Elec.& Cont. Inst.*, Vol. 23, No. 4, November 1976, pp. 391–399.
3. W. Leonhard, *Control of electrical drives*, 2nd Edition, Springer, Berlin, 1996, Section 7.2 and 8.5.
4. R. Kennel et al., Predictive control strategy for converters, in *Proceedings of IFAC Control in Power Electronics and Electrical Drives*, 1983, pp. 415–422.
5. H. Sugimoto et al., *Practical Aspects of Theory and Design of AC Servo System*, Motor Electronics Publications, Tokyo, 1990, Section 3.2 (in Japanese).
6. V. Blasko et al., Sampling of discontinuous voltage and current signals in electrical drives: A system approach, *IEEE Trans. Ind. Appl.*, Vol. 34, No. 5, September/October 1998, pp. 1123–1130.
7. Ahno Yoo et al., Design of current regulator with extended bandwidth for servo motor drive, in *Power Conversion Conference (PCC) Proceedings, JIEE*, 2007.
8. C. Bohn et al., An Analysis Package comparing PID anti-windup strategies, *IEEE Control Systems Mag.*, Vol. 16, No. 2, April 1995, pp. 33–40.
9. A. Visioli, *Modified* Anti-windup scheme for PID controllers, *IEE Proc. Control Theory Appl.*, Vol. 150, January, 2003, pp. 49–54.
10. B. K. Bose, *Power Electronics and Variable Frequency Drives*, IEEE Press, New York, 1997, Section 5.5.
11. J. Jang, *Sensorless control of PMSMs for wide speed operation range*, Ph.D thesis, Appendix B, Seoul National University, Seoul, Korea, 2006 (in Korean).
12. F. Briz et al., Analysis and design of current regulators using complex vectors, *IEEE Trans. Ind. Appl.*, Vol. 36, May/June 2000, pp. 817–825.
13. H. Yoo et al., Anti-windup for complex vector synchronous frame PI current controller, *Trans. Korean Institute of Power Electron.*, Vol. 11, No. 5, 2006, pp. 404–407 (in Korean).
14. Y. Jeong, On line minimum copper loss control of a PMSM considering magnetic saturation, Ph.D thesis, Seoul National University, Seoul, Korea, 2005 (in Korean).
15. *Incremental Rotary Encoders*, Heidenhain, Schaumburg, IL. Available online at http://www.heidenhain.com/products/rotary/dg_i.htm.
16. K. Kubo et al., Disturbance torque compensated speed observer for digital servo drives, in *Proceedings, IEEE-IECON*, 1990, pp. 1182–1187.
17. T. Ohmae et al., A microprocessor-controlled high-accuracy wide-range speed regulator for motor drives, *IEEE Trans. Ind. Electron.*, Vol. IE-29, August 1982, pp. 207–211.
18. H. Kim, Improvement of speed control performance for PMSM using Kalman filter in very low speed range, Ph.D thesis, Seoul National University, Seoul, Korea, 1999. (in Korean).
19. G. Franklin, et al. *Feedback control of dynamic system*, 2nd edition, Addison-Wesley, Reading, MA, 1991, pp. 361–458.
20. H. Kim et al., A new motor speed estimator using kalman filter in low-speed range, *IEEE Trans. Ind. Appl.*, Vol. 43, No. 4, August. 1996. pp. 498–504.
21. R. D. Lorenz, High-resolution velocity estimation for all-digital AC servo drives, *IEEE Trans. Ind. Appl.*, Vol. 27, No. 4, July/August, 1991, pp. 701–705.

22. G. Ellis, *Control System Design Guide*, 3rd edition, Elsevier Academic Press, New York, 2004, Chapter 10.
23. Y. D. Yoon et al., Dual observers for the disturbance rejection of a motion control system, in *IEEE Industrial Application (IAS) Annual Meeting Conference Record*, 2007.
24. J. Jang, *Sensorless control of PMSMs for wide speed operation range*, Ph.D thesis, Appendix C, Seoul National University, Seoul, Korea, 2006 (in Korean).
25. H. Sugimoto et al., *Practical Aspects of Theory and Design of AC Servo System*, Motor Electronics Publications, 1990 (in Japanese).
26. T. Umeno et al., Robust speed control of DC servomotors using modern two degrees-of-freedom controller design, *IEEE Trans. Ind. Electron*, Vol. 38, No. 5, October, 1991, pp. 363–368.
27. P. Schmit, Acceleration feedback implemented through the utilization of an observer as applied to a robotic manipulator, Ph.D. thesis, ECE Department, University of Wisconsin—Madison, 1991.
28. http://www.columbiaresearchlab.com.
29. H. Shim et al., Use of accelerometer for precision motion control of linear motor driven positioning system, in *IEEE Industrial Electronics Society, IECON '98, Proceedings of the 24th Annual Conference of the IEEE,* Vol. 4, 1998, pp 2409–2414.
30. P. Schmit et al., Design principles and implementation of acceleration feedback to improve performance of DC drives, *IEEE Trans. Ind. Appl.*, Vol. 28, No. 3, May. 1992, pp. 594–599.
31. J. K. Kim et al., High precision position control of linear permanent magnet synchronous motor for surface mount device placement system, in Conference Record PCC—Osaka, Vol. 1, 2002, pp. 37–42.
32. S. Lee, New PLL method and voltage controller of series compensator for voltage sag compensation, Ph.D thesis, Seoul National University, Seoul, Korea, 2003 (in Korean).
33. M. Bollen, *Understanding Power Quality Problems-Voltage Sags and Interruptions*, IEEE Press, New York, 2000.
34. S. J. Lee et al., A new phase detecting method for power conversion systems considering distorted conditions in power system, in *IEEE Industrial Application (IAS) Annual Meeting Conference Record*, 1999, pp. 2167–2172.

Chapter 5

Vector Control [1 and 2]

The electric machines convert the mechanical power to electrical power as a generator or convert electrical power to mechanical power as a motor. The mechanical power is presented as the instantaneous product of the rotating speed and the torque. The electric machines basically convert the current (or torque) to the torque (or current) under the excitation flux as a motor (or generator). The speed is decided by the mechanical system connected to the electric machine and by the torque. Usually, the mechanical system includes inertia components, and the speed is the low-pass filtered form of the torque as presented in (2.22) and in Fig. 2.12. The cutoff angular frequency of the low-pass filter is B/J. In many industrial drive systems, where the speed of the electric machine is the concern even though the torque of the electric machine has some fluctuations instantaneously, if the frequency of the fluctuation is high enough compared to the cutoff frequency, the speed can be regulated satisfactorily. But, in a rolling mill drive system shown in Fig. 1.20 or in a high-speed elevator system shown in Fig. P1.3, the torque should be controlled instantaneously to regulate tension of the web or to regulate the acceleration of the cage. In many high-precision motion control systems, where the acceleration, speed, and position are all regulated instantaneously according to their references, the instantaneous torque control is a prerequisite. In these systems, by controlling the torque instantaneously, the acceleration is controlled, and the instantaneous speed is the time integral of the acceleration and it is decided by the acceleration. Also, the instantaneous position is that of the speed, and it is decided by the speed. The instantaneous torque of the electric machine comes from the cross product of the flux linkage vector and the line vector, where the current flows. Therefore, to control the torque instantaneously, the flux linkage and the line vector should be controlled instantaneously. Hence, not only the magnitudes of the current and flux linkage, but also the relative angle between two vectors, should be controlled instantaneously. The angle can be oriented from the excitation flux, where the d axis of

Control of Electric Machine Drive Systems, by Seung-Ki Sul

the rotor reference frame is in the case of AC machine. Therefore, the vector control of the electric machine can be understood from the above context. Also, field (flux)-oriented control can be understood in the same context. Therefore, both the field-oriented control and the vector control signify the instantaneous regulation of the torque of the electric machine. In this chapter, the principles of instantaneous torque control, referred to as vector control, are described in the case of several electric machines. And to extend the operating region of the electric machine, the capability curve of the machine mentioned in Chapter 2 is further investigated using the flux weakening control concept based on *d–q* reference frame theory in Chapter 3.

5.1 INSTANTANEOUS TORQUE CONTROL

The instantaneous torque of the electric machine comes from the cross product of the flux linkage vector and the line vector, where the current flows. And to control the torque instantaneously, the flux linkage and the line vector should be controlled instantaneously. In this section, from the modeling of a DC machine, it is described how the instantaneous torque control is possible in a DC machine. Also, the principle of the torque control is extended to permanent magnet AC machines. And, finally, I describe how the principle is implemented in the induction machine.

5.1.1 Separately Excited DC Machine

As shown in Fig. 5.1a, the excitation (field) flux is regulated by the current of the field winding, and the armature current, from which the torque comes through the interaction with the flux, is regulated by the armature voltage. The armature current may distort the field flux by the armature reaction. However, under the assumption that the armature reaction is fully prevented by the commutating poles and the compensation winding as shown in Fig. 2.5, the flux linkage to the armature current can be fully regulated only by the field winding current. The angle between the armature current and the excitation flux is kept as 90° spatially by means of the commutator and the brush regardless of the rotation of the rotor. Hence, the instantaneous torque of DC machine is simply the product of the magnitude of the armature current and that of the excitation flux, where the polarity of the torque is decided by the polarity of the flux and by that of the current. If the magnitude and the direction of the flux are kept constant, then torque is solely proportional to the armature current as seen from (2.20) in Section 2.3. Hence, by controlling the armature current instantaneously, the torque can be regulated instantaneously. The same torque control capability can be obtained if the relative position of the armature winding and field winding is maintained even if both windings are rotating as shown in Fig. 5.1b. The outer field winding and the inner armature winding may be rearranged as shown in Fig. 5.1c. In this arrangement, if the relative position of the two windings and the magnitude of magnetic motive force (MMF) of both windings are maintained, then the same torque control capability can be achieved. As described in Section 2.5, the equivalent MMF of outer armature winding in Fig. 5.1c can be obtained by three-phase symmetry winding as shown in Fig. 5.1d, which is a structure of the AC synchronous machine in Section 3.3. Now the armature winding is stationary, the MMF by the outer armature windings is rotating. Also, the instantaneous position of the equivalent MMF of the winding is the decided by the instantaneous three-phase AC current in the winding. In Fig. 5.1d, to have the same position of MMF in Fig. 5.1c, the current in a-phase winding is null and the magnitude of current in b- and c-phase winding is identical, but the polarity of the current in b-phase winding is positive but that in c-phase winding is negative. In this way, the instantaneous torque control principle of a DC machine can be applied to an AC machine if the relative position and the magnitude of magnetic motive force (MMF) of both windings are maintained. As seen in Fig. 5.1, the instantaneous torque control of a synchronous machine is exactly the same as in the case of a DC machine, except the rotation of MMF of armature winding and MMF of the field winding. However, in the case of the induction machine, the instantaneous torque control is rather difficult to understand, where there is no separate field winding and no

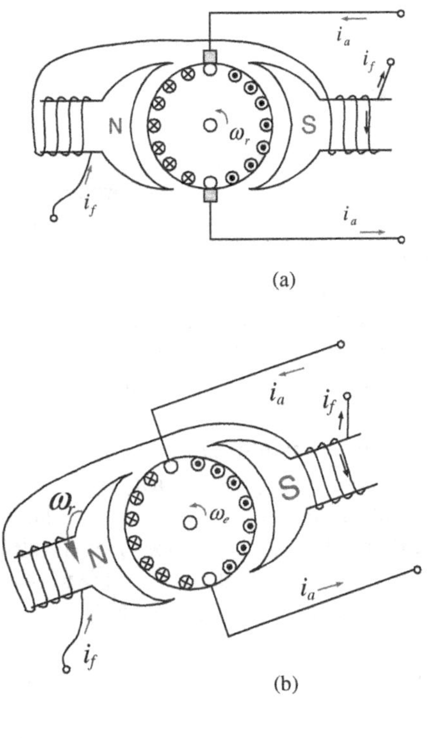

Figure 5.1 Relative position of field flux and armature current. (a) Separately excited DC machine; Stationary field and armature windings (b) Rotating field and armature winding: outer field winding, inner armature winding. (c) Rotating field and armature winding: outer armature winding and inner field winding. (d) Rotating field and armature winding: stationary outer armature winding by three-phase symmetry winding, as well as rotating inner field winding.

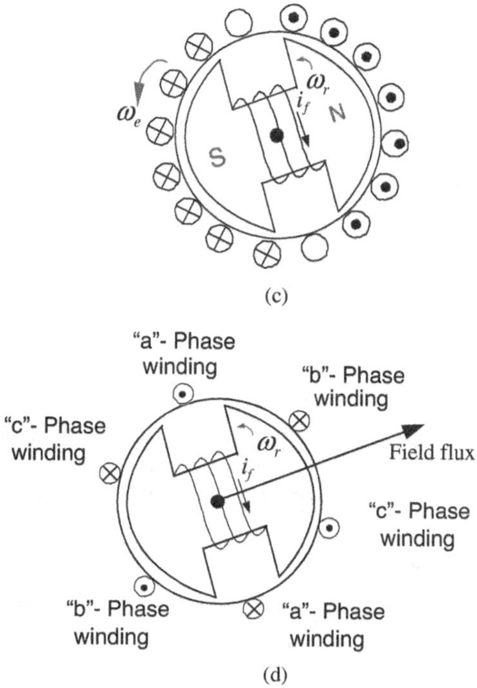

Figure 5.1 *(Continued)*

permanent magnet. But, the rotating flux linkage, whether it is rotor flux linkage, air gap flux linkage, or stator flux linkage, can be defined. Also, part of the stator current can be classified as the armature current with regard to the defined rotating flux linkage, which generates an MMF whose position is perpendicular to the rotating flux linkage. Through the interaction of the rotating flux linkage and MMF by the equivalent armature current, torque can be generated as like a DC machine. At first, the principles of the instantaneous torque control of AC machine can be understood easily from SMPMSM as follows.

5.1.2 Surface-Mounted Permanent Magnet Synchronous Motor (SMPMSM)

The torque of an SMPMSM in Fig. 5.2 can be represented as $T_e = \frac{3}{2}\frac{P}{2}\lambda_f i_{qs}^r$ derived in Section 3.3.3.1. Because the pole number of SMPMSM in Fig. 5.2 is four, the d axis and q axis are apart by 45° spatially. After measuring the rotor position instantaneously, if the q-axis current in the measured rotor reference frame is controlled instantaneously as

$$i_{qs}^{r*} = \frac{T_e^*}{\frac{3}{2}\frac{P}{2}\lambda_f}$$

where T_e^* means the instantaneous torque reference, the torque of SMPMSM can be regulated instantaneously because the flux is kept constant by the permanent magnet.

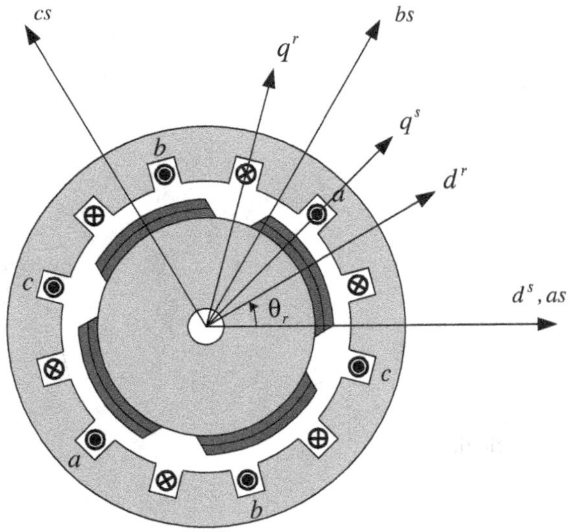

Figure 5.2 Four-pole surface-mounted permanent magnet synchronous machine (SMPMSM) and its rotor reference *d–q* frame.

The current in the rotor reference *d–q* frame, i^r_{ds} and i^r_{qs}, can be obtained through the calculation in (5.1) using the measured rotor position, θ_r, and measured phase currents as explained in Section 3.1.

$$\mathbf{i}^r_{dqs} = i^r_{ds} + ji^r_{qs} = \tfrac{2}{3}(i_{as} + \mathbf{a}i_{bs} + \mathbf{a}^2 i_{cs})e^{-j\theta_r} \tag{5.1}$$

To get maximum torque per ampere (MTPA), the current at the *d* axis should be set to zero as $i^r_{ds} = 0$ because *d*-axis current, i^r_{ds}, does not contribute to the torque. Then, the *q*-axis current, i^r_{qs}, which is now equal to the total current, $\mathbf{i}^r_{dqs}$, is always perpendicular to the field flux by the magnet. Hence, the torque is regulated instantaneously by regulating only i^r_{qs}. The instantaneous torque regulation can be represented as a control block diagram as shown in Fig. 5.3.

By the instantaneous torque control principles of a DC machine and an SMPMSM, the following conditions can be deduced to control the torque of the electric machine instantaneously.

1. The current, which interacts with the field flux, should be regulated instantaneously regardless of the variation of back EMF, leakage inductance, and winding resistance.
2. The field flux should be regulated regardless of the variation of the current, which interacts with the field flux.
3. The flux and the current should be kept as perpendicular instantaneously either by mechanical means or by angle measurement and control.

If above three conditions are fulfilled simultaneously, then the torque of the electric machine can be regulated instantaneously.

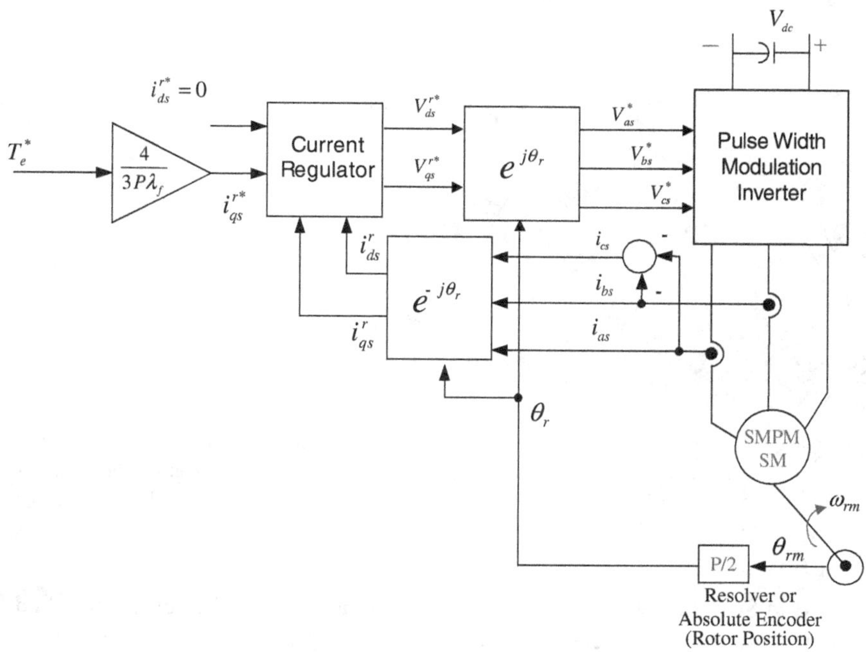

Figure 5.3 Control block diagram for instantaneous torque control of SMPMSM.

5.1.3 Interior Permanent Magnet Synchronous Motor (IPMSM)

The torque of the IPMSM in Fig. 5.4 can be represented as $T_e = \frac{3}{2}\frac{P}{2}\{(L_d - L_q)i_{ds}^r i_{qs}^r + \lambda_f i_{qs}^r\}$ at rotor reference d–q frame derived at Section 3.3.3.2. It is the sum of the reluctance torque and the field torque. And, even if the field flux, λ_f, by the magnet is kept constant, the pair of i_{ds}^r and i_{qs}^r is enormous to generate the given torque reference, T_e^*. However, if the total losses are minimized at the given torque, the pair is uniquely decided. If the iron loss can be neglected, the minimization of the magnitude of the stator current vector, $\mathbf{i}_{dqs}^r$, is the minimization of the copper loss, which is the only loss now considered. Such an operation is called a maximum torque per ampere (MTPA) operation, which gives the best efficiency while generating the given reference torque. Usually, the iron loss is relatively quite small compared to the copper loss at the rated operating conditions of IPMSM. In this context, the instantaneous torque control with MTPA operation involves (a) finding a pair of i_{ds}^r and i_{qs}^r, which minimizes the magnitude of the current vector and generates the reference torque, and (b) regulating d–q current according to the current reference pair found. All of these controls should be done instantaneously. If the parameters of IPMSM are constant regardless of the operating condition, the pair can be derived analytically at a particular speed and torque. However, as

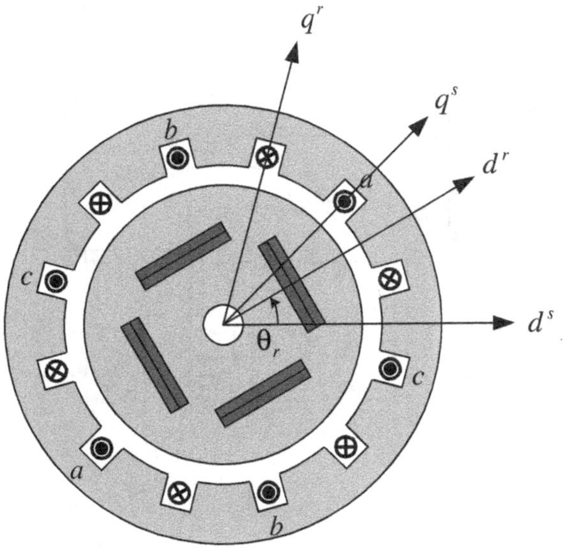

Figure 5.4 Four-pole interior permanent magnet synchronous machine (IPMSM) and its rotor reference d–q frame.

mentioned in Section 2.9.2, if the inductances, L_d and L_q, vary by several hundred percentages, then the analytical solution cannot be an option. The pair can be found through the experiment at various operating points and/or through the parameter data considering the variation of inductances and the field flux by the magnet. After the pair is stored in a look-up table (LUT) with the index of the speed and torque, the current is regulated according to the pair in LUT at the specified speed and torque. The control block diagram of IPMSM can be depicted as Fig. 5.5.

5.2 VECTOR CONTROL OF INDUCTION MACHINE [1]

In the case of an induction machine, unlike the case of previous machines such as the separately excited DC machine and the synchronous machine, both the current for the field flux (field current) and the current for the torque (armature current) should be provided by the stator winding because of no field winding. So, it is quite difficult to understand the separate and independent control of the field flux and the current for the torque by 90° spatially in electrical angle. However, based on the principle of rotating MMF in Section 2.5, the vector control, which means instantaneous torque control, can be understood with the reference frame theory discussed in Chapter 3. The vector control methods of the induction machine can be classified as the indirect method and the direct method. In the early stage of development, the former one is mainly implemented and commercialized. In the indirect method, the slip angular frequency of the induction machine is controlled and the flux and the torque component currents are indirectly adjusted through the slip angular frequency. For this method, though the rotational speed of the machine should be measured, the torque of the induction machine can be controlled instantaneously from the starting of the machine to the top speed of the machine. But the method has shortcomings such as dependency on the

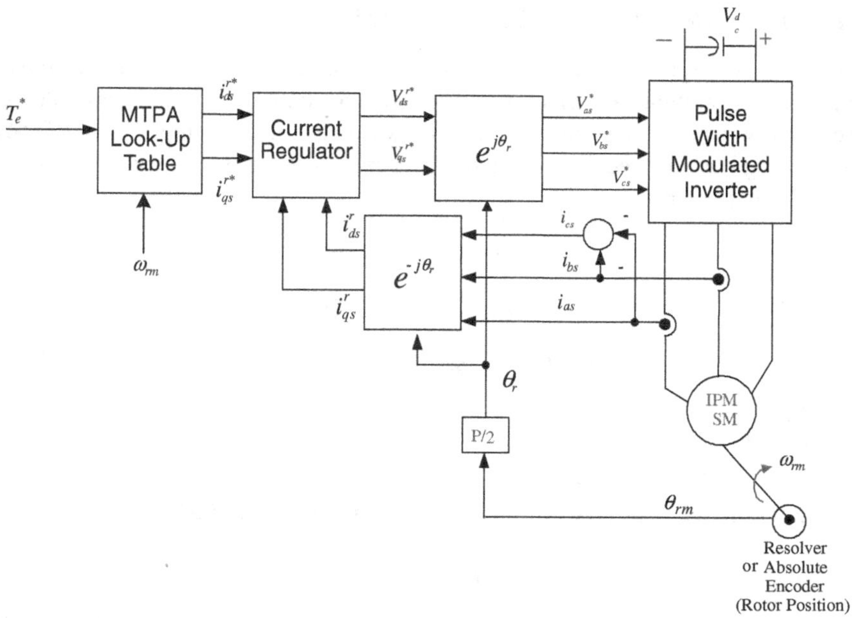

Figure 5.5 Control block diagram for instantaneous torque control of IPMSM.

machine parameters, especially on rotor time constant, namely L_r/R_r. In the direct method, after the magnitude and the position of the field flux, usually rotor flux, is identified by various methods, the stator current is decomposed to the flux component and torque components according to the identified position of field flux. The flux and torque are controlled directly by adjusting the magnitude of the components. The direct method can be easily understood, but the identification of the flux vector is difficult. In particular, in the low speed including the starting of the machine, the identified magnitude and position of the field flux vector may have considerable errors, and the torque control performance would be poor. However, thanks to the development of digital signal processing capability, the direct vector control, which does not require the position sensor on the rotor, has been developed and commercialized recently in the name of "sensorless vector control." The detailed description of the sensorless control is in Chapter 6.

5.2.1 Direct Vector Control

In the direct method, like the case of an SMPMSM, the position of the rotor flux is identified and the torque component current is applied 90° ahead spatially in the direction of the rotation. But unlike the case of an SMPMSM, where the magnitude of the flux is maintained by the permanent magnet, the magnitude of the rotor flux linkage, which is chosen as the orientation of the flux linkage, should be controlled simultaneously by the flux component current together with the torque component current from the stator of the induction machine.

5.2.1.1 Principles of Direct Vector Control

The d–q voltage equations expressed at the d–q reference frame rotating arbitrary speed, ω, can be written in (5.2)–(5.5) as derived in Section 3.2.1. If the neutral point of the stator winding of a squirrel cage rotor is isolated and the impedances of the stator winding are balanced, which is the usual case, then there is no need to consider the n-axis component.

$$V_{ds}^{\omega} = R_s i_{ds}^{\omega} + p\lambda_{ds}^{\omega} - \omega\lambda_{qs}^{\omega} \tag{5.2}$$

$$V_{qs}^{\omega} = R_s i_{qs}^{\omega} + p\lambda_{qs}^{\omega} + \omega\lambda_{ds}^{\omega} \tag{5.3}$$

$$0 = R_r i_{dr}^{\omega} + p\lambda_{dr}^{\omega} - (\omega - \omega_r)\lambda_{qr}^{\omega} \tag{5.4}$$

$$0 = R_r i_{qr}^{\omega} + p\lambda_{qr}^{\omega} + (\omega - \omega_r)\lambda_{dr}^{\omega} \tag{5.5}$$

where p is the differential operator.
Also, the stator and rotor flux linkage can be written in (5.6)–(5.9):

$$\lambda_{ds}^{\omega} = L_s i_{ds}^{\omega} + L_m i_{dr}^{\omega} \tag{5.6}$$

$$\lambda_{qs}^{\omega} = L_s i_{qs}^{\omega} + L_m i_{qr}^{\omega} \tag{5.7}$$

$$\lambda_{dr}^{\omega} = L_m i_{ds}^{\omega} + L_r i_{dr}^{\omega} \tag{5.8}$$

$$\lambda_{qr}^{\omega} = L_m i_{qs}^{\omega} + L_r i_{qr}^{\omega} \tag{5.9}$$

Based on the rotor flux linkage and stator current, the torque can be expressed as

$$T_e = \frac{3}{2}\frac{P}{2}\frac{L_m}{L_r}\left(\lambda_{dr}^{\omega} i_{qs}^{\omega} - \lambda_{qr}^{\omega} i_{ds}^{\omega}\right). \tag{5.10}$$

At the stationary reference frame, where $\omega = 0$, the position of the rotor flux linkage, θ_e, can be derived as

$$\theta_e = \tan^{-1}\left(\frac{\lambda_{qr}^{s}}{\lambda_{dr}^{s}}\right) \tag{5.11}$$

The rotor flux linkage at the d–q axis in the stationary reference frame can be obtained as follows. At first, the stator flux linkage in the stationary d–q reference frame can be calculated by (5.12) and (5.13) from (5.2) and (5.3) by setting $\omega = 0$.

$$\lambda_{ds}^{s} = \int_0^t \left(V_{ds}^{s} - R_s i_{ds}^{s}\right) d\tau \tag{5.12}$$

$$\lambda_{qs}^{s} = \int_0^t \left(V_{qs}^{s} - R_s i_{qs}^{s}\right) d\tau \tag{5.13}$$

Next, the rotor current can be expressed in terms of rotor flux linkage and the stator current from (5.8) and (5.9). Finally, the rotor flux linkage can be represented as (5.14) and (5.15) after eliminating rotor currents by (5.6) and (5.7).

$$\lambda_{dr}^{s} = \frac{L_r}{L_m}\left(\lambda_{ds}^{s} - \sigma L_s i_{ds}^{s}\right) \tag{5.14}$$

$$\lambda_{qr}^{s} = \frac{L_r}{L_m}\left(\lambda_{qs}^{s} - \sigma L_s i_{qs}^{s}\right) \tag{5.15}$$

where σ is the leakage factor and is defined as $\sigma = \frac{L_s L_r - L_m{}^2}{L_s L_r}$. By substituting (5.14) and (5.15) into (5.11), the position of the rotor flux linkage can be obtained. According to the position, θ_e, the stator current can be represented as (5.16) in the synchronously rotating reference frame.

$$\mathbf{i}_{dqs}^{e} = i_{ds}^{e} + j i_{qs}^{e} = \mathbf{i}_{abcs} e^{-j\theta_e} \tag{5.16}$$

In the synchronously rotating reference frame, the rotor flux linkage exists only on the d axis because the d axis is set as the axis where all rotor flux linkage lies. Hence the flux linkage in the q^e axis is inherently null as $\lambda_{qr}^{e} = 0$. By using this condition, (5.4) can be simplified as

$$0 = R_r i_{dr}^{e} + p\lambda_{dr}^{e} \tag{5.17}$$

By using (5.8) and (5.17), i_{dr}^{e} can be eliminated and the rotor flux linkage can be expressed in terms of the machine parameters and d-axis stator current as

$$\lambda_{dr}^{e} = L_m i_{ds}^{e} - \frac{L_r}{R_r} p\lambda_{dr}^{e} \tag{5.18}$$

Equation (5.18) can be rewritten as (5.19) to express the relationship between the current and the flux linkage explicitly.

$$\lambda_{dr}^{e} = \frac{L_m}{1 + p\frac{L_r}{R_r}} i_{ds}^{e} \tag{5.19}$$

Therefore, the flux linkage, λ_{dr}^{e}, can be adjusted by controlling the d-axis current, i_{ds}^{e}. In particular, from (5.19), if i_{ds}^{e} is kept as constant for a while ($t > 4\tau_r = 4\frac{L_r}{R_r}$), then the magnitude of the rotor flux linkage can be approximated as

$$|\lambda_r| = \lambda_{dr}^{e} \approx L_m i_{ds}^{e} \tag{5.20}$$

The magnitude is proportional to the d-axis current, i_{ds}^{e}, under the condition of the constant d-axis current for a while. At the synchronous d–q frame fixed on the rotor flux linkage, because $\lambda_{qr}^{e} = 0$, the torque in (5.10) can be simplified as

$$T_e = \frac{3}{2}\frac{P}{2}\frac{L_m}{L_r}\lambda_{dr}^{e} i_{qs}^{e} \tag{5.21}$$

By substituting (5.19) into (5.21), the torque can be represented in the terms of only the stator current as

$$T_e = \frac{3}{2}\frac{P}{2}\frac{L_m^2}{L_r}\left(\frac{1}{1+p\frac{L_r}{R_r}}i_{ds}^e\right)i_{qs}^e \tag{5.22}$$

If the d-axis current is kept as $i_{ds}^e = \lambda_r/L_m$ for more than $4\tau_r$ and maintained at that value, then the torque of the induction machine can be expressed as (5.23), where the torque is instantaneously controlled by regulating only q-axis stator current.

$$T_e = \frac{3}{2}\cdot\frac{P}{2}\frac{L_m}{L_r}|\lambda_r|i_{qs}^e \tag{5.23}$$

In the direct vector control method mentioned above, after the instantaneous position of the rotor flux linkage, θ_e, is identified, by decomposing the stator current the torque can be controlled instantaneously like the vector control of an SMPMSM. However, identifying the instantaneous position of the rotor flux linkage in real time is quite difficult to achieve, especially at the low rotational speed of the induction machine.

5.2.1.2 Implementation of Direct Vector Control [2]

The instantaneous position of the rotor flux linkage can be identified by the following methods.

1. **Measurement of Air Gap Flux by Hall Effect Sensors.** The Hall effect sensor can measure the magnetic flux density. By installing two sensors on the air gap of the induction machine as shown in Fig. 5.6, which are located 90° apart in the case of a two-pole machine. Then, from the measured air-gap flux, λ_{dm}^s and λ_{qm}^s, the rotor flux linkage can be calculated as follows.
At first, the rotor flux linkage can be expressed in terms of the air-gap flux and the rotor current as

$$\begin{aligned}\lambda_{dqr}^s &= L_{\ell r}\mathrm{i}_{dqr}^s + L_m\left(\mathrm{i}_{dqr}^s + \mathrm{i}_{dqs}^s\right)\\ &= L_{\ell r}\mathrm{i}_{dqr}^s + \lambda_{dqm}^s\end{aligned} \tag{5.24}$$

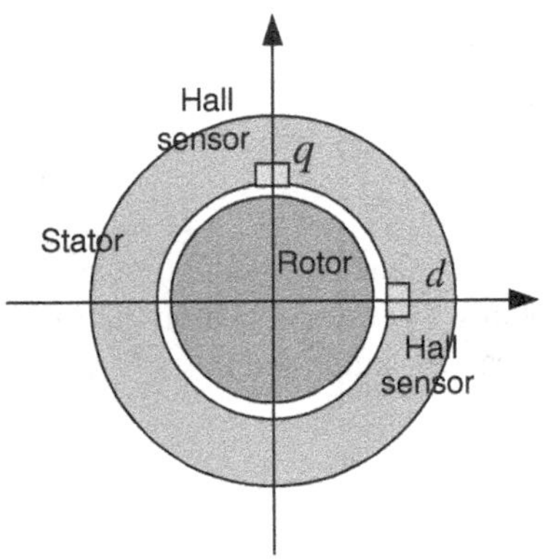

Figure 5.6 Installation of Hall effect sensors in a two-pole induction machine.

And, the rotor current can be represented in terms of stator current and the air-gap flux as

$$\mathrm{i}_{dqr}^{s} = \frac{\lambda_{dqm}^{s} - L_m \mathrm{i}_{dqs}^{s}}{L_m} \tag{5.25}$$

Finally, the rotor flux can be calculated by the measured air-gap flux and the stator current as

$$\begin{aligned} \lambda_{dqr}^{s} &= \left(\frac{\lambda_{dqm}^{s}}{L_m} - \mathrm{i}_{dqs}^{s}\right) L_{\ell r} + \lambda_{dqm}^{s} \\ &= \frac{L_r}{L_m}\lambda_{dqm}^{s} - L_{\ell r}\mathrm{i}_{dqs}^{s} \end{aligned} \tag{5.26}$$

The shortcomings of this method are (a) the sensitivity of Hall effect sensors to the temperature variation and (b) the dissection of the induction machine to install the sensors in the air gap of the machine. Moreover, because the leakage inductance, $L_{\ell r}$, varies according to the magnitude of the rotor current, i_{dqr}^{e}, as mentioned in Section 2.12.2, the accuracy of the identification of the instantaneous position of the rotor flux may be degraded with the variation of the operating condition.

2. **Measurement of Air Gap Voltage by Sensing Coils.** If two multiturn coils are installed on the surface of the stator apart by 90° in electrical angle, then the induced voltage at the coil is proportional to the time differentiation of the air gap flux. By integrating the voltage as (5.27), air-gap flux can be deduced.

$$\lambda_{dqm}^{s} = \int_0^t \mathrm{e}_{dqm}^{s} d\tau \tag{5.27}$$

With these sensing coils, the shortcomings from the Hall effect sensor can be circumvented. But, to install the coils, the induction machine should be dissected. Furthermore, the magnitude of the sensed voltage, e_{dqm}^{s}, by the coils is getting smaller as the frequency, ω_e, of the flux decreases, as in (5.28), and signal-to-noise (S/N) ratio is getting worse. Hence, at low speed, because of poor S/N ratio and the offset in the sensed voltage, the calculation of air-gap flux linkage by (5.27) is impossible.

$$|\mathbf{e}_{dqm}^{s}| \propto \omega_e \tag{5.28}$$

Thus, the practical lower limit of the speed for the vector control of the induction machine by sensing the air gap voltage would be 1/20 of the rated speed of the machine in the case of a few kilowatts or above power range. Even after getting the air gap flux by (5.27), the problem of the dependency on the rotor leakage inductance mentioned in method 1 in this section would still remain.

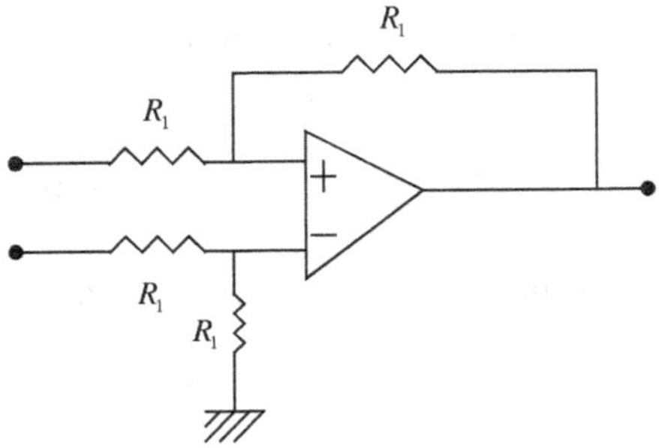

Figure 5.7 Differential amplifier.

3. **Estimation of the Rotor Flux Linkage with Terminal Voltages and Line Current.** By the differential amplifier shown in Fig. 5.7, the terminal voltages to the induction machine can be measured instantaneously. After measuring the voltages, the stator flux linkage can be calculated by (5.29) based on the measured voltage, measured phase current, and the stator resistance. Finally, the rotor flux linkage can be calculated by (5.30).

$$p\lambda^s_{dqs} = \mathrm{V}^s_{dqs} - R_s \mathrm{i}^s_{dqs} \tag{5.29}$$

$$\lambda^s_{dqr} = \frac{L_r}{L_m}(\lambda^s_{dqs} - \sigma L_s \mathrm{i}^s_{dqs}) \tag{5.30}$$

where σ is the leakage factor defined as $\sigma \equiv \frac{L_s L_r - L_m{}^2}{L_s L_r}$.

In addition to all the problems of the previous methods, this method has other problems such as (a) the limited bandwidth and the voltage insulation level of the differential amplifier and (b) the offset of the measured current signal. Thus, the method is simple but the lower limit of the speed range would be around 1/10 and the method can be applied to a less than 500-V class induction machine drive system. The problems due to the differential amplifier for the measurement of the terminal voltage can be circumvented by the estimation of the terminal voltage based on the gating signals of the power semiconductor switches and DC link voltage of the power converter, which is usually the inverter shown in Fig. 5.8. The phase voltage to the induction machine can be described as (5.31) under the assumption of Y connection of the stator winding. In (5.31), S_a, S_b, and S_c are the switching functions of the

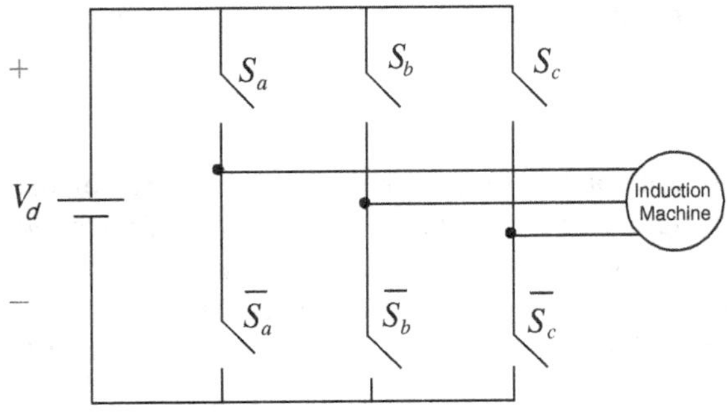

Figure 5.8 Operation of three-phase voltage source inverter with switching function, S_a, S_b, and S_c.

semiconductor switch of each phase, respectively. If the value of the switching function is 1, this represents the conduction of the upper switch of the corresponding phase of the inverter in Fig. 5.8. Otherwise, the lower switch of the corresponding phase is conducting. In the figure, $\bar{S}_a$, $\bar{S}_b$, and $\bar{S}_c$ represent the inverse logic states of S_a, S_b, and S_c. Hence, whenever upper switches turn on, lower switches turn off and vice versa. After estimating the phase voltage, the rotor flux linkage can be calculated by (5.29) and (5.30).

$$\begin{aligned} V_{as} &= \frac{V_d}{3}(2S_a - S_b - S_c) \\ V_{bs} &= \frac{V_d}{3}(2S_b - S_c - S_a) \\ V_{cs} &= \frac{V_d}{3}(2S_c - S_a - S_b) \end{aligned} \tag{5.31}$$

where V_d stands for DC link voltage of the inverter.

Though this method is simple, because of the dead time and nonlinearity of the inverter switches, the estimated phase voltages by (5.31) have some errors especially at low speed. Hence, the performance of the lower speed operation, which is less than 1/10 of the rated speed, is poor with this method. However, because of its simplicity of the implementation, this method is widely used and the operating range can be extended down to a few percentages of the rated speed with the careful dead-time compensation and the real-time machine parameter estimation techniques.

5.2.2 Indirect Vector Control

The indirect vector control method exploits the inherent characteristics of the induction machine as follows: By adjusting the slip angular frequency and the magnitude of the stator current the rotor flux and the torque component current can be controlled separately. In this control method, there is no need to identify the position of the rotor flux linkage. However, to control the slip angular frequency, the instantaneous rotor speed should be measured. In particular, in the case of the larger-size high-efficiency general-purpose induction machines over several hundred kilowatts, the rated slip is less than 1%. So, for 5% accuracy of torque regulation, the rotor speed should be measured with accuracy better than 0.05%. Hence, the accurate speed measurement with an optical encoder is a prerequisite to apply this method. Most of the commercialized induction machine drive systems regulating the torque instantaneously have been implemented with this method based on the rotor speed measured by an optical incremental encoder introduced in Section 4.3.2.

5.2.2.1 Principles of Indirect Vector Control

The voltage and flux linkage equations at the synchronously rotating *d–q* reference frame can be obtained by substituting $\omega = \omega_e$ in (5.2)–(5.9). In these equations, under the assumption that the rotor flux linkage exists only on d axis like (5.32), then $\boldsymbol{\lambda}_r^e = \lambda_{dr}^e + j\lambda_{qr}^e = \lambda_{dr}^e$.

$$\lambda_{qr}^e = 0 \tag{5.32}$$

By substituting (5.32) into (5.9), we can derive (5.33)

$$i_{qs}^e = -\frac{L_r}{L_m} i_{qr}^e \tag{5.33}$$

Also, by substituting (5.32) into (5.10), we can derive (5.34), which is the torque equation:

$$T_e = \frac{3}{2}\frac{P}{2}\frac{L_m}{L_r}\lambda_{dr}^e i_{qs}^e \tag{5.34}$$

Again, by substituting (5.32) into (5.5), the slip angular frequency can be represented in terms of the rotor flux linkage and q-axis current as

$$\omega_e - \omega_r = \omega_{s\ell} = -\frac{R_r i_{qr}^e}{\lambda_{dr}^e} \tag{5.35}$$

From the rotor flux linkage equation in (5.19) and the relationship between the q-axis rotor and the stator current, the slip angular frequency in (5.35) can be derived as (5.36) in terms of only the d–q axis stator current.

$$\omega_e - \omega_r = \omega_{s\ell} = -\frac{R_r i_{qr}^e}{\lambda_{dr}^e} = \frac{R_r}{\lambda_{dr}^e}\frac{L_m}{L_r} i_{qs}^e = \left[\left(1 + p\frac{L_r}{R_r}\right)\frac{1}{i_{ds}^e}\right]\frac{R_r}{L_r} i_{qs}^e \tag{5.36}$$

As seen from the (5.34), the instantaneous torque of the induction machine is directly proportional to the q-axis current, i_{qs}^e, under the condition of the constant rotor flux linkage. Hence, the q-axis current is called the torque component current. And the rotor flux linkage, as seen in (5.19), is regulated solely by the d-axis current, i_{ds}^e. Hence, the d-axis current is called the flux component current. After calculating slip frequency by (5.36) with the required torque and flux component current, the instantaneous angle of the rotor flux can be set by (5.37) with the angular speed of the rotor in electrical angle, ω_r.

$$\theta_e = \int_0^t \omega_e \, d\tau = \int_0^t (\omega_r + \omega_{s\ell}) \, d\tau \tag{5.37}$$

And, the position of the rotor flux is indirectly controlled by adjusting the slip angular frequency through (5.37). Using these processes, the stator current is decomposed to the torque component and the flux component current by the slip angular frequency.

5.2.2.2 Implementation of Indirect Vector Control

A control block diagram of indirect vector control of an induction machine is shown in Fig. 5.9, where a current-regulated PWM inverter is used to control the

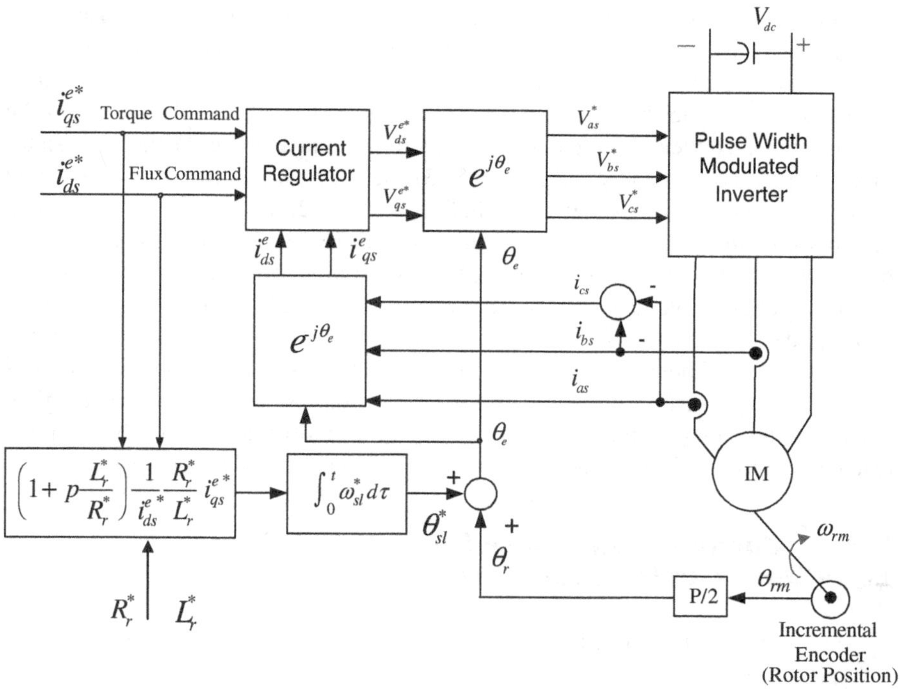

Figure 5.9 Control block diagram of indirect vector control of induction machine based on a current-regulated PWM inverter.

d–q components current of the machine. As seen from Fig. 5.9 and also from (5.36), in the implementation of the indirect vector control, the rotor time constant, $\tau_r = L_r/R_r$, is inevitably involved. The rotor self-inductance (which is the sum of the mutual inductance and the rotor leakage inductance) and the rotor resistance vary according to the operating conditions of the induction machine as mentioned in Section 2.12. Hence, the slip angular frequency calculated by (5.36) may have considerable error due to the machine parameter variations. If there is a real parameter adaptation method, the performance of the indirect vector control could be improved regardless of the variations of the parameters [3, 4]. Some methods to identify the parameters of the induction machine for the vector control at a certain operating condition are also described in Appendix A.

5.3 ROTOR FLUX LINKAGE ESTIMATOR

5.3.1 Voltage Model Based on Stator Voltage Equation of an Induction Machine

The rotor flux linkage can be calculated with the stator voltage and the current of the machine as described in Section 5.2.1.2. Also, the linkage can be calculated from the rotor circuit voltage equations in (5.4) and (5.5). The method to calculate rotor flux linkage using the stator voltage and current is called the voltage model method, whereas the method to calculate rotor flux linkage using rotor voltage equation, where the voltage is zero in the case of the squirrel cage rotor and only the current is the variable, is called the current model method. In the voltage model method, the stator flux linkage is calculated by (5.12) and (5.13) and the rotor flux linkage is calculated from (5.14) and (5.15). As mentioned in Section 5.2.1.2, the voltage model method has shortcomings in the lower operating frequency region, where the magnitude of back EMF is comparable to the voltage drop due to the stator impedance, $R_s + j\omega_e L_{\ell s}$. In this region, the calculated stator flux linkage can be easily contaminated with the measurement noise and offsets in the measured current and voltage. To circumvent the saturation of the integrator due to the measurement offset, the first-order low-pass filter can be used instead of the integrator. But, in this case, in the low-speed region, where the frequency of the stator current is near the cutoff frequency of the filter, the calculated rotor flux linkage may have too much error to be used for the vector control.

5.3.2 Current Model Based on Rotor Voltage Equation of an Induction Machine

At zero speed or low-speed operation of the induction machine, the rotor flux linkage can be calculated based on the rotor speed and the stator current in the rotor reference frame. The rotor voltage equation in the rotor reference d–q frame can be described as

$$\frac{d\lambda^r_{dr}}{dt} = -\frac{R_r}{L_r}\lambda^r_{dr} + R_r\frac{L_m}{L_r}i^r_{ds} \tag{5.38}$$

$$\frac{d\lambda^r_{qr}}{dt} = -\frac{R_r}{L_r}\lambda^r_{qr} + R_r\frac{L_m}{L_r}i^r_{qs} \tag{5.39}$$

where i^r_{ds} and i^r_{qs} is the stator current at the rotor reference frame. Using $\mathbf{i}^r_{dqs} = \mathbf{i}^s_{dqs}e^{-j\theta_r}$, i^r_{ds} and i^r_{qs} can be represented as

$$i^r_{ds} = i^s_{ds}\cos\theta_r + i^s_{qs}\sin\theta_r \tag{5.40}$$

$$i^r_{qs} = -i^s_{ds}\sin\theta_r + i^s_{qs}\cos\theta_r \tag{5.41}$$

By solving the first-order differential equations in (5.38) and (5.39), the rotor flux linkage at the rotor reference frame can be calculated. And then, the rotor flux linkage at the stationary reference d–q frame can be deduced by

$$\lambda^s_{dr} = \lambda^r_{dr}\sin\theta_r - \lambda^r_{qr}\sin\theta_r \tag{5.42}$$

$$\lambda^s_{qr} = \lambda^r_{dr}\sin\theta_r + \lambda^r_{qr}\cos\theta_r \tag{5.43}$$

Because this method relies on the machine parameters such as rotor resistance and rotor inductance, the estimated rotor flux linkage may have considerable errors if there are some errors in these parameters. Moreover, to implement this method, the instantaneous rotor angle should be measured accurately. Thus, the speed/position sensorless control is not compatible with this method.

5.3.3 Hybrid Rotor Flux Linkage Estimator [5]

In general, in the high-speed region where the magnitude of back EMF is large enough, the voltage model method is preferable because of its robustness to the parameter error. But in the low-speed region, the current model method is preferable because of its robustness to the measurement error. Hence a hybrid system exploiting the merits of both methods according to the operating speed had been developed. In Fig. 5.10, a block diagram of such a hybrid estimator is shown, where in the low-speed region the current model method works and in the high-speed region the voltage model works.

The estimator in Fig. 5.10 can be understood as shown in Fig. 5.11, where $\lambda^s_{dqr\_cm}$ stands for the rotor flux linkage by the current model and $\lambda^s_{dqr\_vm}$ stands for that by the voltage model. The gains of proportional and integral (PI) regulator have the following relationship with the gains in Fig. 5.10:

$$K_p = K_1 \frac{L_r}{L_m} \tag{5.44}$$

$$K_i = K_2 \frac{L_r}{L_m} \tag{5.45}$$

From Fig. 5.11, the transfer function between the estimated rotor flux linkage and the flux linkages by two methods can be deduced as

$$\lambda^s_{dqr} = \frac{s^2}{s^2 + K_p s + K_i}\lambda^s_{dqr\_vm} + \frac{K_p s + K_i}{s^2 + K_p s + K_i}\lambda^s_{dqr\_cm} \tag{5.46}$$

Hence, the estimated flux linkage follows the linkage by the current model in a low-frequency region, where the magnitude of the second term of the right-hand side of (5.46) is much larger than that of the first term. In a high-frequency region, the linkage follows that by the voltage model. The transition from current model to voltage model is decided by the gains of a PI regulator. However, by only gains of the regulator, it is difficult to understand where the transition occurs. In (5.46), the first

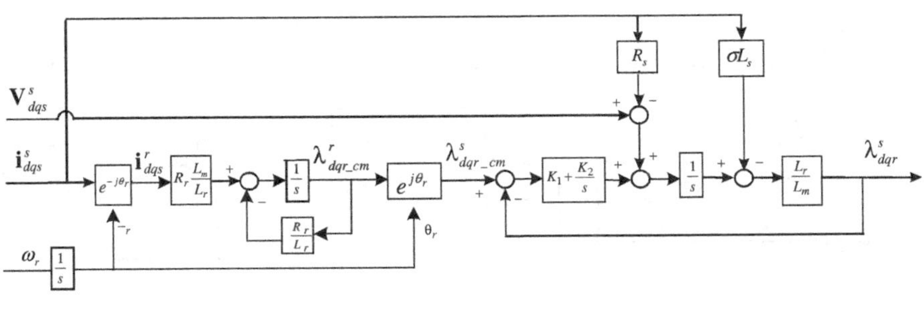

Figure 5.10 Block diagram of a hybrid rotor flux linkage estimator.

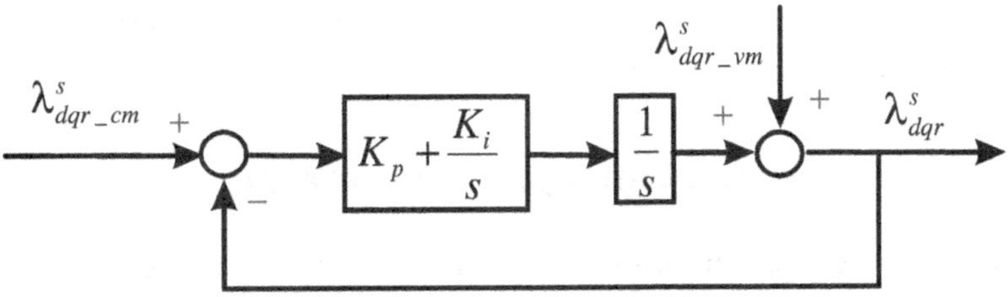

Figure 5.11 Simplified block diagram of a hybrid flux estimator.

term on the right-hand side is the transfer function of the second-order high-pass filter, and the second term on the right-hand side is that of the second-order low-pass filter including a band-pass filter too. And, the transfer function of the hybrid estimator can be understood as the sum of the low-pass filtered flux linkage by the current model and the high-pass filtered flux linkage by the voltage model. Hence, the current model is dominant in the low-frequency region and the voltage model is in high frequency. The transition point can be set by the cutoff frequency of the filter as (5.47) and (5.48) if Butterworth-type filters are assumed.

$$K_p = \sqrt{2}\omega_c \tag{5.47}$$

$$K_i = {\omega_c}^2 \tag{5.48}$$

where ω_c is the cutoff angular frequency of the second-order Butterworth filter.

5.3.4 Enhanced Hybrid Estimator [6]

Compared to the performance of the estimator by voltage model or the current model, the performance of the hybrid estimator in the previous section may be worse at the transition frequency, ω_c, if there are errors in the machine parameters such as R_r and L_m. In particular, the phase of the estimated rotor flux linkage has much error at the transition frequency. This problem can be lessened by compensating the phase of the estimated flux linkage as shown in Fig. 5.12.

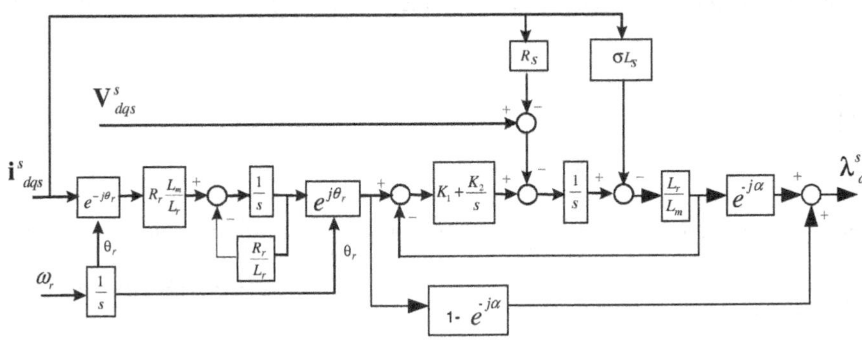

Figure 5.12 Block diagram of enhanced hybrid estimator.

In Fig. 5.12, the phase compensation can be done by an exponential function, $e^{-j\alpha}$, where α is set by

$$\alpha = \left[\pi - \tan^{-1}\left[\frac{K_p\omega_e}{K_i - \omega_e^2}\right]\right] \tag{5.49}$$

With this enhanced hybrid estimator, the estimation of the rotor flux linkage could be reasonably accurate in both phase and magnitude regardless of the errors in the machine parameters and errors in the measurement.

5.4 FLUX WEAKENING CONTROL

The torque and speed of the variable-speed drive system controlled by a PWM inverter is limited by current and voltage rating of inverter and the electric machine. In this section the optimal flux weakening method, which lets the electric machine generate maximum torque under the given current and voltage constraints, is described for an AC machine driven by a three-phase PWM inverter.

5.4.1 Constraints of Voltage and Current to AC Machine [7]

The inverter, which provides variable-voltage and variable-frequency electric power to an AC machine, has limited voltage and current ratings because of the components of the inverter itself and input voltage to the inverter. Also, even if the inverter has large enough voltage and current ratings, the AC machine itself has constraints in current and voltage rating because of insulation, magnetic saturation, and thermal limit. Because the thermal time constant of an AC machine is usually much larger than that of the inverter, several hundred percentage of rated current can flow into the AC machine for a short time. Thus, the torque at a constant torque region, where the torque of the AC machine is only limited by the current constraint, can be increased by several times of the rated torque. Usually, the voltage rating of the inverter is set to equal to the rated voltage of the AC machine. However, the current rating of the inverter is sometimes set as several times that of the AC machine to get higher acceleration and deceleration torque, especially in servo application.

5.4.1.1 Voltage Constraints

As mentioned in Section 2.18.5, the maximum phase voltage, $V_{s\max}$, is decided by DC link voltage, V_d, of a PWM inverter and the PWM method. If the space vector PWM (SVPWM) method is used, $V_{s\max}$ obtained in the linear control range is $V_d/\sqrt{3}$. With the consideration of some margins because of the dead time of the inverter and the control voltage for the current regulation, the maximum phase voltage, $V_{s\max}$, can be set as (5.50), where η can be 0.9–0.95.

$$V_{s\max} = \frac{V_d}{\sqrt{3}}\eta \tag{5.50}$$

If the inverter is controlled in a six-step mode without any PWM, the voltage, $V_{s\max}$, can be increased up to $2V_d/\pi$, and the AC machine can generate the maximum torque at a flux weakening region under the given voltage constraint. However, in this six-step mode operation, current ripple and torque ripple, acoustic noise, and ripple current to a DC link capacitor of inverter are much larger than those in PWM mode operation. Furthermore, because the instantaneous control of the phase and the magnitude of the phase voltage is impossible in the six-step mode operation, the instantaneous torque control cannot be achieved. Also, the dynamics of the AC machine drive system would be degraded conspicuously. If the maximum phase voltage, $V_{s\max}$, is decided by the inverter, then the d–q axis stator voltage should satisfy (5.51) regardless of the reference frame.

$$V_{ds}^{s*2} + V_{qs}^{s*2} = V_{ds}^{r*2} + V_{qs}^{r*2} = V_{ds}^{e*2} + V_{qs}^{e*2} \leq V_{s\max}^{2} \tag{5.51}$$

5.4.1.2 Current Constraint

The maximum current to an AC machine, $I_{s\max}$, is usually decided by the thermal limit of the inverter or the AC machine itself. If the constraint is decided by the inverter, then the limiting condition of the current is set by the heat dissipation of switching and conduction losses of the switching power semiconductors. If the constraint is decided by the AC machine itself, then the limiting condition of the current is set by the heat dissipation from the iron and copper losses of the AC machine. The thermal time constant of the AC machine, which may be from several seconds in the case of less than kilowatts to minutes in the case of a several-hundred-kilowatt machine, is much larger than that of the inverter, which may be several tens of milliseconds. After $I_{s\max}$ is decided, the reference current should satisfy (5.52) regardless of the reference frame.

$$i_{ds}^{s*2} + i_{qs}^{s*2} = i_{ds}^{r*2} + i_{qs}^{r*2} = i_{ds}^{e*2} + i_{qs}^{e*2} \leq I_{s\max}^{2} \tag{5.52}$$

5.4.2 Operating Region of Permanent Magnet AC Machine in Current Plane at Rotor Reference Frame [8]

5.4.2.1 Operating Region Under Current and Voltage Constraints

The voltage and current constraint in (5.51) and (5.52) was expressed in the d–q reference frame, but the constraints are presented in different planes, where one is voltage plane and the other is current plane. Hence, it is difficult to consider both constraints simultaneously. The constraints can be simultaneously depicted in the voltage plane or in the current plane by using the stator voltage equations of an AC machine. If the constraints are depicted in the voltage plane, though the voltage margin and the phase of commanded voltage can be easily understood, the torque, which is usually expressed in terms of current, cannot be easily demonstrated in the voltage plane. Moreover, because most high-performance AC drive systems have an inner current regulation loop, the current plane is a natural choice in the viewpoint of the control loop design [9]. Thus, the current plane is usually used to consider the constraints simultaneously. In this case the voltage constraint can be represented in terms of the current by using the stator voltage equation of the permanent magnet AC machine in (5.53) and (5.54).

$$V_{ds}^r = R_s i_{ds}^r + L_d \frac{d}{dt} i_{ds}^r - \omega_r L_q i_{qs}^r \tag{5.53}$$

$$V_{qs}^r = R_s i_{qs}^r + L_q \frac{d}{dt} i_{qs}^r + \omega_r L_d i_{ds}^r + \omega_r \lambda_f \tag{5.54}$$

From the above equations, like (5.55), the voltage constraint can be expressed in terms of the currents under the assumption of the steady-state operation or slow enough variation of the currents.

$$Z_{ds}^2 \left(i_{ds}^r + \frac{\omega_r^2 L_d \lambda_f}{Z_{ds}^2} \right)^2 + Z_{qs}^2 \left(i_{qs}^r + \frac{\omega_r R_s \lambda_f}{Z_{qs}^2} \right)^2 + 2\omega_r R_s (L_d - L_q)\, i_{ds}^r i_{qs}^r \le V_{s\max}^2 \tag{5.55}$$

where Z_{ds} and Z_{qs} are defined as

$$Z_{ds} \equiv \sqrt{R_s^2 + (\omega_r L_d)^2},\ Z_{qs} \equiv \sqrt{R_s^2 + (\omega_r L_q)^2} \tag{5.56}$$

Inequality by (5.55) can be rewritten as (5.57), and the area satisfying (5.57) is the interior of the dotted ellipse as shown in Fig. 5.13a. The center of the ellipse is

$$\left(-\frac{\omega_r^2 L_{qs} \lambda_f}{R_s^2 + \omega_r^2 L_d L_q},\ -\frac{\omega_r R_s \lambda_f}{R_s^2 + \omega_r^2 L_d L_q} \right)$$

in the current plane. And the major axis of the ellipse is apart from the d axis by $\frac{1}{2} \tan^{-1} \frac{2R_s}{\omega_r (L_d + L_q)}$.

$$\begin{aligned} &Z_{ds}^2 \left(i_{ds}^r + \frac{\omega_r^2 L_{qs} \lambda_f}{R_s^2 + \omega_r^2 L_d L_q} \right)^2 + Z_{qs}^2 \left(i_{qs}^r + \frac{\omega_r R_s \lambda_f}{R_s^2 + \omega_r^2 L_d L_q} \right)^2 \\ &+ 2\omega_r R_s (L_d - L_q) \left(i_{ds}^r + \frac{\omega_r^2 L_q \lambda_f}{R_s^2 + \omega_r^2 L_d L_q} \right) \left(i_{qs}^r + \frac{\omega_r R_s \lambda_f}{R_s^2 + \omega_r^2 L_d L_q} \right) \le V_{s\max}^2 \end{aligned} \tag{5.57}$$

In the case of a surface-mounted permanent magnet synchronous machine (SMPMSM), because $L_d = L_q = L_s$, (5.57) can be simplified as (5.58). The area satisfying (5.58) is the interior of the dotted circle in Fig. 5.13b, and its center is $\left(-\frac{\omega_r^2 L_s \lambda_f}{R_s^2 + \omega_r^2 L_s^2}, -\frac{\omega_r R_s \lambda_f}{R_s^2 + \omega_r^2 L_s^2} \right)$ in the current plane.

$$\left(i_{ds}^r + \frac{\omega_r^2 L_s \lambda_f}{R_s^2 + \omega_r^2 L_s^2} \right)^2 + \left(i_{qs}^r + \frac{\omega_r R_s \lambda_f}{R_s^2 + \omega_r^2 L_s^2} \right)^2 \le \frac{V_{s\max}^2}{R_s^2 + \omega_r^2 L_s^2} \tag{5.58}$$

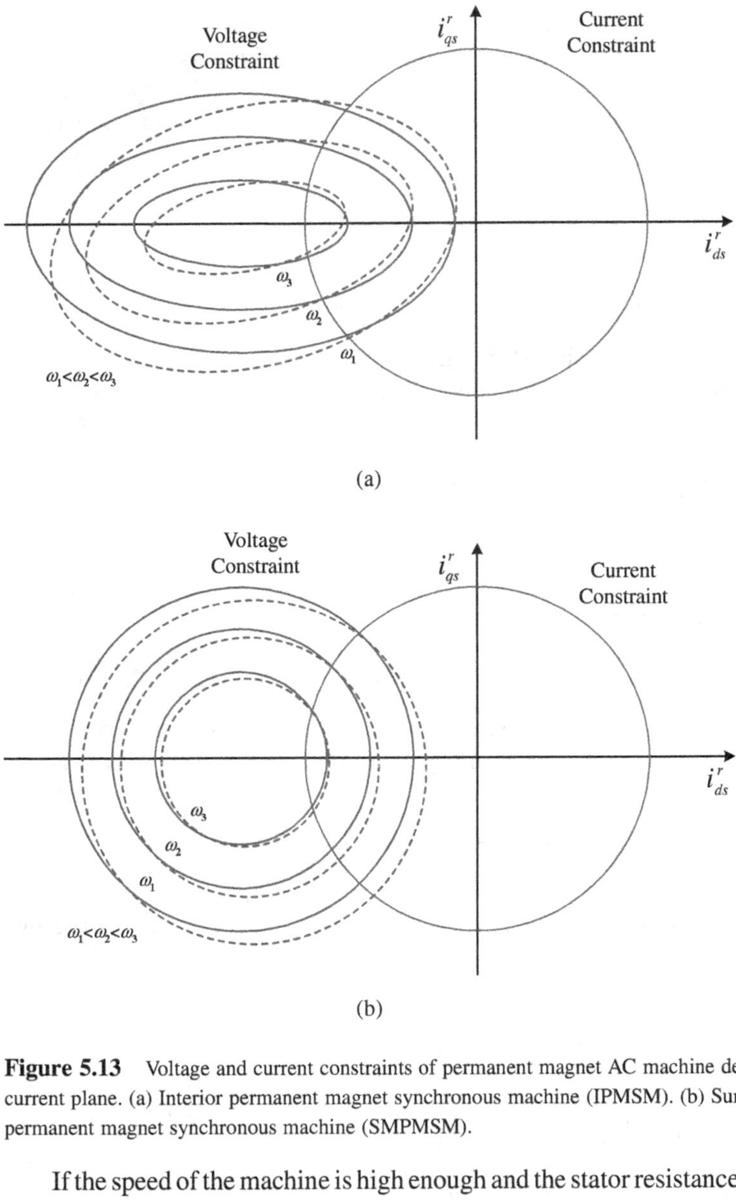

Figure 5.13 Voltage and current constraints of permanent magnet AC machine depicted in a current plane. (a) Interior permanent magnet synchronous machine (IPMSM). (b) Surface-mounted permanent magnet synchronous machine (SMPMSM).

If the speed of the machine is high enough and the stator resistance voltage drop is small enough, then the voltage drop by the stator resistance can be neglected. Hence, (5.57) and (5.58) can be approximated as (5.59) and (5.60):

$$(\omega_r L_d)^2 \left(i_{ds}^r + \frac{\lambda_f}{L_d} \right)^2 + (\omega_r L_q)^2 i_{qs}^{r2} \leq V_{s\max}^2 \tag{5.59}$$

$$\left(i_{ds}^r + \frac{\lambda_f}{L_s} \right)^2 + i_{qs}^{r2} \leq \frac{V_{s\max}^2}{\omega_r^2 L_s^2} \tag{5.60}$$

The area satisfying (5.59) is the interior of the solid ellipse shown in Fig. 5.13a and the area satisfying (5.60) is that of the solid circle in Fig. 5.13b.

As mentioned before, the current constraint is expressed as the inner part of a circle, whose center is the origin of the current plane. The voltage constraint is expressed as the inner part of an ellipse. If the voltage drop in the stator resistance is neglected, the major axis of the ellipse lies on the d axis of the current plane. And its center is constant regardless of the operating speed of the permanent magnet machine. However, the length of major and minor axes of the ellipse decrease as the operation speed increases. If the voltage drop in the stator resistance is considered, the major axis has an offset angle with the d axis, and the center of the ellipse also varies according to the speed. Under the given constraints, the possible operating area in the current plane is the common inner part of the interior of both ellipse and circle. As the operating speed increases, the area by the voltage constraint shrinks and the common area also shrinks. Above a certain speed, there is no common inner area, and the operation of the electric machine at that speed is impossible, satisfying both current and voltage constraints.

5.4.2.2 Operating Region According to the Parameters of the Permanent Magnet AC Machine

The output characteristics of a permanent magnet synchronous machine are decided by the relative location of the center of the ellipse by the voltage constraint to the circle by the current constraint. The center is set by the parameters of the electric machine [10–12]. As shown in Fig. 5.14, the permanent magnet AC machine drive system can be classified as a finite-speed drive system and an infinite speed drive system whether the center of the ellipse is inside of the circle by the current constraint or outside of the circle. In the case of the infinite-speed drive system where the center of ellipse lies inside of the current constraint circle, the maximum speed is limited not by the current and voltage constraints but only by the mechanical reasons such as centrifugal forces or the stress in the bearing. However, in the case of the finite-speed drive system, where the center of ellipse lies outside of the current constraint circle, the voltage constraint limits the maximum operating speed even if there is no mechanical reason to limit the speed. In Fig. 5.14, the voltage and current constraints of the permanent magnet AC machine with the trajectory of current for MTPA operation for both infinite- and finite-speed drive systems neglecting the stator resistance voltage drop are shown. In Fig. 5.14a, for the case of IPMSM, the trajectory of the current for MTPA operation is shown. In the figure, the center of the ellipse is outside of the circle satisfying (5.61), and as described previously, the maximum speed is limited by the ellipse from the voltage constraint. For MTPA operation, up to the speed ω_{1f}, the current moves on the curve by "OA" and above ω_{1f} the current moves on the boundary of the current constraint circle, which is the curve "AB". Above a certain speed, ω_{3f}, there is no more common area of the ellipse and the circle, and the machine cannot be operated above that speed. In Fig. 5.14b, the trajectory of the current for the infinite-speed drive system is shown. In the figure, the center of the ellipse is inside of the circle satisfying (5.62) and the maximum speed is not limited by electrical constraints. For MTPA operation, up to the speed ω_{1i}, the current moves on

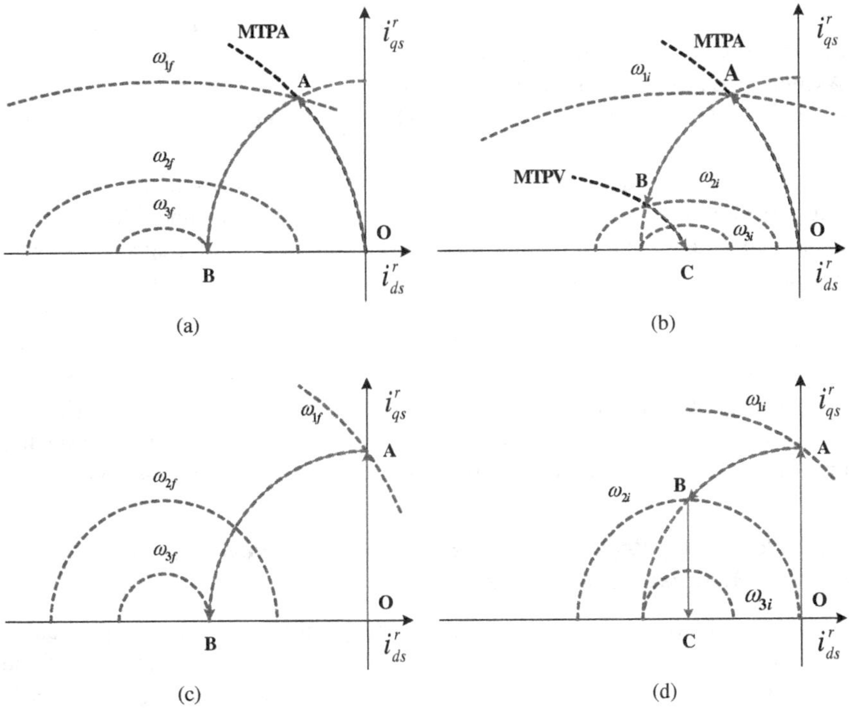

Figure 5.14 Operating regions according to the parameters of the permanent magnet AC machines. (a) IPMSM (Finite-speed drive system). (b) IPMSM (infinite-speed drive system). (c) SMPMSM (finite-speed drive system). (d) SMPMSM (infinite-speed drive system).

the curve by "OA." Also, above ω_{1i} the current moves on the boundary of the current constraint circle, which is the curve "AB" like the case of the finite-speed drive system. However, above ω_{2i} the current moves not on the boundary of the current constraint circle but on the curve for maximum torque per voltage (MTPV) operation, which is the curve "BC" unlike the case of the finite-speed drive system. By following the curve, up to infinite speed the machine can operate if there are no mechanical problems to limit the operating speed.

$$\lambda_f > L_d I_{s\max} \tag{5.61}$$

$$\lambda_f \leq L_d I_{s\max} \tag{5.62}$$

In Fig. 5.14c, for the case of an SMPMSM, the trajectory of the current is shown. In the figure, the center of the voltage constraint circle is outside of the current constraint circle satisfying (5.63), and the maximum speed is limited by the voltage constraint. For MTPA operation, up to the speed ω_{1f}, the current moves along the

q axis represented as "OA" in the figure. And above ω_{1f} the current moves on the boundary of the current constraint circle, which is the curve "AB". Above a certain speed, ω_{3f}, there is no more common area of both circles, and the machine cannot be operated above that speed. In Fig. 5.14d, the trajectory of the current for the infinite-speed drive system is shown. In the figure, the center of the voltage constraint circle is inside of the current constraint circle satisfying (5.64) and the maximum speed is not limited by electrical constraints. For MTPA operation, up to the speed ω_{1i}, the current moves the q axis, "OA". Also, above ω_{1i} the current moves on the boundary of the current constraint circle, which is the curve "AB", like the case of finite-speed drive system. However, above ω_{2i} the current moves not on the boundary of the current constraint circle, but on the line parallel to q axis for maximum torque per voltage (MTPV) operation, which is the line "BC", unlike the case of the finite-speed drive system. By following the line, up to infinite speed the machine can operate if there are no mechanical limitations.

$$\lambda_f > L_s I_{s\max} \tag{5.63}$$

$$\lambda_f \le L_s I_{s\max} \tag{5.64}$$

In the case of an SMPMSM, the flux linkage by the magnet is usually much larger than the flux linkage by the stator current. So, (5.63) holds in the most case of SMPMSM operation unless $I_{s\max}$ is larger than the several times of the rated current of SMPMSM.

The speed up to which the machine is operated in the MTPA mode is called the base speed, ω_b. In Fig. 5.14, it is represented as ω_{1f} and ω_{1i}. Up to the base speed, the speed region is called the constant torque region, and above the base speed the region is called the flux weakening region. In the infinite-speed drive system, the speed where only voltage constraint limits the operating speed is called the critical speed, ω_c. After neglecting the stator resistance voltage drop, the base speed, the maximum possible operating speed in the finite-speed drive system, and the critical speed in the infinite-speed drive system can be derived for both an SMPMSM and an IPMSM as (5.65)–(5.70).

At the base speed, in the case of SMPMSM, both circles by the constraints meet on the q axis. In the case of an IPMSM, the ellipse meets the circle along the MTPA line at ω_b. And it can be presented as (5.65) and (5.66), respectively.

$$\omega_{b,\mathrm{SMPMSM}} = \frac{V_{s\max}}{\sqrt{\lambda_f^2 + (L_s I_{s\max})^2}} \tag{5.65}$$

$$\omega_{b,\mathrm{IPMSM}} = \frac{V_{s\max}}{\sqrt{(L_d i^r_{ds1} + \lambda_f)^2 + \left(L_q i^r_{qs1}\right)^2}} \tag{5.66}$$

where

$$i^r_{ds1} = \frac{\lambda_f - \sqrt{\lambda_f^2 + 8(L_q - L_d)^2 I_{s\max}^2}}{4(L_q - L_d)} \quad \text{and} \quad i^r_{qs1} = \pm\sqrt{I_{s\max}^2 - i^{r2}_{ds1}}$$

In the finite-speed drive system, the maximum speed is the speed where the ellipse and the circle meet at one point in the current plane. Above the speed, there is no common point or area of the ellipse and the circle. The maximum speed can be derived as (5.67) and (5.68) for an SMPMSM and an IPMSM, respectively.

$$\omega_{\max,\text{SMPMSM}} = \frac{V_{s\max}}{\lambda_f - L_s I_{s\max}} \tag{5.67}$$

where $\lambda_f > L_s I_{s\max}$.

$$\omega_{\max,\ \text{IPMSM}} = \frac{V_{s\max}}{\lambda_f - L_d I_{s\max}} \tag{5.68}$$

where $\lambda_f > L_d I_{s\max}$.

In the infinite-speed drive system, the critical speed is the speed where only voltage constraint limits the operating speed. Under the condition of the maximum torque with both voltage and current constraints until at the speed, ω_c, and above that speed where only voltage constraint works, the critical speed can be derived as (5.69) and (5.70) for an SMPMSM and an IPMSM, respectively.

$$\omega_{c,\text{SMPMSM}} = \frac{V_{s\max}}{\sqrt{(L_s I_{s\max})^2 - \lambda_f^2}} \tag{5.69}$$

where $\lambda_f \leq L_s I_{s\max}$.

In the case of an IPMSM, the critical speed can not be easily derived analytically. The speed can be obtained by solving (5.70) under the condition of maximum torque per voltage (MTPV) operation such as $L_q^2\left(I_{s\max}^2 - i_{ds}^{r2}\right)^2 - L_d^2\left(i_{ds}^r + \frac{\lambda_f}{L_d - L_q}\right)\left(i_{ds}^r + \frac{\lambda_f}{L_d}\right) = 0$:

$$\frac{V_{s\max}^2}{\omega_{c,\text{IPMSM}}^2} = \left(L_d i_{ds}^r + \lambda_f\right)^2 + L_q^2\left(I_{s\max}^2 - i_{ds}^{r2}\right) \tag{5.70}$$

The solution can be found numerically by solving (5.70) and MTPV condition simultaneously.

From (5.67) and (5.68), it can be seen that if the center of ellipse is lies on the circle, then the system is the infinite-speed drive system with only one flux weakening region, while in general the infinite-speed drive system has two flux weakening regions. In the flux weakening region I, both voltage and current constraints apply to the drive system. In flux weakening region II, only voltage constraint applies. The base speed, the maximum speed, and the critical speed can be represented by the parameters of the machine as derived previously. But, because of the variation of parameters mentioned in Section 2.9, the specified speeds vary according to the operating conditions.

5.4.3 Flux Weakening Control of Permanent Magnet Synchronous Machine

If the current reference is outside of the possible operating area shown in Fig. 5.13 or Fig. 5.14, then actual current cannot follow the current reference. Hence, the drive system is out of control. By a proper flux weakening method, the current reference can be set to achieve maximum available torque under the voltage and current constraints. In this section, several flux weakening methods are introduced with its merits and demerits.

5.4.3.1 Flux Weakening Control with Feed-Forward Compensation [13,14]

To weaken the flux, the feed-forward compensation technique can be implemented based on the steady-state voltage equation of the machine. The torque command is limited within the available maximum value at that operating speed. With the limited torque command, the optimal current reference in the rotor reference *d–q* frame is calculated considering voltage and current constraints from the steady-state voltage equations in (5.53) and (5.54). As mentioned before, the voltage constraint is decided by DC link voltage and the PWM method. In Fig. 5.15, a control block diagram of this flux weakening method is shown. In this method, because the voltage is decided under the steady-state operating condition, the voltage margin for the regulation of the current should be considered. The method can be easily implemented with the nominal machine parameters, and it is simple because of no gains to set for the flux weakening control. But the performance would be degraded with the variation of the parameters because the compensation is done in the open loop manner. Also, if the speed and torque varys rapidly and if the current varies suddenly, then the performance would be poor because of the deficiency of the voltage margin for the current regulation.

5.4.3.2 Flux Weakening Control with Feedback Compensation [15]

As shown in Fig. 5.16, this flux weakening method is keeping the output voltage of the current regulator within the voltage constraint by feedback of the output voltage to the

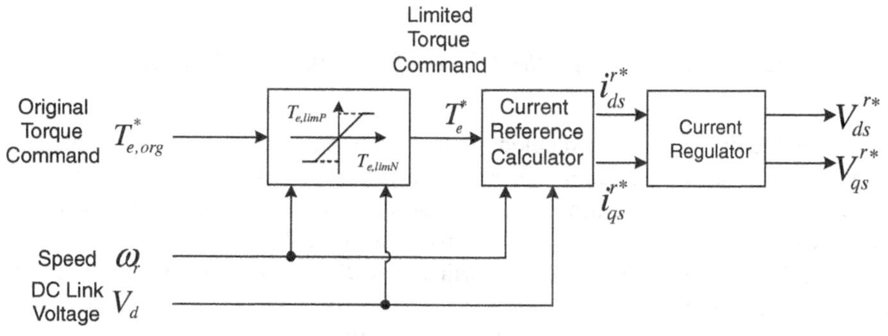

Figure 5.15 Flux weakening method using feed-forward compensation.

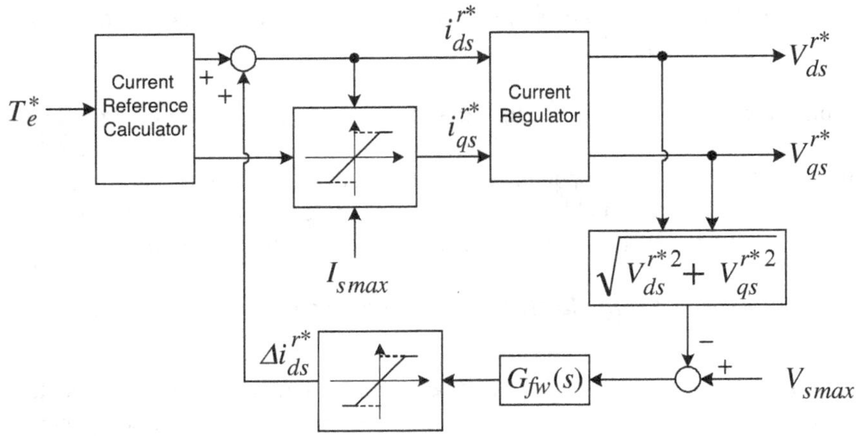

Figure 5.16 Flux weakening method by feedback of the output voltage.

flux weakening regulator. The input to the flux weakening regulator is the difference between the output of the current regulator and the available maximum voltage, $V_{s,\max}$. The output of the flux weakening regulator is d-axis current to adjust the flux level of the electric machine. Because of the feedback of the output voltage of the current regulator, this method is robust to the variation of the parameters according to the operating conditions. However, because the output of the current regulator includes not only the term to cancel the current error but also the term to decouple the back EMF and the cross-coupling voltage in rotor reference frame as mentioned in Section 4.2.5, the parameter variations affect the performance of the current regulator, which is a part of the flux weakening control. Also, the performance of the flux weakening regulator is affected by the parameter variations through the current regulator. Moreover, the gain setting of the flux weakening controller, $G_{fw}(s)$, in Fig. 5.16 is not straightforward and the gains should be set by trial and error. The bandwidth of the flux weakening control loop should be low enough to prevent the interference with the current control loop. Hence, if the speed varies rapidly, the performance of flux weakening control is degraded severely, and the current of the machine, especially the d-axis current, may reveal oscillatory responses.

5.4.3.3 Flux Weakening Method Including Nonlinear Modulation Region

The above two flux weakening methods are formulated under the assumption that the output of the current regulator can be synthesized exactly by a PWM inverter. Therefore, the voltages are limited in the inscribed circle of a hexagon of voltage plane, which is $V_d/\sqrt{3}$ in the case of space vector PWM as described at Section 5.4.1.1. Actually, the voltage is furthermore limited to consider the control margin of the current and the dead time effect of the inverter as (5.50). In some application field such as an electric/hybrid vehicle, the torque of the electric machine should be maximized to get maximum acceleration under varying DC link voltage from the battery. In this case, overmodulation is inevitable and PWM is extended to nonlinear modulation region, where the output voltage includes low-order harmonics. The fundamental component of output voltage in the nonlinear modulation region can be maximized at the cost of low-order harmonic currents, and the available maximum torque increases [16]. But in this region the flux weakening control method is complicated due to the nonlinearity of PWM.

The stator flux linkage of the permanent magnet AC machine is given by the sum of the flux of the magnet and the current in the stator windings. Because there is no rotor circuit, the stator flux linkage is instantaneously decided by the current in the stator windings. Exploiting that, the flux weakening control can be incorporated with the anti-wind-up controller of the current regulator [17]. Unlike the previous flux weakening method based on the feedback compensation, which modifies the current reference according to the voltage error, the input of the anti-wind-up controller can be used directly to modify the current reference. And in this method, not only linear modulation region but also nonlinear modulation region, approximately up to the six-step operation region, can be used to regulate the current. Hence the available maximum torque or the maximum operating speed could be extended by at least 10% in the flux weakening region, where the voltage constraints limit the torque. Furthermore, in this method, there is no need to set voltage margin for the current regulation and for the nonlinear effects of the inverter such as dead time effects and voltage drop of switching devices, and the available voltage of the inverter can be maximally exploited by the flux weakening control loop itself.

In Fig. 5.17, a control block diagram of the flux weakening method incorporating an anti-wind-up controller is shown for SMPMSM [17]. In the figure, the d-axis current is modified according to the voltage error in the q-axis, which was originally the input of the anti-wind-up controller in Figs. 4.18 and 4.19. The q-axis current is also modified to keep the total current within the maximum current, $I_{s\max}$ with the consideration of the modified d-axis current reference. The cutoff frequency, ω_c, of

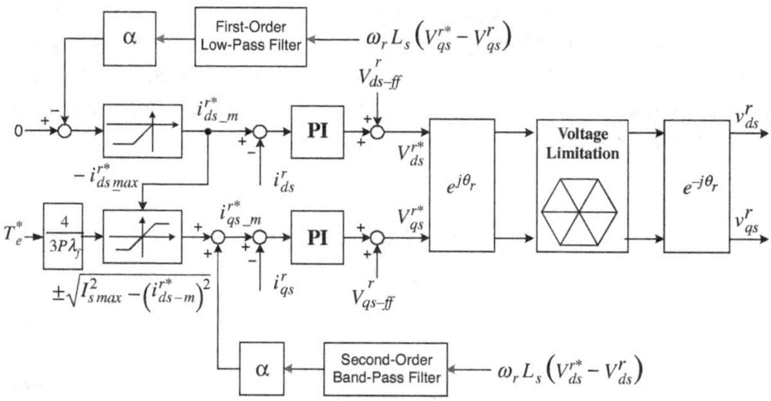

Figure 5.17 Control block diagram of a flux weakening method incorporating an anti-wind-up controller for a surface-mounted permanent magnet synchronous machine (SMPMSM).

the first-order low-pass filter in the block diagram can be set as (5.71) if PI gains of the current regulators are given by (4.23).

$$\omega_c = \frac{K_i}{K_p} \tag{5.71}$$

The α in the figure is a gain deciding the level of flux weakening to resolve the deficiency of the voltage in the q axis. If α is getting large, the flux is weakened more intensively and current is better regulated but the voltage utilization factor defined by (5.72) would be less. Hence, the available maximum torque or the maximum operating speed would decrease.

$$\text{Voltage utilization factor} \equiv \frac{\sqrt{V_{ds}^{r2} + V_{qs}^{r2}}}{\frac{2}{\pi} V_d} \tag{5.72}$$

The second-order band-pass filter can be considered as the cascade connection of the first-order low-pass filter and the first-order high-pass filter as (5.73), and the cutoff frequency of the filters may be set differently, but for easy tuning the cutoff frequency of both filters can be set as (5.71). When the machine is operated in the steady state at the flux weakening region, the output of the second-order band-pass filter is always zero and it does not affect to the operation of the machine. However, in the transient state, especially the machine is running as a generator, the band-pass filter improves the control performance remarkably.

$$\frac{\omega_{cL}}{s + \omega_{cL}} \frac{s}{s + \omega_{cH}} \tag{5.73}$$

The flux weakening method for SMPMSM in Fig. 5.17 can be modified as Fig. 5.18 to accommodate the flux weakening control of IPMSM, where q-axis current should also be adjusted according to d-axis voltage error [18].

Like the SMPMSM case, the voltage errors, ΔV_{ds}^r, ΔV_{qs}^r, in the d–q axis, which were originally the input of anti-wind-up controller, fed to the first-order low-pass filters, LPFd and LPFq. But, unlike the SMPMSM case, the cutoff frequency of the filters should be set separately as (5.74) and (5.75) even if the bandwidths of the current regulator of d–q axis are identical to ω_{bw}.

$$\omega_{cd} = K_i / K_{pd} \tag{5.74}$$

$$\omega_{cq} = K_i / K_{pq} \tag{5.75}$$

where K_{pd} and K_{pq} are given by (5.76) and (5.77) as mentioned in Section 4.2.5.

$$K_{pd} = L_d \omega_{bw} \tag{5.76}$$

$$K_{pq} = L_q \omega_{bw} \tag{5.77}$$

The reference of the stator flux linkage is modified according to the voltage errors. The d–q current references come from the three-dimensional look-up table,

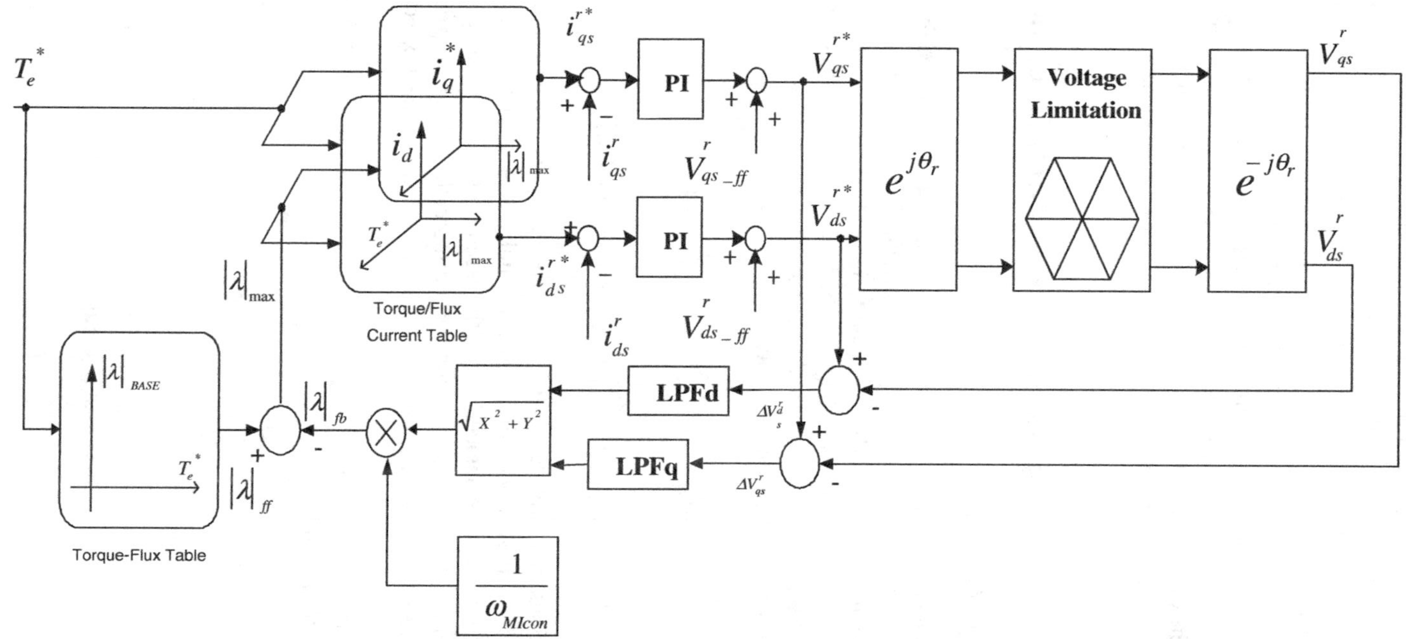

Figure 5.18 Control block diagram of a flux weakening method incorporating an anti-wind-up controller for an interior permanent magnet synchronous machine (IPMSM).

which is prepared through the off-line test of an IPMSM. In the table, the optimal set of d–q current according to the given flux linkage reference and torque command is prepared for MTPA operation or for flux weakening control. ω_{MIcon} in Fig. 5.18 is the gain deciding the level of flux weakening to resolve the deficiency of the voltage, like α in Fig. 5.17. But, unlike α, as ω_{MIcon} increases, the voltage utilization factor increases and the current regulation is getting worse. Usually, ω_{MIcon} can be set as a certain speed between base speed and the maximum speed or it can simply be set as a variable gain such as the motor speed, ω_r. If ω_r is used as ω_{MIcon}, there is no need to tune ω_{MIcon}, but the voltage utilization factor would vary according to the speed. In this case, at higher operating speed, the system would be operated like in the quasi six-step operation mode.

5.4.4 Flux Weakening Control of Induction Machine [7]

5.4.4.1 Voltage Equations

At the given voltage and current constraints, an inequality satisfied by torque component current, i_{qs}^{e*}, and the flux component current, i_{ds}^{e*}, for an induction machine can be derived as follows, similarly to the case of a PMSM. In the synchronously rotating reference frame, the stator voltage can be represented in terms of the stator current and rotor flux linkage as (5.78) and (5.79) under the assumption of the vector control based on the rotor flux linkage orientation.

$$V_{ds}^{e*} = R_s i_{ds}^{e*} + \sigma L_s \frac{di_{ds}^{e*}}{dt} + \frac{L_m}{L_r}\frac{d\lambda_{dr}^{e*}}{dt} - \omega_e \sigma L_s i_{qs}^{e*} \tag{5.78}$$

$$V_{qs}^{e*} = R_s i_{qs}^{e*} + \sigma L_s \frac{di_{qs}^{e*}}{dt} + \frac{L_m}{L_r}\omega_e \lambda_{dr}^{e*} + \omega_e \sigma L_s i_{ds}^{e*} \tag{5.79}$$

In the equations, the term due to the flux variation can be neglected under the assumption of relatively slow variation of the flux linkage. Also, the term due to the variation of the current can be neglected with the assumption of slow enough current variation or under the steady-state operation. Furthermore, the voltage drop due to the stator resistance can be neglected at higher operating speed, where the flux weakening occurs.

With above assumptions, the stator voltage equation in (5.78) and (5.79) can be approximated as (5.80) and (5.81).

$$V_{ds}^{e*} \approx -\omega_e \sigma L_s i_{qs}^{e*} \tag{5.80}$$

$$V_{qs}^{e*} \approx \frac{L_m}{L_r}\omega_e \lambda_{dr}^{e*} + \omega_e \sigma L_s i_{ds}^{e*} \tag{5.81}$$

where the rotor flux linkage reference can be set as (5.82) under the assumption of the slow enough variation of the flux linkage compared to the rotor time constant.

$$\lambda_{dr}^{e*} = L_m i_{ds}^{e*} \tag{5.82}$$

Then, (5.81) can be further approximated as

$$V_{qs}^{e*} \approx \omega_e L_s i_{ds}^{e*} \tag{5.83}$$

By substituting (5.80) and (5.83) to the voltage constraint in (5.51), the following inequality can be obtained, which is the interior of an ellipse in the synchronous reference current plane. Also, the possible operating point of *d–q* current of the induction machine should lie in the common part of (5.52) and (5.84), which is the cross section of the ellipse and the circle.

$$\left(\omega_e \sigma L_s i_{qs}^{e*}\right)^2 + \left(\omega_e L_s i_{ds}^{e*}\right)^2 \leq V_{s\max}^2 \tag{5.84}$$

5.4.4.2 Optimal Current to Maximize Torque

Under the assumption of the slow enough variation of the rotor flux linkage and the precise vector control, the current maximizing the torque of the induction machine, which is represented as (5.85), can be derived from the constraints given by (5.52) and (5.84). Like the case of PMSM drive system, the size of the ellipse by the voltage constraint decreases as the operating speed increases as shown in Fig. 5.19.

$$T_e = \frac{3}{2}\frac{P}{2}\frac{L_m{}^2}{L_r} i_{ds}^{e*} i_{qs}^{e*} \tag{5.85}$$

The current constraint in (5.52) can be depicted as a circle in the current plane as shown in Fig. 5.20. And the possible operating region is the shaded area in Fig. 5.21, which is the cross section of the ellipse and the circle. And the torque is depicted as a reciprocal proportion curve in the current plane as shown in Fig. 5.21.

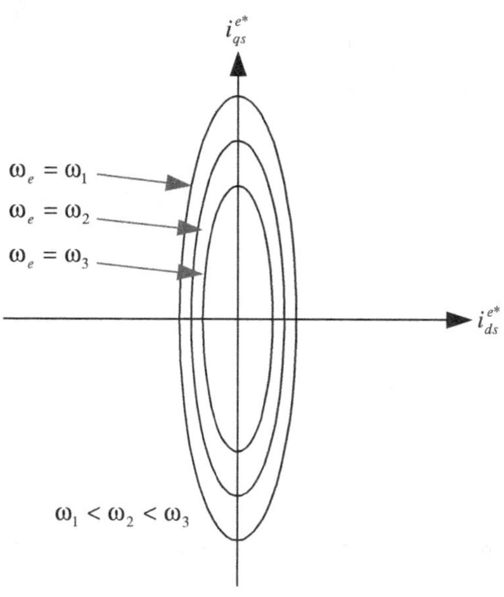

Figure 5.19 Ellipses at different operating speed.

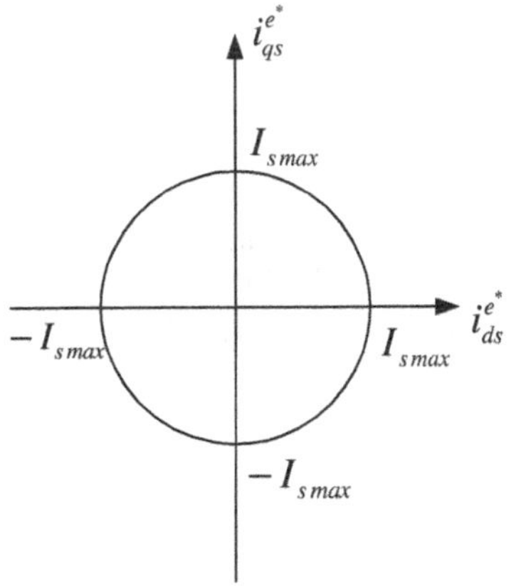

Figure 5.20 Current constraint circle.

5.4.4.3 Constant Torque Region ($\omega_e \le \omega_b$)

If d-axis current for the maximum torque, which is the current at the crossing point of the ellipse and the circle, is larger than the rated value of d-axis current of the induction machine, then the d-axis current reference should be set as the rated value to prevent the magnetic saturation of the induction machine. That is the case of point A in Fig. 5.22, where the torque, T_{e1} may be obtained by i_{ds1}^e, but it is larger than $i_{ds\_rate}^e$ and it would result in the severe saturation of the magnetic circuit of the machine. In this case, i_{ds}^* should be set as $i_{ds\_rate}^e$ as (5.86) and the operating point should be B in the figure generating torque, T_{e2}. Also, the maximum available torque is decided only by the maximum q-axis current, which is given by the current constraint as (5.87), and the maximum torque is always the same as T_{e2} in this region. Hence, the region is called the constant torque region.

$$i_{ds}^{e*} = i_{ds\_rate}^e \tag{5.86}$$

$$i_{qs}^{e*} = \sqrt{I_{s\max}^2 - i_{ds\_rate}^{e2}} \tag{5.87}$$

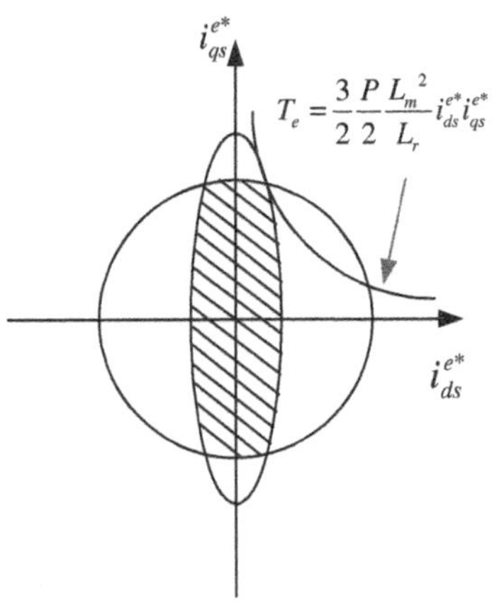

Figure 5.21 Voltage constraint ellipse, current constraint circle, and constant torque locus.

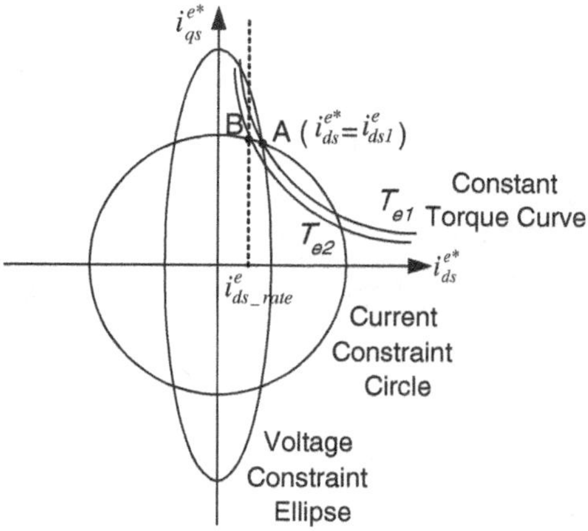

Figure 5.22 Constant torque region.

The rotor flux linkage reference can be derived as (5.88) from (5.82). As the speed increases and the size of the ellipse decreases, the constant torque region ends when the d-axis current at the crossing point of the ellipse and the circle coincides with the rated d-axis current, $i_{ds\_rate}^{e}$. The angular frequency where the constant torque operation region ends is defined as the base frequency, ω_b, and it can be deduced as (5.89).

$$\lambda_{dr}^{e*} = \lambda_{dr\_rate}^{e} = L_m i_{ds}^{e*} = L_m i_{ds\_rate}^{e} \tag{5.88}$$

where $\lambda_{dr\_rate}^{e}$ is the rated rotor flux linkage and the unit is Wb-t.

$$\omega_b = \sqrt{\frac{(V_{s\max})^2}{\lambda_{dr\_rate}^{e2}\frac{L_s^2-(\sigma L_s)^2}{L_m^2} + (\sigma L_s I_{s\max})^2}} \tag{5.89}$$

Unlike a synchronous machine, because of the slip angular frequency of the induction machine, the operating speed of the induction machine at the base frequency is not the base frequency, ω_b. To find the operating speed, the slip angular frequency should be subtracted or added from ω_b according to the operation modes of the induction machine, which are motoring mode or generating mode.

5.4.4.4 Flux Weakening Region I ($\omega_{base} < \omega_e \leq \omega_1$)

This operation region starts from the base speed, ω_b, and ends at ω_1, above which there is no more crossing point between the ellipse and the circle. In this region the d-axis current reference is always smaller than the rated d-axis current as shown in Fig. 5.23. The d-axis current to maximize the torque can be derived as (5.90) from the crossing point of the ellipse and the circle, and the q axis is simply derived as (5.91) from the current constraint.

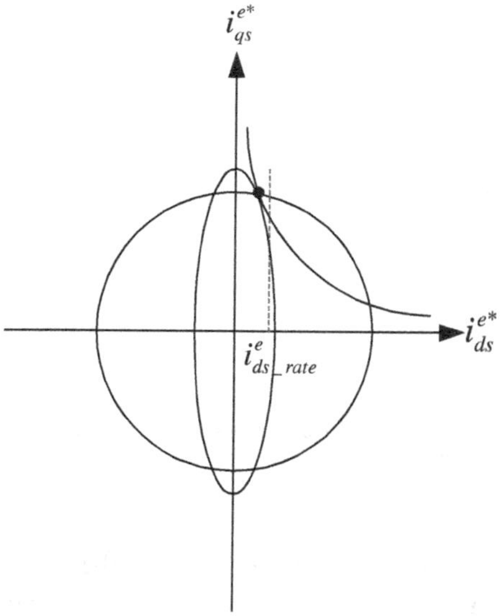

Figure 5.23 Flux weakening region I.

$$i_{ds}^{e*} = \sqrt{\frac{\left(\frac{V_{s\max}}{\omega_e}\right)^2 - (\sigma L_s I_{s\max})^2}{L_s^2 - (\sigma L_s)^2}} \tag{5.90}$$

$$i_{qs}^{e*} = \sqrt{I_{s\max}^2 - \left(i_{ds}^{e*}\right)^2} \tag{5.91}$$

Also, the rotor flux linkage reference can be obtained as (5.92) from (5.90).

$$\lambda_{dr}^{e*} = L_m i_{ds}^{e*} = L_m \sqrt{\frac{\left(\frac{V_{s\max}}{\omega_e}\right)^2 - (\sigma L_s I_{s\max})^2}{L_s^2 - (\sigma L_s)^2}} \tag{5.92}$$

5.4.4.5 Flux Weakening Region II ($\omega_e > \omega_1$)

If the operating frequency of the induction machine further increases from the flux weakening region I and the ellipse shrinks furthermore, then the ellipse would be included in the circle as shown in Fig. 5.24. Hence, there is no more crossing point between the ellipse and the circle. In this case, the torque is limited only by the voltage constraint. This operating region is referred to as flux weakening region II. And it is sometimes called the characteristic region of the induction machine. The frequency, ω_1, where the flux weakening region II starts, can be derived as (5.93) from the fact that at that frequency the circle meets the ellipse at single point.

$$\omega_1 = \sqrt{\frac{L_s^2 + (\sigma L_s)^2}{2(L_s \sigma L_s)^2}} \times \frac{V_{s\max}}{I_{s\max}} \tag{5.93}$$

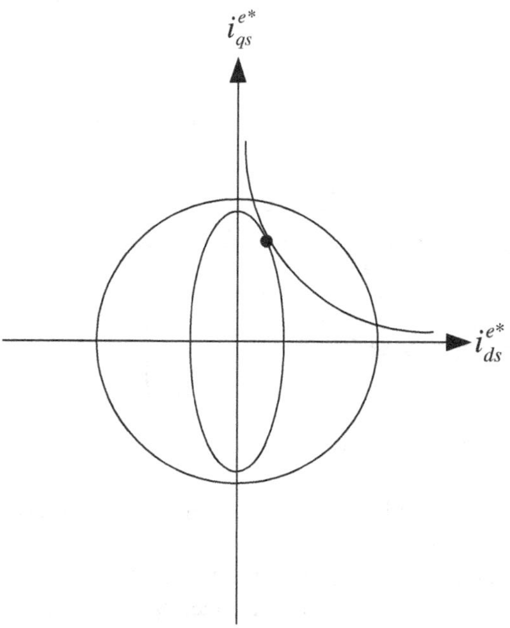

Figure 5.24 Flux weakening region II.

In this region, the reference current to maximize the torque can be represented in terms of the maximum voltage and the machine parameters as (5.94) and (5.95), which is the contact point of the torque curve and the ellipse as shown in Fig. 5.24.

$$i_{ds}^{e*} = \frac{V_{smax}}{\sqrt{2}\omega_e L_s} \tag{5.94}$$

$$i_{qs}^{e*} = \frac{V_{smax}}{\sqrt{2}\omega_e \sigma L_s} \tag{5.95}$$

The rotor flux linkage reference at the point can be obtained as (5.96).

$$\lambda_{dr}^{e*} = L_m i_{ds}^{e*} = \frac{L_m V_{smax}}{\sqrt{2}\omega_e L_s} \tag{5.96}$$

5.4.5 Flux Regulator of Induction Machine

The flux weakening control method described in Section 5.4.4 is based on the assumption of slow enough variation of the rotor flux holding (5.82). However, if the acceleration or deceleration rate is high and the flux variation is fast, then the flux of the induction machine should be controlled instantaneously by (5.19). In this case for the reasonable performance of the flux regulation, the flux regulator should be incorporated in the control loop of the induction machine. At the rapid variation of the speed, if the flux is too small, then the available maximum torque cannot be obtained. However, if the flux is too much, then the current cannot be regulated due to

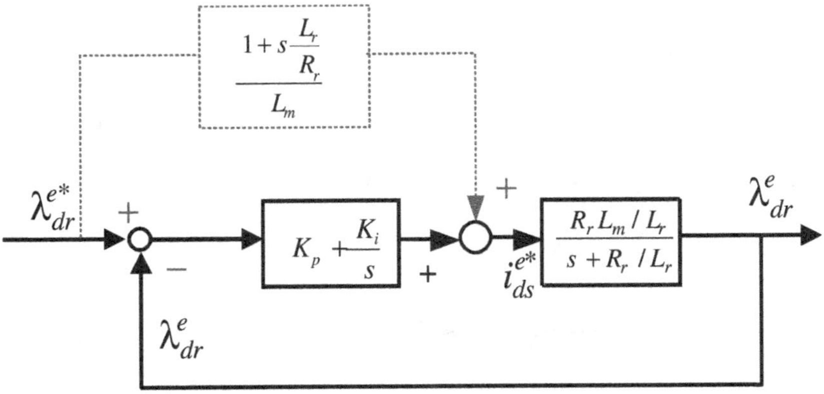

Figure 5.25 Flux regulator of an induction machine.

the deficiency of the voltage. A proportional and integral (PI) flux regulator can solve these problems with a proper flux weakening regulator.

If the machine is controlled in vector control mode precisely, the dynamics of rotor flux linkage can be represented as (5.97). From (5.97), it can be seen that by adjusting the *d*-axis current the rotor flux linkage can be regulated. Also, a simple PI regulator can be employed to control the rotor flux linkage as shown in Fig. 5.25. For reasonable performance of the flux regulation the bandwidth of the current regulator should be high enough compared to that of the flux regulator—for example, at least five times high. Furthermore, in this case the current regulator loop can be assumed to be an ideal one as its transfer function be unity.

$$\frac{d\lambda_{dr}^{e}}{dt} = R_r \frac{L_m}{L_r} i_{ds}^{e} - \frac{R_r}{L_r} \lambda_{dr}^{e} \tag{5.97}$$

The transfer function between the flux reference and the actual flux can be derived as

$$\frac{\lambda_{dr}^{e}(s)}{\lambda_{dr}^{e*}(s)} = \frac{R_r \frac{L_m}{L_r}(K_p s + K_i)}{s^2 + \frac{R_r}{L_r}(1 + L_m K_p)s + R_r \frac{L_m}{L_r} K_i} \tag{5.98}$$

where K_p is a proportional gain of the regulator and K_i, is an integral gain. If the proportional and integral gains are set as (5.99) and (5.100), respectively, then the transfer function in (5.98) can be simplified as (5.101).

$$K_p = \frac{L_r}{L_m} \frac{\omega_c}{R_r} \tag{5.99}$$

$$K_i = \frac{\omega_c}{L_m} \tag{5.100}$$

$$\frac{\lambda_{dr}^{e}(s)}{\lambda_{dr}^{e*}(s)} = \frac{\omega_c}{s + \omega_c} \tag{5.101}$$

It can be seen from (5.101) that the transfer function of the flux regulator is that of the first-order low-pass filter and its cutoff frequency is ω_c, which is the bandwidth of the flux regulator. Hence after setting the bandwidth of the flux regulator, the gains are easily decided and the flux can be regulated without any overshoot even in the case of step change of the flux reference. Because the output of the flux regulator is also limited due to the maximum d-axis current, the anti-wind-up controller, mentioned in Section 4.2.4, should be incorporated in this flux regulator too. The performance of the flux regulator can be enhanced by adding the dotted part in Fig. 5.25 to the regulation loop. With this added block, the flux can be set as desired without any time delay if there is no parameter error and no time delay in the signal processing in the regulation loop. Even if there are delays and parameter errors, the performance of the flux regulation loop can be improved conspicuously with this feed-forwarding term shown as the dotted part in Fig. 5.25.

PROBLEMS

1. An SMPMSM with the following parameters is running in vector control mode. At $t = 0$ s, the direction of the magnetic axis of the rotor flux linkage coincides with that of magnetic axis of "a" phase winding. The speed varies according to the pattern in Fig. P5.1. Answer the following questions.
SMPMSM parameters

Rated voltage : 208 V (line-to-line voltage rms)
Rated current : 11.8 A (phase current rms)
Rated power : 5 Hp
Number of poles : 4
$R_s = 0.423\ \Omega$
$L_s = 4.76$ mH
$\lambda_f = 0.422\ V/(\text{rad/s})$

The total inertia, J_{M+L}, of the drive system is 0.03 kgm^2, and the total friction coefficient, B_{M+L}, of the drive system is 0.04 N-m/(rad/s). The internal connection of stator winding is assumed to be a Y connection.

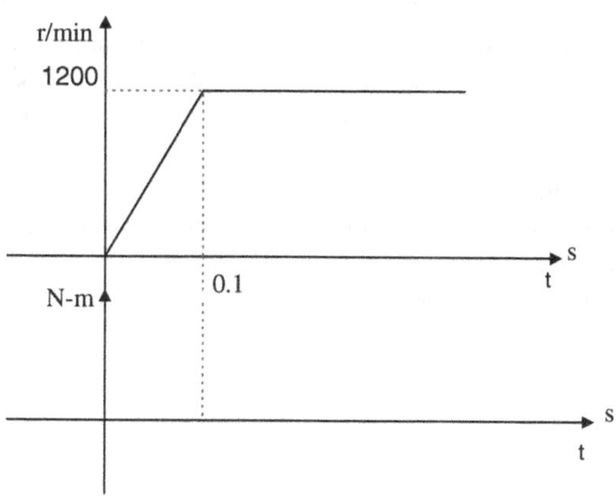

Figure P5.1 Speed pattern.

(1) Calculate the torque to achieve the given speed variation according to time t, and plot the torque of an SMPMSM together with the speed as shown in Fig. P5.1.

(2) Represent "a" phase current in terms of time t.

2. An induction machine with the following parameters is controlled in the rotor flux linkage-oriented vector control mode.

Rated power : 3 Hp
Rated line-to-line rms voltage : 220 V
Rated frequency : 60 Hz
Number of poles : 4
$R_s = 0.435\,\Omega, \quad X_{ls} = 0.754\,\Omega,$
$X_m = 26.13\,\Omega, \quad X_{lr} = 0.754\,\Omega,$
$R_r = 0.816\,\Omega$

The impedances are evaluated at the rated frequency.
The total inertia, J_{M+L}, of the drive system is 1.0 kgm^2, and the total friction coefficient, B_{M+L}, of the drive system is 0.1 N–m/(rad/s). The internal connection of stator winding is assumed to be a Y connection.
The torque component current has been zero and the flux component current has been set to keep the rotor flux linkage as 0.45 V/(rad/s). The drive system has been in the steady state. At $t = 0$ s, the torque command varies in step manner to 20 N-m. Plot the "a" phase current for 0–0.5 s. The current regulator can be assumed as ideal, which means that the transfer function of the current regulation loop is unity. At $t = 0$s, the direction of the rotor flux linkage coincides with that of the magnetic axis of "a" phase winding.

3. The SMPMSM in problem 1 is controlled in the vector control mode. The maximum current of the inverter is 300% of the rated current of the SMPMSM, and the maximum voltage of the inverter is the same as the peak value of rated voltage of the SMPMSM, which is $\sqrt{\frac{2}{3}}$ times of the rated line-to-line rms voltage of the SMPMSM. The current regulator can be assumed to be ideal, which means that the transfer function of the current regulation loop is unity.

(1) Calculate the base speed, ω_b, in r/min.

(2) In the flux weakening control, calculate maximum operation speed (r/min) where the drive system operates within the current constraints without any load torque. In this calculation, the friction of the drive system can be neglected, that is, $B_{M+L} = 0$.

(3) In flux weakening control, calculate the maximum operating speed (r/min) where the drive system operates within the current constraints providing rated torque of SMPMSM. At this operating point, calculate the power factor of SMPMSM at the terminal of SMPMSM.

4. When an induction machine is controlled in stator flux linkage-oriented vector control mode, represent torque, stator flux linkage reference, and slip angular speed in terms of given variables and machine parameters in the following format. In the stator flux linkage oriented vector control, the q-axis stator flux linkage is controlled as null, which is $\lambda_{qs}^e = 0$.

(1) $T_e = f\left(\lambda_{ds}^e, i_{qs}^e\right)$

(2) $\lambda_{ds}^e = g\left(i_{ds}^e, i_{qs}^e, \omega_{sl}\right)$

(3) $\omega_{sl} = h\left(\lambda_{ds}^e, i_{qs}^e, i_{ds}^e\right)$

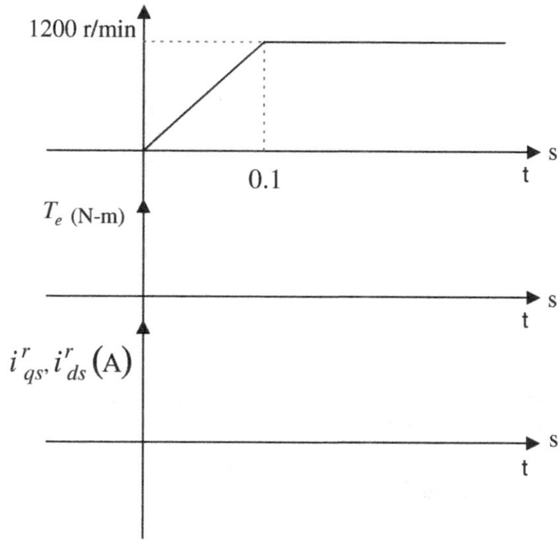

Figure P5.2 Speed pattern.

(4) Describe merits and demerits of the stator flux linkage-oriented vector control compared to those of the rotor flux linkage-oriented vector control.

5. When an eight pole SMPMSM is driven externally at constant speed 1800 r/min by a prime mover, the measured line-to-line rms voltage is 180 V and the waveform is sinusoidal. At this operation, no current flows in the SMPMSM. The same SMPMSM is now controlled under the vector control mode. The speed of the SMPMSM is regulated according to the speed pattern in Fig. P5.2. Represent the "a" -phase current in terms of parameters of the SMPMSM according to the time, t. Along with speed pattern in Fig. P5.2, plot torque and the d–q axes current i^r_{qs} and i^r_{ds}. The total inertia, J_{M+L}, of the drive system is 0.01 kg m^2, and the total friction coefficient, B_{M+L}, of the drive system is 0.1 N-m/(rad/s). The current regulator can be assumed to be ideal, which means that the transfer function of the current regulation loop is unity. At $t = 0$s, the direction of the magnetic axis of the permanent magnet flux linkage coincides with that of the magnetic axis of an "a"-phase winding.

(1) Plot torque, T_e.

(2) Plot current i^r_{qs} and i^r_{ds}.

(3) Represent rotor position, θ_r, in terms of time, t.

(4) Plot "a"-phase current in terms of time, t.

6. A four-pole induction machine with the following parameters is driven by an inverter. The rotor flux linkage is 0.4 Wb-t and the running speed is 1200 r/min, providing 20-N-m torque to the load. The drive system including the induction machine is in the steady state. The inverter is operated in six-step mode, which means no PWM. Find the average DC link current, $<i_{dc}>$, of the invereter and average capacitor voltage, $<V_c>$, in Fig. P5.3 by using the following procedures. The harmonics of the current and voltage due to six-step operation can be neglected, and inductor and capacitor at DC link are ideal. The DC source voltage, V_d, is constant.

$$R_s = 0.4\,\Omega, \quad L_{\ell s} = L_{\ell r} = 2\,\text{mH}, \quad L_m = 70\,\text{mH}, \quad R_r = 0.8\,\Omega$$

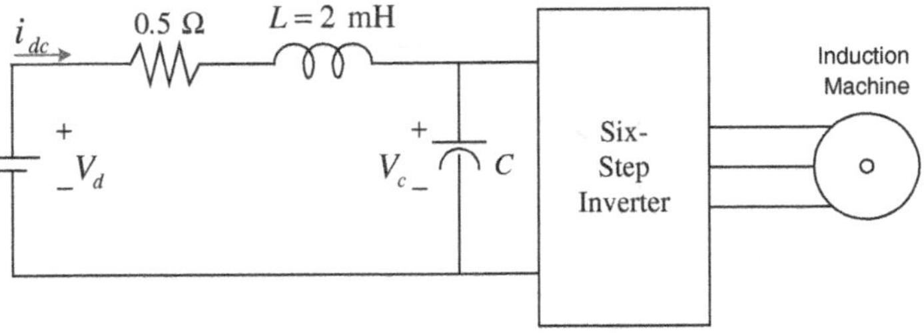

Figure P5.3 Induction machine drive system by a six-step voltage source inverter.

(1) Find the stator current, i_{ds}^e and i_{qs}^e, at a synchronous reference frame for a period of the fundamental frequency in the steady state.

(2) Find the slip angular frequency, $\omega_{s\ell}$, and the angular speed of rotating MMF of the machine in electrical angle.

(3) Calculate input power to the machine neglecting the iron loss and the stray loss.

(4) Find $<V_c>$, $<i_{dc}>$, V_d.

7. The induction machine in problem 2 is controlled in the rotor flux linkage-oriented vector control mode by a PWM inverter with an ideal current regulation loop. The machine is running in the steady state at 1200 r/min and $i_{ds}^e = 6$ A. The torque of the machine is zero. At $t = 0$ s, i_{qs}^e increases linearly with the slew rate, 5000 A/s. And the machine is controlled in the vector control mode continuously. For 0–0.002 s, calculate the line-to-line voltage between the "a" and "b" phase of the machine by using the following procedures. At $t = 0$ s, the direction of the rotor flux linkage coincides with that of the magnetic axis of the "b"-phase winding. Because of a large enough inertia of the drive system, the speed of the induction machine can be assumed to be constant at 1200 r/min for 0–0.002 s. The harmonics in the currents and voltages by the inverter can be neglected.

(1) Find the slip angular frequency, $\omega_{s\ell}$, and the angular speed of rotor flux linkage of the machine in electrical angle for 0–0.002 s.

(2) Find the stator voltage, V_{ds}^e and V_{qs}^e, at synchronously rotating reference frame for 0–0.002 s.

(3) Find the stator voltage, V_{ds}^s and V_{qs}^s, at stationary reference frame for 0–0.002 s.

(4) Find the line-to-line voltage between the "a" and "b" phase, V_{ab}, of the machine for 0–0.002 s.

8. A four-pole induction machine with following parameters is controlled in the rotor flux linkage-oriented vector control mode by a PWM inverter with an ideal current regulation loop. The machine is running in the steady state, $i_{ds}^e = 7$ A. The frequency of the stator current is 55 Hz. At $t = 0$ s, when the direction of the rotor flux linkage coincides with that of the magnetic axis of the "a"-phase winding, the input contactor of the machine is opened as shown in Fig. P5.4. After $t = 0^+$, find the line-to-line voltage between the "a" and "b" phase of the machine by using the following procedures. The current to the machine is immediately zero after the opening of the contactor.

$$R_s = 0.55\,\Omega,\quad R_r = 0.36\,\Omega$$
$$L_{ls} = 1.8\text{ mH},\quad L_{lr} = 1.8\text{ mH}$$
$$L_m = 59\text{ mH},\ \text{(inertia of motor itself)}\quad J_M = 0.04\text{ kg-m}^2$$

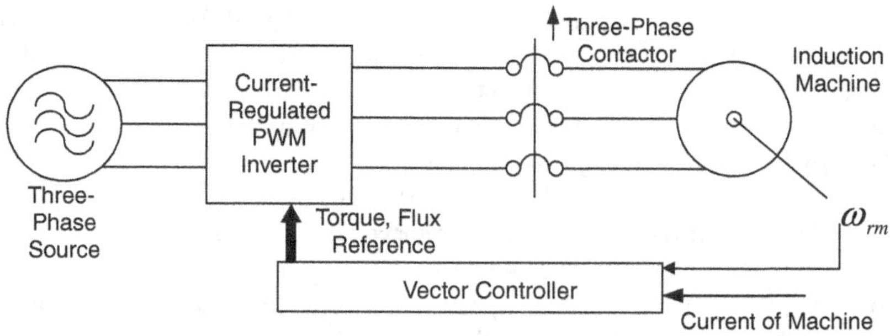

Figure P5.4 Figure P5.4 vector control of induction machine by a current-regulated PWM inverter.

The rated power of the induction machine is 5 Hp, and the inertia of the load is nine times the machine inertia, J_M. And the friction coefficient, B_{M+L}, of the drive system is 0.1 N-m/(rad/s).

(1) Just after the opening of the contactor, find the rotor angular speed, $\omega_r(0^+)$, and the d–q axis stator flux linkage, $\lambda^r_{ds}(0^+)$ and $\lambda^r_{qs}(0^+)$, in the rotor reference frame.

(2) Find the rotor angular speed, $\omega_r(t)$, $t>0^+$, in electric angle.

(3) Find $\lambda^r_{dr}(t), t>0^+$.

(4) Find the "a" phase voltage, $V_{as}(t)$, $t>0^+$.

(5) Find the "b" phase voltage, $V_{bs}(t)$, $t>0^+$.

(6) Find line-to-line voltage, $V_{ab}(t)$, $t>0^+$.

9. A four-pole SMPMSM with following parameters is driven by an inverter in vector control mode according to given speed pattern shown in Fig. P5.5. The friction and load torque of the drive system can be neglected. Find the "a"-phase current, i_{as}, and q-axis voltage at the rotor reference frame, V^r_{qs}, through following procedures. At $t = 0$s, the direction of the magnetic axis of the permanent magnet coincides with that of the magnetic axis of the "a"-phase winding. The current regulator can be assumed to be ideal, which means that the transfer function of the current regulation loop is unity.

SMPMSM Parameters

Rated Power : 200 W
Number of Poles : 8
$R_s = 0.65\ \Omega$
$L_s = 2.35$ mH
$\lambda_f = 0.061$ V/(rad/s)

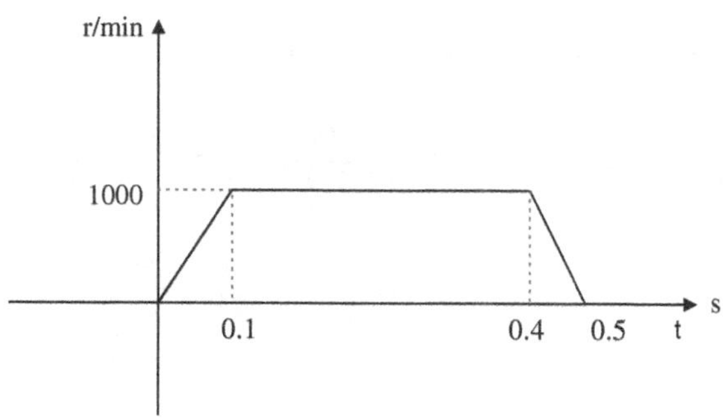

Figure P5.5 Speed pattern.

The total inertia, J_{M+L}, of the drive system is 1.29×10^{-3} kg-m^2. The internal connection of stator winding is assumed to be a Y connection. The harmonics in the currents and voltages by the inverter can be neglected.

(1) Plot the torque of the SMPMSM along with the speed pattern from 0 to 0.5 s.

(2) Derive the expression of q-axis stator current, i^r_{qs}, at rotor reference frame from 0 to 0.5 s in terms of the speed of the SMPMSM and its parameters.

(3) Derive the expression of an "a" -phase stator current, i_{as}, from 0 to 0.5 s in terms of the speed of the SMPMSM and its parameters.

(4) Derive the expression of the q-axis phase voltage, V^r_{qs}, at the rotor reference frame from 0 to 0.5 s in terms of the speed of the SMPMSM and its parameters.

10. The induction machine in problem 2 is controlled in the indirect vector control mode. During the operation, the rotor resistance increases by two times due to the temperature rise of the rotor bar of the squirrel cage rotor. The torque command to the machine is 10 N-m, $i^{e*}_{ds} = 6$ A, and the machine is running in the steady state. The current regulator can be assumed to be ideal, which means that the transfer function of the current regulation loop is unity. The rotor resistance set to the vector controller, $\hat{R}_r$, is 0.36 Ω, but the real rotor resistance is $R_r = 0.72$ Ω. Find the following values. All other parameters except the rotor resistance in the vector controller are matched to the real parameters.

(1) Magnitude of air-gap flux linkage, $|\lambda^e_m| = \sqrt{\lambda^{e2}_{dm} + \lambda^{e2}_{qm}}$.

(2) Magnitude of rotor flux linkage, $|\lambda^e_r| = \sqrt{\lambda^{e2}_{qr} + \lambda^{e2}_{dr}}$.

(3) Actual q-axis rotor flux linkage at synchronously rotating reference frame, λ^e_{qr}, with the parameter error in the rotor resistance.

(4) Actual torque of the induction machine with the parameter error in the rotor resistance.

11. Derive the condition of the vector control for a double cage rotor induction machine based on the equivalent circuit in Fig. P5.6 through following procedures [19].
The flux linkages and torque of a double-cage rotor induction machine are given by the following equations.

$$\text{Stator flux linkage}: \ \lambda^e_{dqs} = (L_m + L_{ls})\mathrm{i}^e_{dqs} + L_m\mathrm{i}^e_{dqr}$$

$$\text{The first cage rotor flux linkage}: \ \lambda^e_{dq1} = L_m\left(\mathrm{i}^e_{dqs} + \mathrm{i}^e_{dqr}\right) + L_1\mathrm{i}^e_{dqr}$$

$$\text{The second cage rotor flux linkage}: \ \lambda^e_{dq2} = L_m\left(\mathrm{i}^e_{dqs} + \mathrm{i}^e_{dqr}\right) + L_1\mathrm{i}^e_{dqr} + L_2\mathrm{i}^e_{dq2}$$

$$\text{Torque of the machine}: \ T_e = \frac{3}{2}\frac{P}{2}\frac{L_m}{L_m + L_1}\mathrm{Im}\left(\lambda^e_{dq1} \cdot \mathrm{i}^e_{dqs}\right)$$

where Im(•) means the imaginary part of the complex number (•).

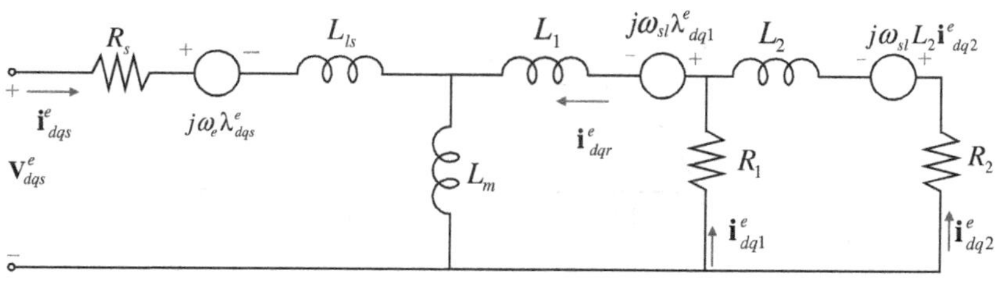

Figure P5.6 The d–q equivalent circuit of a double-cage rotor induction machine.

Answer the following questions under the assumption of the steady-state operation, and the q-axis rotor flux linkage at the first cage is controlled as zero, like $\lambda^e_{q1} = 0$.

(1) Derive ω_{sl} in terms of λ^e_{d1}, R_1, i^e_{q1}.

(2) Derive ω_{sl} in terms of λ^e_{d2}, R_2, i^e_{q2}.

(3) Under the assumption that $\lambda^e_{d1} = L_m i^e_{ds}$ and $\lambda^e_{d2} \approx \lambda^e_{d1}$, derive the relationship between i^e_{q1} and i^e_{qs}.

(4) Derive ω_{sl} in terms of the machine parameters, i^e_{ds} and i^e_{qs}.

12. An eight-pole SMPMSM with an equivalent circuit at the rotor reference frame is driven by the position sensorless control (Fig P5.7). Answer the following questions.

(1) Derive λ^s_{ds}andλ^s_{qs} in terms of V^s_{ds}, V^s_{qs}, i^s_{ds}, and i^s_{qs}.

(2) At a certain instant, $i^r_{ds} = 0A$, $i^r_{qs} = 10$ A, $\lambda^s_{ds} = 0.06$ Wb-t, and $\lambda^s_{qs} = -0.0573$ Wb-t. Find the rotor position, θ_r, at that instant.

(3) In the case of part 2, calculate V^s_{ds} and V^s_{qs}. At that instant, the speed of the SMPMSM was 600 r/min.

(4) From the answers of parts 1 and 2, discuss the expected problems of the position sensorless control where the rotor position is estimated from V^s_{ds}, V^s_{qs}, i^s_{ds}, and i^s_{qs} especially when the machine starts or runs at extremely low speed.

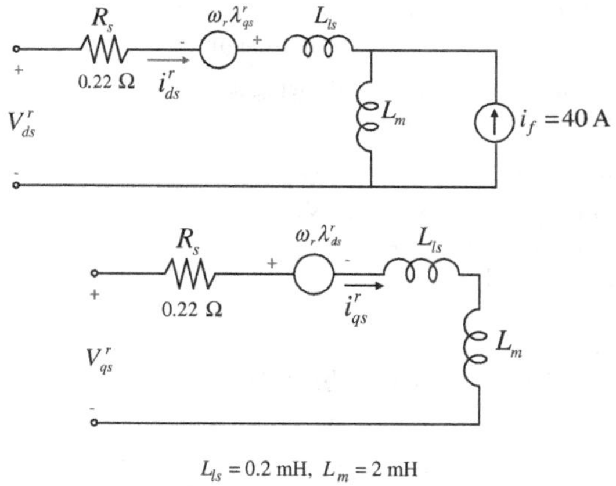

Figure P5.7 Equivalent circuit of an SMPMSM at a rotor reference frame.

13. A synchronous reluctance machine (SynRM) with the following parameters is controlled in the vector control mode achieving maximum torque per ampere (MTPA) operation.

$R_s = 0.232\,\Omega$, $L_d = 43$ mH, $L_q = 3.5$ mH
Total inertia of the drive sytem : $J_{M+L} = 0.026$ kg-m^2
Four-pole, rated frequency : 60 Hz
Rated line-to-line rms voltage : 170 V
Rated Power : 5 Hp
Total friction coefficient of the drive sytem : $B_{M+L} = 0.1$ N-m/(rad/s)
$L_d = L_{\ell s} + L_{md}$, $L_q = L_{\ell s} + L_{mq}$

The current controller shown in Fig. 4.18 is incorporated to the control loop to regulate the current of the SynRM. The current regulator is implemented at the rotor reference frame, and it is assumed that the position of rotor is identified instantaneously. The back EMF of the SynRM is canceled out by the feed-forward voltage of the current controller. The zero of PI regulator cancels out the pole of the SynRM. The parameters for setting of gains of PI regulator and back EMF decoupling are the same as the actual parameters of SynRM. The current regulation bandwidth is set as 500 Hz. It is assumed that the PWM inverter is ideal, and the output of the current regulator is directly applied to the SynRM, but the outputs of the d–q axis current regulator including feed-forward terms are limited within $\frac{300}{\sqrt{3}}$ V at the d–q axis of the rotor reference frame independently, which means that there is no need to consider the hexagon boundary of the voltage of the PWM inverter. Hence, the voltage boundary can be assumed to be a perfect square whose length of one side is $\frac{300}{\sqrt{3}}$ V at the d–q axis of the rotor reference frame voltage plane.

(1) Find proportional gain, K_p, and integral gain, K_i.

(2) If torque command, T_e^*, is given as follows, plot torque command, actual torque, and speed of the SynRM in r/min simultaneously according to the time from zero to 0.5 s. Also plot i_{as}, i_{ds}^{r*}, i_{ds}^{r}, i_{qs}^{r*}, and i_{qs}^{r} according to the time from 0 s to 0.5 s.

$$\text{Torque Command}: \quad T_e^* \begin{cases} 0s \le t \le 0.1\,\text{s} & 20\,\text{N-m} \\ 0.1s < t \le 0.3\,\text{s} & -20\,\text{N-m} \\ 0.3s < t \le 0.5\,\text{s} & 5\,\text{N-m} \end{cases}$$

(3) Compare the vector control of the SynRM with the vector control of an induction machine with regard to the control of the magnitude of flux linkage.

14. A four-pole induction machine with the following parameters is controlled in the rotor flux-oriented direct vector control mode.

Rated power : 5 Hp,
Rated line to line rms voltage : 220 V,
Rated phase rms current : 12 A,
Rated frequency : 60 Hz.
Rated speed : 1739 r/min.
$R_s = 0.55\,\Omega$, $R_r = 0.36\,\Omega$, $L_{\ell s} = 1.8\,\text{mH}$, $L_{\ell r} = 1.8\,\text{mH}$, $L_m = 59\,\text{mH}$
Interia of the induction machine : $J_M = 0.02\,\text{kg-m}^2$
Intertia of total drive system : $J_{M+L} = 0.04\,\text{kg-m}^2$

The friction and load torque can be neglected from t = 0 s to 1 s. From t = 1 s, the load torque comes from friction, and the friction coefficient of the drive system, B_{M+L}, is 10 N-m/(rad/s). The rotor flux linkage reference is shown in Fig. P5.8a The control block diagram including complex vector current regulator is shown in Fig. P5.8b The current regulator is implemented at the synchronous reference frame, and it is assumed that the position of rotor flux linkage is identified instantaneously. In the current regulator, the back EMF of induction machine is canceled out by the feed-forward voltage of the regulator. The zero of PI regulator cancels out the pole of the induction machine. At first, it is assumed that the parameters for setting of gains of PI regulator and back EMF decoupling are the same as the actual parameters of the induction machine. The current regulation bandwidth is set as 500 Hz. It is assumed that the PWM inverter is ideal and the output of the current regulator is directly applied to the induction machine, but the output of the regulator including feed-forward terms is limited within the boundary of the hexagon, where the length of one side of the hexagon is 200 V. The torque command of the induction machine is as follows. Active damping resistance, R_{active}, in the current regulator can be simply set as 10 times the stator resistance.

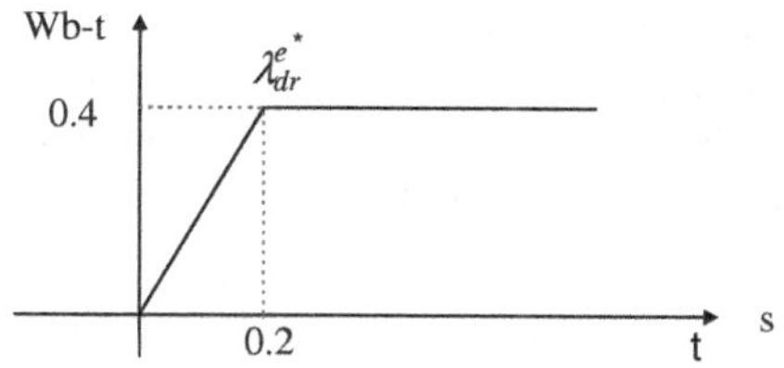

Figure P5.8a Rotor flux linkage reference.

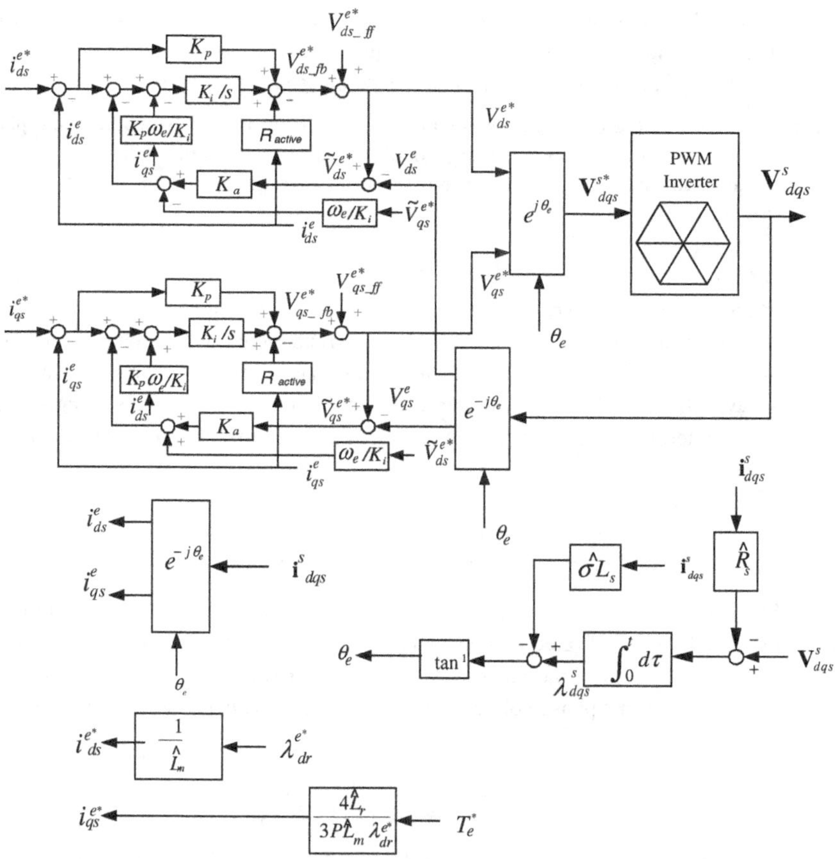

Figure P5.8b Control block diagram of induction machine with direct vector control.

$$\text{Torque command}: \quad T_e^* \begin{cases} 0 \le t \le 0.5\,\text{s} & 0\,\text{N-m} \\ 0.5\,\text{s} < t \le 0.7\,\text{s} & 20\,\text{N-m} \\ 0.7\,\text{s} < t \le 0.9\,\text{s} & -20\,\text{N-m} \\ 0.9\,\text{s} < t \le 1.5\,\text{s} & 20\,\text{N-m} \end{cases}$$

(1) Find current regulator gains, K_p, K_i, and K_a, with the consideration of active damping resistance.

(2) Under the assumption that the phase voltage and the phase current can be measured ideally and that the parameters of the induction machine is accurately known and they are constant regardless of the operating conditions, plot torque command, actual torque, i_{qs}^{e*}, i_{ds}^{e*}, i_{qs}^{e}, i_{ds}^{e}, λ_r, λ_{dr}^{e*}, and the speed in r/min from 0 s to 1.5 s.

(3) Assume that the phase voltage and the phase current can be measured ideally but that the parameters of the induction machine for regulator setting have the following errors:

$$\hat{L}_m = 1.1L_m, \quad \hat{L}_{ls} = 1.2L_{ls}, \quad \hat{L}_{lr} = 1.2L_{lr}, \quad \hat{R}_s = 0.8R_s, \quad \hat{R}_r = 0.7R_r$$

where '^' means the value used to set the gains and decoupling terms of the current regulator and vector controller. Plot torque command, actual torque, i_{qs}^{e*}, i_{ds}^{e*}, i_{qs}^{e}, i_{ds}^{e}, λ_r, λ_{dr}^{e*}, and the speed in r/min from 0 s to 1.5 s.

(4) In addition to the parameter errors in part 3, the measured phase voltage and phase current have the following errors. In the measurement, only "a"-and "b"-phase voltage and current are measured, and "c"-phase voltage and current are calculated as follows:

$$\hat{i}_{cs} = -(\hat{i}_{as} + \hat{i}_{bs}), \quad \hat{V}_{cs} = -(\hat{V}_{as} + \hat{V}_{bs})$$

where the hat (^) denotes the measured value for the current regulator and the vector controller. In the measurement of the phase current, there is white noise, $\eta_i(t)$, whose rms magnitude is 4% of rated rms phase current. The sampling frequency of the white noise is 100 μs and also the measured current has offset, I_{offset}, whose magnitude is 1% of rated rms phase current. And the measured current can be represented as follows:

$$\hat{i}_{as} = i_{as} + \eta_i(t) + I_{\text{offset}}$$
$$\hat{i}_{bs} = i_{bs} + \eta_i(t) + I_{\text{offset}}$$

In the measurement of the phase voltage, there is white noise, $\eta_v(t)$, whose rms magnitude is 1% of rated rms phase voltage. The sampling frequency of the white noise is 100 μs, also, the measured voltage has offset, V_{offset}, whose magnitude is 0.5% of rated rms phase voltage. And the measured voltage can be represented as follows:

$$\hat{V}_{as} = V_{as} + \eta_v(t) + V_{\text{offset}}$$
$$\hat{V}_{bs} = V_{bs} + \eta_v(t) + V_{\text{offset}}$$

Plot torque command, actual torque, i_{qs}^{e*}, i_{ds}^{e*}, i_{qs}^{e}, i_{ds}^{e}, λ_r, λ_{dr}^{e*}, and the speed in r/min from 0 s to 1.5 s.

(5) Through the results from parts 2, 3, and 4, discuss the problem of the direct vector control of the induction machine.

15. Using the indirect vector control shown in Fig. P5.9, repeat problem 14. Parts 1–4. In the case of part 2, it is assumed that the speed measurement is ideal. In the case of parts 3 and 4, the measured speed, ω_{rm}, includes the white noise, $\eta_\omega(t)$, whose rms magnitude is 1 r/min and the sampling time of the noise is 100 μs. Also, the noise has offset, ω_{offset}, whose magnitude is 0.01 r/min. And the measured speed can be represented as follows:

$$\hat{\omega}_{rm} = \omega_{rm} + \eta_\omega(t) + \omega_{\text{offset}}$$

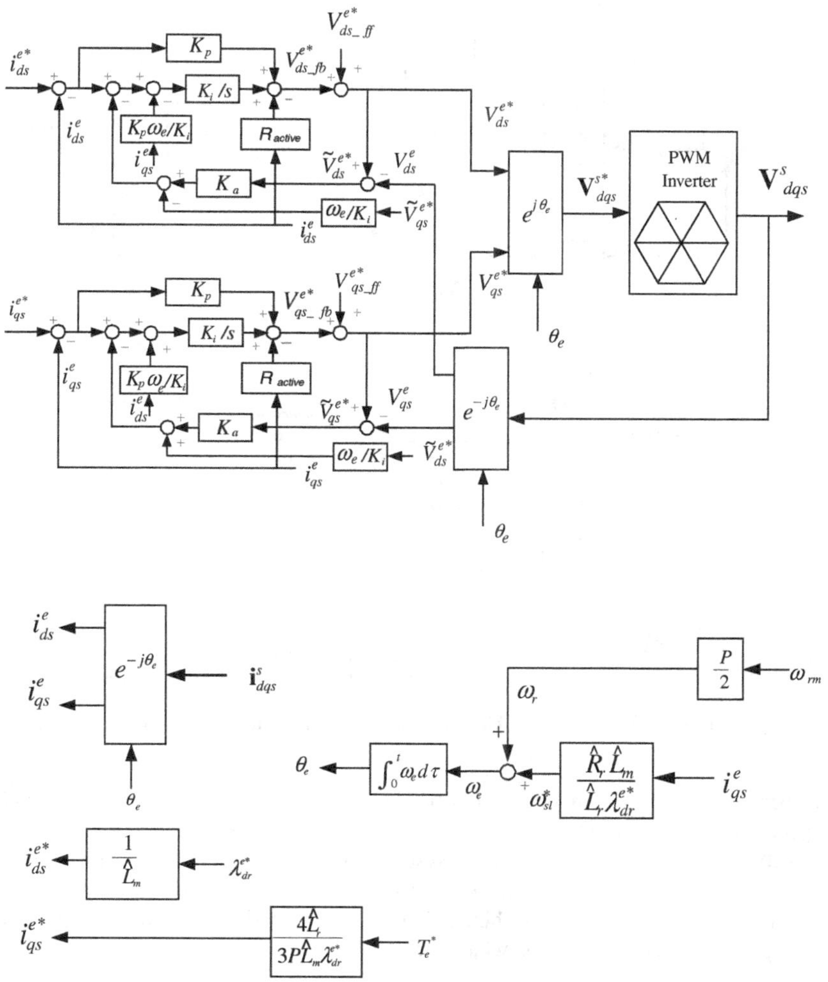

Figure P5.9 Control block diagram of indirect vector control of induction machine.

16. In problem 14, a flux regulator with the flux estimator is incorporated as shown in Fig. P5.10. The bandwidth of the flux regulator is 10 Hz. The maximum value of the flux component current is limited to the magnitude of the rated current of induction machine.

(1) Find the gains, K_p and K_i, of the flux regulator.

(2) Repeat problem 14, parts 2–4. The measured speed in the flux estimator is the same one in the measured speed in problem 15. Use the flux estimator in Fig. 5.12 at this problem.

(3) Compare the results of this problem with the results of problem 14, parts 2–4, with regard to flux regulation and torque control accuracy.

17. The induction machine given in problem 14 is controlled in the rotor flux-oriented indirect vector control mode. The control loop and condition for the vector control is the same as the loop and the condition in problem 15, part 2. The speed of the induction machine is

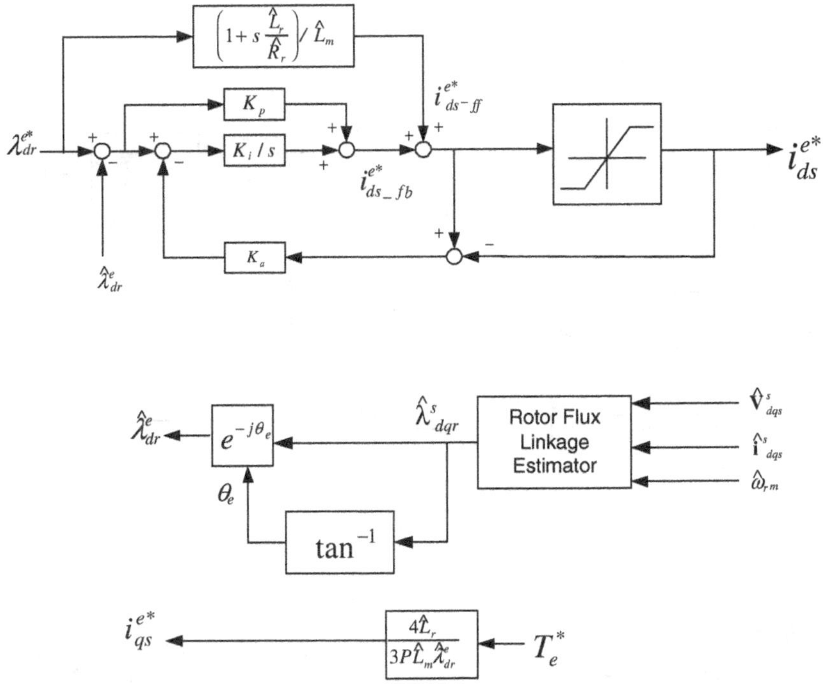

Figure P5.10 Control block diagram of flux regulator and estimator.

controlled by a PI speed regulator. The bandwidth of the speed regulator is 40 Hz and the speed command is as follows. The output of the regulator is limited within ±150 N-m. The load torque and friction of the drive system can be neglected. The total inertia, J_{M+L}, of the drive system is 0.15 kgm^2. The rotor flux linkage reference is 0.4 Wb-t up to base speed, and above the base speed the machine is controlled to provide maximum torque within the voltage and current constraints, which are

$$V_{s\max} = \frac{300}{\sqrt{3}}V, \quad I_{s\max} = 1.6\sqrt{2}I_{\text{rated},rms}$$

$$\omega^*_{rm}\begin{pmatrix} 0\,\text{s} \le t \le 0.3\,\text{s} & 0\,\text{r/min} \\ 0.3\,\text{s} < t & 6000\,\text{r/min} \end{pmatrix}$$

(1) Find the gains of the speed regulator, K_p and K_i.

(2) Plot torque command, actual torque, i^{e*}_{qs}, i^{e*}_{ds}, i^{e}_{qs}, i^{e}_{ds}, $|\lambda^e_r|$, $|i^e|$, $I_{s\max}$, $|V^e_{dqs}|$, $V_{s\max}$, and speed in r/min from 0 s utill the speed reaches 6000 r/min.

18. The synchronous reluctance machine (SynRM) in Fig. 3.11 and Fig. 3.12 in Section 3.3.4 is driven in MTPA mode. The magnetic saturation of the SynRM can be neglected.

(1) Derive the relationship between i^r_{ds} and i^r_{qs}.

(2) Under the constraints $V_s = \sqrt{V^{r2}_{ds} + V^{r2}_{qs}} \le V_{s\max}$ and $i_s = \sqrt{i^{r2}_{ds} + i^{r2}_{qs}} \le I_{s\max}$, plot the capability curve of the SynRM in the first quadrant of the torque–speed plane. If the operating region of SynRM is classified as the constant torque region where only current constraint applies, then both (a) the flux weakening region I where both current and voltage constraints apply and (b) the flux weakening region II where only voltage constraint applies represent (a) the speed ω_{base} where the flux weakening region I starts and (b) the speed ω_1 where the flux weakening region II starts in terms of the parameters of the SynRM and the constraints, $V_{s\max}$ and $I_{s\max}$. In this derivation the voltage drop by the stator resistance can be neglected. At each flux weakening region, represent a pair i_{ds}^r and i_{qs}^r where the maximum torque occurs, in terms of the speed, ω_r, in electrical angle, the parameters of the SynRM, and the constraints, $V_{s\max}$ and $I_{s\max}$.

19. An 11-kW, 6-pole IPMSM with following parameters is controlled in the vector control mode. The maximum phase voltage, $V_{s\max}$, is 170 V and the maximum phase current, $I_{s\max}$, is 60 A. Plot four-quadrant capability curve in the torque–speed plane with given current and voltage constraints by using the following procedures. In the following questions, the voltage drop at the stator resistance can be neglected.

$$R_s = 0.109\,\Omega, \quad L_d = 3.6\,\text{mH}, \quad L_q = 4.5\,\text{mH}, \quad \lambda_f = 0.2595\,\text{Wb-t}$$

(1) Find maximum torque in the region where only current constraint applies.

(2) What is the maximum speed (r/min) where the torque found at part 1 can be maintained?

(3) At 5000 r/min, find the maximum torque while keeping the constraints.

(4) Find the maximum operating speed in r/min while keeping the constraints. In this calculation, load torque and the friction torque can be neglected.

20. Derive (5.65) and (5.69) in Section 5.4.2.2.

21. Derive (5.94) and (5.96) in Section 5.4.4.5.

REFERENCES

1. D. W. Novotny et al., *Vector Control and Dynamics of AC Drives*, Clarendon Press, Oxford, 1996, Chapters 5 and 6.
2. W. Leonhard, *Control of Electrical Drives*, 2nd edition, Springer, Berlin, 1996, Section 12.2.2.
3. P. Vas, *Parameter Estimation, Condition Monitoring, and Diagnosis of Electrical Machines*, Clarendon Press, Oxford, 1993.
4. T. M. Rowan et al., A simple on-line adaptation for indirect field orientation of an induction machine, *IEEE Trans. Ind. Appl.*, Vol. 27, No. 4, July/August 1991, pp. 720–727.
5. P. L. Jansen et al., A physically insightful approach to the design and accuracy assessment of flux observers for field oriented induction machine drives, *IEEE Trans. Ind. Appl.*, Vol. 30, No. 1, January/February 1994, pp. 101–110.
6. J. Kim, Novel rotor flux observer using observer characteristic function in complex vector space for field orientated induction motor drive, *IEEE Trans. Ind. Appl.*, Vol. 38, September/October 2002, pp. 1334–1343.
7. S. Kim, Maximum torque control of induction machine in field weakening region, *IEEE Trans. Ind. Appl.*, Vol. 31, July/August 1995, pp. 787–794.
8. J. Jang,Sensorless control of PMSMs for wide speed operation range, Ph.D. thesis, Seoul National University, Seoul, Korea, 2006 (in Korean).
9. J. Song,Flux weakening control of surface mounted permanent magnet synchronous motor, Ph.D. thesis, Seoul National University, Seoul, Korea, 1996 (in Korean).

10. R. F. Schiferl et al., Power capability of salient pole permanent magnet synchronous motors in variable speed drive applications, *IEEE Trans. Ind. Appl.*, Vol. 26, January/February 1990, pp. 115–123.
11. A. K. Adnanes, Torque analysis of permanent magnet synchronous motors, in Conference Record of the IEEE-PESC, 1991, pp. 695–701.
12. W. L. Soong et al., Field-weakening performance of brushless synchronous AC motor drives, *IEE Proc.-B*, Vol. 141, November 1994, pp. 331–340.
13. J. Song et al., Torque maximizing control of permanent magnet synchronous motor under voltage and current limitations of PWM inverter, in Conference Record of the IEEE-APEC, Vol. 2, 1996, pp. 758–763.
14. J. Kim et al., Improved dynamic performance of interior permanent magnet synchronous motor drive in flux-weakening operation, in Conference Record of the IEEE-PESC, Vol. 2, 1996, pp. 1562–1567.
15. J. Song et al., A new robust SPMSM control to parameter variations in flux weakening region, in Proceedings, IEEE-IECON, Vol. 2, 1996, pp. 1193–1198.
16. J. Holtz et al., On continuous control of PWM inverters in the over-modulation range including the six-step mode, *IEEE Trans. Power Electron.*, Vol. 4, October, 1993, pp. 546–553.
17. T. Kwon et al., Novel anti-windup of a current regulators of a surface-mounted permanent-magnet synchronous motor for flux-weakening control, *IEEE Trans. Ind. Appl.*, Vol. 42, September/October 2006, pp. 1293–1300.
18. G. Choi et al., Novel flux weakening control of an IPMSM for quasi six-step operation, in Conference Record of the IEEE-IAS Annual Meeting, 2007.
19. J. Seok et al., Pseudorotor-flux-oriented control of an induction machine for deep-bar-effect compensation, *IEEE Trans. Ind. Appl.*, Vol. 34, May/June 1998, pp. 429–434.

Chapter 6

Position/Speed Sensorless Control of AC Machines

V/F control for the variable-speed control of AC machines described in Section 2.17.2 has been widely used because of its merits such as no need of accurate machine parameters and no requirement of position/speed sensors. But the performance of the drive system is poor, especially at the start or at the low-speed operation. Moreover, the instantaneous torque regulation by V/F control is almost impossible, and its application is limited to mainly fan and pump drives, where accurate speed and torque control is not a concern. In the case of V/F control with the synchronous machine, the machine can be easily out of synchronization because of no rotor position information, while the vector control of AC machines as described in Chapter 5 can provide instantaneous torque controllability even at zero speed. However, to implement the vector control including zero speed, not only accurate machine parameters but also the position sensor is inevitable [1].

The position/speed sensors described in Section 4.3.1 may result in many practical problems such as complexity of hardware, difficulties in application to hostile environment, increased cost, reduced reliability due to cables and sensor itself, difficulties of mechanical attachment of sensor to the electric machine, increased axial length of the machine, and electromagnetic noise interference. To solve or lessen these problems, for the past 40 years, lots of studies about sensorless control of AC machines have been done [2]. In the 1970s, most studies were based on the steady-state equivalent circuit, mentioned in Chapter 2, and the torque and speed control performance had been limited. And the performance is slightly better than the performance of conventional V/F control drive [3, 4]. In the 1980s, after the development of self-commutated power semiconductors and power electronics technology, the sensorless control algorithm appeared, whose performance is comparable to the performance of sensored AC machine drive system above a certain speed. These algorithms are based on stator voltage and rotor current model of AC machines, but still the performance at the low-speed operation region is poor [5–7]. In

Control of Electric Machine Drive Systems, by Seung-Ki Sul

the 1990s, modern control theory such as adaptive control, state feedback control, and nonlinear control are applied to sensorless control of AC machine and the performance of sensorless control had been remarkably improved and the AC machine drive system employing the above control theories had been commercialized.

In conventional sensorless control algorithms, the rotor position or the rotor flux linkage position has been estimated based on the stator voltage equation of the AC machine [8–13], based on the reference model of the AC machine [9–11], based on the state estimator [14–25], or based on modern control theory such as Kalman filtering, nonlinear feedback, and intelligent control [26–34]. In these algorithms, it should be noted that the accuracy of voltage/current measurement and that of AC machine parameters are critical to the performance of sensorless drive [35–39]. The parameters of AC machines vary according to operating conditions as discussed in Sections 2.9 and 2.12. Hence, the conventional sensorless control algorithm cannot provide satisfactory performance in overall operating conditions. In particular, because the voltage to an AC machine decreases as the speed decreases, the performance of the sensorless control degrades rapidly at low speed. And at zero speed and/or at zero frequency operation, an AC machine cannot run in sensorless operation maintaining torque controllability.

In Fig. 6.1, the operating region of a vector-controlled induction machine drive system is shown. If the vector control is implemented ideally, then the instantaneous torque control would be possible in overall speed range where a physical limit such as voltage and current limits allows. However, even if position/speed sensor is employed, the torque control performance at extremely low speed would be unsatisfactory due to the limited resolution of the position/speed sensor itself. Furthermore, if a sensorless control algorithm is used, then the satisfactory operating region is further shrunk. In particular, in the case of regenerative operation of the induction

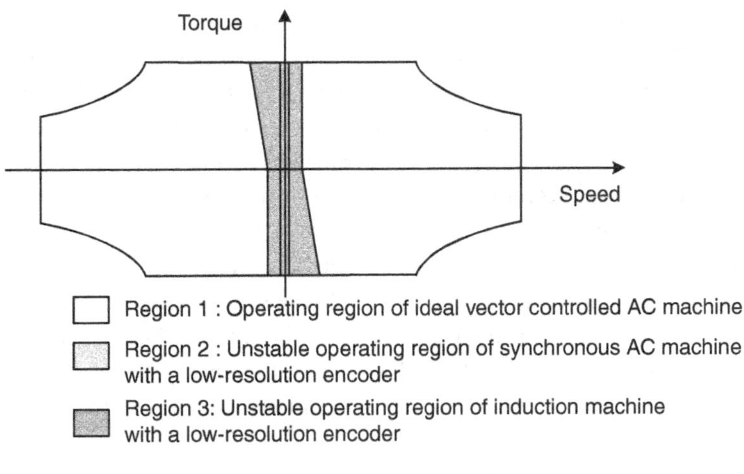

Figure 6.1 Operation region of a vector-controlled AC machine.

machine, the possible operating region of the machine at positive rotational speed is further reduced because of negative slip angular frequency. And the operating region of the induction machines reveals asymmetry in the torque–speed plane, as shown in Fig. 6.1.

Except for the sensitivity to machine parameters, most sensorless control algorithms have two more critical problems as follows. At first, there is a problem due to the integration operation. Usually, the integration operation is used to estimate rotor speed itself or the flux linkage of the machine in the conventional sensorless control algorithm. As mentioned before, the back EMF at low speed is quite low and the integration process of the voltage is vulnerable to the offset of the measured signals. Even with several proposed methods to circumvent this offset problem, it is still very difficult to get reasonable torque control performance at extremely low operating frequency including zero frequency [7, 28]. And next, there is always a mismatch between the reference input voltage, which is used for sensorless control, and the real voltage to the AC machine due to nonlinearity of the drive system itself such as the dead time, voltage drop of switching devices, and stray components of the inverter and the AC machine itself. In particular, the dead time, which is used to prevent an arm short of the inverter, affects severely the mismatch of the voltages as mentioned in Section 7.1. In the range from several-kilowatt to several-hundred-kilowatt IGBT-based inverters, the dead time is several microseconds. If the switching frequency is a few kilohertz, then voltage error due to the dead time alone would be a few percentages of the rated voltage of the AC machine. Hence, if the machine is controlled under a few percentages of the rated speed, then voltage error itself is comparable to the voltage applied to the machine. Hence, the voltage information for sensorless control is useless. With some dead time compensation techniques, the problem can be lessened but cannot be totally removed. The direct measurement of the voltage to the machine may be a solution to the dead time problem, but that results in additional hardware and cost [29].

To escape the above problems inherently, the sensorless control algorithms based on nonideality of the AC machine itself such as eccentricity of rotor, slot harmonics, and electromagnetic unbalance are proposed. But these methods need considerably long calculation time to extract the rotor frequency component from the nonideality, and the signal-to-noise ratio is poor. Hence, the bandwidth of the control loop is too low to be used in most industrial applications. Furthermore, the knowledge about the AC machine such as number of rotor and stator slot or shape of slot is required to apply these algorithms, but that information is not readily accessible [39, 40].

Recently, a novel idea has been proposed for sensorless control of an AC machine, which is based on the magnetic saliency of an AC machine [41–48]. By injecting some signals to an AC machine, the variation of inductance according to the rotor flux position can be measured. From the variation, the position of the rotor flux linkage or rotor position itself can be estimated [49–63, 66–70]. This idea can solve the inherent problem of the conventional sensorless control algorithm based on back EMF of an AC machine. And with the idea, not only speed and torque control of the AC machine but also position control of the AC machine would be possible [60, 61]. In the case of the interior permanent magnet synchronous machine or salient rotor machine, because of inherent saliency the rotor position can be easily identified from the variation of the inductance. Also, the position of the rotor flux linkage of the induction machine and the rotor position of the surface-mounted permanent magnet synchronous machine, where there is no saliency on the rotor, can be identified by injecting proper signals by occurring local magnetic saliency in AC machines [58, 62, 63, 68]. In this chapter, several sensorless control algorithms, which have been used in the industry, are introduced and discussed.

6.1 SENSORLESS CONTROL OF INDUCTION MACHINE

With the specification such as speed control accuracy; ±0.5%, torque control accuracy; ±5%, speed regulation bandwidth; 60 rad/s, minimum controllable speed; 5% of rated speed, several induction machine drive systems employing the position and speed sensorless control with the general purpose inverter have been on the market. With these specifications, in the most of the application, the sensorless drive of induction machine can match to the load requirement. In this section, two sensorless control algorithms which can meet above specifications are introduced.

6.1.1 Model Reference Adaptive System (MRAS) [7]

6.1.1.1 Estimation of Rotor Flux Linkage

The stator voltage and the flux linkage of an induction machine can be described as (6.1)–(6.3) in the complex vector form in the stationary d–q reference frame.

$$\mathbf{V}_s^s = R_s\mathbf{i}_s^s + \frac{d}{dt}\boldsymbol{\lambda}_s^s \tag{6.1}$$

$$\boldsymbol{\lambda}_s^s = L_s\mathbf{i}_s^s + L_m\mathbf{i}_r^s \tag{6.2}$$

$$\boldsymbol{\lambda}_r^s = L_m\mathbf{i}_s^s + L_r\mathbf{i}_r^s \tag{6.3}$$

where R_s stands for the stator winding resistance, L_s stands for the stator self-inductance, L_r stands for the rotor self-inductance, and L_m stands for the mutual inductance.

From (6.3), rotor current can be expressed as

$$\mathbf{i}_r^s = \frac{\boldsymbol{\lambda}_r^s - L_m\mathbf{i}_s^s}{L_r} \tag{6.4}$$

By substituting (6.4) into (6.2), the stator flux linkage can be deduced as

$$\boldsymbol{\lambda}_s^s = L_s\mathbf{i}_s^s + L_m\frac{\lambda_r^s - L_m\mathbf{i}_s^s}{L_r} = \frac{L_m}{L_r}\boldsymbol{\lambda}_r^s + \left(L_s - \frac{L_m^2}{L_r}\right)\mathbf{i}_s^s = \frac{L_m}{L_r}\boldsymbol{\lambda}_r^s + \sigma L_s\mathbf{i}_s^s \tag{6.5}$$

where $\sigma L_s = L_s - L_m^2/L_r$ and it is called the stator transient inductance.

By substituting (6.5) to (6.1), the stator voltage can be expressed in terms of the stator current and rotor flux linkage as

$$\mathbf{V}_s^s = R_s\mathbf{i}_s^s + \frac{d}{dt}\left(\frac{L_m}{L_r}\lambda_r^s + \sigma L_s\mathbf{i}_s^s\right) \tag{6.6}$$

From (6.6), the rotor flux linkage can be deduced as

$$\hat{\boldsymbol{\lambda}}_r^s = \frac{L_r}{L_m}\left(\int_0^t \left(\mathbf{V}_s^s - R_s\mathbf{i}_s^s\right)d\tau - \sigma L_s\mathbf{i}_s^s\right) \tag{6.7}$$

where $\hat{\boldsymbol{\lambda}}_r^s$ represents the estimated value of the rotor flux linkage, $\boldsymbol{\lambda}_r^s$.

If the rotor flux linkage is estimated by (6.7), then even with small offset included in the measured voltage and current the estimated flux linkage would diverge as time passes. To circumvent this problem, a high-pass filter (HPF) can be applied to (6.7). Because of the cascade connection of integrator and HPF, the rotor flux linkage can be approximated as a low-pass filtering of the rotor voltage, $\frac{L_r}{L_m}\left(\mathbf{V}_s^s - (R_s + \sigma L_s s)\mathbf{i}_s^s\right)$.

$$\hat{\boldsymbol{\lambda}}_r^s = \frac{T_c s}{1+T_c s}\frac{L_r}{L_m}\left(\frac{1}{s}\left(\mathbf{V}_s^s - R_s\mathbf{i}_s^s\right) - \sigma L_s\mathbf{i}_s^s\right) = \frac{T_c}{1+T_c s}\frac{L_r}{L_m}\left(\mathbf{V}_s^s - (R_s + \sigma L_s s)\mathbf{i}_s^s\right) \tag{6.8}$$

where T_c is the inverse of the cutoff frequency of HPF, ω_c—that is, $T_c = 1/\omega_c$—and s denotes Laplace operator.

In this case, with constant voltage offset $\left(\mathbf{V}_{s\_o}^s = V_{ds\_o}^s + jV_{qs\_o}^s\right)$ and with constant current offset $\left(\mathbf{i}_{s\_o}^s = i_{ds\_o}^s + ji_{qs\_o}^s\right)$, the steady-state error of the estimated rotor flux linkage due to the offsets can be deduced as

$$\lim_{t\to\infty}\hat{\boldsymbol{\lambda}}_{r\_o}^s(t) = \lim_{s\to 0} s\hat{\boldsymbol{\lambda}}_{r\_o}^s(s) = \lim_{s\to 0}\left(s\frac{T_c}{1+T_c s}\frac{L_r}{L_m}\left(\frac{\mathbf{V}_{s\_o}^s}{s} - R_s\frac{\mathbf{i}_{s\_o}^s}{s}\right)\right)$$

$$= T_c\frac{L_r}{L_m}\left(\mathbf{V}_{s\_o}^s - R_s\mathbf{i}_{s\_o}^s\right) \tag{6.9}$$

It can be seen from (6.9) that with smaller T_c the offsets result in less error in the estimation of the flux linkage. Also, from (6.9), if there is no HPF, which means $T_c = \infty$, then the flux linkage would diverge even with very small offsets. Because the phase error and magnitude characteristics of HPF can be described as (6.10) and (6.11), respectively, and the phase error would decrease as the operating frequency increases and/or T_c of HPF increases. Also, as the operating frequency

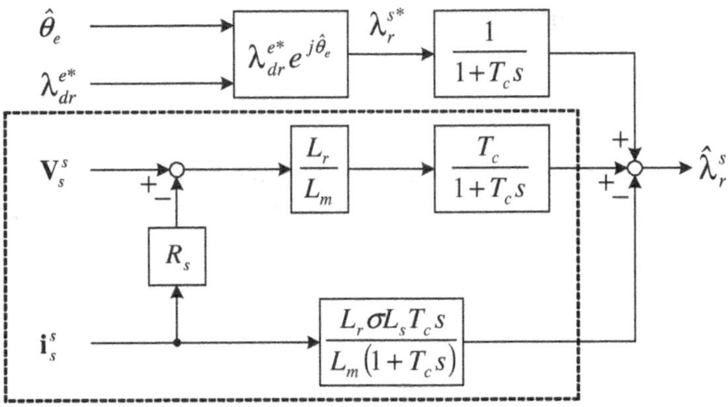

Figure 6.2 Block diagram of estimator of rotor flux linkage of induction machine.

increases and/or T_c of HPF increases, the magnitude approaches to unity.

$$\delta = \tan^{-1}\left(\frac{1}{T_c|\omega_e|}\right) \tag{6.10}$$

$$\text{Magnitude} = \left|\frac{T_c\omega_e}{\sqrt{1+(T_c\omega_e)^2}}\right| \tag{6.11}$$

Hence, T_c should be set as large as possible to reduce estimation error but small enough to keep the estimator from diverging due to the offsets. With this HPF, the performance of sensorless drive in the operation at above a few percentage of rated speed can be improved. Further improvement of the flux estimator is possible with augment of reference of the rotor flux linkage as given by (6.12), which is shown in Fig. 6.2 as a block diagram. With this modification, the starting performance may be improved. But still the torque controllability at lower operating frequency than the cutoff frequency of HPF cannot be improved.

$$\hat{\boldsymbol{\lambda}}_r^s = \frac{T_c}{1+T_c s}\frac{L_r}{L_m}\left(\mathbf{V}_s^s - (R_s + \sigma L_s s)\mathbf{i}_s^s\right) + \frac{1}{1+T_c s}\boldsymbol{\lambda}_r^{s*} \tag{6.12}$$

where $\boldsymbol{\lambda}_r^{s*} = \lambda_{dr}^{s*} + j\lambda_{dr}^{s*} = \lambda_{dr}^{e*}\cos\hat{\theta}_e + j\lambda_{dr}^{e*}\sin\hat{\theta}_e$, and $\hat{\theta}_e$ is the estimated angle of rotor flux linkage.

From (6.12), it can be seen that under the assumption of correct estimated angle and perfect match between $\boldsymbol{\lambda}_r^{s*}$ and $\boldsymbol{\lambda}_r^s = \frac{L_r}{L_m}\left\{\frac{1}{s}\left(\mathbf{V}_s^s - R_s\mathbf{i}_s^s\right) - \sigma L_s\mathbf{i}_s^s\right\}$, the error due to HPF is ideally compensated. That can be easily understood by (6.13) and (6.14).

$$\hat{\boldsymbol{\lambda}}_{r\_vm}^s = \frac{L_r}{L_m}\left(\frac{1}{s}\left(\mathbf{V}_s^s - R_s\mathbf{i}_s^s\right) - \sigma L_s\mathbf{i}_s^s\right) \tag{6.13}$$

where $\hat{\boldsymbol{\lambda}}_{r\_vm}^s$ is the estimated rotor flux linkage by the stator voltage model only.

Then, (6.12) can be rewritten as

$$\hat{\boldsymbol{\lambda}}_r^s = \frac{T_c s}{1+T_c s}\hat{\boldsymbol{\lambda}}_{r\_vm}^s + \frac{1}{1+T_c s}\boldsymbol{\lambda}_r^{s*} \tag{6.14}$$

From (6.14), if $\hat{\boldsymbol{\lambda}}_{r\_vm}^s$ equals $\hat{\boldsymbol{\lambda}}_r^{s*}$, then there is no error due to HPF in the overall frequency range.

The blocks inside of the dotted line in Fig. 6.2 can be implemented using analog hardware to reduce offset and delay due to A/D conversion and digital signal processing.

6.1.1.2 Calculation of Rotor Flux Linkage Angle

To implement the estimator in Fig. 6.2, the control angle, $\hat{\theta}_e$, which is used for coordinate transformation, should be calculated. In Fig. 6.3, $\hat{\theta}_e$ is the angle used in the estimator and $\hat{\theta}_e^*$ is the angle of the rotor flux linkage. For sensorless control, it should be controlled that $\hat{\theta}_e$ tracks $\hat{\theta}_e^*$. In Fig. 6.3, γ^* is the angle between the stator current complex space vector, $\mathbf{i}_s^s$, and the d axis of estimated synchronous reference frame. γ is the angle between estimated rotor flux linkage and the stator current vector. i_{qs}^{e*} is the q-axis component current of the stator current, which is defined as the synchronous reference frame on which the estimator is based. The q-axis component current can be expressed as

$$i_{qs}^{e*} = |\mathbf{i}_s^s|\sin\gamma^* \tag{6.15}$$

In Fig. 6.3, i_{qs}^e stands for the q-axis component current of the stator current defined at the estimated rotor flux axis. And it can be expressed as

$$i_{qs}^{e} = |\mathbf{i}_s^s|\sin\gamma \tag{6.16}$$

To match both angles, $\hat{\theta}_e$ and $\hat{\theta}_e^*$, γ^* should match γ. From (6.15) and (6.16) it means that i_{qs}^{e*} should match to i_{qs}^e.

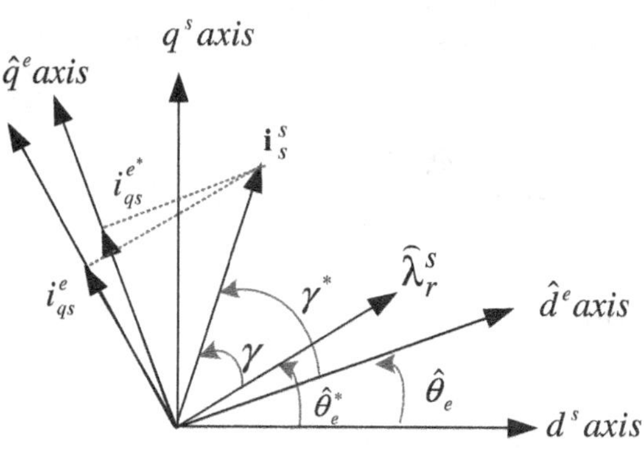

Figure 6.3 Estimated rotor flux linkages and relationship between angles.

Equation (6.16) can be rewritten in terms of d–q components of the stator current and estimated rotor flux linkage at a stationary reference frame as (6.20) by using the following procedures.

$$i_{qs}^{e} = \left|\mathbf{i}_s^s\right|\sin\gamma = \left|\mathbf{i}_s^s\right|\sin\left(\left(\gamma+\hat{\theta}_e^*\right)-\hat{\theta}_e^*\right) = \left|\mathbf{i}_s^s\right|\left(\sin\left(\gamma+\hat{\theta}_e^*\right)\cos\hat{\theta}_e^*-\cos\left(\gamma+\hat{\theta}_e^*\right)\sin\hat{\theta}_e^*\right) \tag{6.17}$$

Also,

$$\begin{aligned}\left|\mathbf{i}_s^s\right|\cos\left(\gamma+\hat{\theta}_e^*\right) &= i_{ds}^s\\ \left|\mathbf{i}_s^s\right|\sin\left(\gamma+\hat{\theta}_e^*\right) &= i_{qs}^s\end{aligned} \tag{6.18}$$

where

$$\begin{aligned}\cos\left(\hat{\theta}_e^*\right) &= \frac{\hat{\lambda}_{dr}^s}{\sqrt{\left(\hat{\lambda}_{dr}^s\right)^2+\left(\hat{\lambda}_{qr}^s\right)^2}}\\ \sin\left(\hat{\theta}_e^*\right) &= \frac{\hat{\lambda}_{qr}^s}{\sqrt{\left(\hat{\lambda}_{dr}^s\right)^2+\left(\hat{\lambda}_{qr}^s\right)^2}}\end{aligned} \tag{6.19}$$

By substituting (6.18) and (6.19) into (6.17), (6.20) can be obtained.

$$i_{qs}^{e} = \frac{i_{qs}^s\hat{\lambda}_{dr}^s - i_{ds}^s\hat{\lambda}_{qr}^s}{\sqrt{\left(\hat{\lambda}_{dr}^s\right)^2+\left(\hat{\lambda}_{qr}^s\right)^2}} \tag{6.20}$$

In Fig. 6.4, a block diagram is shown to calculate the control angle from the estimated rotor flux linkage and the stator current in the stationary reference frame. From the difference between i_{qs}^{e*} and i_{qs}^{e}, the estimated rotor speed is obtained through a proportional and integral (PI) regulator. The synchronous speed is calculated by addition of estimated slip speed, $\hat{\omega}_{sl}$, and the estimated rotor speed. Then, the control angle, $\hat{\theta}_e$, is derived by integrating the synchronous speed. In these calculations, the

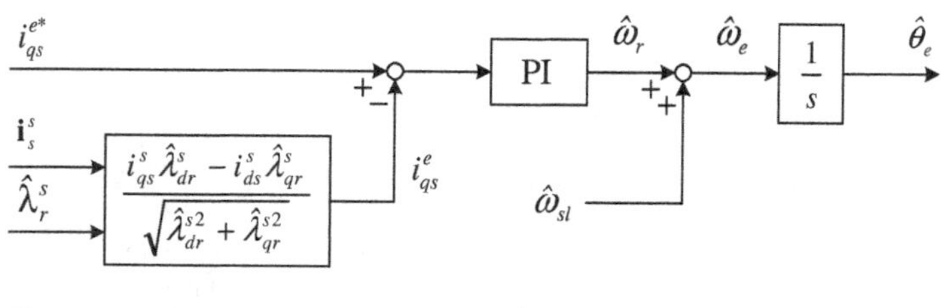

Figure 6.4 Calculation of control angle, $\hat{\theta}_e$.

estimated slip speed can be deduced as (6.21), which comes from slip relationship of indirect vector control of the induction machine as discussed in Section 5.2.2.1.

$$\hat{\omega}_{sl} = \frac{1}{\hat{\tau}_r}\frac{\hat{L}_m}{\lambda_{dr}^{e*}} i_{qs}^{e*} \tag{6.21}$$

where $\hat{\tau}_r$ is the estimated rotor time constant, $\hat{L}_r/\hat{R}_r$.

6.1.2 Adaptive Speed Observer (ASO) [22]

6.1.2.1 State Equation of an Induction Machine

If the rotor current equation in (6.4) expressed in terms of the stator current and the rotor flux linkage is substituted into the rotor voltage equation in (6.22), the rotor voltage equation can be expressed in terms of the stator current and the rotor flux linkage as (6.23).

$$0 = R_r \mathbf{i}_r^s + \frac{d}{dt}\boldsymbol{\lambda}_r^s - j\omega_r \boldsymbol{\lambda}_r^s \tag{6.22}$$

$$0 = R_r\left(\frac{1}{L_r}\boldsymbol{\lambda}_r^s - \frac{L_m}{L_r}\mathbf{i}_s^s\right) + \frac{d}{dt}\boldsymbol{\lambda}_r^s - j\omega_r \boldsymbol{\lambda}_r^s \tag{6.23}$$

From (6.23), the time differentiation of the rotor flux linkage can be represented in terms of stator current and rotor flux linkage itself as

$$\frac{d}{dt}\boldsymbol{\lambda}_r^s = R_r\frac{L_m}{L_r}\mathbf{i}_s^s - \frac{R_r}{L_r}\boldsymbol{\lambda}_r^s + j\omega_r \boldsymbol{\lambda}_r^s \tag{6.24}$$

By substituting (6.4) into (6.2), the stator flux linkage can be represented in terms of the stator current and rotor flux linkage as

$$\begin{aligned}\boldsymbol{\lambda}_s^s &= L_s\mathbf{i}_s^s + L_m\mathbf{i}_r^s = L_s\mathbf{i}_s^s + L_m\left(-\frac{L_m}{L_r}\mathbf{i}_s^s + \frac{1}{L_r}\boldsymbol{\lambda}_r^s\right)\\ &= \left(L_s - \frac{L_m^2}{L_r}\right)\mathbf{i}_s^s + \frac{L_m}{L_r}\boldsymbol{\lambda}_r^s = \sigma L_s\mathbf{i}_s^s + \frac{L_m}{L_r}\boldsymbol{\lambda}_r^s\end{aligned} \tag{6.25}$$

By substituting (6.25) into (6.1), the stator voltage can be expressed in terms of the stator current and rotor flux linkage as

$$\begin{aligned}\mathbf{V}_s^s &= R_s\mathbf{i}_s^s + \frac{d}{dt}\boldsymbol{\lambda}_s^s = R_s\mathbf{i}_s^s + \sigma L_s\frac{d}{dt}\mathbf{i}_s^s + \frac{L_m}{L_r}\frac{d}{dt}\boldsymbol{\lambda}_r^s\\ &= R_s\mathbf{i}_s^s + \sigma L_s\frac{d}{dt}\mathbf{i}_s^s + \frac{L_m}{L_r}\left(R_r\frac{L_m}{L_r}\mathbf{i}_s^s - \frac{R_r}{L_r}\boldsymbol{\lambda}_r^s + j\omega_r\boldsymbol{\lambda}_r^s\right)\\ &= \left\{R_s + R_r\left(\frac{L_m}{L_r}\right)^2\right\}\mathbf{i}_s^s + \sigma L_s\frac{d}{dt}\mathbf{i}_s^s - R_r\frac{L_m}{L_r^2}\boldsymbol{\lambda}_r^s + \left(\frac{L_m}{L_r}\right)j\omega_r\boldsymbol{\lambda}_r^s\end{aligned} \tag{6.26}$$

Equation (6.26) can be rewritten in the form of a differential equation regarding the stator current as

$$\frac{d}{dt}\mathbf{i}_s^s = -\frac{\left\{R_s + R_r\left(\frac{L_m}{L_r}\right)^2\right\}}{\sigma L_s}\mathbf{i}_s^s + \frac{\left(R_r\frac{L_m}{L_r^2}\right)}{\sigma L_s}\boldsymbol{\lambda}_r^s - \frac{L_m}{\sigma L_s L_r}j\omega_r\boldsymbol{\lambda}_r^s + \frac{1}{\sigma L_s}\mathbf{V}_s^s \tag{6.27}$$

By the leakage factor, σ, defined as (6.28), and the rotor time constant given by (6.29), the coefficients of the two differential equations, (6.24) and (6.27), can be rewritten as (6.30), (6.31), and (6.32).

$$\sigma = 1 - \frac{L_m^2}{L_s L_r} \tag{6.28}$$

$$\tau_r = \frac{L_r}{R_r} \tag{6.29}$$

$$\frac{\left\{R_s + R_r\left(\frac{L_m}{L_r}\right)^2\right\}}{\sigma L_s} = \frac{R_s}{\sigma L_s} + \frac{\frac{L_m^2}{L_s L_r}\frac{R_r}{L_r}}{\sigma} = \frac{R_s}{\sigma L_s} + \frac{(1-\sigma)\frac{1}{\tau_r}}{\sigma} = \frac{R_s}{\sigma L_s} + \frac{(1-\sigma)}{\sigma\tau_r} \tag{6.30}$$

$$\frac{\left(R_r\frac{L_m}{L_r^2}\right)}{\sigma L_s} = \frac{L_m\frac{R_r}{L_r}}{\sigma L_s L_r} = \frac{L_m\frac{1}{\tau_r}}{\sigma L_s L_r} = \frac{L_m}{\sigma L_s L_r}\frac{1}{\tau_r} \tag{6.31}$$

$$R_r\frac{L_m}{L_r} = L_m\frac{R_r}{L_r} = \frac{L_m}{\tau_r} \tag{6.32}$$

By the coefficients, the differential equation regarding two states, stator current and rotor flux linkage, can be expressed as (6.33) and (6.34).

$$\frac{d}{dt}\mathbf{i}_s^s = -\left\{\frac{R_s}{\sigma L_s} + \frac{(1-\sigma)}{\sigma\tau_r}\right\}\mathbf{i}_s^s + \frac{L_m}{\sigma L_s L_r}\frac{1}{\tau_r}\boldsymbol{\lambda}_r^s - \frac{L_m}{\sigma L_s L_r}j\omega_r\boldsymbol{\lambda}_r^s + \frac{1}{\sigma L_s}\mathbf{V}_s^s \tag{6.33}$$

$$\frac{d}{dt}\boldsymbol{\lambda}_r^s = \frac{L_m}{\tau_r}\mathbf{i}_s^s - \frac{1}{\tau_r}\boldsymbol{\lambda}_r^s + j\omega_r\boldsymbol{\lambda}_r^s \tag{6.34}$$

Equations (6.33) and (6.34) can be represented in state space equation as (6.35) and (6.36) if the state vector, $\mathbf{X}$, the stator current vector, $\mathbf{i}_s$, and the rotor flux linkage vector, $\boldsymbol{\lambda}_r$, is expressed in vector form as (6.36).

$$\frac{d}{dt}\begin{bmatrix}\mathbf{i}_s\\ \boldsymbol{\lambda}_r\end{bmatrix} = \begin{bmatrix}\mathbf{A}_{11} & \mathbf{A}_{12}\\ \mathbf{A}_{21} & \mathbf{A}_{22}\end{bmatrix}\begin{bmatrix}\mathbf{i}_s\\ \boldsymbol{\lambda}_r\end{bmatrix} + \begin{bmatrix}\mathbf{B}_1\\ 0\end{bmatrix}\mathbf{V}_s = \mathbf{AX} + \mathbf{BV}_s \tag{6.35}$$

$$\mathbf{i}_s = \mathbf{CX}$$

$$\mathbf{X} = \begin{bmatrix} i_{ds}^s & i_{qs}^s & \lambda_{dr}^s & \lambda_{qr}^s \end{bmatrix}^T \tag{6.36}$$

where, $\mathbf{i}_s = \begin{bmatrix} i_{ds}^s & i_{qs}^s \end{bmatrix}^T$ and $\boldsymbol{\lambda}_r = \begin{bmatrix} \lambda_{ds}^s & \lambda_{qr}^s \end{bmatrix}^T$.

$$\mathbf{A}_{11} = -\left\{ \frac{R_s}{\sigma L_s} + \frac{(1-\sigma)}{\sigma \tau_r} \right\} \mathbf{I} = a_{r11}\mathbf{I}$$

$$\mathbf{A}_{12} = \frac{L_m}{\sigma L_s L_r} \left\{ \left(\frac{1}{\tau_r} \right) \mathbf{I}\text{-}\omega_r \mathbf{J} \right\} = a_{r12}\mathbf{I} + a_{i12}\mathbf{J}$$

$$\mathbf{A}_{21} = \frac{L_m}{\tau_r}\mathbf{I} = a_{r21}\mathbf{I}$$

$$\mathbf{A}_{22} = -\left(\frac{1}{\tau_r} \right) \mathbf{I} + \omega_r \mathbf{J} = a_{r22}\mathbf{I} + a_{i22}\mathbf{J}$$

$$\mathbf{B}_1 = \frac{1}{\sigma L_s}\mathbf{I} = b_1 \mathbf{I}$$

$$\mathbf{C} = [\mathbf{I} \quad \mathbf{0}]$$

In the above matrix equations, we have

$$\mathbf{I} = \begin{bmatrix} 1 & 0 \\ 0 & 1 \end{bmatrix}$$

$$\mathbf{J} = \begin{bmatrix} 0 & -1 \\ 1 & 0 \end{bmatrix}$$

Also, the elements in the matrix are as follows.

$$a_{r11} = -\left\{ \frac{R_s}{\sigma L_s} + \frac{(1-\sigma)}{\sigma \tau_r} \right\}, \qquad a_{r12} = \frac{L_m}{\sigma L_s L_r} \cdot \frac{1}{\tau_r} = \frac{1}{c} \cdot \frac{1}{\tau_r},$$

$$a_{i12} = -\frac{L_m}{\sigma L_s L_r}\omega_r = -\frac{1}{c}\omega_r$$

$$a_{r21} = \frac{L_m}{\tau_r}, \quad a_{r22} = -\frac{1}{\tau_r}, \quad a_{i22} = \omega_r, \quad b_1 = \frac{1}{\sigma L_s}, \quad c = \frac{\sigma L_s L_r}{L_m}$$

Through the state observer based on the state equation in (6.35), the rotating speed of the induction machine can be estimated.

6.1.2.2 State Observer

The closed-loop stator observer to estimate both the stator current and rotor flux linkage can be deduced as

$$\frac{d}{dt}\hat{\mathbf{X}} = \hat{\mathbf{A}}\hat{\mathbf{X}} + \mathbf{B}\mathbf{V}_s + \mathrm{G}(\hat{\boldsymbol{i}}_s - \boldsymbol{i}_s) \tag{6.37}$$

where the hat (^) stands for the estimated value and **G** means an observer gain matrix represented as

$$\mathbf{G} = \begin{bmatrix} g_1 & g_2 & g_3 & g_4 \\ -g_2 & g_1 & -g_4 & g_3 \end{bmatrix}^T \tag{6.38}$$

Equation (6.37) can be expressed elementwise as

$$\begin{aligned} \frac{d}{dt}\begin{bmatrix} \hat{i}^s_{ds} \\ \hat{i}^s_{qs} \\ \hat{\lambda}^s_{dr} \\ \hat{\lambda}^s_{qr} \end{bmatrix} &= \begin{bmatrix} a_{r11} & 0 & a_{r12} & -\hat{a}_{i12} \\ 0 & a_{r11} & \hat{a}_{i12} & a_{r12} \\ a_{r21} & 0 & a_{r22} & -\hat{a}_{i22} \\ 0 & a_{r12} & \hat{a}_{i22} & a_{r22} \end{bmatrix}\begin{bmatrix} \hat{i}^s_{ds} \\ \hat{i}^s_{qs} \\ \hat{\lambda}^s_{dr} \\ \hat{\lambda}^s_{qr} \end{bmatrix} + \begin{bmatrix} b_1 & 0 \\ 0 & b_1 \\ 0 & 0 \\ 0 & 0 \end{bmatrix}\begin{bmatrix} V^s_{ds} \\ V^s_{qs} \end{bmatrix} \\ &+ \begin{bmatrix} g_1 & -g_2 \\ g_2 & g_1 \\ g_3 & -g_4 \\ g_4 & g_3 \end{bmatrix}\begin{bmatrix} \hat{i}^s_{ds} & -i^s_{ds} \\ \hat{i}^s_{qs} & -i^s_{qs} \end{bmatrix} \end{aligned} \tag{6.39}$$

From (6.35) and (6.37), the dynamics of the state error can be deduced as

$$\begin{aligned} \frac{d}{dt}(\mathbf{X}-\hat{\mathbf{X}}) &= \left(\mathbf{A}\mathbf{X}-\hat{\mathbf{A}}\hat{\mathbf{X}}\right) + \mathbf{G}\left(\mathbf{i}_s - \hat{\mathbf{i}}_s\right) \\ \frac{d}{dt}\mathbf{e} &= (\mathbf{A}\mathbf{X}-(\mathbf{A}+\Delta\mathbf{A})\hat{\mathbf{X}}) + \mathbf{G}\mathbf{C}(\mathbf{X}-\hat{\mathbf{X}}) \\ &= (\mathbf{A}+\mathbf{G}\mathbf{C})\mathbf{e} - \Delta\mathbf{A}\hat{\mathbf{X}} \end{aligned} \tag{6.40}$$

where $\mathbf{e} = \mathbf{X}-\hat{\mathbf{X}}$, $\Delta\mathbf{A} = \hat{\mathbf{A}}-\mathbf{A} = \begin{bmatrix} \mathbf{0} & -\Delta\omega_r\mathbf{J}/c \\ 0 & \Delta\omega_r\mathbf{J} \end{bmatrix}$, $\mathbf{C} = [\mathbf{I} \ \ \mathbf{0}]$, $c = \frac{\sigma L_s L_r}{L_m}$, and $\Delta\omega_r = \hat{\omega}_r - \omega_r$. If a Lyapunov function is defined as (6.41), then the differentiation of the function **V** regarding time, t, can be calculated as

$$\mathbf{V} = \mathbf{e}^T\mathbf{e} + (\hat{\omega}_r - \omega_r)^2/\gamma \tag{6.41}$$

where γ is a positive real number.

$$
\begin{aligned}
\frac{d}{dt}\mathbf{V} &= \frac{d}{dt}\mathbf{e}^T\mathbf{e} + \mathbf{e}^T\frac{d}{dt}\mathbf{e} + 2(\hat{\omega}_r - \omega_r)\frac{d}{dt}\hat{\omega}_r/\gamma \\
&= \mathbf{e}^T\left\{(\mathbf{A}+\mathbf{GC})^T + (\mathbf{A}+\mathbf{GC})\right\}\mathbf{e} - \hat{\mathbf{X}}^T\Delta\mathbf{A}^T\mathbf{e} - \mathbf{e}^T\Delta\,\mathbf{A}\hat{\mathbf{X}} + 2\Delta\omega_r\frac{d}{dt}\hat{\omega}_r/\gamma \\
&= \mathbf{e}^T\left\{(\mathbf{A}+\mathbf{GC})^T + (\mathbf{A}+\mathbf{GC})\right\}\mathbf{e} - 2\Delta\omega_r\left(e_{ids}\hat{\lambda}^s_{qr} - e_{iqs}\hat{\lambda}^s_{dr}\right)/c + 2\Delta\omega_r\frac{d}{dt}\hat{\omega}_r/\gamma
\end{aligned}
\tag{6.42}
$$

where $e_{ids} = i^s_{ds} - \hat{i}^s_{ds}$, and $e_{iqs} = i^s_{qs} - \hat{i}^s_{qs}$.

The adaptive law can be set as (6.43) by nullifying the sum of the second and the third terms on the right-hand side of (6.42).

$$
\frac{d}{dt}\hat{\omega}_r = \frac{\gamma\left(e_{ids}\hat{\lambda}^s_{qr} - e_{iqs}\hat{\lambda}^s_{dr}\right)}{c} \tag{6.43}
$$

If (6.43) holds and the gain matrix, **G**, is set in order that the first term on the right-hand side of (6.42) is nonpositive, then the differentiation of Lyapunov function, **V**, is always nonpositive. This means that the state error would decrease and that the estimated states track the real states. Because the speed of the induction machine may vary rapidly, instead of the integral controller by (6.43) the speed can be estimated through a PI regulator as (6.44) to increase convergence speed of the observer.

$$
\hat{\omega}_r = K_p\left(e_{ids}\hat{\lambda}^s_{qr} - e_{iqs}\hat{\lambda}^s_{dr}\right) + K_i\int\left(e_{ids}\hat{\lambda}^s_{qr} - e_{iqs}\hat{\lambda}^s_{dr}\right)dt \tag{6.44}
$$

6.1.2.3 Gain Matrix, G

The induction machine itself is a stable system, and its all poles lie in the left half-plane of the Laplace domain. And, the poles of overall system can be set as proportional to the poles of the induction machine. To simplify the description of equations the stator equation can be represented in complex form as

$$
\frac{d}{dt}\mathbf{X} = \hat{\mathbf{A}}\mathbf{X} + \mathbf{B}\mathbf{V}_s + \mathbf{G}\left(\hat{\mathbf{i}}_s - \mathbf{i}_s\right) \tag{6.45}
$$

where

$$
\mathbf{X} = \begin{bmatrix}\hat{\mathbf{i}}^s_s & \hat{\boldsymbol{\lambda}}^s_r\end{bmatrix}^T
$$

$$
\hat{\mathbf{i}}^s_s = \hat{i}^s_{ds} + j\hat{i}^s_{qs}
$$

$$
\hat{\boldsymbol{\lambda}}^s_r = \hat{\lambda}^s_{dr} + j\hat{\lambda}^s_{qr}
$$

$$
\mathbf{i}^s_s = i^s_{ds} + ji^s_{qs}
$$

$$
\mathbf{V}^s_s = V^s_{ds} + jV^s_{qs}
$$

$$\hat{\mathbf{A}} = \begin{bmatrix} \hat{a}_{r11} & \hat{a}_{r12} + j\hat{a}_{i12} \\ \hat{a}_{r21} & \hat{a}_{r22} + j\hat{a}_{i22} \end{bmatrix}$$

$$\mathbf{B} = \begin{bmatrix} b_1 \\ 0 \end{bmatrix}$$

$$\mathbf{G} = \begin{bmatrix} g_1 + jg_2 \\ g_3 + jg_4 \end{bmatrix}$$

The poles of the system including the observer can be found by solving the following characteristic equation of the system.

$$\det\left[s\mathbf{I} - \left(\hat{\mathbf{A}} + \mathbf{GC}\right)\right] = \det\begin{bmatrix} s-(\hat{a}_{r11} + g_1 + jg_2) & -(\hat{a}_{r12} + j\hat{a}_{i12}) \\ -(\hat{a}_{r21} + g_3 + jg_4) & s-(\hat{a}_{r22} + j\hat{a}_{i22}) \end{bmatrix} = 0 \quad (6.46)$$

where det[•] means the determinant of the matrix [•].

By using the relationships as $\hat{a}_{r22} = -c\hat{a}_{r12}$, $\hat{a}_{i22} = -c\hat{a}_{i12}$, the characteristics equation in (6.46) can be deduced as

$$s^2 - \{(\hat{a}_{r11} + \hat{a}_{r22} + g_1) + j(\hat{a}_{i22} + g_2)\}s - (\hat{a}_{r12} + j\hat{a}_{i12}) \times \{c\hat{a}_{r11} + cg_1 + \hat{a}_{r21} + g_3 + j(cg_2 + g_4)\} = 0 \quad (6.47)$$

The characteristic equation of the induction machine can be obtained simply making all elements of gain matrix be zero ($g_1 = g_2 = g_3 = g_4 = 0$), and it is given in

$$s^2 - \{(\hat{a}_{r11} + \hat{a}_{r22}) + j\hat{a}_{i22}\}s - (\hat{a}_{r12} + j\hat{a}_{i12})(c\hat{a}_{r11} + \hat{a}_{r21}) = 0. \quad (6.48)$$

From (6.48) the poles of the overall system including the observer, which are proportional to the poles of the induction machine itself, can be easily calculated by solving (6.49).

$$s^2 - k\{(\hat{a}_{r11} + \hat{a}_{r22}) + j\hat{a}_{i22}\}s - k^2(\hat{a}_{r12} + j\hat{a}_{i12})(c\hat{a}_{r11} + \hat{a}_{r21}) = 0 \quad (6.49)$$

where k is a proportional constant. The elements of the gain matrix can be obtained by comparing the coefficients of (6.47) and (6.49). And the elements can be expressed as following equations:

$$g_1 = (k-1)(\hat{a}_{r11} + \hat{a}_{r22}) \quad (6.50)$$

$$g_2 = (k-1)\hat{a}_{i22} \quad (6.51)$$

$$g_3 = (k^2-1)(c\hat{a}_{r11} + \hat{a}_{r21}) - cg_1 = (k^2-1)(c\hat{a}_{r11} + \hat{a}_{r21}) - c(k-1)(\hat{a}_{r11} + \hat{a}_{r22}) \quad (6.52)$$

$$g_4 = -cg_2 = -c(k-1)\hat{a}_{i22} \quad (6.53)$$

By employing the above adaptive speed observer, the reasonable performance of torque/speed control without speed/position sensor can be obtained down to the operating speed, which is 2–3% of the rated speed in the case of power rating less than a few tens of kilowatts. To get better performance at lower operating speed, the real-time machine parameter adaptation and the direct measurement of the machine terminal voltage can be incorporated into the sensorless control algorithm.

6.2 SENSORLESS CONTROL OF SURFACE-MOUNTED PERMANENT MAGNET SYNCHRONOUS MACHINE (SMPMSM)

Many sensorless control algorithms for an SMPMSM had been developed and applied in the industry. In this section, a simple but robust algorithm is introduced, which exploits *d*-axis current error to estimate the rotor angle. In Fig. 6.5, a block diagram of the sensorless control algorithm is shown [64]. This method also relies on the back EMF and the machine model, and torque control at the zero frequency is impossible. Also, the control performance at low speed is directly affected by the accuracy of the machine parameters and that of voltage and current measurement. Hence, the separate starting algorithm is required to start the machine smoothly from the zero speed. In this section, an open-loop-type starting algorithm is augmented to the sensorless control algorithm.

In Fig. 6.5, feed-forwarding terms, $V^r_{ds\_ff}$ and $V^r_{gs\_ff}$, can be represented as (6.54) and (6.55).

$$V^r_{ds\_ff} = \hat{R}_s i^{r*}_{ds} - \hat{L}_s \hat{\omega}_r i^{r*}_{qs} \tag{6.54}$$

$$V^r_{qs\_ff} = \hat{R}_s i^{r*}_{qs} + \hat{L}_s \hat{\omega}_r i^{r*}_{ds} + \hat{\omega}_r \hat{\lambda}_f \tag{6.55}$$

where ω_r is the rotor angular speed in electrical degree and $L_s = L_m + L_{ls}$. In the equations, the hat (^) means the estimated value of the machine parameters, the speed, or angle.

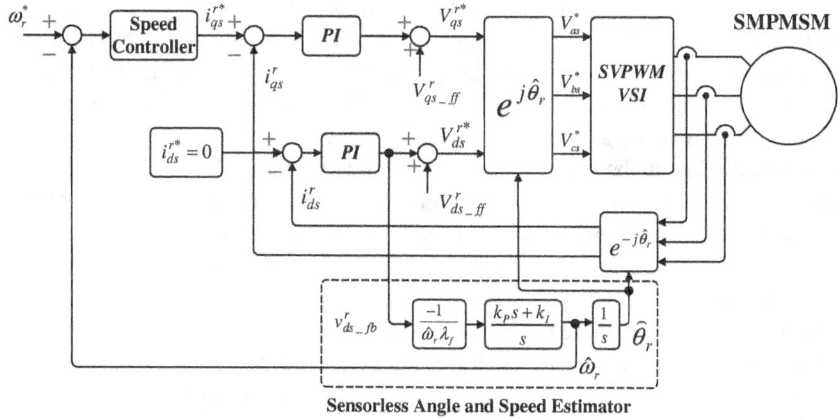

Figure 6.5 Block diagram of sensorless control of an SMPMSM.

If the transient state is neglected, the voltage equations at estimated rotor reference frame can be described as (6.56) and (6.57).

$$V_{ds}^{r} = R_s i_{ds}^{r} - \omega_r L_s i_{qs}^{r} + \omega_r \lambda_f \sin \theta_{err} \tag{6.56}$$

$$V_{qs}^{r} = R_s i_{qs}^{r} + \omega_r L_s i_{ds}^{r} + \omega_r \lambda_f \cos \theta_{err} \tag{6.57}$$

where θ_{err} stands for the angle error definded by

$$\theta_{err} = \hat{\theta}_r - \theta_r \tag{6.58}$$

where θ_r stands for real rotor angle and $\hat{\theta}_r$ for estimated rotor angle.

With the angle error, θ_{err}, the voltage error at the estimated d axis can be deduced as (6.59) under the assumption of the negligible errors for estimated machine parameters and speed such as $\hat{R}_s \approx R_s$, $\hat{L}_s \approx L_s$, and $\hat{\omega}_r \approx \omega_r$. In the case of an SMPMSM control, for MTPA operation the d-axis current is usually controlled as zero and the error in the stator resistance, R_s, does not affect the control performance, whereas the error in the synchronous inductance, L_s, does. However, L_s of an SMPMSM is almost constant regardless of the operating condition because of the equivalent large air gap of the machine due to the surface-mounted permanent magnet. Hence, the error voltage in (6.59) is robust to operating conditions.

$$V_{ds\_error}^{r} = R_s (i_{ds}^{r} - i_{ds}^{r*}) - \omega_r L_s (i_{qs}^{r} - i_{qs}^{r*}) + \omega_r \lambda_f \sin \theta_{err} \tag{6.59}$$

If the currents at d–q axes is well-regulated and θ_{err} is small enough, then the output voltage of PI regulator, which is the d-axis current regulator, can be approximated as

$$V_{ds\_fb}^{r} \approx V_{ds\_error}^{r} \approx \omega_r \lambda_f \sin \theta_{err} \approx \omega_r \lambda_f \theta_{err} \tag{6.60}$$

Based on (6.60), the angle error can be obtained after dividing the output of PI regulator by the estimated speed, $\hat{\omega}_r$, and permanent magnet flux linkage, λ_f. The error can be used as an input to the angle controller, which is implemented as a PI regulator. The output of the angle controller can be the estimated rotor speed; also, the integral of the speed is the estimated rotor angle, and it is used for vector control of an SMPMSM. The PI regulator controls its output to keep the error, which is the input of the PI regulator, at zero. With this sensorless control algorithm, an SMPMSM can be driven without any position/speed sensor above a few percent of the rated speed. The speed regulation bandwidth can be extended to a couple of tens rad/s. Also, even if there are some errors in the estimated rotor speed and the flux linkage, the angle regulator track the rotor angle with reasonable accuracy. However, because of the limited dynamics of the angle regulator, if the rotor angle varies too rapidly, the regulator would fail. As mentioned before, for the starting, a rotating current vector, whose magnitude is fixed as $I_{\max}$, is applied to the machine through the current regulator shown in Fig. 6.6. The speed of the rotating vector can vary as shown in Fig. 6.7. In the figure, for period I the current vector rotates at constant speed, $\omega_{r\_start}^{*}$; and after the rotor flux linkage is synchronized to the rotating current vector, the rotating speed of the current vector increases in the period II. If the rotor speed is above the threshold speed, $\omega_{r\_thres}^{*}$, then

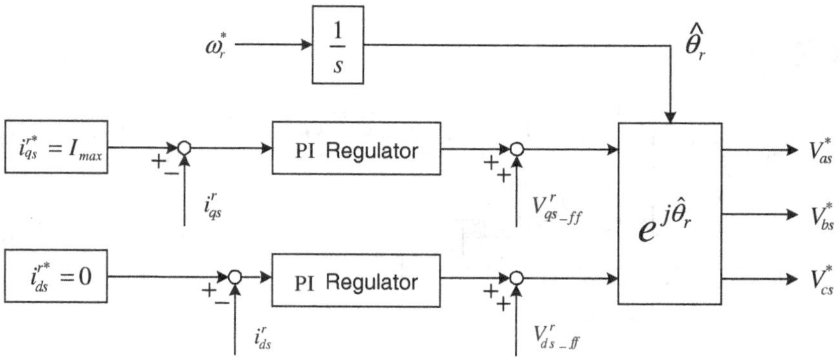

Figure 6.6 Block diagram of current regulator for starting and initial acceleration of an SMPMSM.

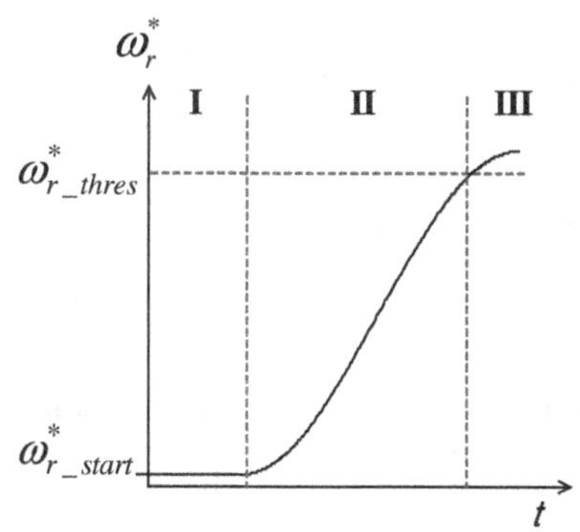

Figure 6.7 Variation of reference speed of the rotating current vector for the starting of an SMPMSM.

the sensorless control algorithm may be engaged in period III, where the magnitude of back EMF is large enough to identify the rotor position. During periods I and II, if sudden impact load is applied to SMPMSM, the rotor can be stalled. And the machine could be out of control.

To incorporate the flux weakening control, the d-axis current reference should be regulated as a negative value. Then the control block diagram in Fig. 6.5 should be modified properly.

6.3 SENSORLESS CONTROL OF INTERIOR PERMANENT MAGNET SYNCHRONOUS MACHINE (IPMSM) [65]

For sensorless control of IPMSM, both d- and q-axis currents should be regulated to achieve maximum torque per ampere (MTPA) operation. Like sensorless control of an SMPMSM, there have been many methods for sensorless control of an IPMSM. One of them is shown in Fig. 6.8, where a closed loop observer is used for the estimation of the rotor position.

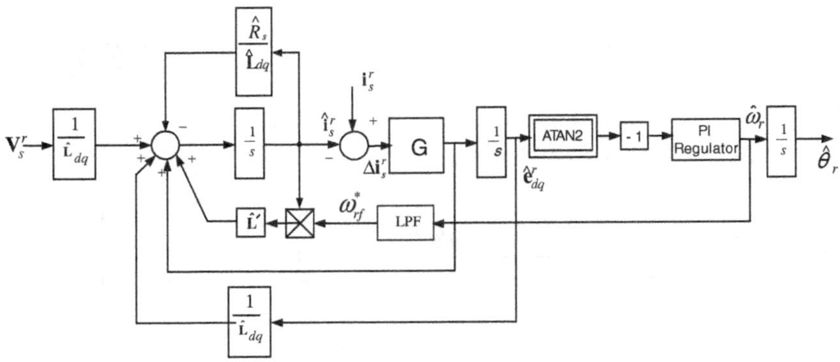

Figure 6.8 Block diagram of the sensorless control of an IPMSM.

In the figure,

$$\frac{1}{\hat{\mathbf{L}}_{dq}} = \begin{bmatrix} \frac{1}{\hat{L}_d} & 0 \\ 0 & \frac{1}{\hat{L}_q} \end{bmatrix}, \quad \hat{\mathbf{L}}' = \begin{bmatrix} 0 & \frac{\hat{L}_q}{\hat{L}_d} \\ -\frac{\hat{L}_d}{\hat{L}_q} & 0 \end{bmatrix}$$

By setting the stator current in estimated rotor reference frame as a state variable, the state-space equation of an IPMSM, whose equivalent circuit is in Fig. 3.10, can be deduced as

$$\frac{d}{dt}\begin{bmatrix} i^r_{ds} \\ i^r_{qs} \end{bmatrix} = \begin{bmatrix} -\frac{R_s}{L_d} & \frac{\omega_r L_q}{L_d} \\ -\frac{\omega_r L_d}{L_q} & -\frac{R_s}{L_q} \end{bmatrix} \begin{bmatrix} i^r_{ds} \\ i^r_{qs} \end{bmatrix} + \begin{bmatrix} \frac{1}{L_d} & 0 \\ 0 & \frac{1}{L_q} \end{bmatrix} \begin{bmatrix} V^r_{ds} \\ V^r_{qs} \end{bmatrix} + \begin{bmatrix} 0 \\ -\frac{\omega_r \lambda_f}{L_q} \end{bmatrix} \tag{6.61}$$

where λ_f is the rotor flux linkage by the permanent magnet.

If the angle error defined as (6.58) is small enough, then the state equation in (6.61) can be rewritten in terms of the angle error as

$$\frac{d}{dt}\begin{bmatrix} i^r_{ds} \\ i^r_{qs} \end{bmatrix} = \begin{bmatrix} -\frac{R_s}{L_d} & \frac{\omega_r L_q}{L_d} \\ -\frac{\omega_r L_d}{L_q} & -\frac{R_s}{L_q} \end{bmatrix} \begin{bmatrix} i^r_{ds} \\ i^r_{qs} \end{bmatrix} + \begin{bmatrix} \frac{1}{L_d} & 0 \\ 0 & \frac{1}{L_q} \end{bmatrix} \begin{bmatrix} V^r_{ds} \\ V^r_{qs} \end{bmatrix} + \begin{bmatrix} -\frac{\omega_r \lambda_f}{L_d} \sin\theta_{err} \\ -\frac{\omega_r \lambda_f}{L_q} \cos\theta_{err} \end{bmatrix} \tag{6.62}$$

If the voltage due to the angle error is set as the new state variable as (6.63) and if the new state is augmented to (6.62), then a closed-loop observer including the augmented state can be described as (6.64) with an observer gain matrix, $\mathbf{G} = \begin{bmatrix} g_{11} & g_{12} \\ g_{21} & g_{22} \\ g_{31} & g_{32} \\ g_{41} & g_{42} \end{bmatrix}$.

$$\mathbf{e}^r_{dq} = \begin{bmatrix} \hat{e}^r_d \\ \hat{e}^r_q \end{bmatrix} = \begin{bmatrix} -\hat{\omega}_r\hat{\lambda}_f \sin\theta_{err} \\ -\hat{\omega}_r\hat{\lambda}_f \cos\theta_{err} \end{bmatrix} \tag{6.63}$$

$$\frac{d}{dt}\begin{bmatrix} \hat{i}^r_{ds} \\ \hat{i}^r_{qs} \\ \hat{e}^r_d \\ \hat{e}^r_q \end{bmatrix} = \begin{bmatrix} -\frac{\hat{R}_s}{\hat{L}_d} & \frac{\hat{\omega}_r\hat{L}_q}{\hat{L}_d} & \frac{1}{\hat{L}_d} & 0 \\ -\frac{\hat{\omega}_r\hat{L}_d}{\hat{L}_q} & -\frac{\hat{R}_s}{\hat{L}_q} & 0 & \frac{1}{\hat{L}_q} \\ 0 & 0 & 0 & 0 \\ 0 & 0 & 0 & 0 \end{bmatrix}\begin{bmatrix} \hat{i}^r_{ds} \\ \hat{i}^r_{qs} \\ \hat{e}^r_d \\ \hat{e}^r_q \end{bmatrix}$$

$$+\begin{bmatrix} \frac{1}{\hat{L}_d} & 0 \\ 0 & \frac{1}{\hat{L}_q} \\ 0 & 0 \\ 0 & 0 \end{bmatrix}\begin{bmatrix} V^r_{ds} \\ V^r_{qs} \end{bmatrix} + \begin{bmatrix} g_{11} & g_{12} \\ g_{21} & g_{22} \\ g_{31} & g_{32} \\ g_{41} & g_{42} \end{bmatrix}\begin{bmatrix} i^r_{ds} - \hat{i}^r_{ds} \\ i^r_{qs} - \hat{i}^r_{qs} \end{bmatrix} \tag{6.64}$$

If all poles of the observer are located at the fixed position on the negative real axis regardless of the rotating speed of an IPMSM as (6.65), then the characteriastic equation of the state equation (6.64) can deduced as (6.66). And the observer gain matrix, **G**, can be found as (6.67).

$$\omega_{1,2} = -\omega_{cd}\left(\zeta_d \pm \sqrt{\zeta_d^2 - 1}\right), \qquad \omega_{3,4} = -\omega_{cq}\left(\zeta_q \pm \sqrt{\zeta_q^2 - 1}\right) \tag{6.65}$$

where ω_{cd}, ω_{cq}, ζ_d, and ζ_q are design variables.

$$\left(s^2 + 2\zeta_d\omega_{cd}s + \omega_{cd}^2\right)\left(s^2 + + 2\zeta_q\omega_{cq}s + \omega_{cq}^2\right) = 0 \tag{6.66}$$

$$\mathbf{G} = \begin{bmatrix} g_{11} & g_{12} \\ g_{21} & g_{22} \\ g_{31} & g_{32} \\ g_{41} & g_{42} \end{bmatrix} = \begin{bmatrix} -\dfrac{\hat{R}_s}{\hat{L}_d} + 2\zeta_d\omega_{cd} & \dfrac{\omega_r\hat{L}_q}{\hat{L}_d} \\ -\dfrac{\omega_r\hat{L}_d}{\hat{L}_q} & -\dfrac{\hat{R}_s}{\hat{L}_q} + 2\zeta_q\omega_{cq} \\ \omega_{cd}^2\hat{L}_d & 0 \\ 0 & \omega_{cq}^2\hat{L}_q \end{bmatrix} \tag{6.67}$$

The estimated rotor speed, $\hat{\omega}_r$, can be calculated from

$$\hat{\omega}_r = \text{sign}\left(-\hat{e}_q^r\right)\sqrt{\left(\hat{e}_d^r\right)^2 + \left(\hat{e}_q^r\right)^2}/\hat{\lambda}_f \tag{6.68}$$

where $\text{sign}\left(-\hat{e}_q^r\right)$ is a signum function, whose value is either 1 or–1 according to the sign of $-\hat{e}_q^r$.

As seen from (6.63), the augumented state, $\mathbf{e}_{dq}^r$, includes the information regarding angle error, and the angle error can be calculated from (6.69). And this angle error could be used directly to compensate the estimated rotor angle.

$$\theta_{err} = \arctan\left(\frac{e_d^r}{e_q^r}\right) \tag{6.69}$$

But to enhance immunity against the measurement noise, as with the case of the sensorless control of SMPMSM, the angle can be obtained indiretly from the integration of the output of a PI regulator, whose input is θ_{err} in (6.69). In Fig. 6.8, this method is shown together with the closed-loop observer. As seen from (6.63), like an SMPMSM, the magnitude of the error voltage, $\mathbf{e}_{dq}^r$, decreases as the speed decreases even at the constant angle error. At zero speed, there is no information regarding angle error in the observer. However, unlike SMPMSM, in the case of an IPMSM the initial rotor position can be identified exploiting the difference of inductance according to the rotor position. By using this inherent phenomenon of IPMSM, an IPMSM can be easily started and can be accelerated slowly up to a certain speed, where the sensorless control algorithm based on back EMF could work [66].

6.4 SENSORLESS CONTROL EMPLOYING HIGH-FREQUENCY SIGNAL INJECTION

The sensorless control methods based on back EMF cannot work at zero frequency, where there is no back EMF. However, by injecting signals to an AC machine, the magnetic saliency according to the rotor position and/or the rotor flux linkage can be detected. And the detected saliency can be used to identify the rotor position. To enhance the control bandwidth, the injected signal should be high enough, but low enough to be manipulated through PWM inverter and its associated controller. The magnetic saliency according to the rotor position is inherently available in the case of a salient rotor AC machine such as IPMSM, synchronous reluctance machine (SynRM), and wound rotor salient synchronous machine, while the induction machine and the SMPMSM have no magnetic saliency. However, an induction machine that has the closed rotor slot or semi-closed rotor slot reveals the magnetic saliency according to the rotor flux linkage due to the local magnetic saturation. And some of the SMPMSM, also, reveals the magnetic saliency according to the rotor position, where the flux linkage of the permanent magnet directs.

The injected signal can be classified as a voltage signal or a current signal as shown in Fig. 6.9. In Fig. 6.9a, a control block diagram to inject a high-frequency voltage signal to an AC machine is shown. In the figure, the injected voltage signal is added to the output of the current regulator. For current regulation the measured current is filtered out through a low-pass filter (LPF) to remove the high-frequency component due to the injected voltage. For identification of rotor position or rotor flux

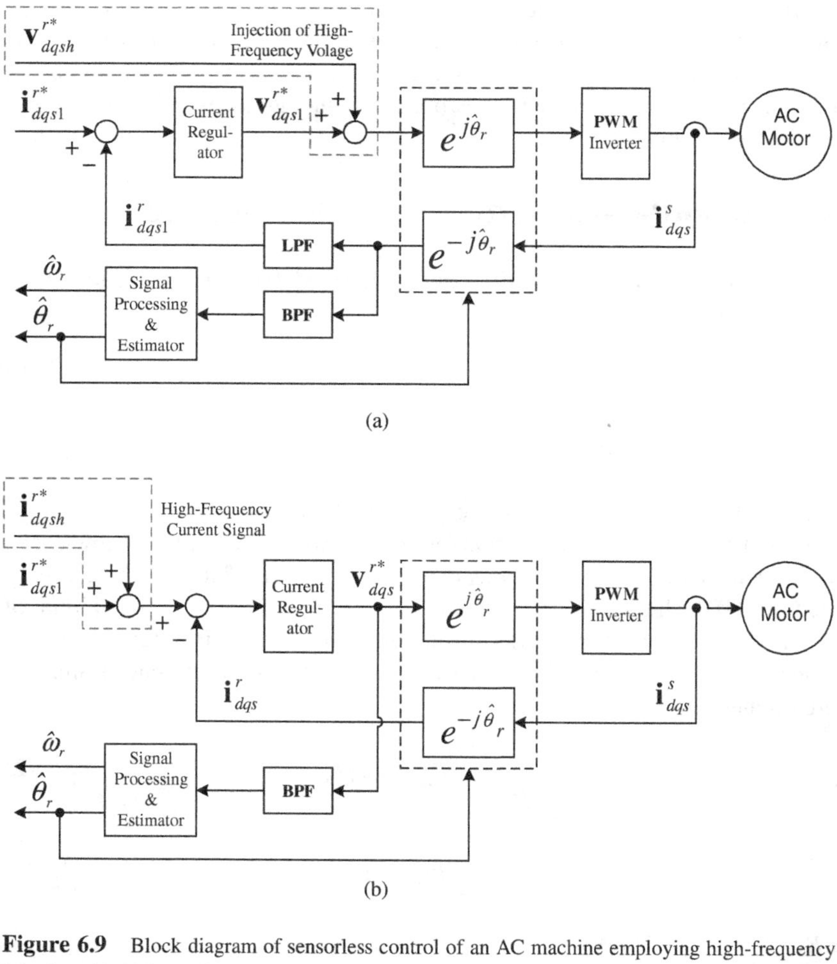

Figure 6.9 Block diagram of sensorless control of an AC machine employing high-frequency signal injection. (a) Voltage signal injection. (b) Current signal injection.

linkage position, the high-frequency components of the current are extracted through a band-pass filter (BPF). In Fig. 6.9b, a control block diagram to inject a high-frequency current signal to an AC machine is shown. Here the high-frequency signal is added to the current reference of the current regulator. For the injection of the high-frequency current signal, the regulation bandwidth of the current regulator should be much higher than the frequency of the injected signal. And it results in burden to hardware and software design of the drive system. Hence, usually the voltage signal injection is preferred.

The high-frequency signal can be classified as a rotating one and a fluctuating one. The former rotates on the stationary reference frame with the frequency of the injected signal. The latter fluctuates on the estimated d or q axis of an AC machine. Though each injection method has its own merits and shortcomings, the fluctuating signal injection is more advantageous in robustness to the measurement noise and multiple saliencies of AC machine [67]. In the case of the fluctuating signal injection, the axis where the signal is injected can be chosen arbitrarily. However, injecting the signal to the estimated d axis of the rotor reference frame in the case of a synchronous machine or estimated d axis of synchronous reference frame in the case of an induction machine is preferable in the viewpoint of reducing losses and torque ripples from the injected signal.

6.4.1 Inherently Salient Rotor Machine [60,61,66]

The voltage equation of both the SynRM and the IPMSM described in Sections 3.3.4 and 3.3.3.2 can be written as

$$V_{ds}^r = R_s i_{ds}^r - \omega_r \lambda_{qs}^r + p\left(L_d i_{ds}^r + \lambda_f\right) \tag{6.70}$$

$$V_{qs}^r = R_s i_{qs}^r + \omega_r \lambda_{ds}^r + p\left(L_q i_{qs}^r\right) \tag{6.71}$$

In the case of the SynRM, because of no permanent magnet, $\lambda_f = 0$ in (6.70).

If the high-frequency signal injection for sensorless control is done at a low-speed region, the voltages in above equations due to the rotation of the machine can be neglected. And if the high-frequency signal is sinusoidal and the equivalent circuit of an AC machine is in the steady state in the viewpoint of the injected signal, the impedances at the high frequency on the d and q axis of the rotor reference frame can be approximated as

$$Z_{dh} \approx \frac{v_{dh}}{i_{dh}} = \sqrt{R_{dh}^2 + (\omega_h L_{dh})^2} \tag{6.72}$$

$$Z_{qh} \approx \frac{v_{qh}}{i_{qh}} = \sqrt{R_{qh}^2 + (\omega_h L_{qh})^2} \tag{6.73}$$

where $R_{dh} \equiv R_s|_{\omega=\omega_h}$, $L_{dh} \equiv L_d|_{\omega=\omega_h}$ at the d axis, $R_{qh} \equiv R_s|_{\omega=\omega_h}$, $L_{qh} \equiv L_q|_{\omega=\omega_h}$ at the q axis and ω_h is the angular frequency of the injected signal.

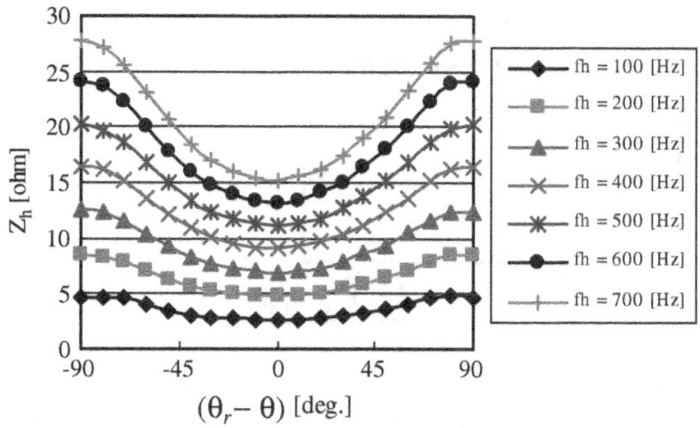

Figure 6.10 High-frequency impedance of a six-pole, 11-kW IPMSM at various injection frequencies (fh $= \omega_h/2\pi$) according to the injected angle oriented to the rotor angle with no load.

From (6.72) and (6.73), if the signal is injected to arbitrary angle, θ, of the *d–q* axis of the rotor reference frame when the rotor position is θ_r, then the high-frequency impedance at the angle, θ, can be described as

$$Z_h(\theta) = Z_{ha} - \tfrac{1}{2}Z_{hp}\cos 2(\theta_r - \theta) \tag{6.74}$$

where Z_{ha} stands for the average value of the *d–q* impedance; that is, $Z_{ha} = \frac{Z_{dh}+Z_{qh}}{2}$ and Z_{hp} stands for the difference of the impedance, $Z_{hp} = Z_{qh} - Z_{dh}$.

In Fig. 6.10, the impedance at the injected high-frequency signal according to the rotor position of a six-pole, 11-kW IPMSM is shown. As the frequency increases, the impedance difference increases. The average impedance, Z_{ha}, decreases as the magnitude of the fundamental frequency component of the stator current of an IPMSM increases, while the difference, Z_{hp}, increases as the current at the *d* axis increases. This comes from the saturation of the magnetic circuit of the IPMSM. At the rotor reference frame, the magnitude of the high-frequency impedance is minimum at the *d* or –*d* axes. However, in the case of the SynRM, the impedance is maximum at the *q* or –*q* axis. For the SynRM, there is no differentiation in the *d* and –*d* axis because of no permanent magnet. In the IPMSM in Fig. 3.9 or SMPMSM in Fig. 3.7, where the flux of the magnet is directed to the *d* axis, the impedance at the *d* axis is smaller than that at the –*d* axis because of the local magnetic saturation by the flux from the permanent magnet. And, thanks to this local saturation, the *d* and –*d* axes can be differentiated using the hysterisis phenomenon [66].

6.4.2 AC Machine with Nonsalient Rotor

6.4.2.1 Induction Machine [58, 59, 68, 69]

The rotor of the induction machine is symmetry to the rotating axis, and there is no variation of the stator impedance according to the rotor position. However, in a closed

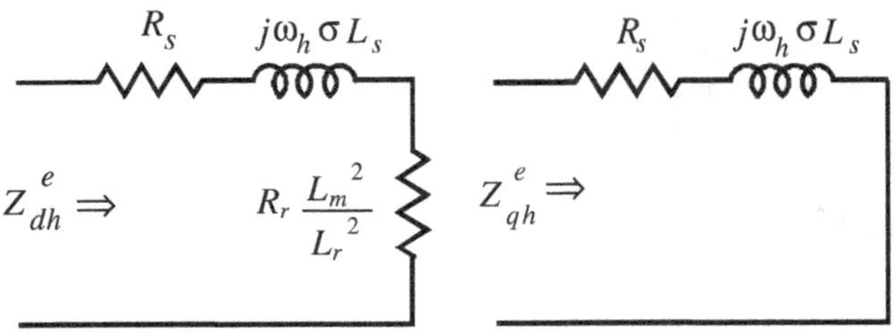

Figure 6.11 High-frequency equivalent circuit at a synchronous reference frame.

slot rotor or a semi-closed slot rotor, there are the leakage impedance variations according to the rotor flux linkage because of the local magnetic saturation of the leakage inductance path due to the high-frequency rotor flux linkage from the injected signal. With the vector control of the induction machine oriented to the rotor flux linkage, the stator voltage equation can be described as (6.75) as derived in Section 4.2.5.2.

$$
\begin{aligned}
V^e_{ds} &= \left(R_s + R_r\frac{L_m^2}{L_r^2}\right) i^e_{ds} + \sigma L_s \frac{di^e_{ds}}{dt} - \omega_e \sigma L_s i^e_{qs} - R_r\frac{L_m}{L_r^2}\lambda^e_{dr} - \omega_r \frac{L_m}{L_r}\lambda^e_{qr} \\
V^e_{qs} &= \left(R_s + R_r\frac{L_m^2}{L_r^2}\right) i^e_{qs} + \sigma L_s \frac{di^e_{qs}}{dt} + \omega_e \sigma L_s i^e_{ds} + \omega_r\frac{L_m}{L_r}\lambda^e_{dr} - R_r \frac{L_m}{L_r^2}\lambda^e_{qr}
\end{aligned} \tag{6.75}
$$

If the stator frequency, ω_e, is small enough, (6.75) can be approximated as (6.76) at the injected high frequency. If the fluctuating signal is injected to the d-axis of the rotor reference frame, the equivalent circuit of the induction machine at the frame can be simplified as the circuit in Fig. 6.11 at the high frequency.

$$
\begin{aligned}
V^e_{dsh} &\approx \left(R_s + R_r\frac{L_m^2}{L_r^2} + j\omega_h \sigma L_s\right) i^e_{dsh} \equiv (R_{dh} + j\omega_h L_{dh}) i^e_{dsh} \\
V^e_{qsh} &\approx (R_s + j\omega_h \sigma L_s) i^e_{qsh} \equiv \left(R_{qh} + j\omega_h L_{qh}\right) i^e_{qsh}
\end{aligned} \tag{6.76}
$$

where $R_{dh} \equiv \left(R_s + R_r\frac{L_m^2}{L_r^2}\right)\big|_{\omega=\omega_h}$, $R_{qh} \equiv R_s\big|_{\omega=\omega_h}$, and $L_{dh} \equiv \sigma L_s\big|_{\omega=\omega_h}$ at the d axis, and $L_{qh} \equiv \sigma L_s\big|_{\omega=\omega_h}$ at the q axis.

By the skin effect due to the injected high-frequency signal, the rotor resistance at the injected frequency could be several times the magnitude of the rotor resistance at nominal frequency of the induction machine. Also, by the skin effect, the transient stator leakage inductance, σL_s, at high frequency is considerably smaller than that at nominal frequency. And, if the signal is injected to the d axis, then d axis resistance in Fig. 6.11 is larger than q axis resistance as

$$
R_{dh} > R_{qh} \tag{6.77}
$$

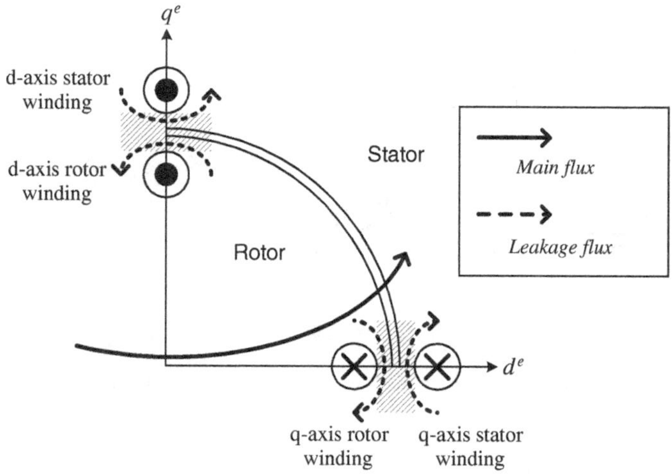

Figure 6.12 Conceptual diagram of the path of the main flux and the leakage flux in the case of two-pole induction machine.

However, even if there is some difference between R_{dh} and R_{qh}, still the impedance difference mainly comes from the difference between leakage inductance, L_{dh} and L_{qh}. The difference between L_{dh} and L_{qh} occurs due to the local saturation of the leakage flux path of the induction machine as shown in Fig. 6.12, and it is mainly affected by the main flux linkage. From the figure, it can be seen that the high-frequency signal injected to the d axis, where the main flux lies, results in more severe saturation due to the main flux. The physical location of windings and its associated main flux linkage is perpendicular. However, the leakage flux path lies on the neighborhood of the winding. Hence, the d axis high-frequency signal results in local magnetic saturation on the q axis of the synchronous reference frame as shown in Fig. 6.12, while the q axis high-frequency signal results in local magnetic saturation on the d axis of the synchronous reference frame. These saturation effects result in the difference of the leakage inductance of the d and the q axes. And because (6.78), the high-frequency inductance, which mainly comes from the leakage inductance, at the d axis is larger than that at the q axis if the high-frequency signal is injected to the d axis, where the main flux linkage lies. The main flux linkage along the rotor position is shown in Fig. 6.13, where the leakage inductance due to the high-frequency signal at the d axis is larger than that at the q axis. Hence, the difference between L_{dh} and L_{qh} reveals the information of the rotor flux position.

$$L_{dh} > L_{qh} \tag{6.78}$$

And, the high-frequency impedance at d axis is larger than that at the q axis as

$$Z_{dh} > Z_{qh} \tag{6.79}$$

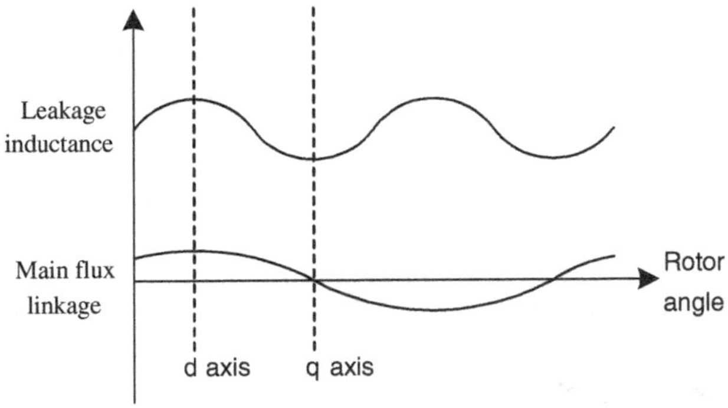

Figure 6.13 Distribution of the main flux linkage and the variation of leakage inductance due to the high-frequency signal.

The high-frequency impedance according to the angle where the high-frequency fluctuating signal is injected can be represented as

$$Z_h(\theta) = Z_{ha} + \frac{1}{2} Z_{hp} \cos 2(\theta_e - \theta) \tag{6.80}$$

In Fig. 6.14, the variation of the high-frequency impedance of a 3.7-kW, 220-V, four-pole, closed rotor slot, induction machine with no load is shown according to the difference between the rotor angle when rotor flux linkage lies and the angle where the high-frequency fluctuating signal is injected. In the figure, Vp stands for the peak value of the injected high-frequency voltage signal.

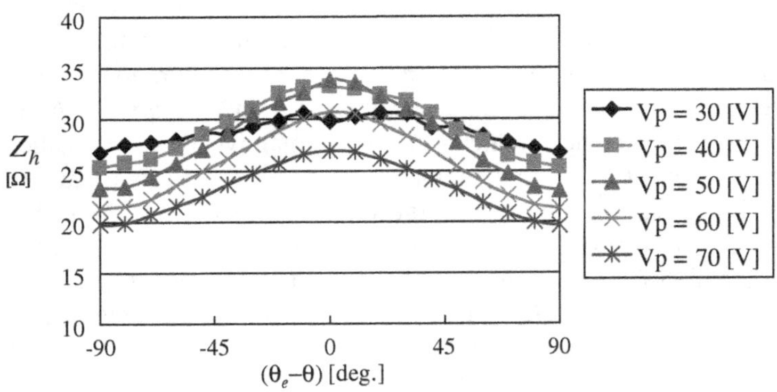

Figure 6.14 Variation of high-frequency impedance of a 3.7-kW, 220-V, four-pole, closed rotor slot, induction machine at various injection voltage magnitudes (Vp) according to the injected angle oriented to the rotor flux linkage angle with no load. Injection of 550-Hz, fluctuating, sinusoidal signal.

6.4.2.2 Surface-Mounted Permanent Magnet Synchronous Machine (SMPMSM) [62, 63, 70]

Like the induction machine, an SMPMSM has no saliency in the rotor. And at the operating frequency, there is no impedance variation due to rotor flux linkage position. However, according to the structure of the machine, with the high-frequency signal injection some SMPMSMs reveal saliency according to the rotor position, where the magnet flux linkage is directed. In Fig. 6.15, the variation of high-frequency impedance of an 11-kW, concentrated winding, eight-pole SMPMSM with the injection of fluctuating sinusoidal signal whose frequency is between 400-Hz and 850-Hz is shown. By exploiting this impedance variation, the estimation of rotor position and speed/position control can be done with a proper control algorithm. It can be seen from Fig. 6.15 that the impedance increases as the injection frequency increases. So, it can be concluded that the impedance mainly comes from the inductance. And when the injection angle oriented from rotor position, $\theta_r - \theta$, equals zero, the magnitude of the impedance is maximum. And when $\theta_r - \theta$ equals 90° or −90°, it is minimum. And it can be said that the high-frequency impedance is maximum at d axis, where the magnet flux linkage is directed. And, at q or $-q$ axis, the impedance is minimum. The physical reason why the largest impedance is at the d axis can be understood like the case of an induction machine described in the previous section. In the case of an SMPMSM, main flux comes from the permanent magnet. If the large current flows in the stator because of torque component current, then the direction of the main flux differs with the flux from the magnet because the main flux linkage is the sum of the magnet flux and the flux by the winding current. Hence, this difference between the position of the main flux linkage and the rotor position should be considered in the sensorless control of an SMPMSM.

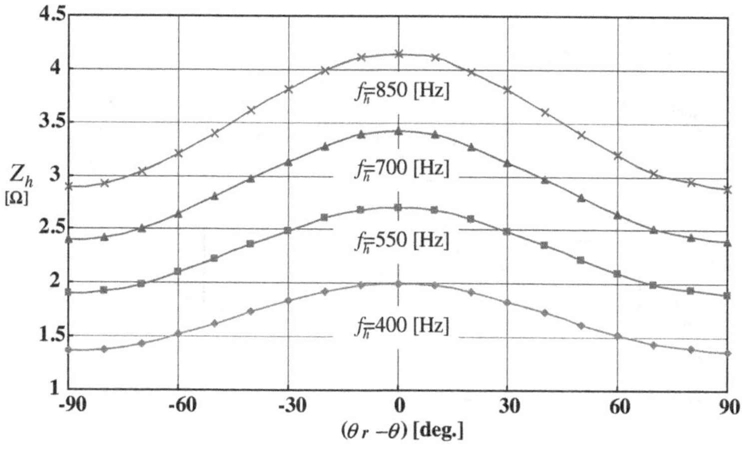

Figigure 6.15 Variation of high-frequency impedance of an 11-kW, eight-pole SMPMSM at various injection frequencies (f_h) with no load according to the injected angle oriented to the rotor angle with no load.

6.4.2.3 Estimation of Position of Rotor Flux Linkage/Rotor Position

Though there are many signal processing methods to estimate rotor flux linkage of the induction machine or the rotor position of the synchronous machine using the high-frequency impedance difference due to the injected fluctuating high-frequency sinusoidal signal, two typical methods are introduced in this section.

1. A measurement axis can be set, which offsets by 45° with the estimated *d–q* axis where vector control and signal injection is done. By the signal processing of the current at the measurement axis the position can be estimated [58, 59, 66].
2. The high-frequency component current at the *q* axis of the estimated *d–q* axis, where vector control and signal injection is done, is used for the estimation of the position [51, 62, 63].

These two methods are basically the same. But because of the difference in the implementations, they reveal some differences in control performance. For a salient rotor machine such as IPMSM and SynRM, the former would be better. For the machine with no saliency on the rotor such as an SMPMSM and an induction machine, the latter is preferable because of its simplicity of the implementation. In Fig. 6.16a, the control block diagram of the former method is shown. And in Fig. 6.16b, the relationship between the measurement axis and control axis is displayed. In Fig. 6.16c, a signal processing block diagram of an extractor to get the error signal, which is proportional to angle error, $\tilde{\theta}_r \equiv \theta_r - \hat{\theta}_r$, from the current in the measurement *d–q* axis is shown. In the control block diagram, a band-pass filter (BPF) is used to get the injected high-frequency component current. And the current is used to drive the extractor, which reveals the angle error between the estimated rotor angle (estimated

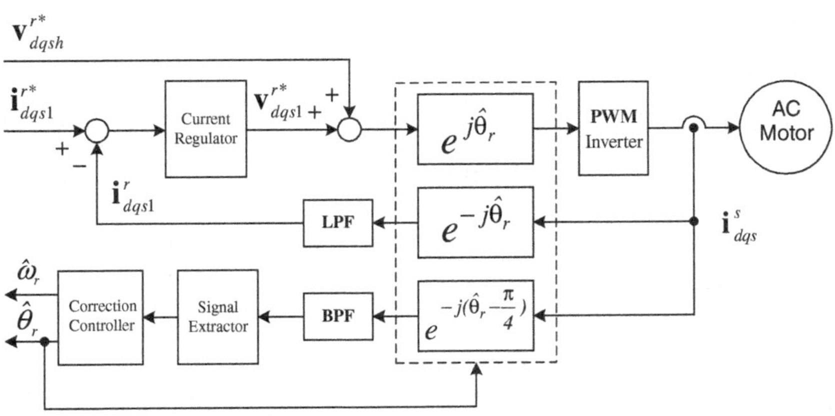

(a)

Figure 6.16 Sensorless control method employing a measurement axis with a high-frequency fluctuating signal injection to the estimated *d* axis. (a) Control block diagram. (b) Relationship between control axis and measurement axis. (c) Error signal extractor. (d) Correction controller.

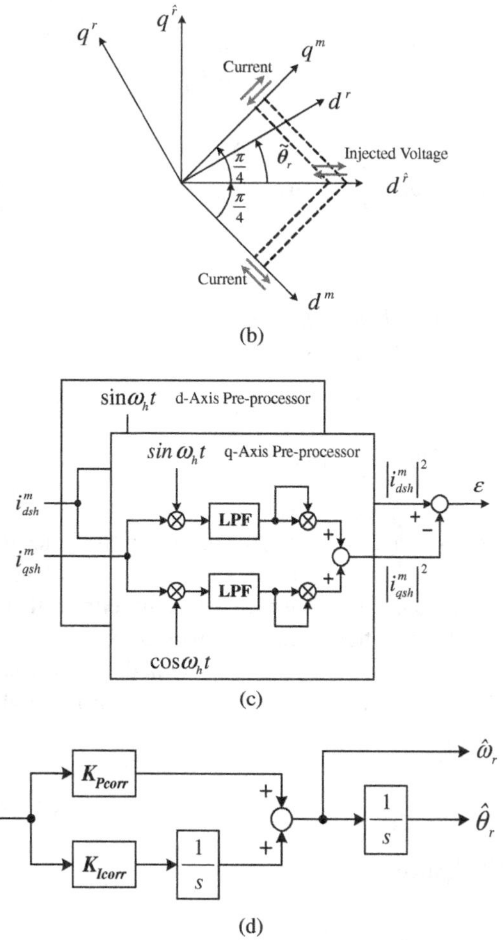

Figure 6.16 (*Continued*)

angle of rotor flux linkage) and real rotor position (real angle of rotor flux linkage), based on the impedance difference. The correction controller may be a bang bang-type regulator, a PI regulator, or a PID regulator, and it generates the estimated position and speed. In Fig. 6.16d, a PI type regulator is used as a correction controller. The output of the PI regulator is the estimated rotor speed, and the position can be calculated through the integration of the estimated speed.

If the high-frequency fluctuating signal given by (6.81) is injected to d axis of the estimated rotor reference frame, then the current in the measurement axis can be described as (6.82) and (6.83) [62].

$$\mathbf{v}_{dqsh\hat{r}} = \begin{bmatrix} V_p \cos \omega_h t \\ 0 \end{bmatrix} \tag{6.81}$$

$$i_{dsh}^{m} = -\frac{V_p}{\sqrt{2}\omega_h^2 L_{dh} L_{qh}} \left\{ \left(\frac{R_{dh}+R_{qh}}{2} + \frac{R_{dh}-R_{qh}}{2}\sin\left(2\tilde{\theta}_r - \frac{\pi}{4}\right)\right)\cos\omega_h t \right.$$
$$\left. -\omega_h \left(\frac{L_{dh}+L_{qh}}{2} + \frac{L_{dh}-L_{qh}}{2}\sin\left(2\tilde{\theta}_r - \frac{\pi}{4}\right)\right)\sin\omega_h t \right\} \tag{6.82}$$

$$i_{qsh}^{m} = -\frac{V_p}{\sqrt{2}\omega_h^2 L_{dh} L_{qh}} \left\{ \left(\frac{R_{dh}+R_{qh}}{2} - \frac{R_{dh}-R_{qh}}{2}\sin\left(2\tilde{\theta}_r + \frac{\pi}{4}\right)\right)\cos\omega_h t \right.$$
$$\left. -\omega_h \left(\frac{L_{dh}+L_{qh}}{2} - \frac{L_{dh}-L_{qh}}{2}\sin\left(2\tilde{\theta}_r + \frac{\pi}{4}\right)\right)\sin\omega_h t \right\} \tag{6.83}$$

It can be seen from (6.82) and (6.83) that the angle error information comes from the difference of the impedance. The error signal extractor in Fig. 6.16c calculates the difference of the magnitude of d–q axis current at the measurement axis. If there is no angle error, which means $\tilde{\theta}_r = 0$, then the difference would be zero. From Fig. 6.16c, (6.82), and (6.83), the output of the extractor can be deduced as (6.84). In Fig. 6.16 and (6.84), LPF stands for low-pass filter, and it filters out the higher-frequency components from the measured current.

$$\begin{aligned}
\varepsilon &\equiv \left|i_{dsh}^{m}\right|^2 - \left|i_{qsh}^{m}\right|^2 \\
&= \left\{LPF\left(i_{dsh}^{m}\cos\omega_h t\right)\right\}^2 + \left\{LPF\left(i_{dsh}^{m}\sin\omega_h t\right)\right\}^2 \\
&\quad - \left\{LPF\left(i_{qsh}^{m}\cos\omega_h t\right)\right\}^2 - \left\{LPF\left(i_{qsh}^{m}\sin\omega_h t\right)\right\}^2 \\
&= -\frac{V_p^2}{4\omega_h^2 L_{dh} L_{qh}}\left\{2\left(R_{avg}\Delta R + \omega_h^2 L_{avg}\Delta L\right) - \sqrt{2}\left((\Delta R)^2 + (\omega_h \Delta L)^2\right)\cos 2\tilde{\theta}_r\right\}\sin 2\tilde{\theta}_r
\end{aligned} \tag{6.84}$$

where

$$R_{avg} \equiv \frac{R_{dh}+R_{qh}}{2}, \quad \Delta R = \frac{R_{dh}-R_{qh}}{2}, \quad L_{avg} \equiv \frac{L_{dh}+L_{qh}}{2}, \quad \Delta L = \frac{L_{dh}-L_{qh}}{2} \tag{6.85}$$

In (6.84), the value inside the curly brackets is independent of the angle error and its polarity is decided by ΔR or ΔL. Hence, by regulating the output of the error signal from the extractor as null, the rotor position can be tracked. In Fig. 6.16d a regulator to track the position is shown. If the output of the extractor is approximated as (6.86), then the transfer function between the real rotor position and the estimated position can be deduced as (6.87) with the correction controller. From the frequency response of the transfer function it can be seen that the angle can be tracked in a low-frequency region very well, but there should be some error in a higher-frequency region.

$$\varepsilon = K'_{err} \sin 2\tilde{\theta}_r \approx K_{err}\, \tilde{\theta}_r \tag{6.86}$$

$$\frac{\hat{\theta}_r}{\theta_r} = \frac{K_{err}K_{Pcorr}s + K_{err}K_{Icorr}}{s^2 + K_{err}K_{Pcorr}s + K_{err}K_{Icorr}} \tag{6.87}$$

where

$$K'_{err} = -\frac{V_p^2}{4\omega_h^2 L_{dh}L_{qh}}\left\{2\left(R_{avg}\,\Delta R + \omega_h^2 L_{avg}\Delta L\right) - \sqrt{2}\left((\Delta R)^2 + (\omega_h \Delta L)^2\right)\cos 2\tilde{\theta}_r\right\}$$

$$\approx -\frac{V_p^2\, L_{avg}\, \Delta L}{2\, L_{dh}L_{qh}} \quad \text{and} \quad K_{err} \approx 2K'_{err} \approx -\frac{V_p^2 L_{avg}\Delta L}{L_{dh}L_{qh}}$$

The test results of the sensorless control with typical AC machines are shown in Figs. 6.17–6.22. For an IPMSM, which has inherent rotor saliency, the speed control performance at low speed including zero speed is shown in Figs. 6.17 and 6.18. The high-frequency impedance characteristics of the machine under test are shown in Fig. 6.10. In Fig. 6.17, the speed command varies from 0 r/min to 50 r/min and back to 0 r/min in step manner. It can be seen that the estimated speed well matches the measured speed, which is used only for monitoring purpose, and that the speed response is comparable to that of the low-resolution sensored vector control system. In Fig. 6.18, the disturbance torque rejection performance is shown with zero speed reference. In this test, the rated torque of an IPMSM is applied in step manner. Except for a short time interval where the load is engaged or disengaged, the speed is well maintained as the commanded null speed. In Figs. 6.19 and 6.20, the sensorless position control results are shown. In Fig. 6.19, the response of the position control

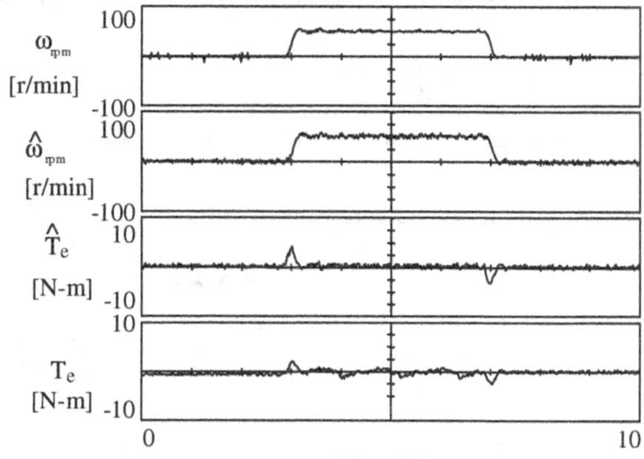

Figure 6.17 Sensorless speed control of an IPMSM with speed reference changes from 0 r/min to 50 r/min and back to 0 r/min.

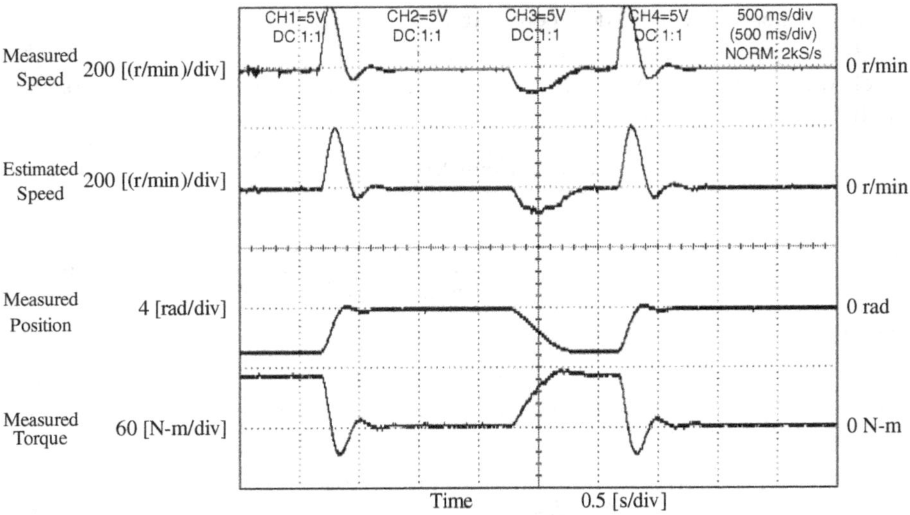

Figure 6.18 Sensorless speed control of an IPMSM with 100% step load torque disturbances at zero speed reference.

with 100% step load torque disturbance is displayed with zero position command. In Fig. 6.20, with 100% load torque, the position response is shown according to the step change of the position reference from $-\pi$ rad to π rad and back to $-\pi$ rad. In these figures, it can be seen that except for a short transient time interval the position error

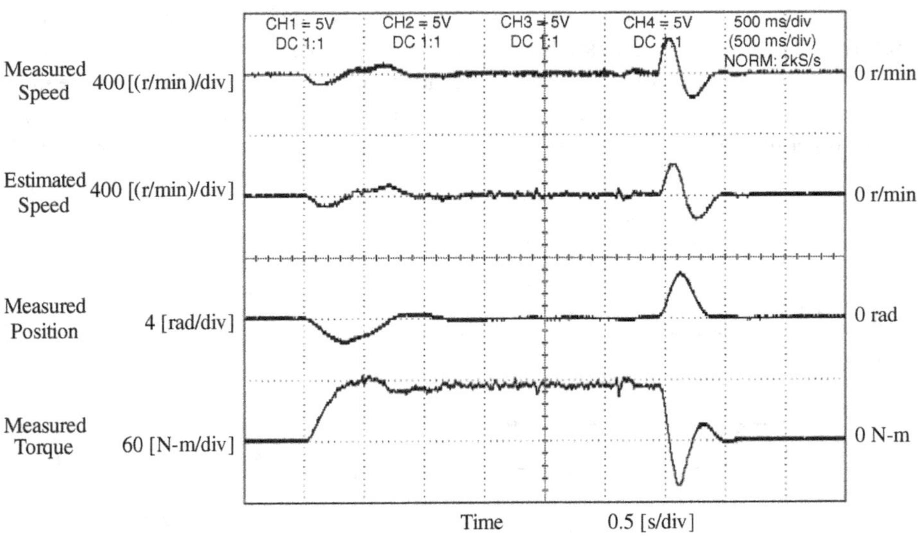

Figure 6.19 Sensorless position control of an IPMSM with 100% step load torque disturbances at zero position reference.

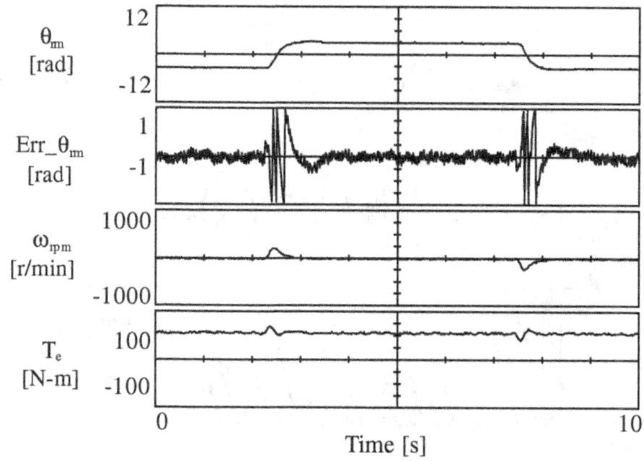

Figure 6.20 Sensorless position control of an IPMSM with position reference changes $-\pi$ rad to π rad and back to $-\pi$ rad.

is maintained at less than 0.1 rad. For an SynRM, similar test results can be obtained [60].

For a four-pole, 220-V, closed-slot rotor, 11-kW induction machine with no inherent saliency, the test results of sensorless speed control employing the high-frequency fluctuating signal injection method are shown in Figs. 6.21 and 6.22. The magnitude of the injected voltage, V_p^*, is 50 Vand the injected frequency, f_h, is 550 Hz. In Fig. 6.21, the speed control performance with 100% load torque at zero speed reference is demonstrated. In this operating speed, the back EMF method cannot work because of too small back EMF and parameter errors. Even with the high-frequency injection method, there are considerable speed ripples due to the dead time and measurement error of the inverter. But, at least, the speed of the induction machine is under control. In Fig. 6.22, the sensorless speed control performance against 100%

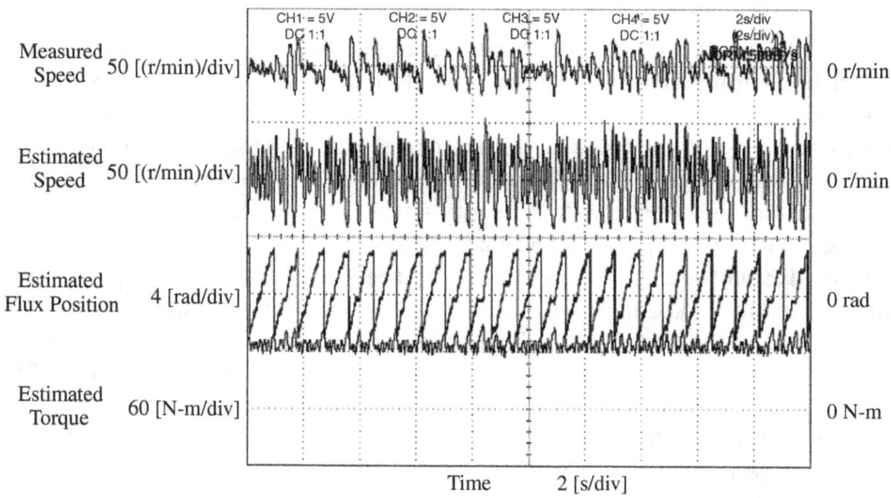

Figure 6.21 Sensorless speed control of a squirrel cage induction machine with 100% load torque at zero speed reference.

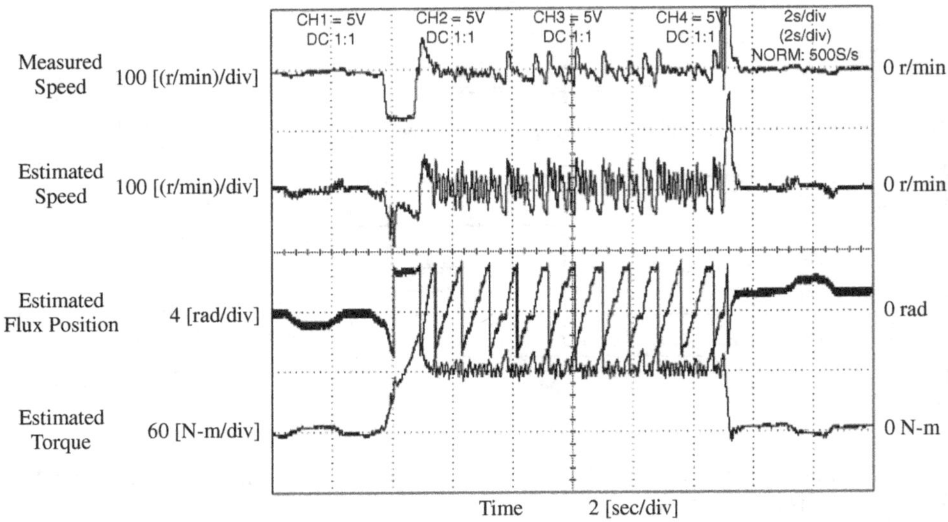

Figure 6.22 Sensorless speed control of a squirrel cage induction machine against 100% load torque disturbances at zero speed reference.

load torque disturbances is shown. At connecting and disconnecting load, though there is relatively large speed variation, in the steady state the speed is well-regulated according to the speed command, which is 0 r/min in this test.

The sensorless control method employing the high-frequency signal injection can be applied at zero frequency, where back EMF signal is unavailable. And it is also robust to the AC machine parameter variations compared to the sensorless control based on the machine model. Furthermore, the method can be implemented without any modification of the hardware of the drive system. However, the sensorless control method can be applied only if there is saliency in impedance characteristics inherently or induced by the injected signal. In some AC machines such as an open-slot rotor induction machine and an SMPMSM with no magnetic saturation, the method cannot be applied because of nonsaliency characteristics. Also, for the injection of the high-frequency signal, there should be extra voltages to synthesize the high-frequency sinusoidal voltage through a PWM inverter. Hence, the method cannot be used in higher operating speed, where back EMF is quite large and comparable to maximum available voltage from the inverter, and all available voltage should be used to regulate the fundamental current. Furthermore, the extra signal results in extra losses and acoustic noise and vibration of the drive system. To mitigate these shortcomings, the method based on high-frequency signal injection can be applied only in the low-speed region, where back EMF is too small to get the satisfactory performance of sensorless control based on back EMF and the AC machine model. In another speed region, the sensorless control described in Sections 6.1, 6.2, and 6.3 can be used successfully [59, 60, 62]. The saliency of the high-frequency impedance could be affected by the saturation not only from the main flux linkage—that is, the flux linkage by the permanent magnet in the case of the machine with the strong permanent magnet and by the main excitation current in the case of the induction machine—but also from the stator current for torque generation or for flux weakening. Hence, the accuracy of the angle tracking might be degraded with the increase of the magnitude of the stator current. For better performance, the distortion of the high-frequency impedance characteristics according to the operating conditions should be carefully considered [59, 62].

PROBLEMS

1. Explain why in the sensorless control of the induction machine based on the equivalent circuit described in Section 3.2, the rotating speed of the rotor of the machine, ω_r, cannot be estimated when the synchronous speed of the machine is null as $\omega_e = 0$ even if the parameters of the induction machine are perfectly known.
2. In problem 14 of Chapter 5, the estimator of the angle of the rotor flux linkage has been changed to the block shown in Fig. 6.4, while other control blocks in Fig. P5.8b are the same.
   (1) Repeat parts 1 to 4 of problem 14 in Chapter 5.
   (2) Describe the expected difficulties of the estimation of the flux linkage, $\widehat{\lambda}_r^s$, when the stator frequency of the induction machine is very low as $\omega_e \approx 0$.
3. In the problem 14 of Chapter 5, the estimator of the angle of the rotor flux linkage has been changed to the estimator described in Section 6.1.2, while other control blocks in Fig. P5.8b are the same. In (6.49), k has been set as 1.5. And the angle is directly calculated by the following equation from the estimated rotor flux linkage, $\hat{\lambda}_r^s$.

$$\theta_e = \arctan\left(\frac{\hat{\lambda}_{dr}^s}{\hat{\lambda}_{qr}^s}\right)$$

   (1) Repeat from parts 1 to 4 of problem 14 in Chapter 5.
   (2) Describe the expected difficulties of the estimation of the flux linkage, $\widehat{\lambda}_r^s$, when the stator frequency of the induction machine is very low as $\omega_e \approx 0$.
4. An eight-pole SMPMSM with the following parameters is driven by an inverter in sensorless vector control mode using the sensorless control method in Fig. 6.5. The current regulator can be assumed to be an ideal one, which means that the transfer function of the regulator is unity. The bandwidth of the speed regulator has been set as 30 rad/s under the assumption of the ideal estimation of the rotor speed and the system parameters. And the speed regulator is a PI-type described in Section 4.3.4.1. The inertia of load is two times of that of the machine itself. From $t = 0$ to 1 s, the SMPMSM starts using the method in Fig. 6.6 and 6.7, where $\omega_{r\_start}^* = 10\,\text{r/min}$ and $\omega_{r\_thres}^* = 150\,\text{r/min}$.

   The duration of period I is 0.3 s, and that of period II, 0.7 s.

   Rated power: 11 kW, Rated speed: 1500 r/min, Rated current: 58.6 A(rms)

   Equivalent circuit parameters: $R_s = 0.0217\,\Omega$, $L_s = 0.7\,\text{mH}$, $\lambda_f = 0.1473\,\text{Wb-t}$

   Inertia of the machine itself: $J_M = 0.0281\,\text{kg-m}^2$

(1) If the speed reference, ω_{rm}^*, and the load torque, T_L, are as listed below, then, plot speed reference, actual speed, actual torque, the *d–q* axis current, i_{ds}^r, i_{qs}^r with their references, and error between estimated rotor position and actual position along with time from 0 to 5 s. It is assumed that the PWM inverter is ideal and the output of the current regulator is directly applied to the SMPMSM without any delay or distortion. The exact parameters of the SMPMSM have been known to the controller. And all control is done in an analog domain.

$$\omega_{rm}^* = \begin{cases} 0\,\text{s} \le t < 1\,\text{s (Starting)} \\ 1\,\text{s} \le t < 3\,\text{s}\ 500\,\text{r/min} \\ 3\,\text{s} \le t < 5\,\text{s}\ 50\,\text{r/min} \\ 3\,\text{s} \le t < 5\,\text{s}\ -50\,\text{r/min} \end{cases} \qquad T_L = \begin{cases} 0\,\text{s} \le t < 2\,\text{s}\ 0\,N\text{-m} \\ 2\,\text{s} \le t < 2.5\,\text{s}\ 60\,N\text{-m} \\ 2.5\,\text{s} \le t < 4\,\text{s}\ 0\,N\text{-m} \\ 4\,\text{s} \le t < 4.5\,\text{s}\ 60\,N\text{-m} \\ 4.5\,\text{s} \le t < 5\,\text{s}\ 0\,N\text{-m} \end{cases}$$

(2) Repeat part 1 under the assumption that the phase voltage and the phase current can be measured ideally, but the parameters of the SMPMSM for the controller setting have errors as follows.

$$\hat{R}_s = 0.9R_s, \qquad \hat{L}_s = 0.9L_s, \qquad \hat{\lambda}_f = 0.95\lambda_f$$

(3) Repeat part 1. In here, in addition to the parameter errors in part 2, the measured phase voltage and phase current have following errors. In the measurement, only "a"- and "b"-phase voltage and current are measured and "c"-phase voltage and current are calculated as follows.

$$\hat{i}_{cs} = -(\hat{i}_{as} + \hat{i}_{bs}), \quad \hat{V}_{cs} = -(\hat{V}_{as} + \hat{V}_{bs})$$

where the hat ($\wedge$) denotes the measured value for the current regulator and the vector controller.

In the measurement of the phase current, there is white noise, $\eta_i(t)$, whose rms magnitude is 4% of rated rms phase current. The sampling frequency of the white noise is 100 μs and also the measured current has offset, I_{offset}, whose magnitude is 1% of rated rms current. Also, the measured current can be represented as follows.

$$\hat{i}_{as} = i_{as} + \eta_i(t) + I_{\text{offset}}$$
$$\hat{i}_{bs} = i_{bs} + \eta_i(t) + I_{\text{offset}}$$

In the measurement of the phase voltage, there is white noise, $\eta_v(t)$, whose rms magnitude is 3 V. The sampling frequency of the white noise is 100 μs and also the measured voltage has offset, V_{offset}, whose magnitude is 0.5 V. Also, the measured voltage can be represented as follows.

$$\hat{V}_{as} = V_{as} + \eta_v(t) + V_{\text{offset}}$$
$$\hat{V}_{bs} = V_{bs} + \eta_v(t) + V_{\text{offset}}$$

5. An eight-pole IPMSM with following parameters is driven by an inverter in sensorless vector control mode using the sensorless control method in Fig. 6.8. The current regulator can be assumed to be ideal one, which means that the transfer function of the regulator is unity. The bandwidth of the speed regulator has been set as 30 rad/s under the assumption of the ideal estimation of the rotor speed and the system parameters. And the speed regulator is a PI-type described in Section 4.3.4.1. The inertia of load is two times that of the machine itself. At $t=0$, the rotor position of the IPMSM has been known to the sensorless controller, which means $\hat{\theta}_r = \theta_r = 0$.

To achieve maximum torque per ampere (MTPA), the current i^r_{ds} and i^r_{qs} should be properly controlled [71]. It can be assumed that the parameters of an IPMSM are constant regardless of the operating condition.

Rated power: 11 kW, Rated *speed* : 1750 r/min, Rated current : 39.5 A (rms)

Equivalent circuit parameters: $R_s = 0.109\,\Omega$, $L_d = 3.6\,\text{mH}$, $L_q = 4.5\,\text{mH}$,

$\lambda_f = 0.2595$ Wb-t

Inertia of IPMSM itself: $J_M = 0.0281\,\text{kg-m}^2$

(1) If the speed reference, ω^*_{rm}, and the load torque, T_L, are as listed below, then plot speed reference, actual speed, actual torque, the *d–q* axis current, i^r_{ds} and i^r_{qs} with their references, and error between estimated rotor position and actual rotor position along with time from 0 to 5 s. It is assumed that the PWM inverter is ideal and the output of the current regulator is directly applied to the IPMSM without any delay or distortion. The exact parameters of the IPMSM have been known to the controller. And all control is done in analog domain.

$$\omega^*_{rm} = \begin{cases} 0\,\text{s} \le t < 2\,\text{s} & 500\,\text{r/min} \\ 2\,\text{s} \le t < 3\,\text{s} & 50\,\text{r/min} \\ 3\,\text{s} \le t < 5\,\text{s} & -50\,\text{r/min} \end{cases} \qquad T_L = \begin{cases} 0\,\text{s} \le t < 1\,\text{s} & 0\,\text{N-m} \\ 1\,\text{s} \le t < 2.5\,\text{s} & 60\,\text{N-m} \\ 2.5\,\text{s} \le t < 4\,\text{s} & 0\,\text{N-m} \\ 4\,\text{s} \le t < 4.5\,\text{s} & 60\,\text{N-m} \\ 4.5\,\text{s} \le t < 5\,\text{s} & 0\,\text{N-m} \end{cases}$$

(2) Repeat part 1 under the assumption that the phase voltage and the phase current can be measured ideally, but the parameters of the IPMSM for the controller setting have errors as follows.

$$\hat{R}_s = 0.9R_s, \qquad \hat{L}_d = 0.9L_d, \qquad \hat{L}_q = 0.8L_q, \qquad \hat{\lambda}_f = 0.95\lambda_f$$

(3) Repeat part 1. Here, in addition to the parameter errors in part 2, the measured phase voltage and phase current have the following errors. In the measurement, only "a"- and "b"-phase voltage and current are measured and "c" -phase voltage and current are calculated as follows.

$$\hat{i}_{cs} = -(\hat{i}_{as} + \hat{i}_{bs}), \quad \hat{V}_{cs} = -(\hat{V}_{as} + \hat{V}_{bs})$$

where the hat (^) denotes the measured value for the current regulator and the vector controller.

In the measurement of the phase current, there is white noise, $\eta_i(t)$, whose rms magnitude is 4% of rated rms phase current. The sampling frequency of the white noise is 100 μs and also the measured current has offset, I_{offset}, whose magnitude is 1% of rated rms current. Also, the measured current can be represented as follows.

$$\hat{i}_{as} = i_{as} + \eta_i(t) + I_{\text{offset}}$$

$$\hat{i}_{bs} = i_{bs} + \eta_i(t) + I_{\text{offset}}$$

In the measurement of the phase voltage, there is white noise, $\eta_v(t)$, whose rms magnitude is 3 V. The sampling frequency of the white noise is 100 μs; also, the measured voltage has offset, V_{offset}, whose magnitude is 0.5 V. Also, the measured voltage can be represented as follows.

$$\hat{V}_{as} = V_{as} + \eta_v(t) + V_{\text{offset}}$$

$$\hat{V}_{bs} = V_{bs} + \eta_v(t) + V_{\text{offset}}$$

6. In a sensorless control method using the high-frequency fluctuating voltage signal injection to the estimated *d* axis as described in Section 6.4.2.2, derive the high-frequency component current at the *q* axis of the estimated *d–q* axis in terms of SMPMSM parameters and angle error under the assumption of small enough angle error. The injected high-frequency signal is represented as (6.81) [62].

7. Based on (6.86) and control block diagram in Fig. 6.16 d, derive the transfer function, (6.87). If the gains of the correction controller are set to get 200-rad/s bandwidth of the transfer function in (6.87) with $K_{err} = 1$, calculate gains. Here the damping coefficient of the transfer function is unity. In this condition, show the Bode plot of the transfer function. If (6.86) is modified to consider time delay, T_d, in the signal processing as $\varepsilon = K_{err}\tilde{\theta}_r e^{-T_d s}$, where K_{err} is unity and T_d is 2 ms, derive the transfer function in terms of K_{err}, T_d, and gains of the correction controller. Show the Bode plot of the modified transfer function with the same gains in the previous case , no time delay. From the Bode plot, find out the bandwidth of the modified transfer function.

REFERENCES

1. F. Briz et al., Speed measurement using rotary encoders for high performance AC drives, in *Proceedings of the 20th International Conference on Industrial Electronics, Control and Instrumentation*, IECON '94, Vol. 1, 1994, pp. 538–542.
2. J. Holtz, Sensorless control of induction machine—with or without signal injection, *IEEE Trans. Ind. Electron.*, Vol. 53, No. 1, January, 2006, pp. 7–30.
3. A. Abbondanti, *Method of flux control in induction motors driven by variable frequency, variable voltage supplies*, in *Conference Record of IEEE Industry Applications Society Annual Meeting*, 1977, pp. 177–184.
4. R. Joetten et al., Control method for good dynamic performance induction motor drives based on current and voltage as measured quantities, *IEEE Trans. Ind. Appl.*, Vol. IA-19 No. 3, May/June 1983, pp. 356–363.
5. I. Takahashi et al., A new quick response and high efficiency control strategy of an induction machine, in *Conference Record of IEEE Industry Applications Society Annual Meeting*, 1985, pp. 496–502.
6. K. Iizuka et al., Microcomputer control for sensorless brushless motor, *IEEE Trans. Ind. Appl.*, Vol. IA-21, No. 4, May/June 1985, pp. 595–601.
7. T. Ohtani et al., Vector control of induction motor without shaft encoder, *IEEE Trans. Ind. Appl.*, Vol. 28, No. 1, January/February 1992, pp. 157–164.
8. A. Consoli et al., Sensorless vector and speed control of brushless motor drives, *IEEE Trans. Ind. Electron.*, Vol. 41, No. 1, February 1994, pp. 91–96.

9. R. Lagerquist et al., Sensorless-control of the synchronous reluctance motor, *IEEE Trans. Ind. Appl.*, Vol. 30, No. 3, May/June 1994, pp. 673–682.
10. T. Kanmachi et al., Sensor-less speed control of an induction motor, *IEEE Ind. Appl. Mag.*, Vol. 1, No. 1, January/February 1995, pp. 22–27.
11. N. Matsui, Sensorless PM brushless DC motor drives, *IEEE Trans. Ind. Electron.*, Vol. 43, No. 2, April 1996, pp. 300–308.
12. M. G. Jovanovic et al., Sensorless vector controller for a synchronous reluctance motor, *IEEE Trans. Ind. Appl.*, Vol. 34, No. 2, March/April 1998, pp. 346–354.
13. R. B. Sepe et al., Real-time observer-based (adaptive) control of a permanent-magnet synchronous motor without mechanical sensors, *IEEE Trans. Ind. Appl.*, Vol. 28, No. 6, November/December 1992, pp. 1345–1352.
14. L. A. Jones et al., A state observer for the permanent-magnet synchronous motor, *IEEE Trans. Ind. Electron.*, Vol. 36, No. 3, August 1994, pp. 374–382.
15. J.-S. Kim et al., New approach for high-performance PMSM drives without rotational position sensors, *IEEE Trans. Power Electron.*, Vol. 12, No. 5, September 1997, pp. 904–911.
16. K. R. Shouse et al., Sensorless velocity control of permanent-magnet synchronous motors, *IEEE Trans. Control Systems Technol.*, Vol. 6, No. 3, May 1998, pp. 313–324.
17. N. Ertugrul et al., Indirect rotor position sensing in real time for brushless permanent magnet motor drives, *IEEE Trans. Power Electron.*, Vol. 13, No. 4, July 1998, pp. 608–616.
18. M.-H. Shin et al., An improved stator flux estimation for speed sensorless stator flux orientation control of induction motors, *IEEE Trans. Power Electron.*, Vol. 15, No. 2, March 2000, pp. 312–318.
19. K. Tatemarsu et al., New approaches with sensorless drives, *IEEE Ind. Appl. Mag.*, Vol. 6, No. 4, July/August 2000, pp. 44–50.
20. C. Schauder, Adaptive speed identification for vector control of induction motor without rotational transducers, *IEEE Trans. Ind. Appl.*, Vol. 28, No. 5, September/October 1992, pp. 1054–1061.
21. H. Tajima et al., Speed sensorless field-orientation control of the induction machine, *IEEE Trans. Ind. Appl.*, Vol. 29, No. 1, January/February 1993, pp. 175–180.
22. H. Kubota et al., DSP-based adaptive flux observer of induction motor, *IEEE Trans. Ind. Appl.*, Vol. 29, No. 2, March/April 1993, pp. 344–348.
23. L. Tian-Hua et al., Adaptive control for a sensorless permanent-magnet synchronous motor drive, *IEEE Trans. Aerospace Electron. Systems*, Vol. 30, No. 3, July 1994, pp. 900–909.
24. R. Blasco-Gimenez, et al., Dynamic performance limitations for MRAS based sensorless induction motor drives. I. stability analysis for the closed loop drive, *IEE Proc. Electric Power Appl.*, Vol. 143, No. 2, March 1996, pp. 113–122.
25. J. Maes et al., Speed-sensorless direct torque control of induction motors using an adaptive flux observer, *IEEE Trans. Ind. Appl.*, Vol. 36, No. 3, May/June 2000, pp. 778–785.
26. Y.-R. Kim et al., Speed sensorless vector control of induction motor using extended kalman filter, *IEEE Trans. Ind. Appl.*, Vol. 30, No. 5, September/October 1994, pp. 1225–1233.
27. L. Ben-Brahim, Motor speed identification via neural networks, *IEEE Ind. Appl. Mag.*, Vol. 1, No. 1, January/February 1995, pp. 28–32.
28. H.-S. Yoo et al., A Polar coordinate-oriented method of identifying rotor flux and speed of induction motors without rotational transducers, *IEEE Trans. Control Systems Technol.*, Vol. 4, No. 3, May 1996, pp. 230–243.
29. N. Hur et al., Sensorless vector control in the presence of voltage and current measurement errors by dead-time, *Conference Record of 1997 IEEE Industry Applications Society Annual Meeting*, Vol. 1, 1997, pp. 433–438.
30. T.-L. Hsien et al., H infinity control for a sensorless permanent-magnet synchronous drive, *IEE Proc. Electric Power Appl.*, Vol. 144, No. 3, May 1997, pp. 173–181.
31. L. Che-Ming et al., Speed sensorless vector control of induction motor using kalman-filter-assisted adaptive observer, *IEEE Trans. Ind. Electron.*, Vol. 45, No. 2, April 1998, pp. 359–361.
32. S. Bolognani et al., Sensorless full-digital PMSM drive with EKF estimation of speed and rotor position, *IEEE Trans. Ind. Electron.*, Vol. 46, No. 1, February 1999, pp. 184–191.

33. Y.-H. Kim et al., High performance IPMSM drives without rotational position sensors using reduced-order EKF, *IEEE Trans. Energy Conversion*, Vol. 14, No. 4, December 1999, pp. 868–873.
34. F.-Z. Peng et al., Robust speed identification for speed-sensorless vector control of induction motors, *IEEE Trans. Ind. Appl.*, Vol. 30, No. 5, September/October 1994, pp. 1234–1240.
35. R. Blasco-Gimenez et al., Dynamic performance limitations for MRAS based sensorless induction motor drives. II. Online parameter tuning and dynamic performance studies, *IEE Proc. Electric Power Appl.*, Vol. 143, No. 2, March 1996, pp. 123–134.
36. K. D. Hurst et al., A self-tuning closed-loop flux observer for sensorless torque control of standard induction machines, *IEEE Trans. Power Electron.*, Vol. 12, No. 5, September 1997, pp. 807–815.
37. G. S. Garcia et al., Reduced-order observers for rotor flux, rotor resistance and speed estimation for vector controlled induction motor drives using the extended Kalman filter technique, *IEE Proc. Electric Power Appl.*, Vol. 146, No. 3, May 1999, pp. 282–288.
38. K. Akatsu et al., Online rotor resistance estimation using the transient state under the speed sensorless control of induction motor, *IEEE Trans. Power Electron.*, Vol. 15, No. 3, May 2000, pp. 553–560.
39. M. Ishida et al., A new slip frequency detection of an induction motor utilizing rotor slot harmonics, *IEEE Trans. Ind. Appl.*, Vol. IA-20 No. 3, May/June 1984, pp. 575–1233.
40. R. Blasco-Gimenez et al., Performance of FFT-rotor slot harmonic speed detector for sensorless induction motor drives, *IEE Proc. Electric Power Appl.*, Vol. 143, No. 3, May 1996, pp. 258–268.
41. M. S. Arefeen et al., Sensorless position measurement in synchronous reluctance motor, *IEEE Trans. Power Electron.*, Vol. 9, No. 6, November 1994, pp. 624–630.
42. S. Ogasawara et al., Implementation and position control performance of a position-sensorless IPM motor drive system based on magnetic saliency, *IEEE Trans. Ind. Appl.*, Vol. 34, No. 4, July/August 1998, pp. 806–812.
43. S. Ostlund et al., Sensorless rotor-position detection from zero to rated speed for an integrated PM synchronous motor drive, *IEEE Trans. Ind. Appl.*, Vol. 32, No. 5, September/October 1996, pp. 1158–1165.
44. J. S. Lee et al., Stator-flux-oriented sensorless induction motor drive for optimum low-speed performance, *IEEE Trans. Ind. Appl.*, Vol. 33, No. 5, September/October 1997, pp. 1170–1176.
45. K. D. Hurst et al., Zero-speed tacholess IM torque control: simply a matter of stator voltage integration, *IEEE Trans. Ind. Appl.*, Vol. 34, No. 4, July/August 1998, pp. 790–795.
46. R. Mizutani et al., Current model-based sensorless drives of salient-pole PMSM at low speed and standstill, *IEEE Trans. Ind. Appl.*, Vol. 34, No. 4, July/August 1998, pp. 841–846.
47. E. K. K. Sng et al., New observer-based DFO scheme for speed sensorless field-oriented drives for low-zero-speed operation, *IEEE Trans. Power Electron.*, Vol. 13, No. 5, September 1998, pp. 959–968.
48. A. Consoli et al., A new zero-frequency flux-position detection approach for direct-field-oriented-control drives, *IEEE Trans. Ind. Appl.*, Vol. 36, No. 3, May/June 2000, pp. 797–804.
49. F. Blaschke et al., Sensorless direct field orientation at zero flux frequency, *Conference Record of IEEE 1996 Industry Applications Annual Meeting*, Vol. 1, 1996, pp. 189–196.
50. M. Schroedl et al., EMF-based rotor flux detection in induction motors using virtual short circuits, *IEEE Trans. Ind. Appl.*, Vol. 34, No. 1, January/February 1998, pp. 142–147.
51. T. Aihara et al., Sensorless torque control of salient-pole synchronous motor at zero-speed operation, *IEEE Trans. Power Electron.*, Vol. 14, No. 1, January 1999, pp. 202–208.
52. P. L. Jansen et al., Transducerless position and velocity estimation in induction and salient AC machines, *IEEE Trans. Ind. Appl.*, Vol. 31, No. 2, March/April 1995, pp. 240–247.
53. P. L. Jansen et al., Transducerless field orientation concepts employing saturation-induced saliencies in induction machines, *IEEE Trans. Ind. Appl.*, Vol. 32, No. 6, November/December 1996, pp. 1380–1393.
54. J. Holtz, Sensorless position control of induction motors—An emerging technology, *IEEE Trans. Ind. Electron.*, Vol. 45, No. 6, December 1998, pp. 840–851.
55. M. J. Corley et al., Rotor position and velocity estimation for a salient-pole permanent magnet synchronous machine at standstill and high speeds, *IEEE Trans. Ind. Appl.*, Vol. 34, No. 4, July/August 1998, pp. 784–789.

56. S. J. Kang et al., Position sensorless control of synchronous reluctance motor using high frequency current injection, *IEEE Trans. Energy Conversion.*, Vol. 14, No. 4, December 1999, pp. 1271–1275.
57. A. Consoli et al., Low- and zero-speed sensorless control of synchronous reluctance motors, *IEEE Trans. Ind. Appl.*, Vol. 35, No. 5, September/October 1999, pp. 1050–1057.
58. J.-I. Ha et al., Sensorless field-orientation control of an induction machine by high-frequency signal injection, *IEEE Trans. Ind. Appl.*, Vol. 35, No. 1, January/February 1999, pp. 45–51.
59. J. Ha, Sensorless control of AC machines using high frequency signal injection, Ph.D. thesis, Seoul National University, Seoul, Korea, 2001 (in Korean).
60. J.-I. Ha et al., Position controlled synchronous reluctance motor without rotational transducer, *IEEE Trans. Ind. Appl.*, November/December Vol. 35, No. 6, 1999, pp. 1193–1198.
61. J.-I. Ha et al., Sensorless position control and initial position estimation of interior permanent magnet motor, in *Conference Record of the 2001 IEEE Industry Applications Society Annual Meeting*, 2001, Vol. 4, pp. 2607–2613.
62. J. Jang, Sensorless control of PMSMs for wide speed operation range, Ph.D. thesis, Seoul National University, Seoul, Korea, 2006 (in Korean).
63. J. H. Jang et al., Sensorless drive of surface mounted permanent magnet motor by high frequency signal injection based on magnetic saliency, *IEEE Trans. Ind. Appl.*, Vol. 39, No.4, July/August 2003, pp. 1031–1039.
64. B. H. Bae et al., Implementation of sensorless vector control for super-high speed PMSM of turbo-compressor, *IEEE Trans. Ind. Appl.*, Vol. 39, No. 3, May/June 2003, pp. 811–818.
65. Y. C. Son et al., Sensorless operation of permanent magnet motor using direct voltage sensing circuit, in *Conference Record of the 2002 IEEE Industry Applications Society Annual Meeting*, 2002.
66. J. I. Ha et al., Sensorless rotor position estimation of an interior permanent magnet motor from initial states, *IEEE Trans. Ind. Appl.*, Vol. 39, No. 3, May/June 2003, pp. 761–766.
67. J. Holtz, Initial rotor polarity detection and sensorless control of PM synchronous machines, in *Conference Record of the 2006 IEEE Industry Applications Society Annual Meeting*, 2006, Vol. 4, pp. 2040–2047.
68. J. I. Ha et al., Physical understanding of high frequency injection method to sensorless drives of an induction machine, in *Conference Record of the 2000 IEEE Industry Applications Society Annual Meeting*, 2000, Vol. 3, pp. 1802–1808.
69. J. I. Ha et al., Design and selection of AC machines for saliency-based sensorless control, in *Conference Record of 2000 IEEE Industry Applications Society Annual Meeting*, 2002, Vol. 2, pp. 1155–1162.
70. J. H. Jang et al., Analysis of permanent magnet machine for sensorless control based on high frequency signal, *IEEE Trans. Ind. Appl.*, Vol. 40, No. 6, November/December 2004, pp. 1595–1604.
71. T. M. Jahn et al., Interior permanent magnet synchronous motor for adjustable speed drive, *IEEE Trans. Ind. Appl.*, Vol. IA-22 No. 4, July/August 1986, pp. 738–747.

Chapter 7

Practical Issues

In this chapter, several practical problems, which appear to implement the control algorithms described in the previous chapters of this text, are addressed and the possible solutions against the problems are introduced. At the first, to drive the electric machines using the power converter mentioned from Sections 2.18.2 to 2.18.6, because of the finite switching time of the power semiconductor switches and the delays in the signal propagation, a dead time or a blanking time is inserted in the gating signal of the switches. The dead time degrades the control performance of the electric machine drive system, especially at low-speed operation. To lessen the problem due to the dead time, some countermeasures are introduced in this chapter. And next, because the performance of the regulator basically relies on the measurement of the variables, the accurate measurement of the current is crucial to the high-performance electric machine drive system. The offset and scale errors of the current sensors and the delays in the measurement system is inevitable, but their negative effects to the control performance of the drive system can be reduced by some tricks. Finally, because of the digital implementation of the control algorithm, there are delays from the sample and holder, the execution time of the algorithm, and PWM of the power converter. The delays may severely limit the control performance of the drive system, especially for higher control bandwidth system or for super-high-speed operation of the electric machine. Some remedies to cope with the delays from the digital signal processing are discussed in this chapter.

7.1 OUTPUT VOLTAGE DISTORTION DUE TO DEAD TIME AND ITS COMPENSATION [1–3]

A power circuit of a three-phase voltage source PWM inverter with an IGBT (insulated gate bipolar transistor) to drive an AC machine is shown in Fig. 7.1. The output voltage of the inverter is decided by the complementary switching of the power semiconductor switches in each phase. If, at each phase, the turning-on signal of the upper switch of a phase is set as soon as the turning-on signal of the lower switch disappears, then there may be shoot-through because of the signal transfer delay and

Control of Electric Machine Drive Systems, by Seung-Ki Sul

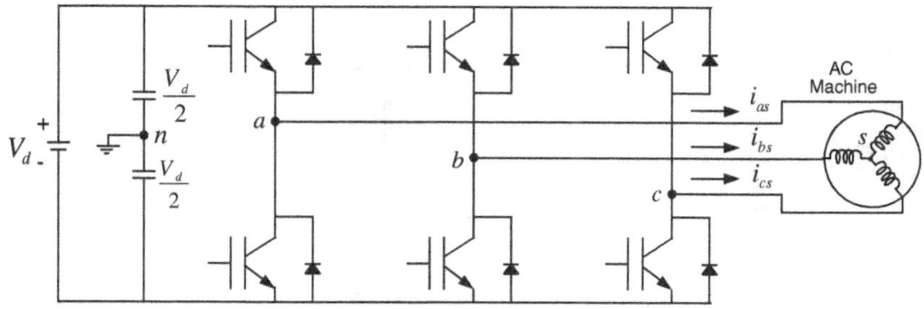

Figure 7.1 Three–phase PWM inverter with an AC machine employing an IGBT as an active switching devices.

turn-on and turn-off time of the semiconductor switches. Hence, for a certain time period, the gating signals of both upper and lower switches are maintained as turning off. This time period is called the dead time or the blanking time. For the dead time, the pole voltage of the inverter at the phase is decided by the polarity of the current flowing at the phase. Also, the pole voltage is out of control for the dead time. Furthermore, the output voltage to the AC machine and the current in the machine would be distorted. The distortion of the current may result in torque ripples and acoustic noises in the drive system. To compensate the voltage and current distortion, the phenomenon due to the dead time should be clearly understood.

For the dead time, because of no turning-on signals to both switches at a phase of three-phase inverter, the current cannot flow through the active switching devices, which is the IGBT in Fig. 7.1. Only the current can flow through the upper diode or the lower diode according to the polarity of the phase current. If the current flows to an AC machine, which is defined as the positive direction of the current flow, then the lower diode is turned on. However, for the negative direction of the current flow, the upper diode is turned on. According to the direction of the current flow, the pole voltage, which is defined as the voltage between the output terminal of a phase of inverter and mid point of DC link noted as n in Fig. 7.1, can be defined. The voltage distortion due to the dead time can be compensated by adjusting the timing of the gating signals for the active switches in the inverter.

7.1.1 Compensation of Dead Time Effect

The basic principle of the compensation is adjusting the timing of the gating signal to make the output pole voltage be the same as the commanded reference pole voltage by considering the polarity of the current and the sequence of on and off signals. In Fig. 7.2, the compensation method according to the polarity of the current is shown. In Fig. 7.2a, since the current polarity is positive, the lower diode conducts for the dead time. Hence, the pole voltage is decided solely by the gating signal of the upper switch. Therefore, the gating signal of the upper switch can be set without considering the dead time to synthesize the commanded reference pole voltage. And the dead time is implemented by shrinking the turn-on interval of the lower switch by the dead time. If the polarity is negative as shown in Fig. 7.2b, then the pole voltage is decided solely by

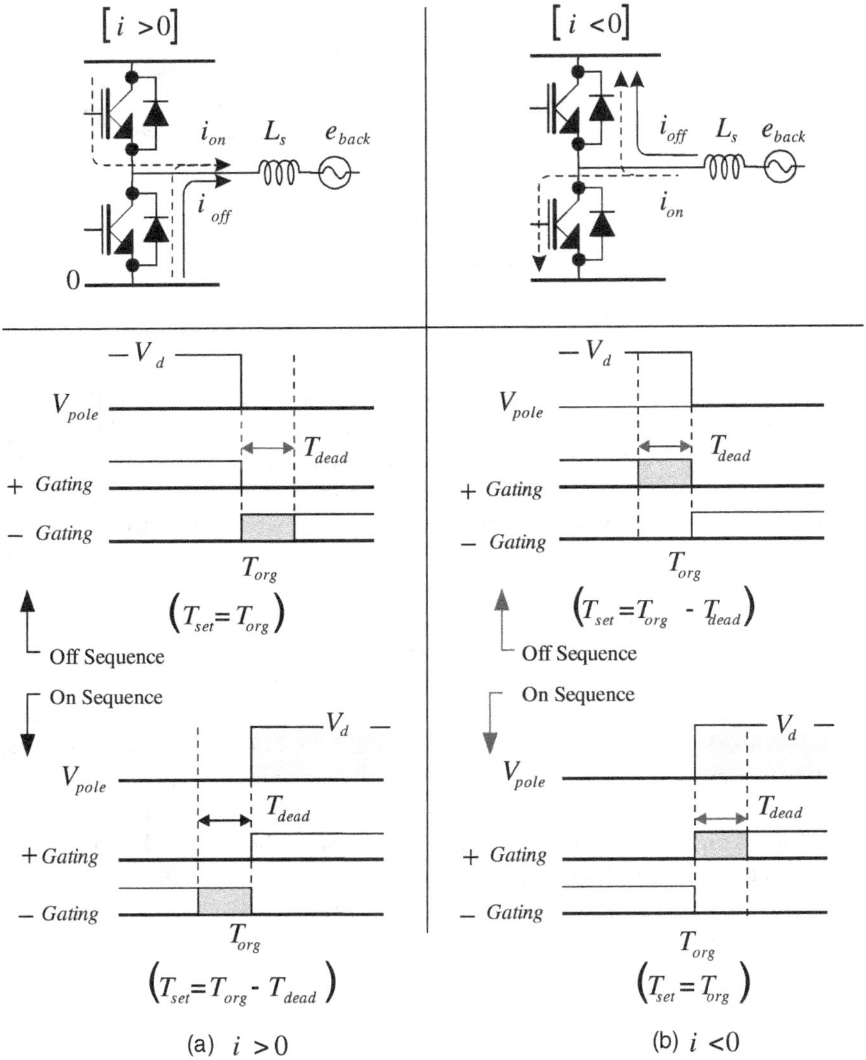

Figure 7.2 Compensation of dead time effects according to current polarity and on–off sequence (i_{on}, phase current before or after dead time; i_{off}, current during dead time; T_{dead}, time interval of dead time; T_{org}, ideal switching time point; T_{set}, switching time point after dead time compensation; L_s, equivalent inductance of an AC machine defined at the stator side; e_{back}, back EMF voltage of AC machine, + Gating: Gating signal for upper switch, where logic "high" means turn-on. – Gating: Gating signal for lower switch).

the gating signal of the lower switch. Therefore, the gating signal of the lower switch can be set without considering the dead time to synthesize the commanded voltage. In the figure, off sequence means that the current is commutated from upper switches to lower switches whether the polarity of the current is positive or negative. And on sequence is vice versa.

The above description can be expressed as (7.1) and (7.2).
For positive polarity of the current

$$\begin{array}{lll}\text{Off sequence} & \Rightarrow & T_{set} = T_{org} \\ \text{On sequence} & \Rightarrow & T_{set} = T_{org} - T_{dead}\end{array} \tag{7.1}$$

For negative polarity of the current

$$\begin{array}{lll}\text{Off sequence} & \Rightarrow & T_{set} = T_{org} - T_{dead} \\ \text{On sequence} & \Rightarrow & T_{set} = T_{org}\end{array} \tag{7.2}$$

If the propagation delay of the gating signal to all six active switches of the inverter and the turn-off and the turn-on time of all switches are the same, the above compensation method would work perfectly. But if there are imbalance in the switching and signal propagation characteristics between the upper and lower switches and imbalance between phases of the inverter, those imbalances should be considered [1].

7.1.2 Zero Current Clamping (ZCC) [2]

The voltage distortion due to the dead time is getting severe as the modulation index of PWM decreases, which is that the magnitude of pole voltage decreases. In particular, when the magnitude of the phase current is almost zero, then the current clamps to zero due to the dead time effects even if there are some voltage command reference in that phase. For the dead time, where both active switches in a phase are turned off, the current can flow through a diode in the phase. If the current flows through the diode, the inductive energy stored on the inductance of an AC machine is transferred to a DC link and the magnitude of the phase current decreases. If the magnitude of the current decreases to zero for the dead time, then the current can flow through neither the active switches nor the diode in the phase during the dead time. In this situation, if the modulation index is continuously small and the frequency of the output voltage is low, then the magnitude of the current increases slightly in a consecutive switching period. Hence, the current again reaches to zero for the dead time and it is kept as zero for the dead time. This situation would be continued until the modulation index is high enough. And for a while, the phase current clamped on zero. This zero current clamping phenomenon results in low-order harmonics to phase current. To prevent this phenomenon, the voltage synthesization considering the dead time is required especially when the magnitude of the current is near zero at switching of the power semiconductors.

7.1.3 Voltage Distortion due to Stray Capacitance of Semiconductor Switches [3]

If the magnitude of the current is small enough, then for the dead time the voltage across the power semiconductor switch varies slowly due to the stray capacitance parallel to the switch. This slow variation of the voltage results in voltage distortion if it is not compensated properly. This phenomenon occurs when the capacitance of the stray capacitor across the switch, C_{st}, is larger than that the threshold value, C_{zcc}, given by the magnitude of the phase current, i, the dead time, T_{dead}, and DC link voltage, V_d, as (7.3). This phenomenon occurs regularly in a region where the current is small enough to satisfy the inequality as $C_{st} > C_{zcc}$. To understand this phenomenon, the pole voltage from an inverter where the active switching devices are MOSFETs can be investigated as follows.

$$C_{zcc} = |i| \frac{T_{dead}}{V_{dc}} \tag{7.3}$$

As shown in Fig. 7.3, if the polarity of the current is positive and the magnitude of the current is small enough, to turn on the upper switch the lower switch should be turned off at the first. In this situation, for the dead time the output current flows through the lower diode. After the dead time period ends, the upper active switch is turned on and the stray capacitance is charged by a DC link instantaneously. And there is no distortion due to the stray capacitance in this switching situation. But, just after the instant the gating signal of the upper switch is changed from logic "1" to logic "0" to turn off the upper active switch, both the upper and lower active switches are nonconducting for the dead time. At that instant if the polarity of the current is positive, then the current should flow through the diode. However, because of the reverse voltage across the lower diode due to the stray capacitors of both switches, the current doesn't flow through the diode. Instead, the current flows through the stray capacitors across the switches, and if the phase current is constant for the dead time, the voltage across the switches vary linearly from the DC link voltage, V_d, to zero for the case of the lower switch. Hence the pole voltage varies from a half of DC link to a negative half of DC link as shown in Fig. 7.4a. At another instant, if the gating signal of

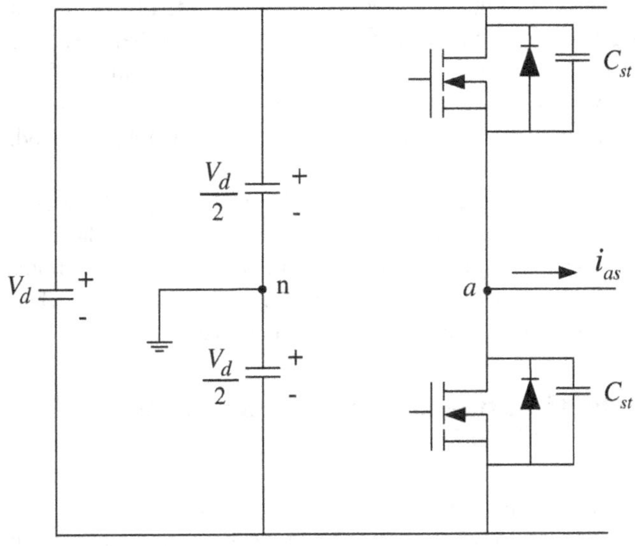

Figure 7.3 One phase of a PWM inverter implemented by MOSFET switches (here, "A" phase is assumed).

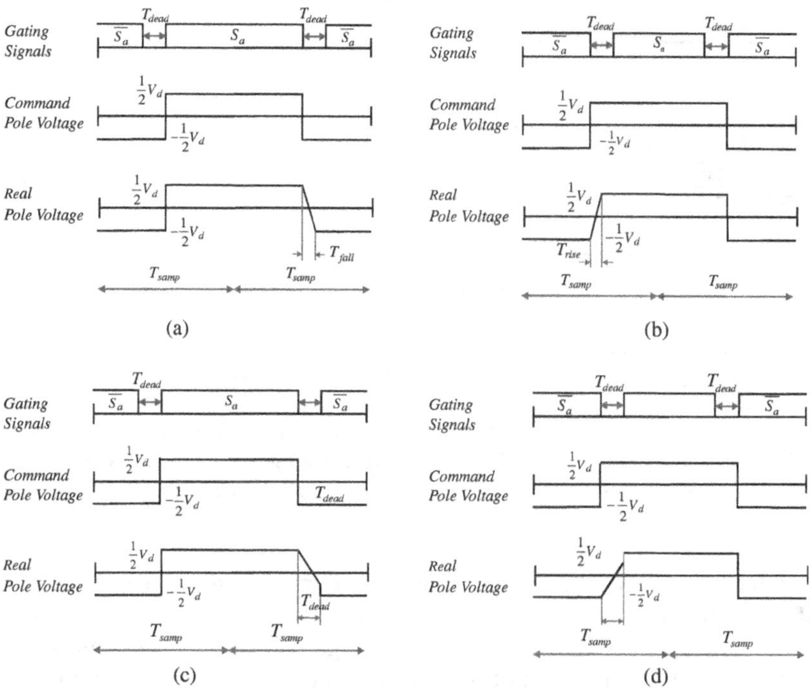

Figure 7.4 Voltage distortion due to the stray capacitor for the dead time. (a) Voltage distortion when $i_{as} > 0$. (b) Voltage distortion when $i_{as} < 0$. (c) Voltage distortion when $i_{as} > 0, i_{as} \cong 0$. (d) Voltage distortion when $i_{as} < 0, i_{as} \cong 0$.

the lower switch is changed from logic "1" to logic "0" to turn off the lower active switch and if the upper active switch is still nonconducting for the dead time, then the pole voltage varies from a negative half of DC link to a positive half of DC link as shown in Fig. 7.4b. The rate of the voltage variation is inversely proportional to the capacitance of the stray capacitors of both switches and proportional to the magnitude of the current. After the dead time, even though the pole voltage still varies linearly toward a negative or a positive half of the DC link, if the active switch is turned on, then the pole voltage jumps instantaneously to the negative half of the DC link as Fig. 7.4c or to the positive half of the DC link as Fig. 7.4d. This voltage distortion phenomenon during the dead time due to the stray capacitances of the power semiconductors is becoming severe as the magnitude of the phase current is becoming small.

In Figs. 7.4a and 7.4b, T_{fall} stands for the time for which the pole voltage varies from a positive half of DC link to a negative half of DC link, and T_{rise} stands for the time for which the pole voltage varies from a negative half of DC link to a positive half of DC link. The average of the commanded reference pole voltage and the actual pole voltage for a sampling period can be deduced as (7.4) and (7.5), respectively. To nullify the voltage distortion due to the stray capacitors for the dead time in the average sense, two average voltages in a switching period should be the same. If the

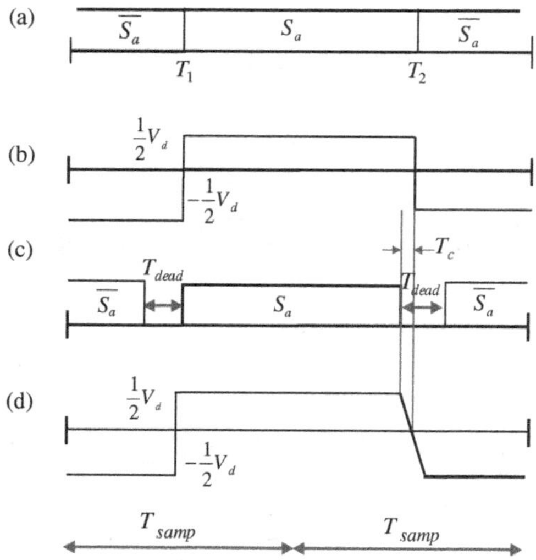

Figure 7.5 Compensation of the distortion voltage due to stray capacitors when $i_{as} > 0$. (a) Ideal gating signals. (b) Command pole voltage. (c) Compensated gating signals. (d) Real pole voltage.

turn-off and turn-on time of the switching devices are neglected, if the conduction voltage of the devices are also neglected, and if the phase current is small enough but its polarity is positive, then the same average voltage can be obtained simply by advancing the turn-off timing point of the gating signal of the upper switch by the compensation time, T_c, as shown in Fig. 7.5.

$$\langle V_{an}^* \rangle = \frac{1}{T_{samp}} \int_0^{T_{samp}} V_{an}^*(t)\, dt \tag{7.4}$$

$$\langle V_{an} \rangle = \frac{1}{T_{samp}} \int_0^{T_{samp}} V_{an}(t)\, dt \tag{7.5}$$

If the polarity of the current is negative, the voltage distortion can be compensated by simply advancing the turn-off timing point of the gating signal of the lower switch by the compensation time, T_c, as shown in Fig. 7.6. The compensation time, T_c, is a nonlinear function of the magnitude of the phase current. It can be obtained through the experimental test or by solving (7.4) and (7.5) considering the rate of the voltage variation for the dead time. The rate is decided by the magnitude of the phase current and total capacitance of the stray capacitors in a phase. If the current is almost zero, then the compensation time is the same with the dead time itself as seen in Fig. 7.7.

7.1.4 Prediction of Switching Instant [4]

To compensate the voltage distortion due to the dead time, the polarity and the magnitude of the phase current at the switching instant should be known to a PWM

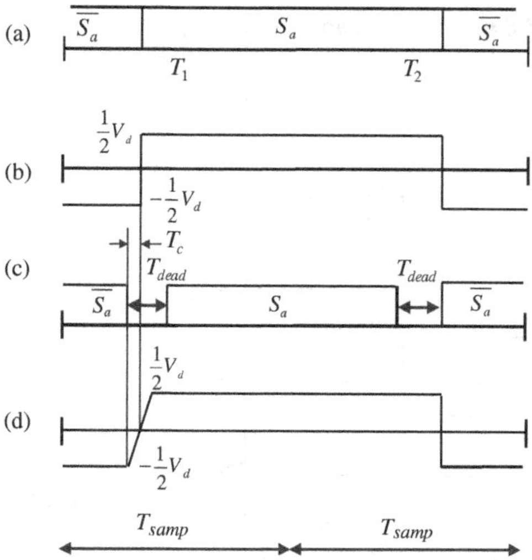

Figure 7.6 Compensation of the distortion voltage due to stray capacitors when $i_{as} < 0$. (a) Ideal gating signals. (b) Command pole voltage. (c) Compensated gating signals. (d) Real pole voltage.

gating signal generator. However, because of the digital implementation of the control algorithm and the generation of PWM gating signals in discrete manner, the current is usually sampled at peak and/or valley of the carrier wave as described in Section 4.2.3.2. At the peak or the valley of the carrier wave, the switching occurs very rarely. And, there are differences between the sampled current and the current at the switching instant of each phase. In particular, if the current magnitude is small and the inductance of AC machine is small, then not only the magnitude but also the polarity would be different between the sampled current and the current at the switching instant. In this case the compensation of the voltage distortion based on the sampled current makes the voltage distortion more severe. In a three-phase AC machine drive system, for a period of fundamental current of an AC machine, the three-phases current changes its polarity six times in total. And at each zero crossing point, when the polarity changes, the magnitude of the current is very small and the voltage distortion occurs. That results in torque ripples whose frequency is six times the fundamental frequency. Hence, for accurate compensation, the current polarity and the magnitude at the switching instant should be identified. The current at the switching instant can be directly measured, or the current at that instant can be predicted based on the sampled current and the parameters of the inverter and AC machine. The direct measurement is robust to the parameter errors of the drive system, but it needs an additional 12 times of the sampling of the current in a switching period. And the 12 sampling instants should be synchronized to the gating signal of the

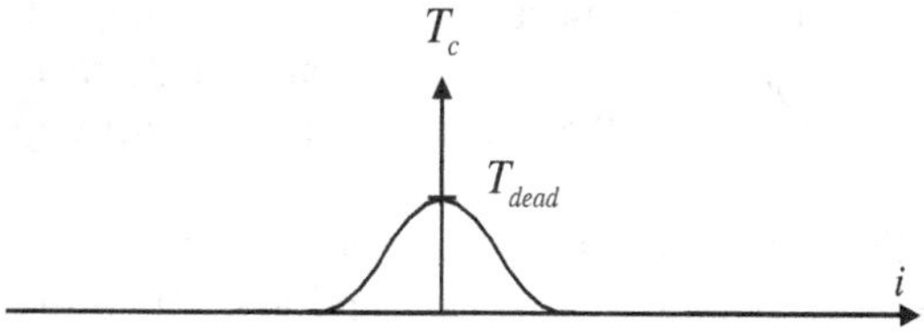

Figure 7.7 Compensation time, T_c.

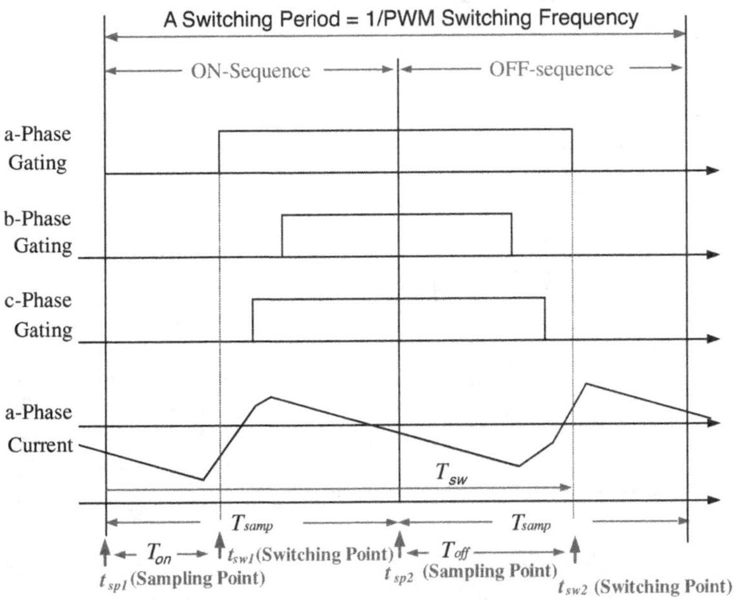

Figure 7.8 Gating signals from a three-phase symmetry space vector PWM and a-phase current in a switching period.

corresponding switch. Also, it results in extra hardware burdens to the drive system. The prediction of the current is sensitive to the parameter errors of the drive system, but it can be implemented by the software without any extra hardware burden. In the case of most AC machines, the inductance is large enough to assume that the current varies linearly regarding to the time in a switching period. The typical current waveform and the gating signals of three phases are shown in Fig. 7.8 under the assumption of double sampling in a switching period as mentioned in Section 4.2.3.2. The prediction process can be easily formulated through the following equations for the case of an induction machine driven by a three-phase symmetry space vector PWM inverter.

The current is sampled at the beginning of the sampling period, which is the peak or valley of the triangular carrier wave in the case of doubling sampling. Because of the digital execution delay of the current regulation algorithm, the current sampled at t_{sp1} should be used to compensate the current at the switching instant t_{sw2}. In Fig. 7.8, the current at t_{sp1} is totally different from the current at the switching instant at t_{sw2}. Hence, if the dead time compensation is done based on the current at t_{sp1}, then the voltage distortion would be severe. To predict the current at t_{sw2} based on the sampled current at t_{sp1}, the time difference between t_{sp1} and t_{sw2} should be calculated. As seen from Fig. 7.8, the difference, which is noted as T_{sw}, is the sum of the sampling period, T_{samp}, and the gating time, T_{off}.

The stator voltage equation of the induction machine at the synchronously rotating reference frame can be written as (7.6) under the assumption of the rotor flux-oriented vector control described in Section 5.2.

$$
\begin{aligned}
V_{ds}^{e} &= R_s i_{ds}^{e} + \sigma L_s p i_{ds}^{e} + \frac{L_m}{L_r} p \lambda_{dr}^{e} - \omega_e \sigma L_s i_{qs}^{e} \\
V_{qs}^{e} &= R_s i_{qs}^{e} + \sigma L_s p i_{qs}^{e} + \omega_e \left(\sigma L_s i_{ds}^{e} + \frac{L_m}{L_r} \lambda_{dr}^{e} \right)
\end{aligned} \tag{7.6}
$$

where p is the differential operator, that is, d/dt. If the rotor flux is kept as constant as $\lambda_{dr}^{e} = L_m i_{ds}^{e}$, then (7.6) can be simplified as follows:

$$
\begin{aligned}
V_{ds}^{e} &= R_s i_{ds}^{e} + \sigma L_s p i_{ds}^{e} - \omega_e \sigma L_s i_{qs}^{e} \\
V_{qs}^{e} &= R_s i_{qs}^{e} + \sigma L_s p i_{qs}^{e} + \omega_e L_s i_{ds}^{e}
\end{aligned} \tag{7.7}
$$

Usually, because the time difference, T_{sw}, is much smaller than the stator transient time constant, defined as $\tau_s = \sigma L_s / R_s$, the current variation can be approximated as a linear function of time. Then the variation of the current for the time difference can be deduced as (7.8) under the assumption of ideal PWM of the inverter.

$$
\begin{aligned}
&V_{ds}^{e} = V_{ds}^{e*}, && V_{qs}^{e} = V_{qs}^{e\,*} \\
&i_{ds}^{e} = \left(i_{ds}^{e\,m} + i_{ds}^{e\,P}\right)/2, && i_{qs}^{e} = \left(i_{qs}^{e\,m} + i_{qs}^{e\,P}\right)/2 \\
&p i_{ds}^{e} \approx \left(i_{ds}^{e\,P} - i_{ds}^{e\,m}\right)/T_{sw}, && p i_{qs}^{e} \approx \left(i_{qs}^{e\,P} - i_{qs}^{e\,m}\right)/T_{sw}
\end{aligned} \tag{7.8}
$$

where V_{ds}^{e*} and V_{qs}^{e*} are the commanded reference d–q voltages, i_{ds}^{em} and i_{qs}^{em} are the sampled d–q currents at t_{sp1}, i_{ds}^{eP} and i_{qs}^{eP} are the predicted d-q currents at t_{sw2}. By substituting (7.8) into (7.7), (7.9) can be obtained with the reference d–q voltages instead of the actual d–q voltages.

$$
\begin{bmatrix} i_{ds}^{eP} \\ i_{qs}^{eP} \end{bmatrix} = \frac{1}{a^2 - bc} \left(\begin{bmatrix} ad + bc & -ab - bd \\ -cd - ac & bc + ad \end{bmatrix} \begin{bmatrix} i_{ds}^{em} \\ i_{qs}^{em} \end{bmatrix} + \begin{bmatrix} a & -b \\ -c & a \end{bmatrix} \begin{bmatrix} V_{ds}^{e*} \\ V_{qs}^{e*} \end{bmatrix} \right) \tag{7.9}
$$

where $a = R_s/2 + \sigma L_s/T_{sw}$, $b = -\omega_e \sigma L_s/2$, $c = \omega_e \sigma L_s/2$, and $d = \sigma L_s/T_{sw} - R_s/2$.

Because T_{sw} is small, and $R_s, \omega_e L_s << \sigma L_s/T_{sw}$. With this approximation, the voltage equation (7.7) can be written as (7.10). Based on the voltage equation, the predicted current can be calculated using (7.11).

$$
\begin{aligned}
V_{ds}^{e*} &= R_s i_{ds}^{em} + \sigma L_s \left(i_{ds}^{eP} - i_{ds}^{em}\right)/T_{sw} - \omega_e \sigma L_s i_{qs}^{em} \\
V_{qs}^{e*} &= R_s i_{qs}^{em} + \sigma L_s \left(i_{qs}^{eP} - i_{qs}^{em}\right)/T_{sw} + \omega_e L_s i_{ds}^{em}
\end{aligned} \tag{7.10}
$$

$$
\begin{aligned}
i_{ds}^{eP} &= i_{ds}^{em} + \left(V_{ds}^{e*} - R_s i_{ds}^{em}\right) T_{sw}/\sigma L_s + \omega_e T_{sw} i_{qs}^{em} \\
i_{qs}^{eP} &= i_{qs}^{em} + \left(V_{qs}^{e*} - R_s i_{qs}^{em}\right) T_{sw}/\sigma L_s - \omega_e T_{sw} i_{ds}^{em}
\end{aligned} \tag{7.11}
$$

After finding d–q currents at t_{sw2}, the a-phase current at t_{sw2} can be calculated through the reference frame transformation described in Section 3.1. With similar process, the b- and c-phase currents at switching instant can be predicted, respectively. The compensations of the voltage distortion due to the dead time, which are described in the previous sections, can be done based on these predicted three-phase currents at switching instants of each phase.

7.2 MEASUREMENT OF PHASE CURRENT

As mentioned in Section 4.2, for the high-performance control of the electric machines and power converters, the current regulation is indispensable. Because the regulation is done on the basis of the measured current, the accurate measurement of the current is the utmost important task for the high-performance current control [5]. Some basic principles of the current measurement are described in Section 4.2.1 and some implementation issues are also discussed in Section 4.2.3.2. However, for the practical AC machine drive system fed by a PWM inverter, due to PWM, the phase current includes ripples inevitably. Because the most of the modern AC machine drive system is controlled by digital microelectronics, the current should be sampled at every sampling point and converted to digital value. To get the corresponding digital value from the ripple-rich current, several signal processing techniques can be used as discussed in Section 4.2.3.2. To achieve maximum regulation bandwidth, the synchronized sampling with the PWM carrier is widely used in the industry [6, 7]. Through this method, the current can be sampled twice per switching period in the center of zero vectors if a three-phase symmetry space vector PWM is used. However, even with this method, there are still some measurement errors because of the hardware of the current sampling, namely, the current sensor itself, the analog low-pass filter to cut off the measurement noise, and the analog-to-digital converter. In particular, the time delays come from the analog filter, the execution of the control algorithm, and the PWM, resulting in the cross-coupling of the current dynamics at the d–q reference frame. And, the current regulation performance would be degraded, and in the extreme case the current regulation loop would be unstable [8]. Furthermore, because of the delay from the analog filter, the ripples of the phase current would be included in the measured current. And the regulation bandwidth cannot be extended as designed due to this ripple component. In following sections, the practical issues regarding current measurement is addressed in detail and the remedies to cope the above problems are discussed.

7.2.1 Modeling of Time Delay of Current Measurement System [9]

A block diagram of a typical current measurement system for digitally controlled electric machine drive system is shown in Fig. 7.9, where the sensor, analog low-pass filter (LPF), and analog-to-digital converter (A/D) are depicted. The A/D and the sensor itself have some time delay, but most of the time delay comes from the analog

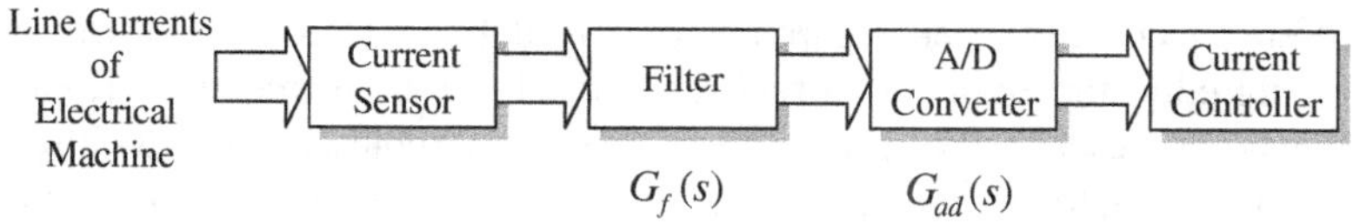

Figure 7.9 Block diagram of current measurement system of digitally controlled drive system.

LPF. The delay varies according to the tolerance and aging of the analog components. After neglecting this variation, the transfer function of the filter can be described as (7.12) and the total delay of the measurement system can be modeled as (7.13).

$$G_f(s) = \frac{(2\pi f_n)^2}{s^2 + 4\pi\zeta f_n s + (2\pi f_n)^2} \tag{7.12}$$

$$T_d = \tau_1 + \frac{1}{2\pi f_x}\tan^{-1}\frac{(2\zeta f_x/f_n)}{\left(1-\left({f_x}/{f_n}\right)^2\right)} \tag{7.13}$$

where $G_f(s)$ is the transfer function of a typical second order LPF, τ_1 is the sum of the delay due to the parts in the measurement system except LPF, f_x stands for input frequency to LPF, f_n stands for the natural undamped frequency of LPF, and ζ stands for damping coefficient of LPF. If f_x is small enough compared to f_n, (7.13) can be approximated as

$$T_d = \tau_1 + \frac{\zeta}{\pi f_n} \tag{7.14}$$

As mentioned in Section 4.2.3.2, if there is no delay between sampled current and actual current, the sampled current at the peak or the valley of PWM would be the average value of the sampling interval. However, as shown in Fig. 7.10, if there is delay, then the sampled current is not the average value. For example, in Fig. 7.10, at the sampling point the actual current, i_{as}^0, is decreasing, but the sampled current, I_{as}^f, which comes from the filtered current, i_{as}^f, is larger than the actual current at the sampling point, I_{as}^0. The difference depends on the slope of the current at the sampling

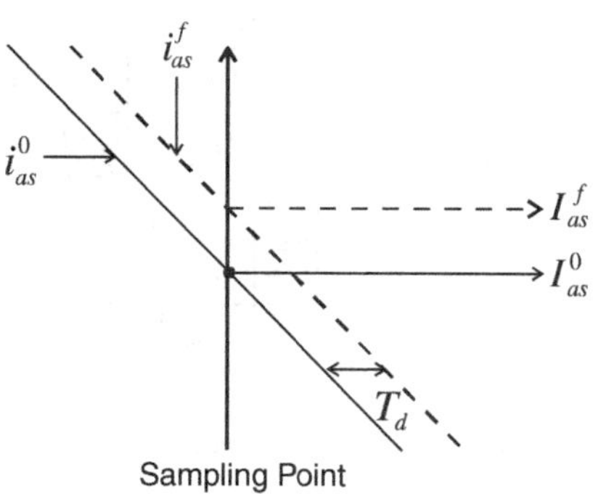

Figure 7.10 Delay in current measurement i_{as}^0, actual current; i_{as}^f, filtered current by LPF; I_{as}^0, sampled current if the actual current is sampled; I_{as}^f, sampled current from the filtered current).

point. The difference occurs at every sampling point. And the sampled current, which is used in the digitally implemented current regulator, has a high-frequency ripple component that is not present in the actual current. Due to the ripples, the bandwidth of the current regulator would be limited and the actual current may include the ripple components due to the feedback regulation.

In Fig. 7.11, with a three-phase symmetry space vector PWM the actual current and the delayed current by the second-order LPF is shown through the computer simulation at each sector of hexagon of space vector plane. In the figure, Sa, Sb, and Sc stand for the switching function of each phase of three-phase PWM inverter in Fig. 7.1. If the value of the switching function is 1, then the upper switch of the

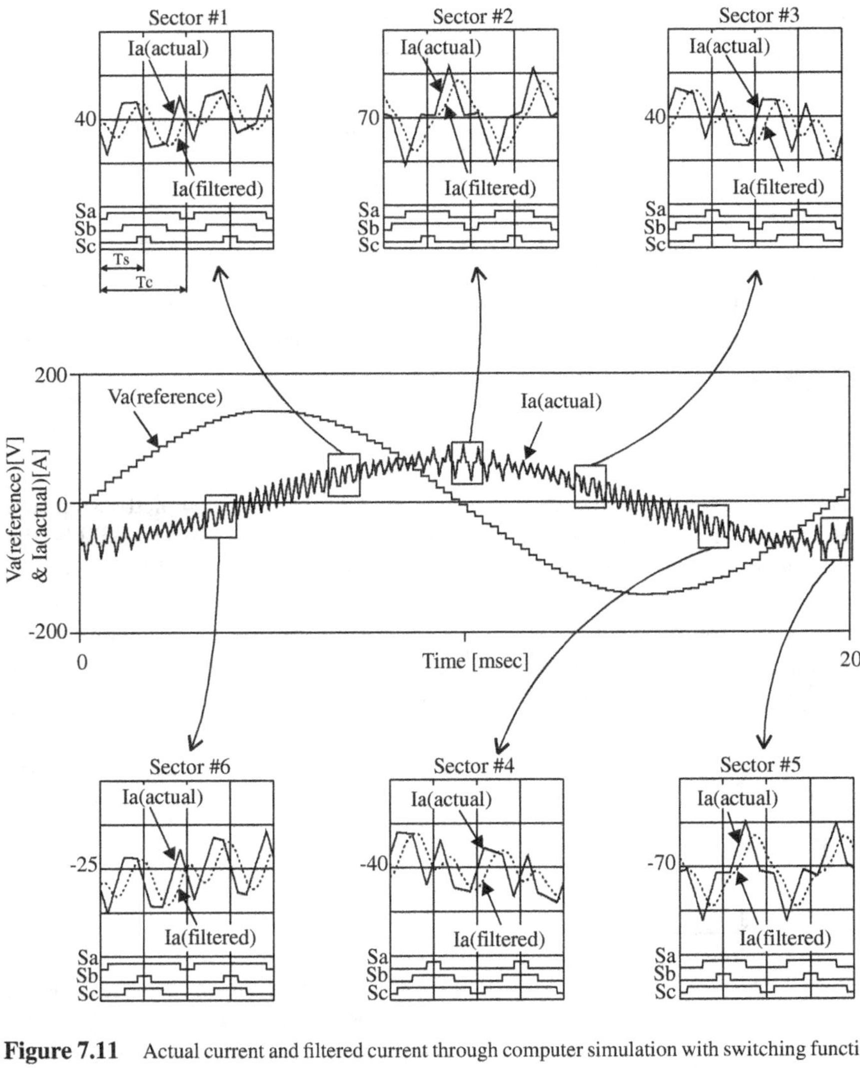

Figure 7.11 Actual current and filtered current through computer simulation with switching functions of a three-phase PWM inverter.

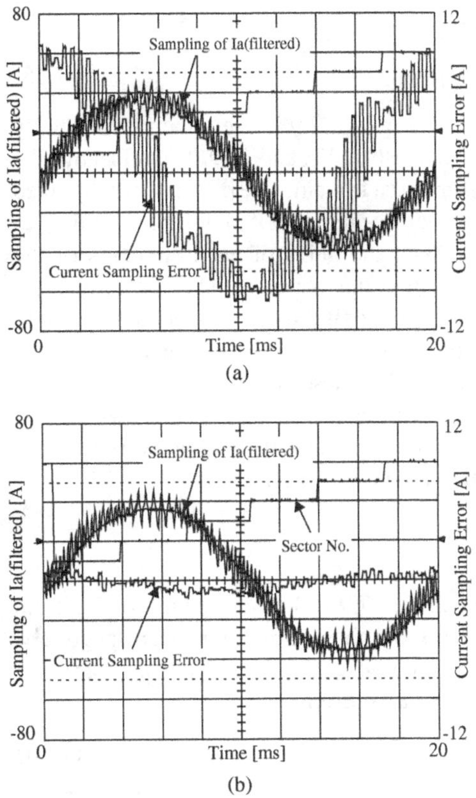

Figure 7.12 Current measurement error in the sampling according to the sampling instant. (a) Sampling at the peak of a PWM carrier. (b) Delayed sampling by T_d from the peak of a PWM carrier.

corresponding phase is turned on and the lower switch is turned off. Otherwise, the lower switch is turned on and upper switch is turned off.

The problem occurs because of the delay of the current measurement, and it can be solved by simply delaying the sampling point by the time delay of the current measurement system. In Fig. 7.12, the experimental result of the delayed sampling is shown. As seen from the figure, the error can be reduced to one twentieth of the error sampled at peak of the carrier. The proper value of delay, T_d, can be easily set by the design parameters of LPF and other parts in the measurement system. However, it should be noted that because of the delayed sampling the execution time of the current regulation loop would increase by T_d.

7.2.2 Offset and Scale Errors in Current Measurement [10]

In the three-phase AC machine drive system shown in Fig. 7.1, usually two phase currents are measured and the other phase current is calculated from the measured current as (7.15) under the assumption that the instantaneous sum of three phase currents is zero. In (7.15), the a- and b-phase currents are measured and the c-phase current is calculated based on the measured a- and b-phase currents.

$$i_{cs} = -(i_{as} + i_{bs}) \tag{7.15}$$

If the current measurement system shown in Fig. 7.9 is used, then the digitally converted current may have offset because of sensor, LPF, and A/D. The offset may be different in each phase of the current measurement system. The offset is usually fixed with the components in the current measurement system, but it may vary according to noise, operating temperature, and aging. Also, the gains of each elements of the measurement system vary, and the overall gain of one phase of the current measurement system may differ from that of other phase. So, the scale from actual current to the final digital value is different in each phase. In summary, each current measurement system has different offset and different scale. These offset and scale differences may result in current ripples and torque ripples of the drive system. In this section the effects of these offset and scale differences with regard to AC machine drive system are discussed and the methods to minimize the effects are introduced.

7.2.2.1 Offset

In the current measurement system shown in Fig. 7.9 and described as (7.15), if the sum of all offsets of the components at each a and b phase is δi_{as} and δi_{bs}, respectively, then the digitally converted current of all three phases, i_{as_AD}, i_{bs_AD}, and i_{cs_AD}, can be expressed as (7.16)–(7.18) in terms of the actual currents, i_{as} and i_{bs}.

$$i_{as\_AD} = i_{as} + \delta i_{as} \tag{7.16}$$

$$i_{bs\_AD} = i_{bs} + \delta i_{bs} \tag{7.17}$$

$$i_{cs\_AD} = -(i_{as\_AD} + i_{bs\_AD}) \tag{7.18}$$

The above three-phase currents can be expressed in the synchronously rotating reference d–q frame, whose rotating speed is ω_e as (7.19) and (7.20).

$$i^e_{ds\_AD} = i^e_{ds} + \delta i^e_{ds} \tag{7.19}$$

$$i^e_{qs\_AD} = i^e_{qs} + \delta i^e_{qs} \tag{7.20}$$

If the offset has fixed DC value, then the offsets in the d–q reference frame can be derived as (7.21) and (7.22).

$$\delta i^e_{ds} = \frac{2}{\sqrt{3}}\sqrt{\delta i^2_{as} + \delta i_{as}\delta i_{bs} + \delta i^2_{bs}}\ \sin(\omega_e t + \alpha) \tag{7.21}$$

$$\delta i^e_{qs} = \frac{2}{\sqrt{3}}\sqrt{\delta i^2_{as} + \delta i_{as}\delta i_{bs} + \delta i^2_{bs}}\ \cos(\omega_e t + \alpha) \tag{7.22}$$

where $\alpha = \tan^{-1}\left(\frac{\sqrt{3}\delta i_{as}}{\delta i_{as} + 2\delta i_{bs}}\right)$.

It can be seen from (7.21) and (7.22) that the offset in the phase current measurement results in AC component current ripple, whose frequency is the synchronous frequency, ω_e. In the case of the vector-controlled induction machine drive system based on the rotor flux linkage, these offsets result in torque ripples. If the synchronous speed, ω_e, is large enough compared to the inverse of the rotor time constant, $\tau_r = L_r/R_r$, then the effects of the offset in the d axis can be neglected in the viewpoint of the flux regulation. If the current regulation is perfect and the digitally converted d–q currents well track their references like $i_{ds}^{e*} = i_{ds\_AD}^{e}$ and $i_{qs}^{e*} = i_{qs\_AD}^{e}$, then the actual d–q-axis current can be represented as (7.23) and (7.24).

$$i_{ds}^{e} = i_{ds\_AD}^{e} - \delta i_{ds}^{e} \tag{7.23}$$

$$i_{qs}^{e} = i_{qs\_AD}^{e} - \delta i_{qs}^{e} \tag{7.24}$$

And from the torque equation (5.21), the actual torque due to the offsets can be deduced as

$$T_e = T_e^{*} - \delta T_e = K_T i_{qs}^{e*} - K_T \delta i_{qs}^{e} \tag{7.25}$$

where $K_T = \frac{3}{2}\frac{P}{2}\frac{L_m}{L_r}\lambda_{dr}^{e}$ and i_{qs}^{e*} is the reference of the q-axis components current, which is the torque component current. The error in the torque can be derived as

$$\delta T_e = K_T \delta i_{qs}^{e} = K_T \frac{2}{\sqrt{3}}\sqrt{\delta i_{as}^{2} + \delta i_{as}\delta i_{bs} + \delta i_{bs}^{2}}\ \cos(\omega_e t + \alpha) \tag{7.26}$$

As seen from (7.26), the torque includes the ripple component and this ripple results in speed ripple in a speed-regulated drive system. If the speed of the induction machine is regulated as constant and the bandwidth of the speed regulation loop is low enough compared to the synchronous speed, ω_e, then there should be speed ripple due to the torque ripple by (7.26). And in the system, by minimizing the speed ripple whose frequency is ω_e, the offset of each phase can be nullified with the adjustment of the offset of each phase of the current measurement system. In the case of an SMPMSM drive system where the d-axis component current has no effect to the torque generation, the torque ripple due to the offsets appears similarly like (7.26). However, in the SMPMSM case, the frequency of the ripple is the rotating speed of the rotor in electrical angle, ω_r. In the case of an IPMSM drive system, where q-axis current and d-axis current are both used for the torque generation to exploit reluctance torque, not only the speed ripple whose frequency is ω_r but also that whose frequency is $2\omega_r$ is minimized to nullify the offset of both phases. However, if the d-axis current reference is set as zero, and then the frequency spectrum of the speed ripple has mainly ω_r component. In this case, the speed ripple whose frequency is ω_r can be minimized for offset adjustment. In the SynRM drive system, not only ω_r component and but also $2\omega_r$ component should be used to nullify the offsets.

7.2.2.2 Scale

In the current measurement system shown in Fig. 7.9 and described as (7.15), if the current regulation is ideal, then the digitally converted current can be described as (7.27) and (7.28) in the steady-state operation of the three-phase AC machine drive system.

$$i_{as\_AD} = I\cos(\theta_e + \phi) \tag{7.27}$$

$$i_{bs\_AD} = I\cos(\theta_e - \tfrac{2}{3}\pi + \phi) \tag{7.28}$$

where I is the magnitude of the phase current, θ_e is the instantaneous angle of the synchronous reference frame defined as $\theta_e = \omega_e t$, and ϕ, is the angle between the current vector and d axis of the reference frame. And it is assumed that because of the orthogonality of the effects from offset and from scale error the two problems can be handled separately. If the overall scale of the measurement system of each phase is k_a and k_b, respectively, then the actual phase current can be expressed as (7.29) and (7.30) considering scale errors.

$$i_{as} = \frac{I\cos(\theta_e + \phi)}{k_a} \tag{7.29}$$

$$i_{bs} = \frac{I\cos(\theta_e - \frac{2}{3}\pi + \phi)}{k_b} \tag{7.30}$$

From (7.29), (7.30), and (3.4), if the current regulation is perfect and the digitally converted d–q currents well track their references like $i_{ds}^{e*} = i_{ds\_AD}^{e}$ and $i_{qs}^{e*} = i_{qs\_AD}^{e}$, then the d–q-axis current error in complex vector form can be described as

$$\begin{aligned}\delta\mathbf{i}_{dqs}^{e} &= \mathbf{i}_{dqs\_AD}^{e} - \mathbf{i}_{dqs}^{e} = (i_{qs\_AD}^{e} - i_{qs}^{e}) + j(i_{qs\_AD}^{e} - i_{qs}^{e}) = \delta i_{ds}^{e} + j\delta i_{qs}^{e} \\ &= \left(\frac{k_a - 1}{k_a}\right) I\cos(\phi) - \left(\frac{k_a - k_b}{k_a k_b}\right)\frac{I}{\sqrt{3}}\left[\sin\left(2\omega_e t - \frac{2\pi}{3} + \phi\right) - \sin\left(\phi - \frac{2\pi}{3}\right)\right] \\ &+ j\left\{\left(\frac{k_a - 1}{k_a}\right) I\sin(\phi) - \left(\frac{k_a - k_b}{k_a k_b}\right)\frac{I}{\sqrt{3}}\left[\cos\left(2\omega_e t - \frac{2\pi}{3} + \phi\right)\right.\right. \\ &\left.\left. + \cos\left(\phi - \frac{2\pi}{3}\right)\right]\right\}\end{aligned} \tag{7.31}$$

where $\mathbf{i}_{dqs\_AD}^{e}$ stands for the synchronous reference frame d–q-axis current in complex vector form, which is used in the digitally implemented current regulator, while $\mathbf{i}_{dqs}^{e}$ stands for the actual d–q axis current flowing through an AC machine. Hence, in the case of the vector-controlled induction machine drive system based on the rotor flux linkage, the torque error can be described as (7.32) under the assumption that the synchronous speed, ω_e, is large enough compared to the inverse of the rotor time constant, $\tau_r = L_r/R_r$.

$$\begin{aligned}\delta T_e &= T_e^* - T_e = \frac{3}{2}\frac{P}{2}\frac{L_m^2}{L_r}\left(i_{ds}^{e*} i_{qs}^{e*} - i_{ds}^{e} i_{qs}^{e}\right) \approx \frac{3}{2}\frac{P}{2}\frac{L_m^2}{L_r}\left(\delta i_{ds}^{e} i_{qs}^{e*} + \delta i_{qs}^{e} i_{qds}^{e*}\right) \\ &= \frac{3}{2}\frac{P}{2}\frac{L_m^2}{L_r}\left\{\delta i_{ds}^{e} I\sin(\phi) + \delta i_{qs}^{e} I\cos(\phi)\right\}\end{aligned} \tag{7.32}$$

After substituting δi_{ds}^{e} and δi_{qs}^{e} given by (7.31) into (7.32), it can be seen that torque error has a DC component due to the scale error and $2\omega_e$ frequency component. If the scales of both phases are the same, then there is no AC component torque ripple. Hence, under the constant speed control, at least the difference of the scale of each current measurement system can be nullified by minimizing the speed ripples, whose frequency is $2\omega_e$, with the adjustment of the scale of only one phase of the current measurement system. With this scale adjustment, torque ripple due to the scale difference can be removed. But, due to the scale error there are still magnitude errors in the current regulation after this adjustment, though the scales of both phases are the same. However, this DC components torque error can be rejected by the integral term of the PI regulator in the speed-regulated AC machine drive system. In an SMPMSM drive system, like the induction machine drive system, the scale difference results in $2\omega_r$ frequency component ripple torque. Hence by minimizing $2\omega_r$ frequency component speed ripple, the difference can be resolved. In the case of an IPMSM drive system, by setting $i_{ds}^{e*} = 0$ for the scale adjustment purpose, the scale difference can be adjusted by minimizing $2\omega_r$ frequency component speed ripples like the case of an SMPMSM. In an SynRM drive system, not only the $2\omega_r$ component and but also the $4\omega_r$ component should be used to nullify the difference of the scale.

For the offset and the scale adjustments, the speed ripple at a certain frequency is used in the constant-speed AC machine drive system. To apply this technique, it should be noted that the speed ripples resulting from other reasons, such as eccentricity of the rotor, unbalance of rotating magnetic motive force (MMF), load characteristics, and the dead time, should be decoupled to the ripples from offset and scale error. The eccentricity and the load characteristics usually result in integer multiples of the rotating frequency of the rotor. But the dead time results in six times the synchronous speed, which is $6\omega_e$ or $6\omega_r$. And the speed ripples due to the dead time can be differentiated with ripples due to other reasons. However, if there are differences in the actual dead time for each switch due to the difference of the propagation delay of gating signals and turn-on and turn-off time of switches, then speed ripples, whose frequency is the synchronous speed or two times of the synchronous speed, may occur. Also in some drive systems, the speed ripple may be too small for the adjustment of the offset and the scale difference because of the large inertia. In these systems, the synchronous speed should be set low enough to detect the speed ripples due to the offset and the scale difference. However, the synchronous speed should be large enough compared to the inverse of the rotor time constant in the case of an induction machine drive system. For the adjustments, the bandwidth of the speed regulation loop should be low enough compared to the frequency components of speed ripple to minimize. Otherwise, the speed regulator responds to the speed ripples, and the speed ripple would be reduced at the cost of the severe ripples in torque component current. In this case the output of the integral term of a PI speed regulator can be used for the adjustments. The output of the integral term of a PI regulator includes the frequency components due to the offset and the scale difference. The detailed description of the parameter adjustment using regulator output is in Appendix A, Section A.2.

7.3 PROBLEMS DUE TO DIGITAL SIGNAL PROCESSING OF CURRENT REGULATION LOOP

In most AC machine drive systems the control algorithms are executed digitally in a microelectronics controller such as a digital signal processor (DSP) or a microcomputer. The current regulation loop, which is usually the innermost regulation loop in the drive system, is also implemented and executed digitally. However, because of the nature of the serial execution of the software, the digital execution delay is inevitable. The delay is usually one sampling time to guarantee the maximum utilization of the modulation index of PWM inverter. As mentioned in Section 4.2.3.2, because of PWM, half of the sampling interval is further delayed in addition to the digital execution delay [8]. Hence, a 1.5 sampling interval delay occurs in a digitally controlled PWM inverter system. If the synchronous reference frame current regulator discussed in Section 4.2.5 is employed for the current regulation of an AC machine, the reference frame rotates for a 1.5 sampling interval, $1.5T_{samp}$, and the output voltage of the regulator has errors in the magnitude and angle. These errors can be neglected when the synchronous speed, ω_e, is low enough compared to the sampling frequency—for example, if $\omega_e \leq \frac{1}{40}\frac{2\pi}{T_{samp}}$. Otherwise, the errors may result in the stability problem of the current regulation loop, and overcurrent fault may occur. Furthermore, if the synchronous frequency of an AC machine $f_e = \frac{\omega_e}{2\pi}$ increases above one-twentieth of the sampling frequency, then in addition to the error discussed in Section 7.3.1, the magnitude error in the sampled current occurs. Because of this error, the magnitudes of the flux and the torque of the AC machine reveal the differences with their reference values.

7.3.1 Modeling and Compensation of Current Regulation Error due to Digital Delay [11]

In Fig. 7.13, a current regulator for a three-phase AC load described in Section 4.2.5 is depicted in the complex vector form. In the figure, a PI-type regulator with

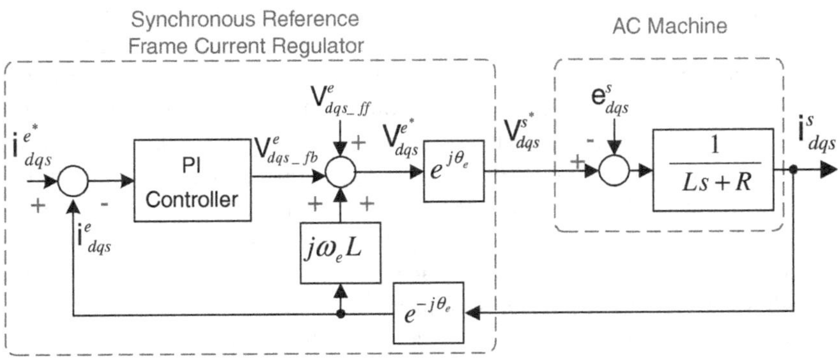

Figure 7.13 Block diagram of three-phase AC current regulator in complex vector form.

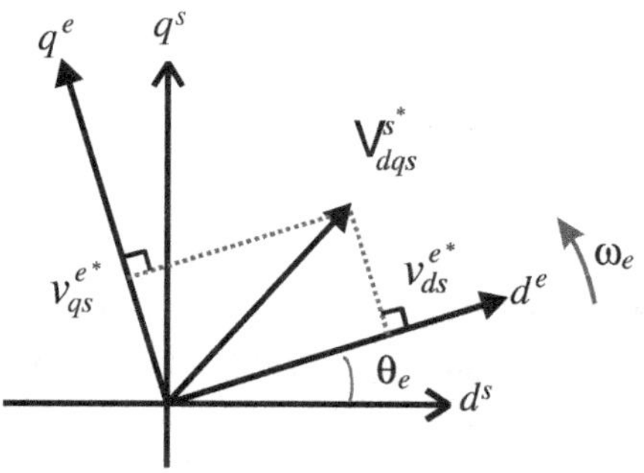

Figure 7.14 Voltage reference in synchronous and stationary reference frames

feed-forward decoupling terms is employed. From the stator voltage equation of the induction machine in synchronous reference frame as (4.31), the feed-forward term can be deduced as

$$\mathbf{V}^e_{dqs\_ff} = \mathbf{e}^e_{dqs} = -\frac{L_m}{L_r}\omega_{br}\boldsymbol{\lambda}^e_{dqr} \tag{7.33}$$

where $\omega_{br} = R_r/L_r - j\omega_r$, and for the induction machine, the resistance and inductance of the load in Fig. 7.13 can be expressed as

$$L = L_\sigma \equiv L_s - \frac{L_m^2}{L_r}, \qquad R = R_s + \left(\frac{L_m}{L_r}\right)^2 R_r \tag{7.34}$$

In the case of an SMPMSM, the feed-forward term can be expressed as

$$\mathbf{V}^r_{dqs-ff} = \mathbf{e}^r_{dqs} = j\omega_r\lambda_f \tag{7.35}$$

where λ_f is the flux linkage by the permanent magnet. Also, the resistance and inductance of the load in Fig. 7.13 can be expressed as

$$L = L_s, \qquad R = R_s \tag{7.36}$$

As shown in Fig. 7.14, if the d axis of the synchronous reference frame is apart from the d axis of the stationary reference frame by θ_e, the reference voltage and current at the synchronous reference frame can be expressed as (7.37) and (7.38) as discussed in Section 3.1, respectively.

$$\mathbf{V}^{s*}_{dqs} = \mathbf{V}^{e*}_{dqs}e^{j\theta_e} \tag{7.37}$$

$$\mathbf{i}^{e*}_{dqs} = \mathbf{i}^{s*}_{dqs}e^{-j\theta_e} \tag{7.38}$$

As shown in Fig. 7.15, if the current is sampled at the starting point of every sampling interval, the current regulation algorithm is executed after the current sampling, and the gating signals based on the results of the algorithm are updated at the end of the sampling interval, then as mentioned in Section 4.2.3.2 there should be delay between the current sampling and the execution of PWM based on the sampled current by 1.5 sampling interval, $1.5T_{samp}$.

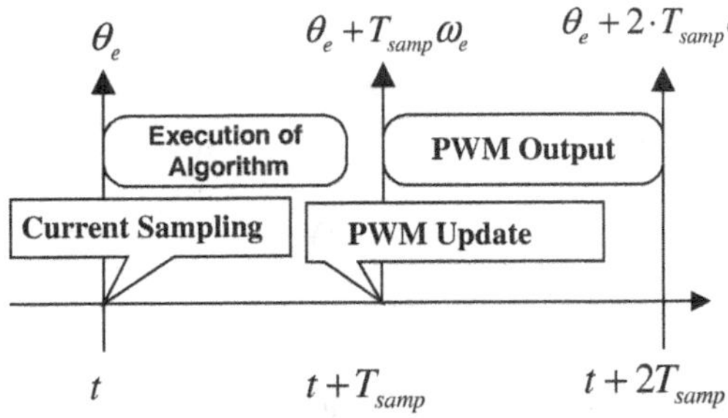

Figure 7.15 Time line of current sampling and PWM.

The rotation of the synchronous reference frame during the sampling interval is shown in Fig. 7.16 together with the outputs of the current regulator, $\mathbf{V}_{dqs}^{e*}$. At the time point, t, the current is sampled and the regulator algorithm is executed from time point, t to time point $t + T_{samp}$. After then, the updated PWM signals are kept for the next sampling period, which is from $t + T_{samp}$ to $t + 2T_{samp}$. During algorithm calculation and PWM, the synchronous reference frame rotates from θ_e to $\theta_e + 2T_{samp}\omega_e$ under the assumption of the constant synchronous speed, ω_e. Because of this rotation, the output of the regulator, which is supposed to be applied to an AC machine at θ_e, is applied to the machine from $\theta_e + T_{samp}\omega_e$ to $\theta_e + 2T_{samp}\omega_e$. Hence, the output of the regulator has error due to the rotation of the axis. The average voltage applied to an AC machine from $\theta_e + T_{samp}\omega_e$ to $\theta_e + 2T_{samp}\omega_e$ can be deduced as (7.39) from the original reference voltage, $\mathbf{V}_{dqs}^{e*}$, after considering the rotation of axis. The average voltage can be rewritten as (7.40) from (7.39).

$$\left\langle \mathbf{V}_{dqs-digital}^{s*} \right\rangle = \frac{1}{T_{samp}} \int_{T_{samp}}^{2T_{samp}} \mathbf{V}_{dqs}^{e*} e^{j(\omega_e \tau + \theta_e)} d\tau \tag{7.39}$$

$$\left\langle \mathbf{V}_{dqs-digital}^{s*} \right\rangle = K(\omega_e, T_{samp}) e^{j(1.5T_{samp}\omega_e + \theta_e)} \mathbf{V}_{dqs}^{e*} \tag{7.40}$$

where

$$K(\omega_e, T_{samp}) = \frac{2}{\omega_e T_{samp}} \sin\left(\frac{\omega_e T_{samp}}{2}\right) \tag{7.41}$$

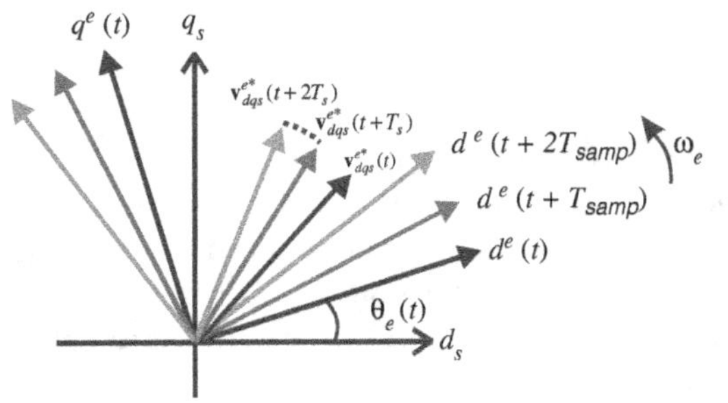

Figure 7.16 Rotation of synchronous reference frame as time passes.

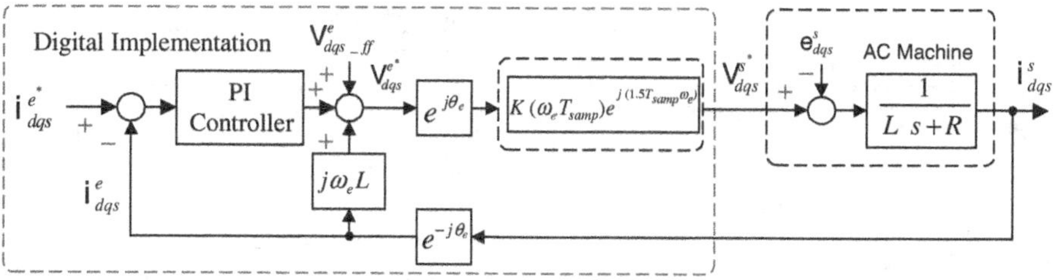

Figure 7.17 Block diagram of the current regulator for a three-phase AC load with the compensation of the error due to the digital control delay.

It can be seen from (7.40) and (7.41) that the voltage has magnitude error by $K(\omega_e, T_{samp})$ and phase angle error by $1.5T_{samp}\omega_e$. These errors can be compensated by multiplying the compensation function as (7.42) to the output of the regulator as shown in Fig. 7.17.

$$K(\omega_e, T_{samp})e^{j(1.5T_{samp}\omega_e)} \tag{7.42}$$

If the synchronous frequency, $f_e = \omega_e/(2\pi)$, is less than one-twentieth of the sampling frequency, $f_s = 1/T_{samp}$, the magnitude error is less than 5%. And the magnitude error may be neglected. However, the phase angle error is 27° even at $f_e = f_s/20$. Because of this angle error, the current regulation performance would be degraded conspicuously.

Experimental results of the induction machine drive system without and with the compensation of the phase angle error are shown in Figs. 7.18, respectively. As shown in Fig. 7.18a, when the ratio between the sampling frequency and

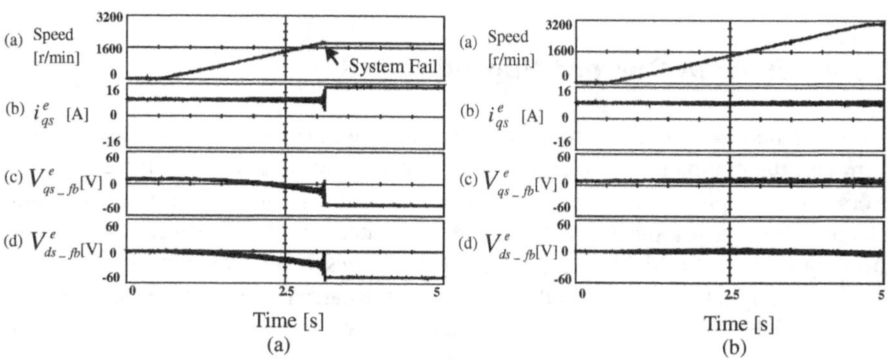

Figure 7.18 Acceleration test of a 7.5-kw, four-pole, 220-V induction machine drive system without and with the compensation of errors due to the digital delay (sampling interval: 400 μs). *From the top*: Speed of the machine, q-axis current, q-axis output of PI current regulator, d-axis output of PI current regulator. (a) Without compensation block. (b) With compensation block.

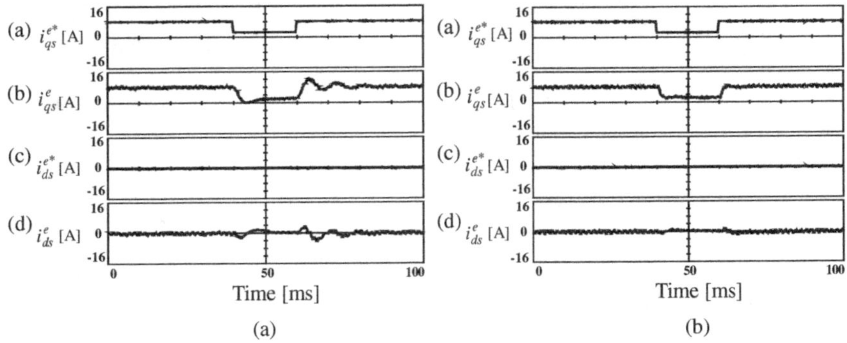

Figure 7.19 Response of current regulator to step change of reference. (*From the top*: q-axis current reference, q-axis actual current, d-axis current reference, d-axis actual current.) (a) Without compensation block. (b) With compensation block.

synchronous frequency is around 37, even at this large ratio, without the compensation, the current regulation loop is unstable and overcurrent fault occurs. However, with the compensation, as shown in Fig. 7.18b, the current is well-regulated until the ratio is less than 25.

The current regulation performance to the step change of the reference can be also improved with the compensation. It can be seen from Figs. 7.19a and 7.19b that with the compensation the oscillations of the current are reduced remarkably. Even with this compensation block, the current regulation loop could be unstable if the ratio is getting smaller and less than 15. In this case the complex vector current regulator with the active damping resistance, which is introduced in Section 4.2.5.4, could enhance the stability of the regulation loop conspicuously [13]. By further decreasing the ratio such as less than 12, the stability of the regulation loop could be lost again. Then, the predictive current in (4.17) can be used for active damping, and the stability could be recovered.

7.3.2 Error in Current Sampling [13]

If the ratio between sampling frequency and the output frequency is getting smaller, even with the compensation of the delays discussed in Section 7.3.1 there are still errors in the sampling of average current in the sampling period because of PWM and discrete sampling of the current. At the computer simulation results of current sampling system of a 65-kW, 20-pole, IPMSM drive system shown in Fig. 7.20, the current is sampled at the peak and the valley of the carrier wave whose frequency is 5 kHz. It can be seen that the sampled current at rotor reference d–q frame is far from the average value for a sampling period, T_{samp}.

The sampled current and the instantaneous current can be displayed in rotor reference d–q axis current plane as shown in Fig. 7.21. Again, it can be confirmed that the sampled current is far from the average value of the trajectory of the instantaneous current.

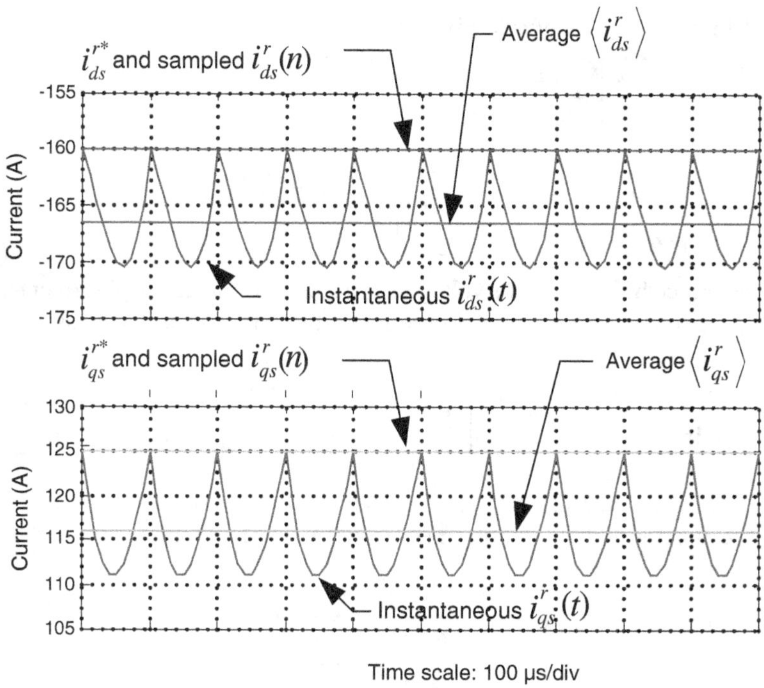

Figure 7.20 Computer simulation results of current sampling system of an IPMSM machine drive system ($f_{samp} = 10$ kHz, $f_e = 1/(\omega_r/2\pi) = 1500\,\text{Hz}$).

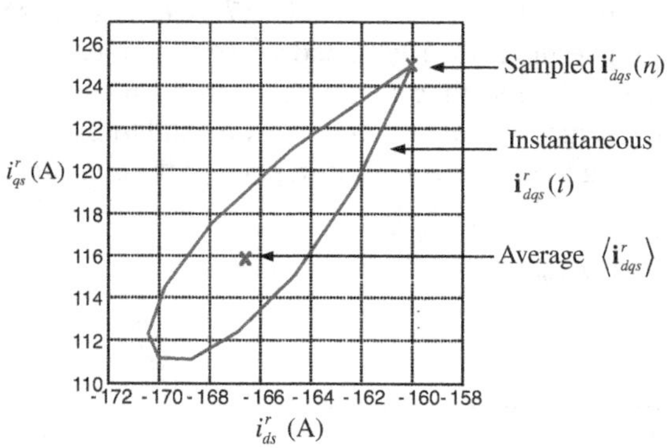

Figure 7.21 Trajectory of the instantaneous current and sampled current in a d–q-axis current plane for a sampling period ($f_{samp} = 10$ kHz, $f_e = 1500$ Hz).

The method to compensate the error can be derived as follows as with an example of an IPMSM drive system, whose voltage equation is in (7.43) as mentioned Section 3.3.3.2. If there is delay in the angle by $1.5T_{samp}$, then the average current for the sampling period, which has no error for the period, can be calculated from the sampled current as (7.50) through the following procedures.

$$\begin{bmatrix} V^r_{ds} \\ V^r_{qs} \end{bmatrix} = R_s \cdot \begin{bmatrix} i^r_{ds} \\ i^r_{qs} \end{bmatrix} + \begin{bmatrix} L_d & -L_q \\ L_d & L_q \end{bmatrix} \cdot \frac{d}{dt} \begin{bmatrix} i^r_{ds} \\ i^r_{qs} \end{bmatrix} + \begin{bmatrix} 0 & -\omega_r \cdot L_q \\ \omega_r \cdot L_d & 0 \end{bmatrix} \cdot \begin{bmatrix} i^r_{ds} \\ i^r_{qs} \end{bmatrix} + \begin{bmatrix} 0 & 0 \\ \omega_r & 0 \end{bmatrix} \cdot \begin{bmatrix} \lambda_f \\ 0 \end{bmatrix} \tag{7.43}$$

The rotor reference frame d–q-axis voltages, which have the angle displacement, θ, in a sampling period, can be expressed in terms of the reference voltages as

$$\begin{bmatrix} V^r_{ds} \\ V^r_{qs} \end{bmatrix} = \begin{bmatrix} \cos\theta & \sin\theta \\ -\sin\theta & \cos\theta \end{bmatrix} \cdot \begin{bmatrix} V^{r*}_{ds} \\ V^{r*}_{qs} \end{bmatrix} = \begin{bmatrix} V^{r*}_{ds} \cdot \cos\theta + V^{r*}_{qs} \cdot \sin\theta \\ -V^{r*}_{ds} \cdot \sin\theta + V^{r*}_{qs} \cdot \cos\theta \end{bmatrix} \tag{7.44}$$

The angle displacement, θ, has the trajectory as time passes as shown in Fig. 7.22.

The average voltage for a sampling period can be deduced in terms of the displacement angles as

$$\begin{bmatrix} \langle V^r_{ds} \rangle \\ \langle V^r_{qs} \rangle \end{bmatrix} = \frac{1}{(\theta_1 - \theta_2)} \cdot \begin{bmatrix} \int_{\theta_2}^{\theta_1} V^r_{ds} \cdot d\theta \\ \int_{\theta_2}^{\theta_1} V^r_{qs} \cdot d\theta \end{bmatrix} = \frac{1}{(\theta_1 - \theta_2)} \cdot \begin{bmatrix} V^{r*}_{ds} \cdot (\sin\theta_1 - \sin\theta_2) - V^{r*}_{qs} \cdot (\cos\theta_1 - \cos\theta_2) \\ V^{r*}_{ds} \cdot (\cos\theta_1 - \cos\theta_2) + V^{r*}_{qs} \cdot (\sin\theta_1 - \sin\theta_2) \end{bmatrix} \tag{7.45}$$

If the sampling frequency, f_{samp} is finite, then, as seen from Fig. 7.22 θ_1 is always larger than θ_2. Also, the cross-coupling terms of the right-hand side of (7.45) can be zero only if $\theta_2 = -\theta_1$. That means that in discrete time control the angle for reference frame transformation should be in the center of the sampling interval. And θ_1 can be set as (7.46) to nullify the cross-coupling terms under the assumption of the constant rotor speed.

$$\theta_1 = \frac{T_{samp}}{2} \cdot \omega_r \tag{7.46}$$

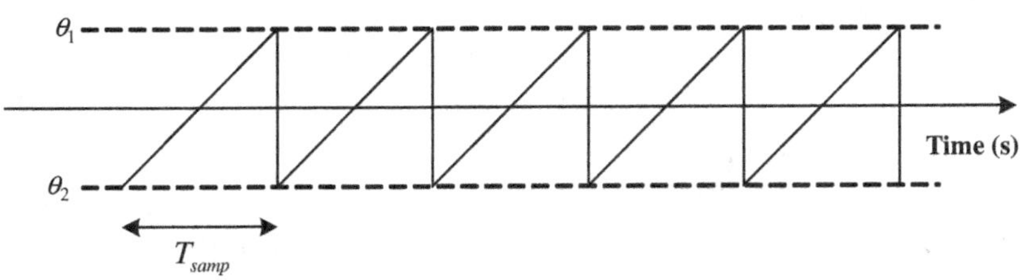

Figure 7.22 Trajectory of angle displacement with time.

The average voltage, $\langle V_{ds}^r \rangle$ and $\left\langle V_{qs}^r \right\rangle$, with the average current, $\langle i_{ds}^r \rangle$ and $\left\langle i_{qs}^r \right\rangle$, for a sampling period can be deduced as (7.47) from (7.43).

$$\begin{bmatrix} \langle V_{ds}^r \rangle \\ \left\langle V_{qs}^r \right\rangle \end{bmatrix} = \begin{bmatrix} R_s & -\omega_r \cdot L_q \\ \omega_r \cdot L_d & R_s \end{bmatrix} \cdot \begin{bmatrix} \langle i_{ds}^r \rangle \\ \left\langle i_{qs}^r \right\rangle \end{bmatrix} + \begin{bmatrix} 0 & 0 \\ \omega_r & 0 \end{bmatrix} \cdot \begin{bmatrix} \lambda_f \\ 0 \end{bmatrix} \tag{7.47}$$

Also, the average voltage can be calculated as (7.48) from (7.45) under the condition of $\theta_2 = -\theta_1$.

$$\begin{aligned} \begin{bmatrix} \langle V_{ds}^r \rangle \\ \left\langle V_{qs}^r \right\rangle \end{bmatrix} &= \frac{1}{2 \cdot \theta_1} \cdot \begin{bmatrix} \int_{-\theta_1}^{\theta_1} V_{ds}^r \cdot d\theta \\ \int_{-\theta}^{\theta_1} V_{qs}^r \cdot d\theta \end{bmatrix} = \frac{1}{2 \cdot \theta_1} \cdot \\ &\begin{bmatrix} V_{ds}^{r*} \cdot (\sin \theta_1 + \sin \theta_1) - V_{qs}^{r*} \cdot (\cos \theta_1 - \cos \theta_1) \\ V_{ds}^{r*} \cdot (\cos \theta_1 - \cos \theta_1) + V_{qs}^{r*} \cdot (\sin \theta_1 + \sin \theta_1) \end{bmatrix} = \frac{\sin \theta_1}{\theta_1} \cdot \begin{bmatrix} V_{ds}^{r*} \\ V_{qs}^{r*} \end{bmatrix} \end{aligned} \tag{7.48}$$

From (7.46–7.48), the average current in a sampling period can be expressed as

$$\begin{aligned} \begin{bmatrix} \langle i_{ds}^r \rangle \\ \left\langle i_{qs}^r \right\rangle \end{bmatrix} &= \begin{bmatrix} R_s & -\omega_r \cdot L_q \\ \omega_r \cdot L_d & R_s \end{bmatrix}^{-1} \cdot \left(\begin{bmatrix} \langle V_{ds}^r \rangle \\ \left\langle V_{qs}^r \right\rangle \end{bmatrix} - \begin{bmatrix} 0 & 0 \\ \omega_r & 0 \end{bmatrix} \cdot \begin{bmatrix} \lambda_f \\ 0 \end{bmatrix} \right) \\ &= \frac{1}{R_s^2 + \omega_r^2 \cdot L_d \cdot L_q} \cdot \begin{bmatrix} R_s & \omega_r \cdot L_q \\ -\omega_r \cdot L_d & R_s \end{bmatrix} \cdot \\ &\left(\frac{2}{\omega_r \cdot T_{samp}} \cdot \sin\left(\frac{\omega_r \cdot T_{samp}}{2} \right) \cdot \begin{bmatrix} V_{ds}^{r*} \\ V_{qs}^{r*} \end{bmatrix} - \begin{bmatrix} 0 & 0 \\ \omega_r & 0 \end{bmatrix} \cdot \begin{bmatrix} \lambda_f \\ 0 \end{bmatrix} \right) \end{aligned} \tag{7.49}$$

Finally, the average current in (7.49) can be represented in terms of the sampled current and parameters of IPMSM as

$$\begin{aligned} \langle i_{ds}^r \rangle &= a_1 \cdot i_{ds\_samp}^r + a_2 \cdot v_{ds}^{r*} + a_3 \cdot v_{qs}^{r*} + a_4 \cdot \omega_r \cdot \lambda_f \\ \left\langle i_{qs}^r \right\rangle &= b_1 \cdot i_{ds\_samp}^r + b_2 \cdot v_{ds}^{r*} + b_3 \cdot v_{qs}^{r*} + b_4 \cdot \omega_r \cdot \lambda_f \end{aligned} \tag{7.50}$$

where the coefficients of terms in (7.50) can be expressed as follows under the assumption of $R_s << \omega_r \cdot L_d$, $R_s << \omega_r \cdot L_q$.

$$a_1 = \alpha \cdot \frac{1}{\cos \theta_1}, \quad a_2 = 0, a_3 = \frac{1}{\omega_r \cdot L_d} \cdot \left(\alpha - \frac{1}{\cos \theta_1} \right), \quad a_4 = \frac{1}{\omega_r \cdot L_d} \cdot \left(\alpha \cdot \frac{1}{\cos \theta_1} - 1 \right)$$

$$b_1 = \alpha \cdot \frac{1}{\cos \theta_1}, \quad b_2 = -\frac{1}{\omega_r \cdot L_q} \cdot \left(\alpha - \frac{1}{\cos \theta_1} \right), \quad b_3 = 0, \quad b_4 = 0$$

where $\alpha = \sin\theta_1 / \theta_1$.

From (7.50), the average current in a sampling period can be evaluated in real time based on the sampled current and parameters of an IPMSM. Similarly, the expressions of the average current in a sampling period for other AC machines can be derived.

PROBLEMS

1. In the inverter shown in Fig. 7.1, the gating signals of the upper switch and the lower switch of each phase have the dead time by 3 μs as shown in Fig. P7.1a. The switching time of IGBT and diode can be neglected. The exact time point of the pole voltage reference transition is in the center of the dead time period as shown in the figure.

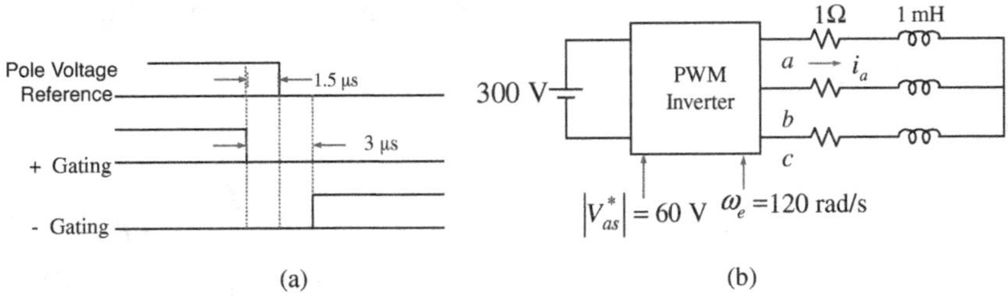

Figure P7.1 Pole voltage reference and gating signals of upper and lower switches of one phase of a PWM inverter. (a) PWM inverter with series R-L load.

As shown in Fig. P7.1b, the inverter supplies a 60-Hz, balanced three-phase voltage, whose peak of a phase voltage is 60 V, to R-L series load. The load is in Y connection and $R = 1\ \Omega$, $L = 1$ mH. The switching frequency of the PWM inverter is 10 kHz, and the DC link voltage is a 300-V constant. The system is under the steady state. Answer the following questions [12].

(1) If there was no dead time, calculate the magnitude of the fundamental component of a phase current and the displacement power factor which is defined as the cosine value of the angle between the fundamental frequency components of the phase voltage and the phase current.

(2) Calculate the magnitude of the voltage error in the fundamental frequency component of the phase voltage due to the dead time by 3 μs.

(3) Calculate the peak of the fundamental frequency component of a phase voltage, and find out the phase angle error of the phase voltage from the reference voltage due to the dead time.

2. In the circuit shown in Fig. 7.3, the stray capacitance of each switch is 3 nF and the switching time of the active switches and diodes can be neglected. Also, the conduction voltage drop of the semiconductor switches can be neglected. Answer the following questions. In the circuit, $T_{\text{dead}} = 3\ \mu\text{s}$, and $V_d = 300$ V.

When the phase current is a –2 A constant, the lower switch is turned off, which is defined as time point zero ($t = 0$ s).

(1) For $t = 0$–$3\ \mu$s, plot the pole voltage and the phase current together.

(2) Calculate compensation time, T_c, to compensate the voltage distortion due to the stray capacitors.

(3) Repeat parts 1 and 2 for the case where the phase current is a –0.5 A constant at switching.

3. As shown in Fig. P7.2, to prevent aliasing phenomenon in digital sampling of the phase current, a first-order low-pass filter (LPF), whose cutoff frequency is set as ω_c, has been added in front of the current measurement system. The fundamental frequency of the phase

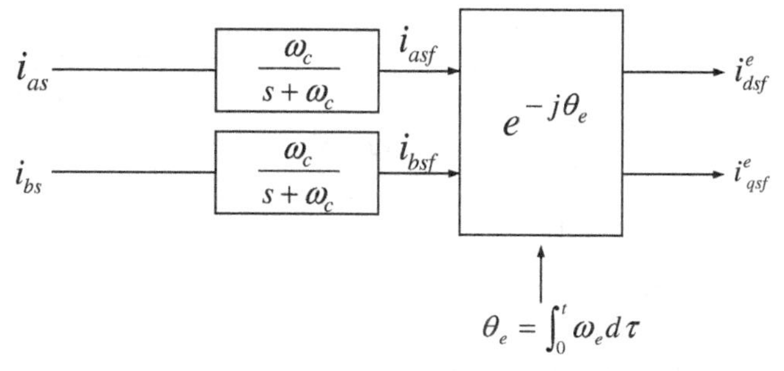

Figure P7.2 Current measurement system with LPF to prevent the aliasing phenomenon.

current is ω_e, and the current is transformed to the value in the synchronous reference frame as mentioned in Section 3.1 [8].

(1) Derive the relationship between the synchronous reference frame current with no LPF (i^e_{ds}, i^e_{qs}), and that with LPF (i^e_{dsf}, i^e_{qsf}), as the following equation. Derive a_{11}, a_{12}, a_{21}, and a_{22}.

$$\begin{bmatrix} i^e_{dsf} \\ i^e_{qsf} \end{bmatrix} = \begin{bmatrix} a_{11} & a_{12} \\ a_{21} & a_{22} \end{bmatrix} \begin{bmatrix} i^e_{ds} \\ i^e_{qs} \end{bmatrix}$$

(2) When $\omega_c = 1000\pi$ (rad/s), $\omega_e = 120\pi$ (rad/s), the bandwidth of the synchronous reference frame current regulator in Fig. 4.18 is set as $\omega_{bw} = 500\pi$ (rad/s). Here, active damping resistance, R_{active}, is set as zero. At a given ω_{bw}, calculate the attenuation rate of the LPF at the synchronous reference frame, which is given by $\left|\frac{i^e_{dsf}}{i^e_{ds}}\right|_{\omega=\omega_e} = |a_{11}|$, and calculate the magnitude of the cross-coupling term, $|a_{12}| = |a_{21}|$.

(3) Describe the effects of the cross-coupling terms, a_{12}, and a_{21}, to the performance of the vector control of an AC machine drive system.

4. A three-phase PWM inverter, shown in Fig. 7.1, supplies voltages to a series-connected R-L load. The three-phase symmetry space vector PWM is used to generate gating signals of the inverter. The load is connected in Y. And, $R = 1\,\Omega, L = 1$mH, DC link voltage is a 300-V constant, and the switching frequency is 5 kHz. The reference of the each phase voltage is given as follows.

$$V^*_{as} = 100\cos(200\pi t) \quad [\text{V}]$$

$$V^*_{bs} = 100\cos\left(200\pi t - \frac{2\pi}{3}\right) \quad [\text{V}]$$

$$V^*_{cs} = 100\cos\left(200\pi t + \frac{2\pi}{3}\right) \quad [\text{V}]$$

All switches of the inverter can be assumed as ideal ones. Hence, their switching time and conduction voltage can be neglected.

By computer simulation for $t = 0$–20 ms answer the following questions.

(1) Plot the instantaneous a-phase current and its sampled one if the current is sampled at the peak and valley of a PWM carrier.

(2) Plot the instantaneous a-phase current and its sampled one if the current is sampled at the peak and valley of PWM carrier through an LPF as shown in Fig. P7.3.

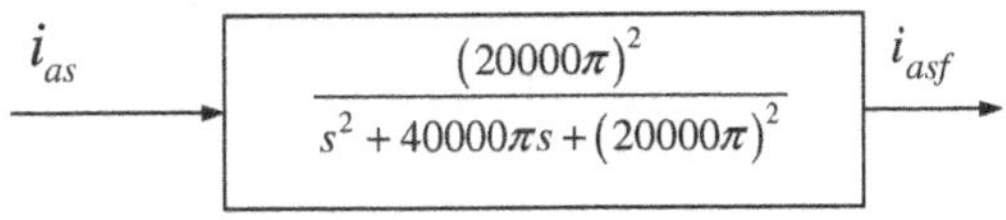

Figure P7.3 Second-order low-pass filter (LPF).

(3) Plot the instantaneous a-phase current and its sampled one if the filtered current, i_{asf}, is sampled with delay by T_d, which is calculated from (7.14), from the peak and valley of a PWM carrier. In (7.14), τ_1 can be neglected.

5. In the current regulator shown in Fig. 7.13, $L = 5$ mH and $R = 0.5\ \Omega$. And the bandwidth of the regulator is set as 100 Hz. The gains of PI controller are calculated as discussed in Section 4.2.5. Here, $R_{\text{active}} = 0\ \Omega$ and the current regulator is implemented in discrete form and its sampling period, T_{samp}, is 300 μs. And, there is $1.5T_{samp}$ delay .by the digital signal processing and PWM [11].

(1) By computer simulation without compensation of the delay, calculate the poles of the transfer function between the synchronous reference current and the actual current, $\mathbf{i}^e_{dqs}/\mathbf{i}^{e*}_{dqs}$ when $\omega_e = 60\pi$ and 600π, respectively.

(2) If the output voltage of the current regulator has been modified to compensate the voltage error due to the delay as shown in following equation, then repeat part 1.

$$\mathbf{V}^{s*}_{dqs} = \frac{2}{\omega_e T_s}\sin\left(\frac{\omega_e T_s}{2}\right)e^{j(1.5T_s\omega_e)}\cdot e^{j\theta_e}\mathbf{V}^{e*}_{dqs}$$

(3) By comparing the results obtained from parts 1 and 2, discuss the effects of the time delay to the performance of the current regulation loop of an AC machine drive system.

6. (1) Derive (7.21), (7.22), and (7.26) [10].

(2) Derive (7.31) and (7.32).

(3) The drive system given in problem 14 in Chapter 5 is used for speed control of the induction machine. The bandwidth of the speed regulator is set as 20 Hz. And the speed command is given as follows. The speed regulator is PI type and the output of the regulator, which is the torque command, is limited within ±150% of the rated torque of the induction machine. The friction of the machine and load can be neglected. The parameters of the induction machine are the same as the ones in problem 14 in Chapter 5 except for the total inertia. The total inertia, J_{M+L}, is four times the inertia of the machine itself. The gains of the speed regulator are set as described in Section 4.3.4.1.

$$t = 0\text{–}1\ \text{s},\qquad \omega^*_{rm} = 0(\text{r/min})$$
$$t \geq 1\ \text{s},\qquad \omega^*_{rm} = 600(\text{r/min})$$

In the measurement of the phase current, there is white noise, $\eta_i(t)$, whose rms magnitude is 4% of the rated rms phase current. Only, a- and b-phase currents are measured and c-phase current is calculated using the measured currents. The sampling frequency of the white noise is 100μs and also the measured currents have offset, δi_{as} and δi_{bs}, whose magnitude is 1% of rated rms current. And a- and b-phase current measurement system had 1% scale difference. Hence, the measured current can be represented as follows.

$$i_{as\_AD} = i_{as} + \delta i_{as} + \eta_i(t)$$
$$i_{bs\_AD} = (i_{bs} + \delta i_{bs} + \eta_i(t))/0.99$$

(a) For $t = 0$–5 s, plot T_e^*(N-m), T_e(N-m), $i_{qs}^{e*}(A)$, $i_{ds}^{e*}(A)$, $i_{qs}^{e}(A)$, $i_{ds}^{e}(A)$, $i_{ds\_AD}^{e}(A)$, $i_{qs\_AD}^{e}(A)$, λ_{dr}^{e}(Wb-t), λ_{dr}^{e*}(Wb-t) and ω_{rm}(r/min).

(b) For $t = 3$–5s, plot the magnitude of the frequency spectrum of T_e^*(N-m), T_e(N-m), $i_{qs}^{e*}(A)$, $i_{ds}^{e*}(A)$, $i_{qs}^{e}(A)$, $i_{ds}^{e}(A)$, $i_{ds\_AD}^{e}(A)$, $i_{qs\_AD}^{e}(A)$, λ_{dr}^{e}(Wb-t), λ_{dr}^{e*}(Wb-t) and ω_{rm}(r/min) using fast fourier transform (FFT).

REFERENCES

1. J. W. Choi et al. Inverter output voltage synthesis using novel dead time compensation, *IEEE trans. Power Electron.*, Vol. 11, No. 2, March 1996, pp. 221–227.
2. J. S. Kim et al. Analysis and compensation of voltage distortion by zero current clamping in voltage-fed PWM inverter, *Conference Record of IPEC-Yokohama*, 1995, pp. 265–270.
3. Y. Murai et al. PWM Strategy for high frequency carrier inverters eliminating current-clamps during switching dead-time, *Conference Record of IEEE Industry Applications Society Annual Meeting*, 1992, pp. 317–322.
4. H.M. Ryu et al. Compensation of voltage distortion in PWM-VSI by prediction of stator currents at switching point, in *IEE-Japan*-D, *Domestic Conference, conference record*, Vol. 3, 1999, pp. 87–90.
5. T. M. Rowan et al. A new synchronous current regulator and an analysis of current-regulated PWM inverters, *IEEE Trans. Ind. Appl.*, Vol. 22, 1986, pp. 678–690.
6. Y. Yamamoto et al. Digital current control method of induction motor using synchronous current detection with PWM signals, *Trans. IEE-Japan*, Vol. 112-D, No. 7, 1992, pp. 613–622 (in Japanese).
7. V. Blasko et al. Sampling of discontinuous voltage and current signals in electrical drives—A system approach, *IEEE Trans. Ind. Appl.*, Vol. 34, No. 5, September/October 1998, pp. 1123–1130.
8. R. B. Sepe et al. Implementation of discrete-time field-oriented current control, *IEEE Trans. Ind. Appl.*, Vol. 30, No. 3, 1994, pp. 723–727.
9. S. H. Song et al. Current measurements in digitally controlled AC drives, *IEEE Ind. Appl. Mag.*, July/August 2000, pp. 51–62.
10. D. W. Chung et al. Analysis and compensation of current measurement error in vector-controlled AC motor drives, *IEEE Trans. Ind. Appl.*, Vol. 34, No. 2, 1998, pp. 340–345.
11. B. H. Bae et al. A compensation method for time delay of full digital synchronous frame current regulator of PWM ac drives, in *Conference Record of the 2001 IEEE Industry Applications Society Annual Meeting*, Vol. 3, 2001, pp. 1708–1714.
12. Y. Murai et al. Waveform distortion and correction circuit for PWM inverters with switching lag-times, *IEEE Trans. Ind. Appl.*, Vol. 23, No. 5, September/October 1987, pp. 881–886.
13. J. S. Yim et al. Modified current control schemes for high-performance permanent-magnet AC drives with low sampling to operating frequency ratio, *IEEE Trans. Ind. Appl.*, Vol. 45, No. 2, March/April 2009, pp. 763–771.

Appendix A

Measurement and Estimation of Parameters of Electric Machinery

To apply the control algorithms in this text to the control of the electric machines and power converters, the parameters of the electric machinery should be identified for setting of gains of the regulators, limiting values of limiters of the controller, reference and feed-forwarding values to the regulator, and so on. The parameters of an electric machine may be known by the locked rotor test or by the no-load test. However, because the parameters from the tests are not from the actual operating condition of the machine, as mentioned in Section 2.9 and 2.12 the parameters may differ widely from the parameters in the real operating condition. The parameters of the electric machinery including power converters may be calculated or estimated from the design data or from the performance test data of the manufacturer, but these data are not easily available in the application field. In this appendix, several practical methods to identify the parameters of the electric machines and drive system are introduced. Most of them do not require any special measurement tool, but relies on the controller of the drive system itself.

A.1 PARAMETER ESTIMATION

In this section, several methods to identify the parameters of electric machines based on the extra tests or name plate data of the machinery are introduced. Though the method based on an extra test provides reasonably accurate parameters, the methods may need some tools to apply the test signals or special setup for the test. Hence, it is difficult to be used generally in the industry site. The parameter estimation from the name plate data is easy and straightforward. However, the estimated parameters are

Control of Electric Machine Drive Systems, by Seung-Ki Sul

only effective at the rated operating condition. Furthermore, they may have considerable errors due to manufacturing tolerance.

A.1.1 DC Machine

A.1.1.1 Armature Winding Resistance, R_a

The armature resistance of a small DC machine can be easily measured by a resistance meter. In this case, to reduce measurement error the armature winding resistance can be measured at several different positions with slow rotation of the rotor. However, the error due to contact resistance of the brush and the commutator cannot be avoided. In the medium- or large-size DC machine, the winding resistance is small and the effect of contact resistance is considerable. In this case, by flow of a constant current to a DC machine using a drive system of DC machine, the armature winding resistance can be calculated by dividing the terminal voltage by the magnitude of the current. In this test the rotor of the machine can rotate slowly to reduce the error due to the contact resistance. But the rotational speed should be low enough in order to prevent back EMF from affecting to the measurement of the resistance.

A.1.1.2 Armature Inductance, L_a

The armature inductance can be identified by the response of the armature current with the step change of the armature voltage as shown in Fig. A.1. In this test, the field current should be set as the actual operating condition of DC machine. After the slope of the armature current at the instant of the step change is measured, the inductance can be calculated by dividing the magnitude of the step voltage by the slope of the current. Because the inductance varies according to the magnitude of the armature current, i_{ao}, the test should be done at several values of i_{ao} within the rated value of the armature current. With this test, the variation of the inductance according to the armature current can be also known.

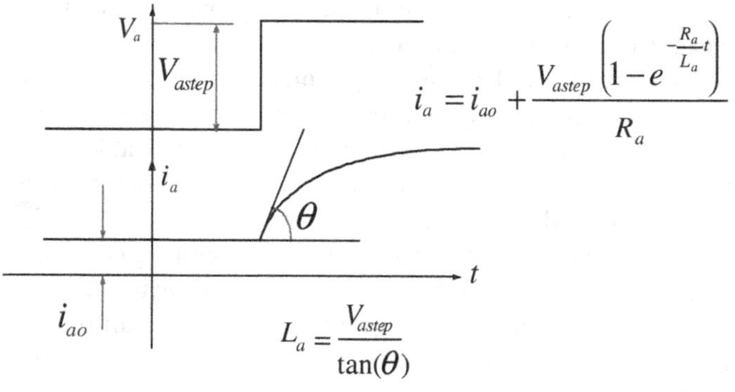

Figure A.1 Estimation of armature inductance by step voltage input.

A.1.1.3 Field Winding Resistance, R_f

The resistance of the field winding can be easily measured by a resistance meter in the case of an under-several-tens of-kilowatts DC machine whose rated voltage of field circuit is around several hundreds volts. For larger machines or lower field voltage machines, the resistance can be calculated by measuring the terminal voltage of the field winding with the constant field current.

A.1.1.4 Field Winding Inductance, L_f

The field winding inductance can be identified like the case of the armature inductance. Because the field inductance varies widely according to the magnitude of the field current, the inductance should be evaluated at several values of the field current within the rated value of the field current. With this test, the magnetic saturation characteristics of the field circuit can be identified.

A.1.1.5 Torque Constant, $K = K_T\lambda_f$

In MKS unit system, as seen from (2.19) and (2.20), K_e is the same with K_T. Hence, identifying the torque constant, $K = K_e\lambda_f$, means to identify the back EMF constant. The back EMF constant can be estimated by measuring the voltage of the armature terminal in the rotation of the rotor by an external prime mover. In this test, to reduce the error due to the brush and the commutator, the rotational speed could be more than half of the rated speed. By measuring the voltage at several different speeds, the accuracy of the estimation of back EMF constant can be enhanced. Because the field flux linkage, λ_f, varies nonlinearly according to the field current, if torque constant, K, is identified at several different field currents, then the nonlinearity of the flux linkage can be also identified. To apply this method to above-several-tens-of-kilowatts machines, rotating the machine is not easy.

Another method to identify torque constant, even without any test, is using the data of the name plate. Usually, in the name plate, there are data regarding rated speed, rated power, rated armature voltage, rated armature current, rated field voltage, and rated field current. If the resistance of the armature winding is known as R_a, then torque constant can be easily estimated with the following simple calculation. For an above-several-tens-of-kilowatts machine, where the armature winding copper loss is relatively small compared to the rated power, even without knowing R_a, the torque constant can be identified within the reasonable error bound. It should be noted that in this method the estimated torque constant is the value at the rated operating condition. From (2.10), if the rated values in the name plate are substituted, then back EMF constant at rated operating condition can be estimated. If the resistance of the armature winding is unknown, the rated armature voltage can be used as a back EMF at a rated operating condition. By substituting the voltage to (2.9), the back EMF constant, $K(= K_e\lambda_f)$, can be guessed. The value from this process can be named as K_1. Also, the torque constant at the rated operating condition can be obtained from the rated power, the rated speed, and the rated current. The rated torque can be calculated by dividing the rated power by the rated speed. The rated torque can be substituted to (2.11) along with the rated armature current, and then the torque constant can be obtained. The value from this process can be called K_2. The torque constant from the former process, K_1, is always larger than the constant from the latter process like $K_1 > K_2$. The reason for this is the following. In the latter process, because the rated power is the power from the shaft of the machine, the friction and windage loss should be added to the rated power to calculate the exact torque constant of the machine itself. And the real torque constant would be a little bit larger than K_2. Also, in the calculation of K_1, the voltage drop in the commutator and the brush is neglected. If the voltage drop is considered as in (2.3), then the real back EMF constant would be a little bit smaller than K_1. Hence, the constant, K, can be set as the average of K_1 and K_2, as $K \approx (K_1 + K_2)/2$.

A.1.2 Estimation of Parameters of Induction Machine

A.1.2.1 Rated Value of Rotor Flux Linkage, λ_{dr}^e

The stator voltage equation for rotor flux-oriented vector-controlled induction machine can be described as (A.1) as discussed in Section 4.2.5.2.

$$\begin{aligned} V_{ds}^e &= \left(R_s + R_r \frac{L_m^2}{L_r^2}\right) i_{ds}^e + \sigma L_s \frac{di_{ds}^e}{dt} - \omega_e \sigma L_s i_{qs}^e - R_r \frac{L_m}{L_r^2} \lambda_{dr}^e \\ V_{qs}^e &= \left(R_s + R_r \frac{L_m^2}{L_r^2}\right) i_{qs}^e + \sigma L_s \frac{di_{qs}^e}{dt} + \omega_e \sigma L_s i_{ds}^e + \omega_r \frac{L_m}{L_r} \lambda_{dr}^e \end{aligned} \tag{A.1}$$

Because $\sigma L_s i_{ds}^e = \left(L_s - \frac{L_m^2}{L_r}\right)\left(\frac{\lambda_{dr}^e}{L_m}\right) = \left(\frac{L_s}{L_m} - \frac{L_m}{L_r}\right)\lambda_{dr}^e \approx \frac{L_{ls}+L_{lr}}{L_m}\lambda_{dr}^e << \lambda_{dr}^e$, we have $\frac{L_m}{L_r} \approx 1$ and $\omega_r = \omega_e$ at no load. At rated operating speed, we have $\left(R_s + R_r \frac{L_m^2}{L_r^2}\right) i_{qs}^e << \omega_r \frac{L_m}{L_r} \lambda_{dr}^e$. Hence, the q-axis voltage in (A.1) can be approximated as (A.2) at near rated operating speed with no load:

$$V_{qs}^e \approx \omega_r \lambda_{dr}^e = \omega_e \lambda_{dr}^e \tag{A.2}$$

In this operating condition, we have $V_{ds}^e \ll V_{qs}^e$. And the peak of the phase voltage is $V_s = \sqrt{V_{ds}^{e2} + V_{qs}^{e2}} \approx V_{qs}^e$. Therefore, from the rated voltage and rated frequency of the induction machine, which are easily obtained from the name plate of the machine, the rated flux can be calculated. The approximated value of the rated flux linkage can be obtained simply by dividing the peak of the rated phase voltage by the rated angular frequency, ω_e, of the machine. For accurate setting of the flux linkage, the voltage drop by the stator winding resistance and stator transient reactance as $\left(\sqrt{R_s^2 |I_s|^2 + (\omega_e \sigma L_s)^2 |I_s|^2}\right)$, where $|I_s|$ is the peak of the rated phase current, can be subtracted from the peak of the phase voltage. Furthermore, after subtraction, the voltage can be divided not by the rated angular frequency but by $\omega_e(L_m/L_r)$ for more accurate flux linkage. For example, the approximated value of the rated rotor flux linkage can be obtained by $\left(\frac{\sqrt{2}}{\sqrt{3}}220\right)/(60 * 2 * \pi) = 0.4765$ Wb-t in the case of a line-to-line three-phase 220-Vrms, 60-Hz induction machine because $\omega_e(L_m/L_r) \approx \omega_e$ and also because there are no data regarding L_m and L_r.

A.1.2.2 Stator Transient Inductance, σL_s [2]

The stator transient inductance of the induction machine can be defined and approximated as

$$\sigma L_s = L_s - \frac{L_m^2}{L_r} \approx L_{ls} + L_{lr} \tag{A.3}$$

As mentioned in Section 2.12, the transient inductance varies with the magnitude and the frequency of the current flowing through the inductance. Because the transient inductance is used for the calculation of the rotor flux linkage as (A.4), the accurate estimation of the inductance at the operating condition of the induction machine is crucial to the performance of the rotor flux-oriented direct vector control and to that of the flux weakening control.

$$\lambda_{dqr}^s = \frac{L_r}{L_m}(\lambda_{dqs}^s - \sigma L_s \mathbf{i}_{dqs}^s) \tag{A.4}$$

If σL_s is not accurately reflected in the estimation of the rotor flux linkage, the torque of the induction machine reveals an oscillatory response; in the extreme case, overcurrent fault may occur. In the flux weakening control, for setting of the frequencies where the flux weakening region I or II begins as mentioned in Section 5.4.4, the value of the transient inductance is crucial. The transient

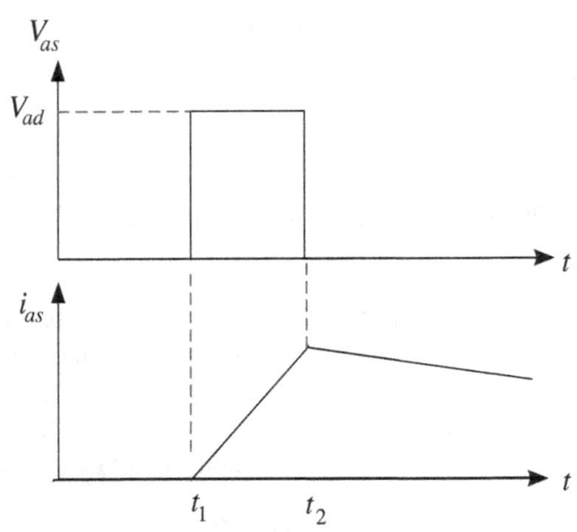

Figure A.2 Short pulse voltage and its associated current to induction machine to estimate stator transient inductance.

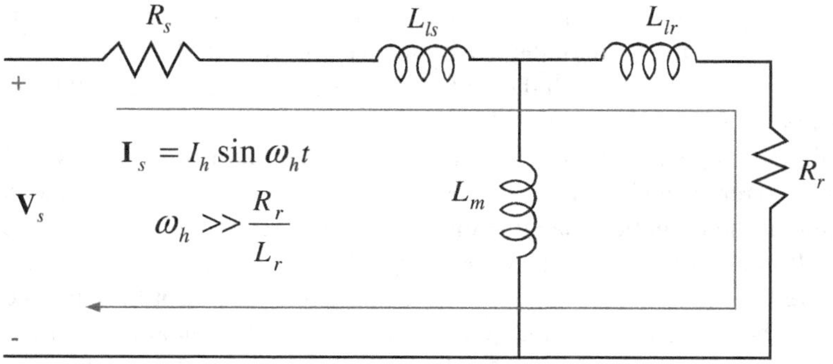

Figure A.3 Current path of an induction machine when high-frequency current is injected to the stator.

inductance can be estimated by applying a short voltage pulse to the induction machine through a PWM inverter as shown in Fig. A.2. Then, the current would flow as shown in Fig. A.3. In this test, the phase voltage equation can be derived as

$$\begin{aligned} V_{as} &= (R_s + R_r)i_{as} + (L_{ls} + L_{lr})\frac{di_{as}}{dt} \\ &\approx (R_s + R_r)i_{as} + \sigma L_s \frac{di_{as}}{dt} \end{aligned} \tag{A.5}$$

If the width of the voltage pulse is small enough compared to the stator time constant defined as $\tau_s \equiv \sigma L_s/(R_s + R_r)$, most of the voltage would be applied to the stator transient inductance and (A.5) can be approximated as

$$V_{as} \approx \sigma L_s \frac{di_{as}}{dt} \tag{A.6}$$

Hence, the transient inductance, σL_s, can be estimated by

$$\sigma L_s = V_{ad} \frac{t_2 - t_1}{i_{as}(t_2) - i_{as}(t_1)} \tag{A.7}$$

The voltage to the machine, V_{ad}, is decided by the switching function of the inverter and the DC link voltage, V_d. When the switching functions are $S_a = 1$, $S_b = 0$, and $S_c = 0$, the voltage of the "a" phase is $2V_d/3$. The duration of the voltage pulse, $t_2 - t_1$, should be set so that the peak of "a"-phase current, i_{as}, is around a rated value. With this pulse voltage, the current shaped as shown in Fig. A.2 would flow. However, because of the skin effect due to the shape of the current, the estimated inductance by (A.7) is smaller than the actual inductance at normal operating condition. Furthermore, the transient inductance may be affected by the rotor current, and the value would decrease as the rotor current increases because of the saturation of rotor leakage flux linkage as mentioned in Section 2.12.2.

Another method to estimate the transient inductance is using a high-frequency component of the stator current. If the high-frequency current, whose angular frequency, ω_h, is high enough compared to the inverse of the rotor time constant, $\tau_r = L_r/R_r$, is injected to the stator of the induction machine, then the high-frequency component current flows not through the mutual inductance branch but through the leakage inductances as shown in Fig. A.3.

By exploiting this phenomenon to estimate the transient inductance, the small high-frequency current is added to the d-axis reference current of the synchronous reference frame of the induction machine under test. The frequency of the current should be small enough to be regulated well with the current regulator but high enough compared to the inverse of the rotor time constant. If the rotor flux linkage is constant, then the d-axis current can be deduced as (A.8). The magnitude of the high-frequency current can be set as several tens percentages of the rated d-axis current [3].

$$i^e_{ds} = I_d + \tilde{i}^e_{ds} = I_d + I_h \sin\omega_h t \tag{A.8}$$

where I_d is the flux component current to set the rated rotor flux linkage in the vector control of the induction machine. The high-frequency component of output of the current regulator can be described as

$$\tilde{V}^e_{ds} = (R_s + R_r)\tilde{i}^e_{ds} + \omega_h(L_{ls} + L_{lr})I_h \cos\omega_h t \tag{A.9}$$

From (A.9), we can derive (A.10).

$$(\tilde{V}^e_{ds} - (R_s + R_r)\tilde{i}^e_{ds})^2 = \omega_h^2(L_{ls} + L_{lr})^2 I_h^2 \frac{1 + \cos 2\omega_h t}{2} \tag{A.10}$$

By applying low-pass filter to (A.10), the DC component of the left-hand term can be used to estimate the transient inductance as

$$\sigma L_s \approx (L_{ls} + L_{lr}) = \sqrt{\frac{2LPF[(\tilde{V}^e_{ds} - (R_s + R_r)\tilde{i}^e_{ds})^2]}{\omega_h^2 I_h^2}} \tag{A.11}$$

Here, if the rotor resistance is unknown, the sum of both resistances can be approximated as $R_s + R_r \approx 2R_s$ for the estimation of the inductance. This method is simple to implement with the current-regulated PWM inverter. Also, the signal processing is quite easy. However, if the frequency of the injected high-frequency current is too high, then because of the skin effect the estimated inductance is smaller than the real value. This method can be used with the normal operation of the induction machine where the rotor current flows, and the variation of the inductance with the rotor current can be identified. Also, the variation of the transient inductance according to the magnitude of the rotor flux linkage can be considered with this method by adjusting the DC component of (A.8), which is the flux component current in the rotor flux-oriented vector-controlled induction machine drive system. This variation of the inductance can be used for accurate setting of the frequencies where the flux weakening region I or II begins as mentioned in Section 5.4.4.

A.1.2.3 Mutual Inductance, L_m

The mutual inductance of the induction machine can be easily identified if no-load operation is possible. In this case, by simply the measuring the phase current and the phase voltage, the sum of the stator leakage reactance and the mutual reactance can be obtained by dividing the voltage by the current. Then the mutual inductance can be approximated by dividing the sum by the operating frequency, ω_e, under the assumption that the stator leakage inductance is much smaller than the mutual inductance. In this no-load test, the operating frequency should be near the rated value to reduce error due to the resistance voltage drop of the stator winding. The mutual inductance of the induction machine varies according to the magnitude of the rotor flux linkage decided by the magnitude of the d-axis current of the synchronous reference frame. If the mutual inductance is not correct, then there should be slip frequency error in the case of indirect vector control system. If the flux linkage varies for the flux weakening control as mentioned in Section 5.4.4, then the variation of the mutual inductance should be identified to set the reference rotor flux linkage as seen from (5.89), (5.92), and (5.96).

A.2 PARAMETER ESTIMATION OF ELECTRIC MACHINES USING REGULATORS OF DRIVE SYSTEM [4]

The parameter estimation methods described in the previous section can estimate parameters of the electric machines. However, the parameters are estimated through extra tests or signals. And the operating condition for the test is quite different with the normal operating condition of the machine. Hence, if the parameters vary according to the operating conditions, the estimated values may be far different from the values in the real operating condition. However, in this section, the parameters of the electric machinery are estimated by the inherent information of the regulators of the system. And the estimated ones are the values at the normal operating conditions. Furthermore, this estimation method does not need any extra tool or setup for the test. And it is compatible to the existing hardware of the drive system and can be applied only by modification of the software. Also, this method may be used as on-line tuning of the parameters against the parameter variation according to the operating condition.

A.2.1 Feedback Control System

A typical single-input and single-output feedback control system can be described as a block diagram shown in Fig. A.4. In the figure, $r(s)$ is the reference input, $y(s)$ is the output, and $\nu(s)$, $d(s)$, and $n(s)$ stand for system uncertainty, disturbance, and sensor noise, respectively. In the figure, if the controlled plant is depicted as $P(s)$ and the

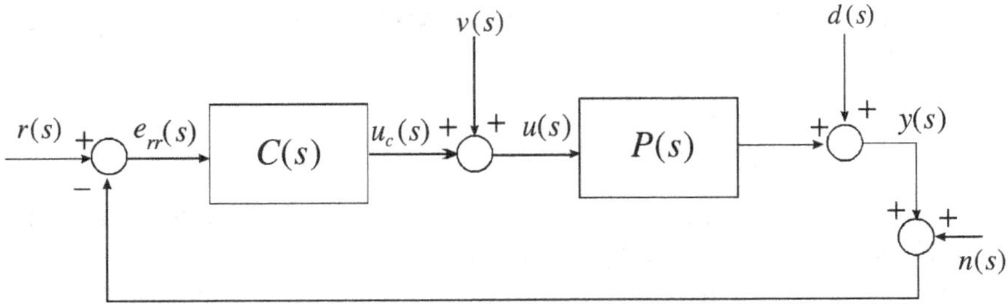

Figure A.4 Block diagram of typical single-input and single-output feedback control system.

controller is represented as $C(s)$, then the transfer function between the reference input and the output can be derived as

$$y(s) = \frac{C(s)P(s)}{1+C(s)P(s)}(r(s)-n(s)) + \frac{P(s)}{1+C(s)P(s)}v(s) + \frac{1}{1+C(s)P(s)}d(s) \tag{A.12}$$

In (A.12), the first term on the right-hand side means the effect of the reference input and noise with regard to the output. The second term means the effect of the uncertainty of the system to the output. And the last term means that of disturbance to the output. The error between the reference input and the output, $e_{rr}(s)$, can be represented as the sum of terms driven by $r(s)$, $v(s)$, $d(s)$, and $n(s)$ as

$$e_{rr}(s) = \frac{1}{1+C(s)P(s)}(r(s)-d(s)-n(s)) - \frac{P(s)}{1+C(s)P(s)}v(s) \tag{A.13}$$

The control target is to nullify the error. To simplify the analysis, with the assumption of negligible disturbance and sensor noise, the error can be nullified under the condition in (A.14).

$$|C(s)P(s)| >> 1 \tag{A.14}$$

With this condition and assumption, the control input to the plant, $P(s)$, can be approximated as

$$\begin{aligned} u_c(s) &= \frac{C(s)}{1+C(s)P(s)}r(s) - \frac{C(s)P(s)}{1+C(s)P(s)}v(s) \\ &\approx \frac{C(s)}{1+C(s)P(s)}r(s) - v(s) \end{aligned} \tag{A.15}$$

From (A.15), the input to the plant can be expressed as

$$u(s) = u_c(s) + v(s) = \frac{C(s)}{1+C(s)P(s)}r(s) \tag{A.16}$$

It can be seen from (A.16) that the system uncertainty, $\nu(s)$, from the modeling error due to parameter variations is added to the output of the controller, $u_c(s)$. Hence, if the input to the physical plant, $u(s)$, is not distorted because of the condition in (A.14), then the output of the controller, $u_c(s)$, would be distorted because of $\nu(s)$. To remove the distortion of the output, $y(s)$, due to the system uncertainty, the controller distorts its output by feedback action. By exploiting this phenomenon, the modeling error, especially parameter error, can be identified.

A.2.2 Back EMF Constant of DC Machine, *K*

To identify a back EMF constant, the feed-forwarding term of the current regulator of a DC machine drive system, whose control block diagram is shown in Fig. 4.12, can be used. By adjusting the feed-forwarding term, $\hat{e}$, which is given by $\hat{e} = K\omega_{rm}$, the integral term of a PI regulator can be kept as constant at the same armature current, i_a^*, regardless of the operating speed of the machine. If the back EMF constant is different with actual value, then the integral term of the regulator would include the voltage due to the error of the back EMF constant. If the integral term is the same regardless of the operating speed, then the feed-forwarding term is correct, and that means the correct back EMF constant. Hence, the back EMF constant, K, can be identified at the given armature current, i_a^*. With this method, the back EMF constant, which is the torque constant, can be easily identified without any special measurement tool or test setup. The reason for using only the integral term instead of the output of PI regulator as whole is that the integral term does not include high-frequency noises because of its inherent low-pass filtering characteristics.

A.2.3 Stator Winding Resistance of Three-Phase AC Machine, R_s [1]

To estimate the stator flux linkage of an AC machine, the accurate value of the stator winding resistance is important. In particular, at the low-speed operation of an AC machine the accuracy is crucial to the performance of the direct vector control of the induction machine. The resistance can be identified by using a current-regulated PWM inverter. In this test, an AC machine can be in a standstill by flowing a DC current only in the *d*-axis of the stationary reference frame of the AC machine. Because of no back EMF, the output voltage of the PWM inverter would be small. Moreover, the voltage distortion due to the dead time effects and voltage drop of the power semiconductor of the PWM inverter may result in errors in the identification of the resistance. However, the errors can be minimized as follows.

The control block diagram of a current-regulated AC machine drive system with the voltage distortion is shown in Fig. A.5. In the figure, $\mathbf{V}'^{s}_{dqs}$ is the distorted voltage due to the dead time and nonlinearity of the inverter. R_{ce} stands for the conducting resistance of the active switch of PWM inverter, and R_d stands for that of the freewheeling diode of the inverter. If the current is well-regulated

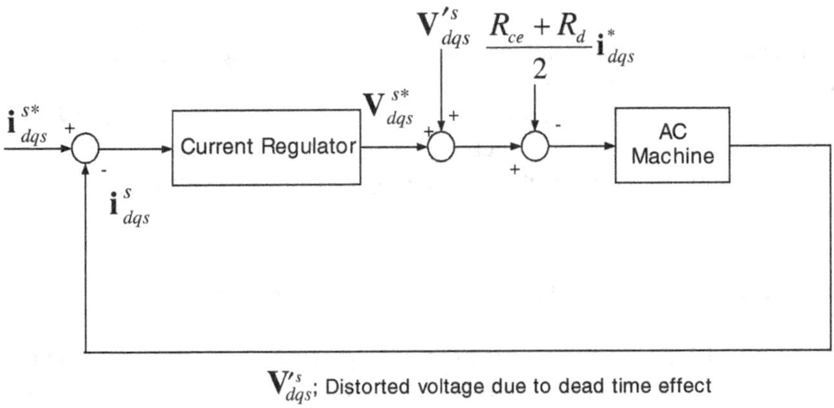

Figure A.5 Block diagram of a current-regulated AC machine drive system.

according to its reference, then the output of the current regulator would be distorted to cancel out the distortion of the voltage due to the dead time and voltage drop, and so on. If the q-axis current is regulated as null and the d-axis current reference is i^{s*}_{ds1}, which is a constant DC value, then the d-axis voltage can be described as

$$V^{s*}_{ds1} = R'_s\, i^{s*}_{ds1} - V'^s_{ds1} \tag{A.17}$$

where R'_s means the total resistance seen from the inverter, which includes not only stator resistance but also cable resistance and resistance of the power semiconductors of the inverter.

For another d-axis current reference, i^{s*}_{ds2}, which is also a DC value, the voltage equation can be set as (A.18) under the assumption of the constant total resistance regardless of the magnitude of the current.

$$V^{s*}_{ds2} = R'_s\, i^{s*}_{ds2} - V'^s_{ds2} \tag{A.18}$$

By solving (A.17) and (A.18) simultaneously, the equivalent resistance can be calculated as (A.19). Here, R_s is the winding resistance and R'_s is the equivalent resistance in the viewpoint of the control system. For the control of the AC machine using the PWM inverter, not R_s but R'_s should be used to estimate the flux linkage and to set the current regulator gain. Hence R'_s is the proper value for the tuning of the regulators.

$$R'_s = R_s + \frac{R_{ce} + R_d}{2} = \frac{V^{s*}_{ds1} - V^{s*}_{ds2}}{i^{s*}_{ds1} - i^{s*}_{ds2}} \tag{A.19}$$

In the above derivation, it is assumed that R'_s and V'^s_{dqs} are independent of the current. In this method, the magnitude of both currents and the difference of the currents, i^{s*}_{ds1} and i^{s*}_{ds2}, should be large enough to enhance nose-to-signal ratio. To sample the voltage, V_{ds1}^{s*} and V_{ds2}^{s*}, the system should be in the steady state to decouple the voltage due to the variation of the stator flux linkage. The bandwidth of the current regulator for this test can be set as low as possible to reduce the effect of measurement noise. To enhance the accuracy of the estimation, the d-axis, where the DC current is injected, can be rotated from one phase to another as follows.

First, $i_a = I_{dc}$, $i_b = -I_{dc}/2$, $i_c = -I_{dc}/2$; next $i_a = -I_{dc}/2$, $i_b = I_{dc}$, and $i_c = -I_{dc}/2$; next $i_a = -\frac{I_{dc}}{2}$, $i_b = -\frac{I_{dc}}{2}$, and $i_c = I_{dc}$; next $i_a = -I_{dc}$, $i_b = I_{dc}/2$, $i_c = I_{dc}/2$; next $i_a = I_{dc}/2$, $i_b = -I_{dc}$, and $i_c = I_{dc}/2$; finally $i_a = I_{dc}/2$, $i_b = I_{dc}/2$, and $i_c = -I_{dc}$. The average value of R'_s through these six tests can be used as the estimated equivalent stator resistance. During the test procedure, because of ohmic loss of the electric machine and cable, the resistance may increase slightly.

A.2.4 Induction Machine Parameters

A.2.4.1 Rotor Time Constant, $\tau_r = L_r/R_r$

The rotor time constant of an induction machine can be identified by exploiting the characteristics of the speed regulator of the indirect vector controlled induction machine drive system. In this test, it is assumed that there is no torque disturbance from the load. If the mechanical system of the drive system can be set as simple inertia as shown in Fig. A.6, then the controlled plant, $P(s)$, and the controller, which is a PI-type speed regulator, $C(s)$, can be described as (A.20) and (A.21), respectively.

$$P(s) = \frac{1}{J_m s} \tag{A.20}$$

$$C(s) = K_p + \frac{K_i}{s} \tag{A.21}$$

where K_p and K_i are the proportional gain and integral gain of PI regulator, respectively.

If the speed of a system shown in Fig. A.6 is well-regulated and the rotor flux-oriented indirect vector control of the induction machine is ideally performed, then

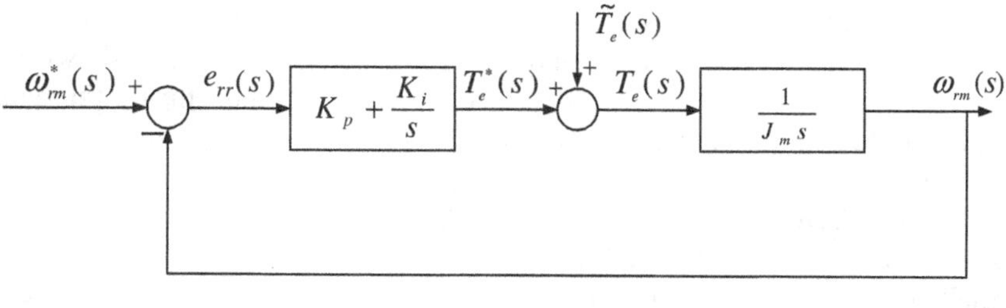

Figure A.6 Control block diagram of a speed regulation system.

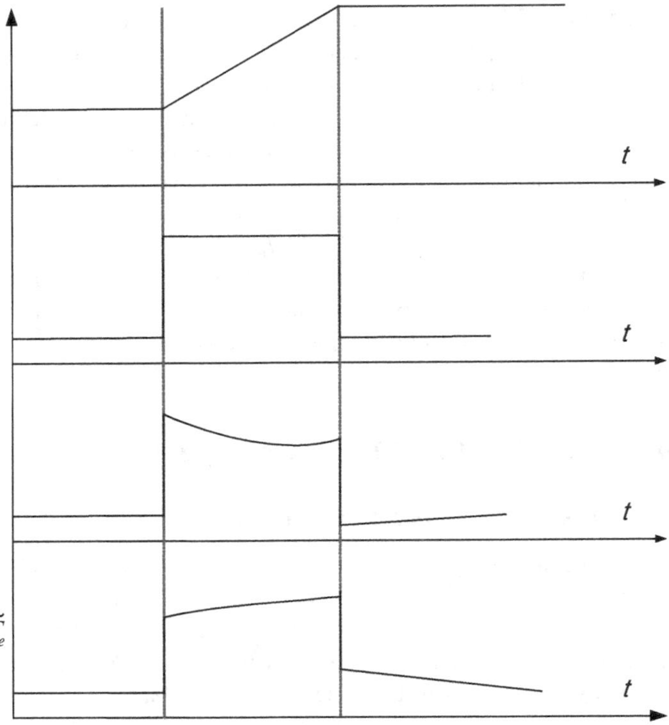

Figure A.7 Output of a speed regulator with a different rotor time constant to the ramp input of a speed reference.

the output of the speed regulator, T_e^*, to the ramp input of the system, shown in Fig. A.7a, would be a square wave as shown in Fig. A.7b. But, if there is an error in the rotor time constant, then the output of the regulator would be distorted as shown in Fig. A.7c or Fig. A.7d. These distortions come from the reflection of the disturbance, $\tilde{T}_e$, to the regulator output, T_e^*. In the indirect vector control of an induction machine drive system, the error in the rotor time constant results in the distortion of the torque of the machine because of the error of the angle of rotor flux linkage. To decouple this distorted torque, $\tilde{T}_e$, the output of the speed regulator would be distorted and the output of the regulator, T_e^*, looks like the waveforms in Fig. A.7c or Fig. A.7d. Hence, by adjusting the rotor time constant of the indirect vector-controlled induction machine drive system to get the square-type output to the ramp input as shown in Fig. A.7a, the rotor time constant can be identified. To use this method, the friction and load torque should be small enough compared to the acceleration torque of the inertia of the system.

In the actual drive system, there are inevitably noises from current and speed sensors. And the output of the speed regulator includes ripples. The ripples of the output make the application of the above method difficult. However, as mentioned before, due to the inherent low-pass filtering characteristics of the integral term of the PI regulator, the integral term, $e_{rr}(s)\cdot\frac{K_i}{s}$ in Fig. A.6, instead of whole output of PI regulator, can be used to identify the rotor time constant.

A.2.4.2 Estimation of Stator Self-Inductance, L_s, and Stator Transient Inductance, σL_s

The inductances of the induction machine can be identified by exploiting the characteristics of the current regulator. First, a method to identify the stator self-inductance is described. It is assumed that the current regulator is implemented as PI type as mentioned in Section 4.2.5 at the synchronous reference frame; also, its transfer function, $C(s)$, is given by (A.22). And, from the stator voltage equation (4.31), the controlled plant, $P(s)$, can be described as (A.23).

$$C(s) = K_p + \frac{K_i}{s} \tag{A.22}$$

$$P(s) = \frac{1}{\left(R_s + R_r\frac{L_m^2}{L_r^2}\right) + \sigma L_s s} \tag{A.23}$$

The integral term of the output of the q axis current regulator can be used to identify the inductance. The control block diagram including plant, $P(s)$, is shown in Fig. A.8, where the feed-forwarding term can be given as (A.24) under the assumption of no rotor flux linkage variation during the test.

$$V_{qs\_ff}^{e*} = \omega_e \hat{L}_s i_{ds}^e - \omega_{sl}\frac{\hat{L}_m^2}{\hat{L}_r} i_{ds}^e \tag{A.24}$$

If the rotor time constant has been set by the method in Section A.2.4.1, then it can be assumed that the d-axis current, i_{ds}^e, and ω_{sl} are well-matched to their actual values.

If there is error in the estimated stator self-inductance, $\hat{L}_s$, by ΔL_s, then feed-forwarding component can be described as (A.25). If this test is done at near the rated speed, then the third term, $\omega_{sl}\frac{\hat{L}_m^2}{\hat{L}_r} i_{ds}^e$, on the right-hand side of (A.25) is small enough

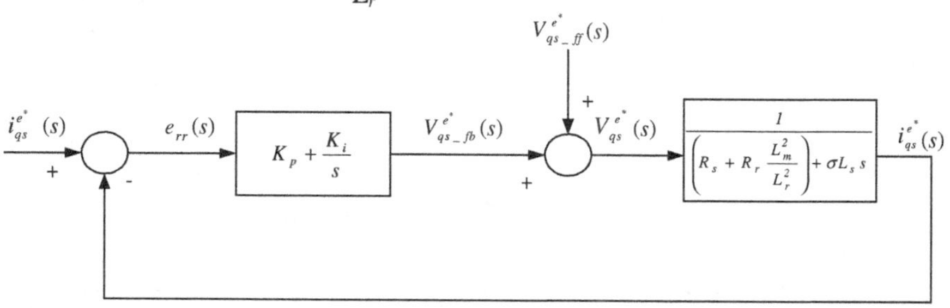

Figure A.8 Control block diagram of a q-axis current regulator of vector-controlled induction machine drive system.

compared to the second term, $\omega_e \Delta L_s i^e_{ds}$. And the third term has nothing to do with the synchronous angular frequency, ω_e, but the magnitude of the second term is proportional to ω_e.

$$\begin{aligned} V^{e*}_{qs\_ff} &= \omega_e \hat{L}_s i^e_{ds} - \omega_{sl} \frac{\hat{L}_m^2}{\hat{L}_r} i^e_{ds} \\ &= \omega_e L_s i^e_{ds} + \omega_e \Delta L_s i^e_{ds} - \omega_{sl} \frac{\hat{L}_m^2}{\hat{L}_r} i^e_{ds} \end{aligned} \tag{A.25}$$

And, when ω_e varies, it can be seen that the error in the feed-forwarding term comes from only the error in the estimation of the self-inductance. Hence, the error in the voltage would be compensated by the integral term of the current regulator under the assumption that q-axis current is well-regulated according to its reference. If there is error in the estimated stator self-inductance, the voltage of the integral term would be proportional to the operating speed, ω_e, with constant d-axis current i^e_{ds}. In Fig. A.9, the waveforms of the integral term of the current regulator

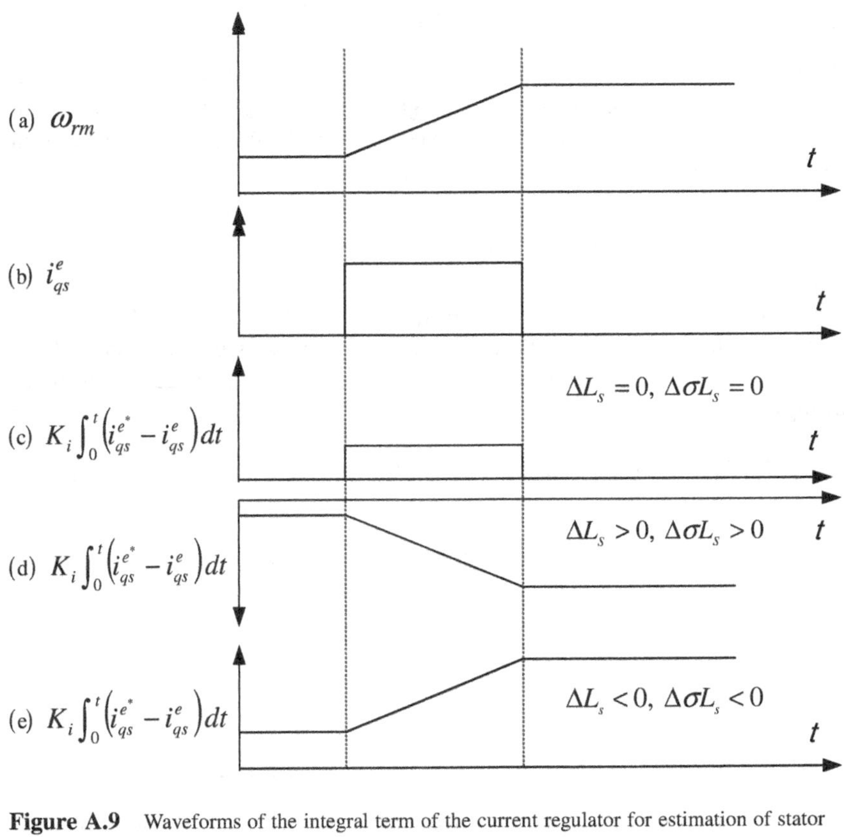

Figure A.9 Waveforms of the integral term of the current regulator for estimation of stator self-inductance.

are shown along with the ramp input of speed reference and its associated q-axis current reference according to different values of the estimated stator self-inductances.

To the ramp speed reference as shown in Fig. A.9a, the q-axis current would be the trace as shown in Fig. A.9a under the assumption that speed is well-regulated and that the mechanical load can be considered as mainly inertia load. If there is no error in the stator transient inductance, the trace of the integral term would be the one shown in Fig. A.9c, while, if there is error, then the trace shown in Fig. A.9d or Fig. A.9e appears according to the polarity of the error. Hence by adjusting, $\hat{L}_s$, in the feed-forwarding term, the integral term can be set until the trace in Fig. A.9c appears. Another reason of voltage distortion of the output of the current regulator would be dead time effects. However, if this test is done at near rated speed, the dead time effects can be neglected. From this identified stator self-inductance, the mutual inductance can be found without much error because the stator leakage inductance is much smaller than the stator self-inductance as seen from (2.75) and (2.76).

After identifying the stator self-inductance, the stator transient inductance, $\sigma\hat{L}_s$, can be identified similarly. For the identification of $\sigma\hat{L}_s$, the feed-forwarding term in (A.24) can be used again. But, the feed-forwarding term can be calculate by (A.26) to reflect the effect of the stator transient inductance. The first term and the third term on the right-hand side of (A.26) can be set reasonably accurately because of the identified stator self-inductance. Also, because the third term has nothing to do with the operating speed, it is only related to the torque of the induction machine, and it is quite small compared to other terms. If there is error, $\Delta\sigma L_s$, in the estimated $\sigma\hat{L}_s$, then the feed-forwarding term can be approximated as (A.27). And the voltage proportional to the speed appears again in the integral term of the current regulator. By keeping the trace of the integral term of the current regulator as the trace shown in Fig. A.9c, the stator transient inductance can be identified.

$$V_{qs\_ff}^{e*} = \omega_e \frac{\hat{L}_m}{\hat{L}_r} \lambda_{dr}^{e*} + \omega_e \sigma \hat{L}_s i_{ds}^e - \omega_{sl} \frac{\hat{L}_m^2}{\hat{L}_r} i_{ds}^e \tag{A.26}$$

$$\begin{aligned} V_{qs\_ff}^{e*} &\approx \omega_e \frac{\hat{L}_m}{\hat{L}_r} \lambda_{dr}^{e*} + \omega_e \sigma \hat{L}_s i_{ds}^e \\ &= \omega_e \frac{\hat{L}_m}{\hat{L}_r} \lambda_{dr}^{e*} + \omega_e \sigma L_s i_{ds}^e + \omega_e \Delta\sigma L_s i_{ds}^e \end{aligned} \tag{A.27}$$

It should be noted that the estimation of the transient inductance may have considerable error if there is still error in the stator self inductance. And it is important to estimate the stator self-inductance as accurately as possible. The mutual inductance can be found by simply subtracting half of the transient inductance from the stator self-inductance under the assumption of the equal stator and rotor leakage inductances.

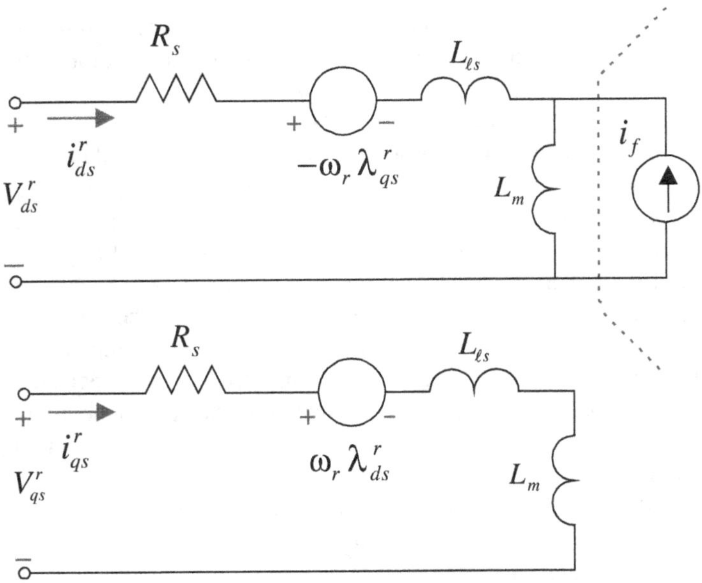

Figure A.10 Equivalent circuit of surface-mounted permanent magnet synchronous machine.

A.2.5 Permanent Magnet Synchronous Machine [2]

A.2.5.1 Surface-Mounted Permanent Magnet Synchronous Machine (SMPMSM)

The stator voltage equation at rotor reference frame of SMPMSM can be derived from the equivalent circuit in Fig. A.10 as mentioned in Section 3.3.3.1.

$$\begin{aligned} V_{ds}^r &= R_s i_{ds}^r + L_s \frac{di_{ds}^r}{dt} - \omega_r L_s i_{qs}^r \\ V_{qs}^r &= R_s i_{qs}^r + L_s \frac{di_{qs}^r}{dt} + \omega_r L_s i_{ds}^r + \omega_r \lambda_f \end{aligned} \tag{A.28}$$

where λ_f represents the flux linkage of the permanent magnet to the stator winding and it can be expressed in terms of the equivalent current source, i_f, and the mutual inductance, L_m, as $\lambda_f = L_m i_f$. And the torque can be represented as

$$T_e = \frac{3}{2} \cdot \frac{P}{2} \cdot \lambda_f i_{qs}^r \tag{A.29}$$

A.2.5.1.1 Rotor Position, θ_r. If the rotor position, $\hat{\theta}_r$, used for the transformation of reference frame, is different from the actual rotor position, θ_r, then not only the performance of the vector control is degraded due to the coupling of the *d–q* axis but also the estimated parameters of SMPMSM would be inaccurate. And, to estimate the parameters, the rotor position should first be estimated accurately. The absolute angle from the absolute encoder or from the resolver may have errors with the real position of the rotor because of the attachment error of the resolver or encoder to the shaft of the rotor. For example, if the attachment error is 3° in mechanical angle and the number of poles in the SMPMSM machine is over 16, the error would be more than 24°, and the torque error would be more than 8% only due to this angle error. The angle error can be compensated by exploiting the characteristics of the current regulator as follows. If the current regulator in Section 4.2.5 is used, then the *d–q*-axis voltage to the SMPMSM can be described as (A.30). The voltage consists of a term from the output of the feedback regulator and a feed-forwarding term.

$$\begin{aligned} V_{ds}^{r*} &= V_{ds\_fb}^{r*} + V_{ds\_ff}^{r*} \\ V_{qs}^{r*} &= V_{qs\_fb}^{r*} + V_{qs\_ff}^{r*} \end{aligned} \tag{A.30}$$

where $V_{ds\_ff}^{r*}$ and $V_{qs\_ff}^{r*}$ can be given by

$$\begin{aligned} V_{ds\_ff}^{r*} &= -\omega_r \hat{L}_s i_{qs}^r \\ V_{qs\_ff}^{r*} &= \omega_r \hat{L}_s i_{ds}^r + \omega_r \hat{\lambda}_f \end{aligned} \tag{A.31}$$

If the angle, $\hat{\theta}_r$, has a difference with regard to the rotor position by θ_{err} as shown in Fig. A.11, then the back EMF would appear not only in the *d* axis but also in the *q* axis as

$$e_{ds\text{-}err}^r = -\omega_r \lambda_f \sin \theta_{err} \tag{A.32}$$

Hence, this voltage appears at the output of the *d*-axis feedback controller, $V_{ds\_fb}^{r*}$. By exploiting this phenomenon, the angle error can be compensated. In Fig. A.12, when the error is 10° and–10° in electrical angle for the case of a 24-pole SMPMSM drive system, which is less than 1° in mechanical angle, the integral term of the *d*-axis current regulator is shown according to the error. The speed of SMPMSM varies from -50r/min to 50 r/min linearly. If the polarity of the angle error is positive, then the trace of the integral term is the reverse of the trace of the speed as shown in Fig. A.12a. If the

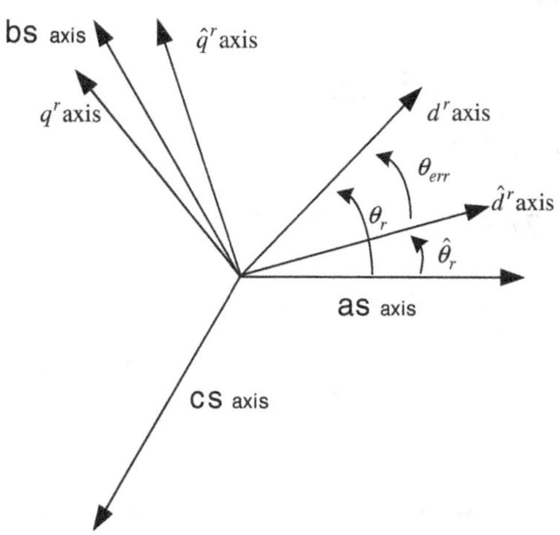

Figure A.11 Angle error of an SMPMSM.

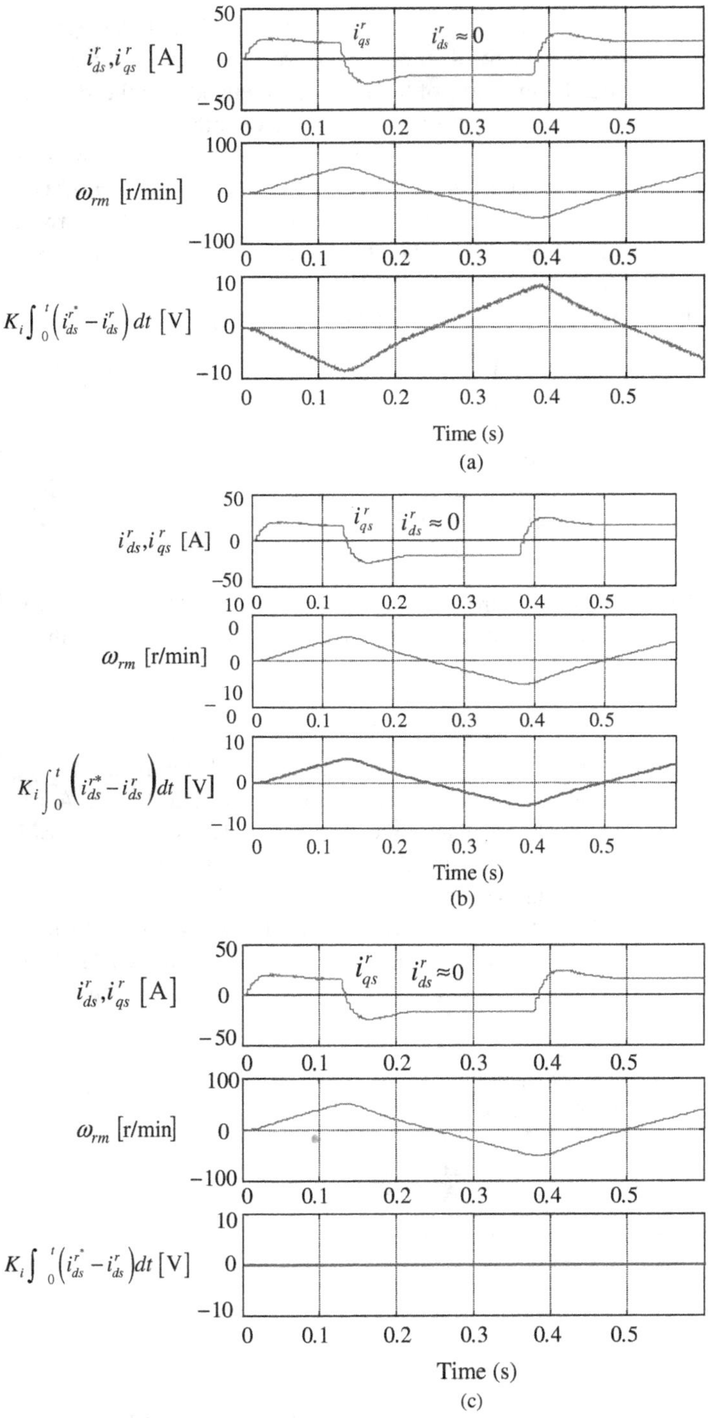

Figure A.12 Integral term of a d-axis current regulator with rotor angle error (*from top*: trace of d–q axis current, speed of rotor, and integral term of d-axis current regulator). (a) With error by 10°. (b) With error by –10°. (c) With correct angle of rotor.

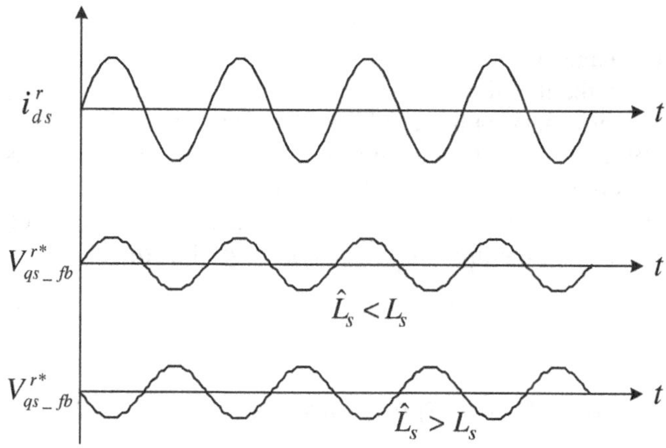

Figure A.13 Estimation of L_s by injecting an AC current to d axis.

polarity of the error is negative, the trace is the one shown in Fig. A.12b. When the angle is correct, then the integral term is almost zero regardless of the variation of the speed as shown in Fig. A.12c. Hence, by keeping the integral term as zero regardless of the speed variation, the correct rotor angle can be identified.

A.2.5.1.2 Synchronous Inductance, L_s. To estimate the synchronous inductance,L_s, as seen from (A.28), the d–q voltage to SMPMSM should be measured under the variation of the d-axis current or the q-axis current. If q-axis current varies, then torque would vary. And the sinusoidal variation of d-axis current can be used to estimate the inductance. If an SMPMSM is running at constant speed with a speed regulator, the q-axis voltage equation can be derived as

$$V^r_{qs} = R_s i^r_{qs} + \omega_r L_s i^r_{ds} + \omega_r \lambda_f \tag{A.33}$$

In (A.33), if i^r_{ds} varies as the first trace in Fig. A.13 and if $\hat{L}_s < L_s$, then the output of q-axis current regulator would be the second trace in Fig. A.13. Or if $\hat{L}_s > L_s$, it would be the third trace in Fig. A.13. Hence, by keeping the output of the q-axis current regulator as a direct line with adjustment of $\hat{L}_s$, the inductance can be identified. In this test the frequency of the d-axis current should be less than a fraction of the bandwidth of the current regulator. And the magnitude should be small enough not to saturate the output of the current regulator, but large enough to enhance the signal-to-noise ratio. And the operating speed of SMPMSM would be near the rated speed to increase the sensitivity of the voltage to the inductance error.

A.2.5.1.3 Flux Linkage of Permanent Magnet, $\lambda_f = L_m i_f$. The flux linkage, λ_f, can be easily identified similar to the back EMF constant of DC machine as mentioned in Section A.2.2. In (A.31), if the d-axis current is regulated as null, the feed-forwarding term has only a back EMF term proportional to the rotor speed like the feed-forwarding term of the current regulator of DC machine in Fig. 4.12. And if there is error in $\hat{\lambda}_f$, then the integral term of a q-axis current regulator varies according to the speed at constant q-axis current reference. By keeping the integral term as constant regardless of speed, the flux linkage can be identified.

Another simple method is observing back EMF without any current through an SMPMSM. The back EMF can be monitored by regulating both d–q-axis currents as zero when the SMPMSM is running at near the rated speed. In this case, the q-axis voltage to the SMPMSM is only the back EMF term as in (A.34). And by dividing the voltage by the rotor speed, ω_r, the flux linkage can be easily identified.

$$V^r_{qs} = \omega_r \lambda_f \tag{A.34}$$

A.2.5.2 Interior Permanent Magnet Synchronous Machine (IPMSM)

The parameters of an IPMSM can be estimated similarly to the parameters of an SMPMSM. The error of rotor angle can be compensated exactly by the same manner as in the case of an SMPMSM. However, unlike an SMPMSM, if the constant power speed ratio (CPSR) of IPMSM is high, then the back EMF voltage of IPMSM would be much smaller compared to that of an SMPMSM. And, the sensitivity of voltage, $-\omega_r\lambda_f\sin(\theta_{err})$, to angle error would be reduced. Hence, tuning of the angle would be difficult. The inductances, L_d and L_q, of IPMSM can be estimated by injecting a sinusoidal current to an IPMSM. Unlike an SMPMSM, to estimate both L_d and L_q, the current should be injected to both d–q axes of the rotor reference frame of the IPMSM. $L_d(L_d = L_{ls} + L_{md})$ can be identified by using the output of the q-axis current regulator and $L_q(L_q = L_{ls} + L_{mq})$ by the output of the d axis. As mentioned in Section 2.9.2, L_q varies widely according to the currents, i^r_{qs} and i^r_{ds}. Hence, L_q should be identified at several pairs of the currents, i^r_{qs} and i^r_{ds}. Unlike an SMPMSM, with sinusoidal current injection to an IPMSM for the estimation of the inductances, the sinusoidal torque would occur and the speed of the IPMSM would vary according to this torque. To minimize the speed variation and to suppress the voltage due to the speed variation, the frequency and the magnitude of the injected sinusoidal current should be set carefully. The frequency should be high enough compared to the inverse of the mechanical time constant, which is defined as $\tau_m = J/B$, of the drive system of the IPMSM to filter out the sinusoidal torque by the mechanical time constant. But, the frequency should be set within the bandwidth of the current regulator. And the frequency could be a fraction of the bandwidth. The operating speed of the IPMSM for inductance estimation should be above or near the base speed of the IPMSM to enhance the accuracy of the estimation.

A.3 ESTIMATION OF MECHANICAL PARAMETERS

A.3.1 Estimation Based on Mechanical Equation

The mechanical parameters of a drive system are the inertia, J_m[kg-m^2], the friction coefficient, B[N-m/(rad/s)], and the stiffness, K_{sh}[N-m/rad]. The friction coefficient,

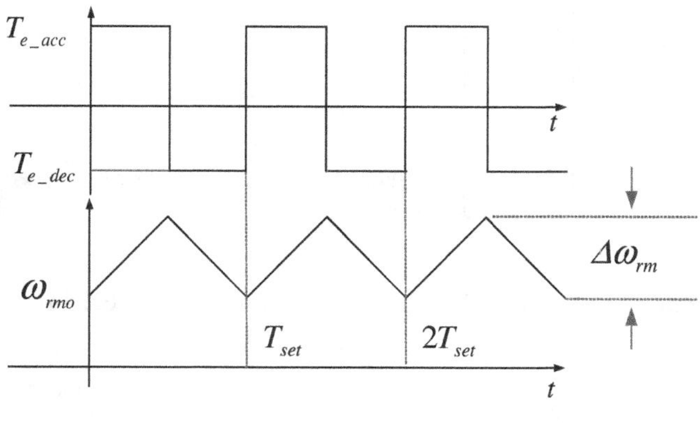

Figure A.14 Estimation of inertia by acceleration and deceleration of rotor.

B, varies widely and nonlinearly according to the direction of the rotation and rotational speed. Also, it may vary according to the operating environment such as temperature, humidity, and aging. If the drive system can be operated without any load, then by measuring or by estimating torque based on the flux linkage and current, the friction coefficient can be evaluated at different speeds as

$$B = T_e/\omega_{rm} \tag{A.35}$$

The friction coefficient due to windage and friction loss has a tendency whereby the coefficient increases as magnitude of speed increases.

The inertia of the drive system can be calculated with the data of the structure of the rotating body and its materials as mentioned in Section 1.2.3. However, such data are not easily available in the industry field. If the friction coefficient is known accurately or if the friction torque is negligible, then the inertia can be estimated by accelerating or decelerating the rotor as shown in Fig. A.14. The inertia can be found as (A.36) from the torque for the acceleration, T_{e_acc}, and deceleration, T_{e_dec}, and the rate of speed variation, $\frac{\Delta\omega_{rm}}{T_{set}/2}$.

$$J_m = \left(\frac{T_{e\_acc} - B \cdot \omega_{rm}}{\frac{\Delta\omega_{rm}}{T_{set}/2}} - \frac{T_{e\_dec} - B \cdot \omega_{rm}}{\frac{\Delta\omega_{rm}}{T_{set}/2}} \right) /2 \tag{A.36}$$

The acceleration and deceleration rate can be set as large as possible to enhance signal-to-noise ratio. In order to minimize parameter variation according to the speed, the period and the magnitude of the square wave can be set to bound the speed variation within several tens of percentages of the rated speed. The minimum speed of the test, ω_{rmo}, can be above several tens of percentages of rated speed to minimize errors in speed and torque measurement.

Another method is exploiting the natural decaying characteristics of speed without mechanical load. At near rated speed or at a certain speed, if the electric power to the machine is turned off, where it is defined as $t = 0$ s, then the speed

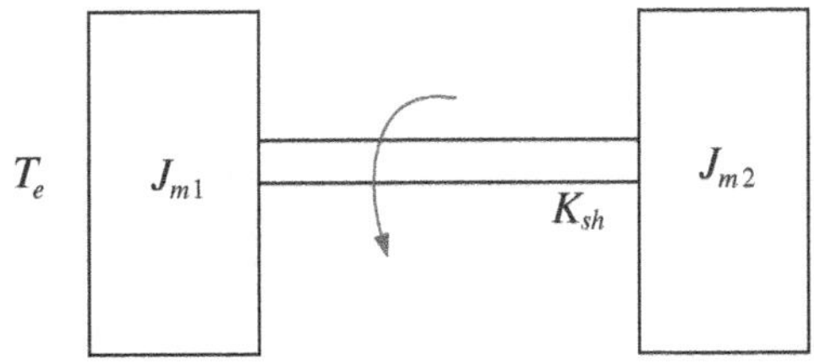

Figure A.15 Two inertia loads connected with a long shaft.

of the rotor would decay as (A.37) under the assumption of no mechanical load.

$$\omega_{rm} = \omega_{rmo} e^{-\frac{B}{J_m}t} \tag{A.37}$$

At $t = 0^+$ s, the slope of the speed variation can be evaluated like

$$\frac{d\omega_{rm}}{dt}_{t=0^+} = -\omega_{rmo} \cdot \frac{B}{J_m} \tag{A.38}$$

where ω_{rmo} is the speed just after the turning-off of the power to the machine. By doing the above-mentioned test at several different speeds, $\omega_{rmo1}, \omega_{rmo2},$ and $\omega_{rmo3}, \ldots$, the inverse of the mechanical time constant, B/J_m, can be evaluated at those speeds. If the tests are successful, then the calculated inertias from the mechanical time constant would be almost same as each other with the consideration of the variation of the friction coefficient.

The stiffness, K_{sh}, can be calculated by the structure and material of coupling shaft between the electric machine and the mechanical load. But such data are not easily available. Another method to identify the stiffness is to measure the resonant frequency of the drive system. As shown in Fig. A.15, if two inertias are connected with a long shaft, at the step torque the rotor accelerates with the mechanical resonance, whose frequency is ω_{sh}, as shown in Fig. A.16.

The resonance frequency can be calculated as (A.39) after neglecting the inertia of the connecting shaft and damping of the system.

$$\omega_{sh} = \sqrt{K_{sh}\left(\frac{1}{J_{m1}} + \frac{1}{J_{m2}}\right)} \tag{A.39}$$

Hence, from (A.39) if ω_{sh}, J_{m1}, and J_{m2} are known, then the stiffness, K_{sh}, can be estimated.

A.3.2 Estimation Using Integral Process [6]

In the parameter estimation methods at the previous section, the estimation of one parameter is related to other parameters. Also, the accuracy of the estimation is related to the accuracy of other parameters. Furthermore, according to load conditions, the estimated error would be large. Also, in some mechanical system with a flexible coupling and/or with small backlash, the estimation would be impossible because of

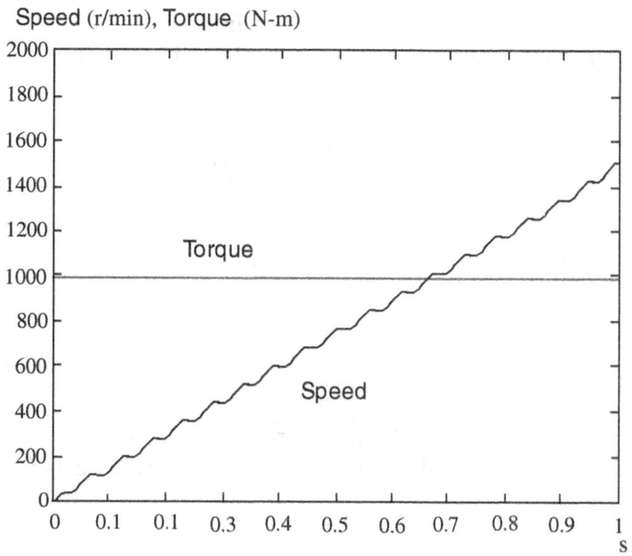

Figure A.16 Speed response of two inertia loads with step torque.

the resonance and nonlinearity of the mechanical system. In this section a method to estimate the parameters is introduced. Because the method relies on the basic mechanical equation and integral process, it is robust to the operating condition of the drive system. Hence it can give reasonably accurate estimation results even with above resonances and nonlinearities. Furthermore, this method can be used in the case that the rotational direction and the operational speed of the drive system are limited with any reason. In Fig. A.17, a mechanical system considering Coulomb friction and disturbance torque is shown. In the figure, C_m is the maximum friction at standstill, representing Columb friction; T_d is the disturbance torque including modeling error of the system, and it is assumed that it varies slowly. From the figure, the torque equation can be derived as

$$T_e = (J_m + J_L)\frac{d\omega}{dt} + B\omega + \text{sign}(\omega)C_m - T_d \tag{A.40}$$

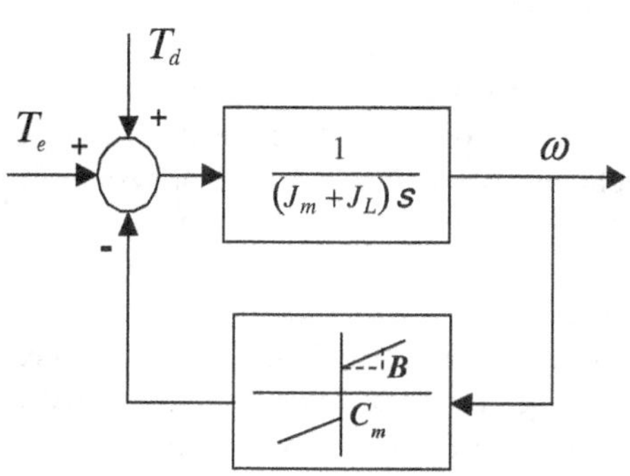

Figure A.17 Control block diagram of a mechanical system considering Coulomb friction and disturbance torque.

where "sign" is sign function, whose value is 1 or –1 according to the sign of the variable. By multiplying angular acceleration, $d\omega/dt$, on both sides of (A.40), we can obtain (A.41).

$$T_e \frac{d\omega}{dt} = (J_m + J_L)\left(\frac{d\omega}{dt}\right)^2 + B\omega\frac{d\omega}{dt} + T'_d \frac{d\omega}{dt} \tag{A.41}$$

where $T'_d = \text{sign}(\omega)C_m - T_d$. If T'_d varies slowly and its differentiation regarding time can be neglected, then by integrating both sides of (A.41), we can derive (A.42).

$$\int_{t_1}^{t_2} T_e \frac{d\omega}{dt}\,dt = (J_m + J_L)\int_{t_1}^{t_2}\left(\frac{d\omega}{dt}\right)^2 dt + B\int_{t_1}^{t_2}\omega\frac{d\omega}{dt}\,dt + + T'_d\int_{t_1}^{t_2}\frac{d\omega}{dt}\,dt \tag{A.42}$$

where t_1 is the initial time point where the estimation process start, and t_2 is the ending time point. The second term on the right-hand side of (A.42) can be expressed as

$$B\int_{t_1}^{t_2}\omega\frac{d\omega}{dt}\,dt = \frac{B}{2}\left(\omega^2_{at\ t=t_2} - \omega^2_{at\ t=t_1}\right) \tag{A.43}$$

If the torque, T_e, of the electric machine is well-regulated according to its reference, T_e^*, then $T_e = T_e^*$. From (A.42), the inertia can be identified by

$$J_T = J_m + J_L = \frac{\int_{t_1}^{t_2} T_e^* \frac{d\omega}{dt}\,dt}{\int_{t_1}^{t_2}\left(\frac{d\omega}{dt}\right)^2 dt} - \frac{B}{2}\frac{\left(\omega^2_{at\ t=t_2} - \omega^2_{at\ t=t_1}\right)}{\int_{t_1}^{t_2}\left(\frac{d\omega}{dt}\right)^2 dt} - T'_d \frac{\left(\omega_{at\ t=t_2} - \omega_{at\ t=t_1}\right)}{\int_{t_1}^{t_2}\left(\frac{d\omega}{dt}\right)^2 dt} \tag{A.44}$$

In (A.41), the angular acceleration, $d\omega/dt$, can be obtained by high-pass filtering or band-pass filtering of the measured speed to suppress the measurement noise. Or the acceleration can be obtained from the acceleration observer mentioned in Section 4.3.4.2. As seen from (A.44), if the estimation time defined as $t_2 - t_1$ is long enough, the denominators on the right-hand side of (A.44) are large enough to neglect the second and third terms on the right-hand side of (A.44). And only the first term can be used to estimate the inertia as (A.45). In particular, if the speeds at t_1 and at t_2 are the same, then the second and the third terms should be zero. And the inertia can be estimated only by the first term without the approximation error. With this method, the inertia can be estimated even under the restriction of operational speed and torque range. Hence, this method can be easily applicable to existing mechanical system. If the operating speed for the test is limited in a narrow speed region, then the estimation error due to the variation of the friction and disturbance torque can be minimized. In particular, if the rotational direction is fixed and the speed region is limited above some speed, then the error from Coulomb friction can be eliminated. Furthermore, because the signal processing relies on the integral process, which is inherently a strong low-pass filter, the estimation process is robust to the measurement noise as long as the average of the noise is zero.

$$J_T = \frac{\int_{t_1}^{t_2} T_e^* \frac{d\omega}{dt} dt}{\int_{t_1}^{t_2} \left(\frac{d\omega}{dt}\right)^2 dt} \tag{A.45}$$

The friction coefficient can be estimated similarly. Under the assumption that the disturbance torque and Coulomb friction are constant during the estimation process, by differentiating (A.40) regarding time and multiplying the angular acceleration, (A.46) can be obtained.

$$\frac{dT_e^*}{dt}\frac{d\omega}{dt} = (J_m + J_L)\frac{d^2\omega}{dt^2}\frac{d\omega}{dt} + B\left(\frac{d\omega}{dt}\right)^2 \tag{A.46}$$

By integrating both sides of (A.46), we can derive (A.47).

$$\int_{t_1}^{t_2} \frac{dT_e^*}{dt}\frac{d\omega}{dt} dt = J_T \int_{t_1}^{t_2} \frac{d^2\omega}{dt^2}\frac{d\omega}{dt} dt + B\int_{t_1}^{t_2} \left(\frac{d\omega}{dt}\right)^2 dt \tag{A.47}$$

The first term on the right-hand side of (A.47) can be expressed as

$$J_T \int_{t_1}^{t_2} \frac{d^2\omega}{dt^2}\frac{d\omega}{dt} dt = \frac{J_T}{2}\left(\left(\frac{d\omega}{dt}\right)^2_{at\ t=t_2} - \left(\frac{d\omega}{dt}\right)^2_{at\ t=t_1}\right) \tag{A.48}$$

By substituting (A.48) into (A.47), the friction coefficient can be estimated as

$$B = \frac{\int_{t_1}^{t_2} \frac{dT_e^*}{dt}\frac{d\omega}{dt} dt}{\int_{t_1}^{t_2} \left(\frac{d\omega}{dt}\right)^2 dt} - \frac{J_T}{2}\frac{\left(\left(\frac{d\omega}{dt}\right)^2_{at\ t=t_2} - \left(\frac{d\omega}{dt}\right)^2_{at\ t=t_1}\right)}{\int_{t_1}^{t_2} \left(\frac{d\omega}{dt}\right)^2 dt} \tag{A.49}$$

Like the estimation of the inertia, in (A.49), if the estimation time is long enough or if the acceleration is the same at $t_1$and at t_2, then (A.49) can be simplified as

$$B = \frac{\int_{t_1}^{t_2} \frac{dT_e^*}{dt}\frac{d\omega}{dt} dt}{\int_{t_1}^{t_2} \left(\frac{d\omega}{dt}\right)^2 dt} \tag{A.50}$$

The mechanical parameter estimation method in this section can be applied even to the system with the mechanical resonance and nonlinearity. By extending estimation time long enough or by limiting the estimation speed region where the resonance or nonlinearity would be less or not exist, the parameters can be estimated within reasonable error bound.

REFERENCES

1. J. W. Choi et al., Inverter output voltage synthesis using novel dead time compensation, *IEEE Trans. Power Electron.*, Vol. 11, No. 2, March 1996, pp. 221–227.
2. J. Seok, High performance drive of deep bar induction machine using pseudo-rotor flux oriented control, Ph. D. thesis, Seoul National University, Seoul, Korea, 1998, (in Korean).
3. J. K. Seok et al., Optimal flux selection of an induction machine for maximum torque operation in flux-weakening region, *IEEE Trans. Power Electron.*, Vol. 14, No. 4, July 1999, pp. 700–708.
4. J. K. Seok et al., Induction motor parameter tuning for high-performance drives, *IEEE Trans. Ind. Appl.*, Vol. 37, No. 1, January/February 2001, pp. 35–41.
5. J. K. Ji et al., Kalman Filter and LQ-Based Speed Controller for Torsional Vibration Suppression in 2-Mass Motor Drive System, *IEEE Trans. Ind. Electron.*, Vol. 42, No. 6, December 1995, pp. 1–8.
6. T. Kwon et al., Identification of the mechanical parameters for servo drive, in *Conference Record IEEE-IAS Annual Meeting*, 2006.

Appendix B

d–q Modeling Using Matrix Equations

B.1 REFERENCE FRAME AND TRANSFORMATION MATRIX

In Fig. B.1, the relationship between a three-phase axis, a stationary reference frame d–q axis, and a rotating d–q axis with arbitrary speed, ω, is shown. Conventionally, in an AC machine, a d axis in the synchronous reference frame or rotor reference frame means a direct axis where the excitation flux linkage is directed. And a q axis, which means a quadrature axis, is located 90° away in the positive rotating direction. In the steady state, the electric phenomenon such as flux linkage, voltage, and current in the d axis appears at the q axis after a quarter of the fundamental time period of the rotating MMF.

At the axes in Fig. B.1, the transformation matrix to convert a variable in a three-phase axis into the variable in a rotating axis with arbitrary speed can be derived as

$$\mathbf{f}_{dqn}^{\omega} = \mathbf{T}(\theta)\mathbf{f}_{abc} \tag{B.1}$$

where $\mathbf{f}_{dqn}^{\omega}$ and $\mathbf{f}_{abc}$ are column vectors of variables given as (B.2) and (B.3).

$$\mathbf{f}_{dqn}^{\omega} = [f_d^{\omega}\, f_q^{\omega}\, f_n^{\omega}]^T \tag{B.2}$$

$$\mathbf{f}_{abc} = [f_a f_b f_c]^T \tag{B.3}$$

$[\ldots]^T$ means the transpose of a matrix or a vector.

Control of Electric Machine Drive Systems, by Seung-Ki Sul

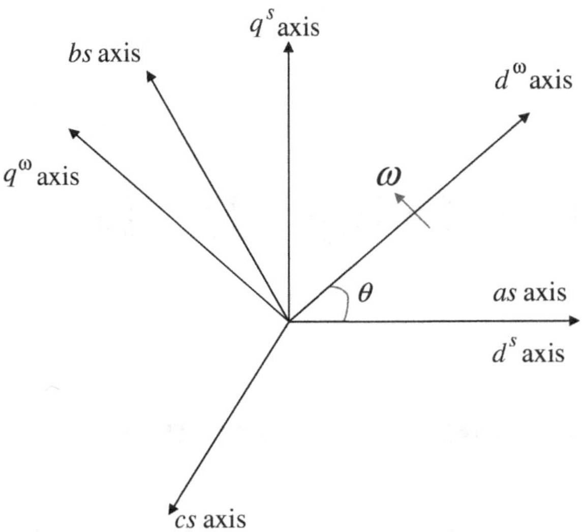

Figure B.1 Relation of reference frames.

The transformation matrix, $\mathbf{T}(\theta)$, can be defined as

$$\mathbf{T}(\theta) \equiv \frac{2}{3}\begin{bmatrix} \cos\theta & \cos\left(\theta-\frac{2}{3}\pi\right) & \cos\left(\theta+\frac{2}{3}\pi\right) \\ -\sin\theta & -\sin\left(\theta-\frac{2}{3}\pi\right) & -\sin\left(\theta+\frac{2}{3}\pi\right) \\ \frac{1}{\sqrt{2}} & \frac{1}{\sqrt{2}} & \frac{1}{\sqrt{2}} \end{bmatrix} \tag{B.4}$$

where θ is the angle between the magnetic axis of an "a" phase (d^s) and an arbitrary speed rotating d axis (d^ω) and is given by

$$\theta = \int_0^t \omega(\zeta)d\zeta + \theta(0) \tag{B.5}$$

where $\theta(0)$ is the initial angle between two axes at time point zero. It is conventionally set as zero, such as $\theta(0) = 0$.

$\mathbf{T}(\theta)$ may be defined differently. If $\mathbf{T}(\theta)$ is defined as (B.6), then the magnitude of phase variables in *d–q* reference frame is larger than that of corresponding variables in a three-phase axis by the ratio of $\sqrt{3}/\sqrt{2}$. But, the power and torque calculated by *d–q–n* variables are the same as those by three-phase variables, which means that the transformation by (B.6) is called a power invariant transformation. The inverse matrix of the transformation matrix in (B.6) is the transpose of the matrix itself—that is, $\mathbf{T}(\theta)^{-1} = \mathbf{T}(\theta)^T$. Hence the matrix in (B.6) is an orthogonal matrix.

$$\mathbf{T}(\theta) = \sqrt{\frac{2}{3}}\begin{bmatrix} \cos\theta & \cos\left(\theta - \frac{2}{3}\pi\right) & \cos\left(\theta + \frac{2}{3}\pi\right) \\ -\sin\theta & -\sin\left(\theta - \frac{2}{3}\pi\right) & -\sin\left(\theta + \frac{2}{3}\pi\right) \\ \frac{1}{\sqrt{2}} & \frac{1}{\sqrt{2}} & \frac{1}{\sqrt{2}} \end{bmatrix} \tag{B.6}$$

However, if $\mathbf{T}(\theta)$ is defined as (B.7), then the magnitude of phase variables in the *d–q* reference frame is the same as the corresponding variables in the three-phase axis like the case with (B.4). But, the power and torque calculated by *d–q–n* variables are smaller than those calculated by three-phase variables by the ratio of 2/3, and the transformation by (B.7) is called a magnitude invariant transformation. Furthermore, with the transformation matrix in (B.7), the neutral component is the same one defined by (3.3). But, the inverse matrix is not the transpose of itself. And, the manipulating of the transformation matrix is getting complicated.

$$\mathbf{T}(\theta) = \frac{2}{3}\begin{bmatrix} \cos\theta & \cos\left(\theta - \frac{2}{3}\pi\right) & \cos\left(\theta + \frac{2}{3}\pi\right) \\ -\sin\theta & -\sin\left(\theta - \frac{2}{3}\pi\right) & -\sin\left(\theta + \frac{2}{3}\pi\right) \\ \frac{1}{2} & \frac{1}{2} & \frac{1}{2} \end{bmatrix} \tag{B.7}$$

In this text, the transformation matrix in (B.4) is used as a nominal transformation matrix. In this case, the magnitude of the *d–q* variables is the same as that for the corresponding variables, but the magnitude of the neutral component is not the same as the one in (3.3). It is larger than that of neutral component in (3.3) by a ratio of $\sqrt{2}$. However, the inverse matrix of (B.4) can be simply obtained as (B.8), though it is not the transpose of itself. It should be noted that the transformation by (B.4) is not power invariant and that the torque and power from *d–q–n* variables should be scaled by 3/2. The matrix in (B.4) has many useful characteristics as (B.8)–(B.11), and those characteristics can be used to simplify the manipulation of the dynamic equation of AC machines.

$$\mathbf{T}(\theta)^{-1} = \frac{3}{2}\mathbf{T}(\theta)^{T} = \begin{bmatrix} \cos\theta & -\sin\theta & \frac{1}{\sqrt{2}} \\ \cos\left(\theta - \frac{2}{3}\pi\right) & -\sin\left(\theta - \frac{2}{3}\pi\right) & \frac{1}{\sqrt{2}} \\ \cos\left(\theta + \frac{2}{3}\pi\right) & -\sin\left(\theta + \frac{2}{3}\pi\right) & \frac{1}{\sqrt{2}} \end{bmatrix} \tag{B.8}$$

$$\mathbf{T}(\theta)\left[\frac{d}{dt}\mathbf{T}(\theta)^{-1}\right] = \begin{bmatrix} 0 & -\omega & 0 \\ \omega & 0 & 0 \\ 0 & 0 & 0 \end{bmatrix} \tag{B.9}$$

$$\left[\frac{d}{dt}\mathbf{T}(\theta)\right]\mathbf{T}(\theta)^{-1} = \begin{bmatrix} 0 & \omega & 0 \\ -\omega & 0 & 0 \\ 0 & 0 & 0 \end{bmatrix} \tag{B.10}$$

$$\mathbf{T}(\theta) = \mathbf{R}(\theta)\mathbf{T}(0) \tag{B.11}$$

where

$$\mathbf{R}(\theta) = \begin{bmatrix} \cos\theta & \sin\theta & 0 \\ -\sin\theta & \cos\theta & 0 \\ 0 & 0 & 1 \end{bmatrix} \tag{B.12}$$

and

$$\mathbf{T}(0) = \frac{2}{3}\begin{bmatrix} 1 & -\frac{1}{2} & -\frac{1}{2} \\ 0 & \frac{\sqrt{3}}{2} & -\frac{\sqrt{3}}{2} \\ \frac{1}{\sqrt{2}} & \frac{1}{\sqrt{2}} & \frac{1}{\sqrt{2}} \end{bmatrix} \tag{B.13}$$

By transformation matrix in (B.4), the variables in the three-phase axis can be converted to the variables in the stationary d–q–n reference frame as

$$\mathbf{f}^s_{dqn} = \mathbf{T}(0)\mathbf{f}_{abc} \tag{B.14}$$

Similarly, the variables in the stationary d–q–n reference frame can be converted to the variables in arbitrary speed rotating d–q–n reference frame as

$$\mathbf{f}^\omega_{dqn} = \mathbf{R}(\theta)\mathbf{f}^s_{dqn} \tag{B.15}$$

For example, the relationship to convert the variables in the three-phase axis to the variables in the stationary d–q–n reference frame can be expressed as (B.16), (B.17), and (B.18) from (B.14).

$$f^s_d = \frac{(2f_a - f_b - f_c)}{3} \tag{B.16}$$

$$f_q^s = \frac{(f_b - f_c)}{\sqrt{3}} \tag{B.17}$$

$$f_n^s = \frac{\sqrt{2}(f_b + f_b + f_c)}{3} \tag{B.18}$$

As mentioned in Chapter 3, if the neutral point of an AC machine is isolated and the three-phase winding of an AC machine is symmetry, then there is no neutral component current.

That is, if $f_b + f_b + f_c = 0$, then

$$f_n^\omega = 0 \tag{B.19}$$

$$f_d^s = f_a^s \tag{B.20}$$

In (B.15), if the neutral component is zero, then the variables in the stationary *d–q* reference frame can be converted to the variables in the synchronously rotating *d–q* reference frame as (B.21) and (B.22).

$$f_d^e = f_d^s \cos\theta_e + f_q^s \sin\theta_e \tag{B.21}$$

$$f_q^e = -f_d^s \sin\theta_e + f_q^s \cos\theta_e \tag{B.22}$$

Also, inversely, the variables in the synchronously rotating reference frame can be converted to the ones in the stationary frame as (B.23) and (B.24).

$$f_d^s = f_d^e \cos\theta_e - f_q^e \sin\theta_e \tag{B.23}$$

$$f_q^s = f_d^e \sin\theta_e + f_q^e \cos\theta_e \tag{B.24}$$

The instantaneous power by three-phase variables can be calculated as

$$\text{Power} = V_{as} i_{as} + V_{bs} i_{bs} + V_{cs} i_{cs} \tag{B.25}$$

And, by using the transformation matrix, the power can be represented in terms of *d–q–n* variables as

$$\begin{aligned}\text{Power} &= \mathbf{V}_{abc}^T \mathbf{I}_{abc} = [\mathbf{T}(\theta)^{-1}\mathbf{V}_{dqn}^\omega]^T [\mathbf{T}(\theta)^{-1}\mathbf{I}_{dqn}^\omega] \\ &= \tfrac{3}{2}\,\mathbf{V}_{dqn}^\omega \mathbf{I}_{dqn}^\omega = \tfrac{3}{2}(V_{ds}^\omega i_{ds}^\omega + V_{qs}^\omega i_{qs}^\omega + V_{ns}^\omega i_{ns}^\omega)\end{aligned} \tag{B.26}$$

where

$$\mathbf{V}_{abc} = \begin{bmatrix} V_{as} & V_{bs} & V_{cs} \end{bmatrix}^T \tag{B.27}$$

$$\mathbf{I}_{abc} = \begin{bmatrix} i_{as} & i_{bs} & i_{cs} \end{bmatrix}^T \tag{B.28}$$

$$\mathbf{V}^{\omega}_{dqn} = \begin{bmatrix} V^{\omega}_{ds} & V^{\omega}_{qs} & V^{\omega}_{ns} \end{bmatrix}^T \tag{B.29}$$

$$\mathbf{I}^{\omega}_{dqn} = \begin{bmatrix} i^{\omega}_{ds} & i^{\omega}_{qs} & i^{\omega}_{ns} \end{bmatrix}^T \tag{B.30}$$

From (B.26), it can be seen that the power by *d–q–n* variables is the same regardless of the rotating speed of the frames.

B.2 *d–q* MODELING OF INDUCTION MACHINE USING TRANSFORMATION MATRIX

The voltage equation of the stator and rotor of a three-phase induction machine can be represented as (B.31) and (B.32), respectively.

$$\mathbf{V}_{abcs} = R_s\mathbf{I}_{abcs} + p\boldsymbol{\lambda}_{abcs} \tag{B.31}$$

$$\mathbf{V}_{abcr} = R_r\mathbf{I}_{abcr} + p\boldsymbol{\lambda}_{abcr} \tag{B.32}$$

where p means $p = d/dt$, and it is a differential operator.

In the above equations, the relationship between stator and rotor axes, angle of rotor, and rotational speed of the rotor is shown in Fig. 3.2. The voltage, current, and flux linkage in the above equations can be expressed as column vectors as (B.33)–(B.38).

$$\mathbf{V}_{abcs} = \begin{bmatrix} V_{as} & V_{bs} & V_{cs} \end{bmatrix}^T \tag{B.33}$$

$$\mathbf{I}_{abcs} = \begin{bmatrix} i_{as} & i_{bs} & i_{cs} \end{bmatrix}^T \tag{B.34}$$

$$\boldsymbol{\lambda}_{abcs} = \begin{bmatrix} \lambda_{as}\ \lambda_{bs}\ \lambda_{cs} \end{bmatrix}^T \tag{B.35}$$

$$\mathbf{V}_{abcr} = \begin{bmatrix} V_{ar}\ V_{br}\ V_{cr} \end{bmatrix}^T \tag{B.36}$$

$$\mathbf{I}_{abcr} = \begin{bmatrix} i_{ar}\ i_{br}\ i_{cr} \end{bmatrix}^T \tag{B.37}$$

$$\boldsymbol{\lambda}_{abcr} = \begin{bmatrix} \lambda_{ar}\ \lambda_{br}\ \lambda_{cr} \end{bmatrix}^T \tag{B.38}$$

The stator and rotor flux linkage can be expressed in terms of inductances and currents as

$$\begin{bmatrix} \boldsymbol{\lambda}_{abcs} \\ \boldsymbol{\lambda}_{abcr} \end{bmatrix} = \begin{bmatrix} \mathbf{L}_s & \mathbf{L}_{sr} \\ (\mathbf{L}_{sr})^T & \mathbf{L}_r \end{bmatrix} \begin{bmatrix} \mathbf{I}_{abcs} \\ \mathbf{I}_{abcr} \end{bmatrix} \tag{B.39}$$

where

$$\mathbf{L}_s = \begin{bmatrix} L_{ls}+L_{ms} & -\frac{1}{2}L_{ms} & -\frac{1}{2}L_{ms} \\ -\frac{1}{2}L_{ms} & L_{ls}+L_{ms} & -\frac{1}{2}L_{ms} \\ -\frac{1}{2}L_{ms} & -\frac{1}{2}L_{ms} & L_{ls}+L_{ms} \end{bmatrix} \tag{B.40}$$

$$\mathbf{L}_r = \begin{bmatrix} L_{lr}+L_{mr} & -\frac{1}{2}L_{mr} & -\frac{1}{2}L_{mr} \\ -\frac{1}{2}L_{mr} & L_{lr}+L_{mr} & -\frac{1}{2}L_{mr} \\ -\frac{1}{2}L_{mr} & -\frac{1}{2}L_{mr} & L_{lr}+L_{mr} \end{bmatrix} \tag{B.41}$$

and

$$\mathbf{L}_{sr} = L_{sr}\begin{bmatrix} \cos\theta_r & \cos\left(\theta_r+\frac{2\pi}{3}\right) & \cos\left(\theta_r-\frac{2\pi}{3}\right) \\ \cos\left(\theta_r-\frac{2\pi}{3}\right) & \cos\theta_r & \cos\left(\theta_r+\frac{2\pi}{3}\right) \\ \cos\left(\theta_r+\frac{2\pi}{3}\right) & \cos\left(\theta_r-\frac{2\pi}{3}\right) & \cos\theta_r \end{bmatrix} \tag{B.42}$$

By (B.1) and (B.4) and by the characteristics of the transformation matrix in (B.8)–(B.10), the stator voltage equation in (B.31) can be expressed in terms of *d–q–n* variables as (B.43), (B.44), and (B.45) at arbitrary speed rotating *d–q–n* reference frame.

$$V_{ds}^{\omega} = R_s i_{ds}^{\omega} + p\lambda_{ds}^{\omega} - \omega\lambda_{qs}^{\omega} \tag{B.43}$$

$$V_{qs}^{\omega} = R_s i_{qs}^{\omega} + p\lambda_{qs}^{\omega} + \omega\lambda_{ds}^{\omega} \tag{B.44}$$

$$V_{ns}^{\omega} = R_s i_{ns}^{\omega} + p\lambda_{ns}^{\omega} \tag{B.45}$$

Similarly, the rotor voltage equation can be expressed in terms of *d–q–n* variables as (B.46), (B.47), and (B.48) at arbitrary speed rotating *d–q–n* reference frame. Here, the rotational speed of the rotor is ω_r as shown in Fig. 3.2.

$$V_{dr}^{\omega} = R_r i_{dr}^{\omega} + p\lambda_{dr}^{\omega} - (\omega-\omega_r)\lambda_{qr}^{\omega} \tag{B.46}$$

$$V_{qr}^{\omega} = R_r i_{qr}^{\omega} + p\lambda_{qr}^{\omega} + (\omega - \omega_r)\lambda_{dr}^{\omega} \tag{B.47}$$

$$V_{nr}^{\omega} = R_s i_{nr}^{\omega} + p\lambda_{nr}^{\omega} \tag{B.48}$$

In particular, in the case of squirrel cage induction machine, because the rotor is short-circuited by the end ring, the rotor voltages at the *d–q–n* axis are zero as $V_{dr}^{\omega} = V_{qr}^{\omega} = V_{nr}^{\omega} = 0$.

Also, the stator flux linkage can be represented in terms of *d–q–n* variables as (B.49), (B.50), and (B.51).

$$\lambda_{ds}^{\omega} = L_s i_{ds}^{\omega} + L_m i_{dr}^{\omega} \tag{B.49}$$

$$\lambda_{qs}^{\omega} = L_s i_{qs}^{\omega} + L_m i_{qr}^{\omega} \tag{B.50}$$

$$\lambda_{ns}^{\omega} = L_{ls} i_{ns}^{\omega} \tag{B.51}$$

Similarly, the rotor flux linkage can be expressed in terms of *d–q–n* variables as (B.52), (B.53), and (B.54).

$$\lambda_{dr}^{\omega} = L_m i_{ds}^{\omega} + L_r i_{dr}^{\omega} \tag{B.52}$$

$$\lambda_{qr}^{\omega} = L_m i_{qs}^{\omega} + L_r i_{qr}^{\omega} \tag{B.53}$$

$$\lambda_{nr}^{\omega} = L_{ls} i_{nr}^{\omega} \tag{B.54}$$

where $L_m = \frac{3}{2}L_{ms}$, $L_s = L_m + L_{ls}$, and $L_r = L_m + L_{lr}$.

The torque of the induction machine can be found by differentiating the coenergy, W_c, regarding the displacement as (B.55). If the magnetic saturation of the induction machine can be neglected, then the coenergy is the same with the field energy, W_f, expressed as (B.56).

$$T_e = \frac{P}{2}\frac{\partial W_c}{\partial \theta_{rm}} \tag{B.55}$$

where P is the number of poles.

$$W_c = W_f = \frac{1}{2}(\mathbf{I}_{abcs})^T(\mathbf{L}_s - L_{ls}\mathbf{I})(\mathbf{I}_{abcs}) + (\mathbf{I}_{abcs})^T\mathbf{L}_{sr}(\mathbf{I}_{abcr}) + \frac{1}{2}(\mathbf{I}_{abcr})^T(\mathbf{L}_r - L_{lr}\mathbf{I})(\mathbf{I}_{abcr}) \tag{B.56}$$

In (B.56), the first term and the last term on the right-hand side of (B.56) is independent of the displacement, θ_{rm}. Hence, the torque can be found by differentiating the second term as

$$T_e = \frac{P}{2}(\mathbf{I}_{abcs})^T\frac{\partial \mathbf{L}_{sr}}{\partial \theta_{rm}}(\mathbf{I}_{abcr}) \tag{B.57}$$

Instead of the stator, rotor current, and mutual inductances of (B.57) in three phase variables, the torque can be expressed as (B.58) in terms of *d–q–n* stator flux linkage and stator current as (B.58) by using the transformation matrix in (B.4) and the characteristics of the matrix in (B.8)–(B.10).

$$T_e = \frac{3}{2}\frac{P}{2}\frac{L_m}{L_r}\left(\lambda_{dr}^{\omega} i_{qs}^{\omega} - \lambda_{qr}^{\omega} i_{ds}^{\omega}\right) \tag{B.58}$$

Based on (B.43)–(B.54), the equivalent circuit of the induction machine at an arbitary speed rotating *d–q–n* reference frame can be depicted as Fig. B.2.

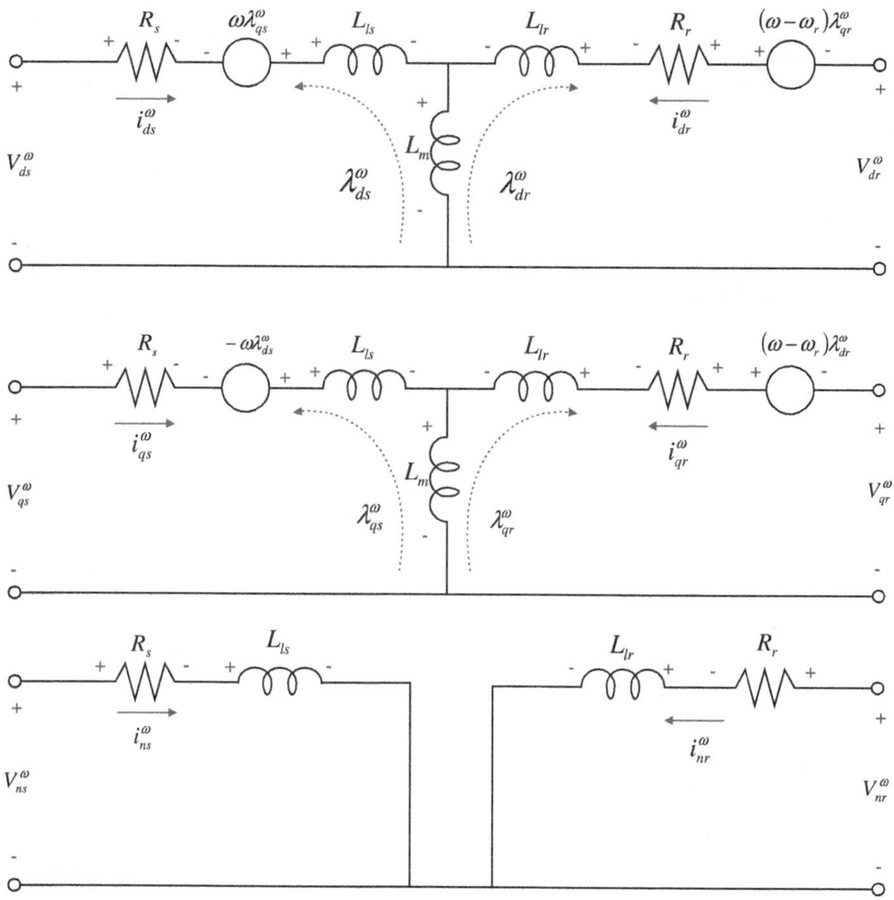

Figure B.2 Equivalent circuit of the induction machine at an arbitary speed rotating *d–q–n* reference frame. (*from the top*: *d* axis, *q* axis, and *n* axis.)

B.3 d–q MODELING OF SYNCHRONOUS MACHINE USING TRANSFORMATION MATRIX

Similaly, the synchronous machine can be modeled in d–q–n reference variables using the transformation matrix. And the torque also can be derived from the differentiation of coenergy regarding the displacement. All the results are the same as those in Section 3.3.

Index

IEEE Press Series on Power Engineering

1. *Principles of Electric Machines with Power Electronic Applications, Second Edition*
M. E. El-Hawary

2. *Pulse Width Modulation for Power Converters: Principles and Practice*
D. Grahame Holmes and Thomas Lipo

3. *Analysis of Electric Machinery and Drive Systems, Second Edition*
Paul C. Krause, Oleg Wasynczuk, and Scott D. Sudhoff

4. *Risk Assessment for Power Systems: Models, Methods, and Applications*
Wenyuan Li

5. *Optimization Principles: Practical Applications to the Operations of Markets of the Electric Power Industry*
Narayan S. Rau

6. *Electric Economics: Regulation and Deregulation*
Geoffrey Rothwell and Tomas Gomez

7. *Electric Power Systems: Analysis and Control*
Fabio Saccomanno

8. *Electrical Insulation for Rotating Machines: Design, Evaluation, Aging, Testing, and Repair*
Greg Stone, Edward A. Boulter, Ian Culbert, and Hussein Dhirani

9. *Signal Processing of Power Quality Disturbances*
Math H. J. Bollen and Irene Y. H. Gu

10. *Instantaneous Power Theory and Applications to Power Conditioning*
Hirofumi Akagi, Edson H. Watanabe, and Mauricio Aredes

11. *Maintaining Mission Critical Systems in a 24/7 Environment*
Peter M. Curtis

12. *Elements of Tidal-Electric Engineering*
Robert H. Clark

13. *Handbook of Large Turbo-Generator Operation Maintenance, Second Edition*
Geoff Klempner and Isidor Kerszenbaum

14. *Introduction to Electrical Power Systems*
Mohamed E. El-Hawary

15. *Modeling and Control of Fuel Cells: Disturbed Generation Applications*
M. Hashem Nehrir and Caisheng Wang

16. *Power Distribution System Reliability: Practical Methods and Applications*
Ali A. Chowdhury and Don O. Koval

17. *Introduction to FACTS Controllers: Theory, Modeling, and Applications*
Kalyan K. Sen and Mey Ling Sen

18. *Economic Market Design and Planning for Electric Power Systems*
James Momoh and Lamine Mili

19. *Operation and Control of Electric Energy Processing Systems*
James Momoh and Lamine Mili

20. *Restructured Electric Power Systems: Analysis of Electricity Markets with Equilibrium Models*
Xiao-Ping Zhang

21. *An Introduction to Wavelet Modulated Inverters*
S.A. Saleh and M. Azizur Rahman

22. *Probabilistic Transmission System Planning*
Wenyuan Li

23. *Control of Electric Machine Drive Systems*
Seung-Ki Sul

24. *High Voltage and Electrical Insulation Engineering*
Ravindra Arora and Wolfgang Mosch

25. *Practical Lighting Design with LEDs*
Ron Lenk and Carol Lenk

Forthcoming Titles

Electricity Power Generation: the Changing Dimensions
Digambar M. Tagere

Power Conversion and Control of Wind Energy Systems
Bin Wu, Yongqiang Lang, Navid Zargan, and Samir Kouro

Electric Distribution Systems
Abdelhay A. Sallam and O. P. Malik

Doubly Fed Induction Machine: Modeling and Control for Wind Energy Generation Applications
Gonzalo Abad, Jesus Lopez, Miguel Rodriguez, Luis Marroyo, and Grzegorz Iwanski

Maintaining Mission Critical Systems, Second Edition
Peter M. Curtis

Printed in the USA
K061860SCI081417 01S29053000000002160